PHYSIOLOGY, BIOCHEMISTRY AND MOLECULAR BIOLOGY
OF PLANT LIPIDS

Physiology, Biochemistry and Molecular Biology of Plant Lipids

Edited by

John Peter Williams

Mobashsher Uddin Khan

and

Nora Wan Lem

KLUWER ACADEMIC PUBLISHERS
DORDRECHT / BOSTON / LONDON

A C.I.P. Catalogue record for this book is available from the Library of Congress

ISBN 0-7923-4379-4

Published by Kluwer Academic Publishers,
P.O. Box 17, 3300 AA Dordrecht, The Netherlands.

Kluwer Academic Publishers incorporates
the publishing programmes of
D. Reidel, Martinus Nijhoff, Dr W. Junk and MTP Press.

Sold and distributed in the U.S.A. and Canada
by Kluwer Academic Publishers,
101 Philip Drive, Norwell, MA 02061, U.S.A.

In all other countries, sold and distributed
by Kluwer Academic Publishers Group,
P.O. Box 322, 3300 AH Dordrecht, The Netherlands.

Printed on acid-free paper

All Rights Reserved
© 1997 Kluwer Academic Publishers
No part of the material protected by this copyright notice may be reproduced or
utilized in any form or by any means, electronic or mechanical,
including photocopying, recording or by any information storage and
retrieval system, without written permission from the copyright owner.

Printed in the Netherlands

TABLE OF CONTENTS

Preface — xvii
Dedication — xix

SECTION 1: FATTY ACID BIOSYNTHESIS

Is the Chloroplast Fatty Acid Synthase a Multienzyme Complex?
    P.G. Roughan and J.B. Ohlrogge. — 3

Structure-Function Studies on Desaturases and Related Hydrocarbon Hydroxylases.
    J. Shanklin, E.B. Cahoon, E. Whittle, Y. Lindqvist, W. Huang, G. Schneider and N. Schmidt. - TERRY GALLIARD LECTURE- — 6

Acetyl-CoA Carboxylase from *Brassica napus*.
    J.E. Markham, K.M. Elborough and A.R. Slabas. — 11

Transgenic Modification of Acetyl-CoA Carboxylase and β-Keto Reductase Levels in *Brassica napus*. Functional and Regulatory Analysis.
    A.J. White, K.M. Elborough, H. Jones and A.R. Slabas. — 14

Enzyme Studies on Isoforms of Acetyl-CoA Carboxylase.
    L.J. Price, D. Herbert, C. Alban, D. Job, D.J. Cole, K.E. Pallett and J.L. Harwood. — 17

Expression of the Acetyl-CoA Carboxylase Multi-Subunit Enzyme in Oilseed Rape.
    D. Burtin and S. Rawsthorne. — 20

Biotin Carboxyl Carrier Protein and Biotin Carboxylase Subunits of the Multi-Subunit Form of Acetyl-CoA Carboxylase from *Brassica napus*.
    K.M. Elborough, J.E. Markham, I.M. Evans, R. Winz, A.J. White and A.R. Slabas. — 23

Acetyl-CoA Carboxylase Activity in the Oil Palm
    R. Sambanthamurthi and J.B. Ohlrogge. — 26

Soybean Chloroplast Acetyl-CoA Carboxylase.
    S.V. Reverdatto, V. Beilenson and N.C. Nielsen. — 29

Two Forms of Acetyl-CoA Carboxylase in Higher Plants and Effects of UV-B on the Enzyme Levels.
    Y. Sasaki and T. Konishi. — 32

Characterization and Cloning of Biotin Holocarboxylase Synthetase from Higher Plants.
    G. Tissot, R. Douce and C. Alban. — 35

Structural Studies on Plant and Bacterial Reductases Involved in Lipid Biosynthesis.
A.R. Slabas, J. Rafferty, D. Rice, C. Baldock, J.W. Simon, N. Thomas, S. Bithell and A.R. Stuitje. 38

A Novel Pathway for the Biosynthesis of Straight and Branched, Odd- and Even-Length Medium Chain Fatty Acids in Plants.
G.J. Wagner and A.B. Kroumova. 42

Does Endogenous Acyl-CoA Synthetase Activity Generate the Primer in Biosynthesis of Very Long Chain Fatty Acids?
A. Hlousek-Radojcic, K.J. Evenson, J.G. Jaworski and D. Post-Beittenmiller. 45

Modification of Plant Epicuticular Waxes.
R.K. Deka and D. Post-Beittenmiller. 48

Biochemistry of Short-Chain Alkanes: Evidence for an Elongation/Reduction/C1 Elimination Pathway.
T.J. Savage, M.K. Hristova and R. Croteau. 51

A Re-examination of Pathways for the Biosynthesis of Medium and Long Chain Fatty Acids in Seeds of Several Plants.
A.B. Kroumova and G.J. Wagner. 54

Biosynthesis of an Acetylenic Fatty Acid in Microsomal Preparations from Developing Seeds of *Crepis alpina*.
A. Banas, M. Bafor, E. Wiberg, M Lenman, U. Ståhl and S. Stymne. 57

Metabolism of Palmitate Differs in *Neurospora crassa* Mutants with Impaired Fatty Acid Synthase.
M. Goodrich-Tanrikulu, A.E. Stafford and T.A. McKeon. 60

Hydroxy Fatty Acid Biosynthesis and Genetic Transformation in the Genus *Lesquerella* (Brassicaceae).
D.W. Reed, J.K. Hammerlindl, C.E. Palmer, D.C. Taylor, W. Keller and P.S. Covello. 63

Developmental Changes in Substrate Utilization for Fatty Acid Synthesis by Plastids Isolated from Oilseed Rape Embryos.
P.J. Eastmond and S. Rawsthorne. 66

β-Ketoacyl-Acyl Carrier Protein [ACP] Synthase II in the Oil Palm (*Elaeis guineensis* Jacq.) Mesocarp.
U.S. Ramli and R. Sambanthamurthi. 69

The Products of the Microsomal Fatty Acid Elongase are Determined by the Expression and Specificity of the Condensing Enzyme.
A. Millar and L. Kunst. 72

Chloroplastic Carnitine Acetyltransferase.
    C. Wood, C. Masterson and J.A. Miernyk.   75

Function of Chloroplastic Carnitine Palmitoyltransferase.
    C. Masterson, C. Wood and J.A. Miernyk.   78

Kinetic Analysis of the Mechanism of Enoyl-ACP Reductase.
    T. Fawcett and C. Overend.   81

Identification and Characterization of 9-*cis*-Hexadecenoic Acid *Cis-Trans* Isomerase of *Pseudomonas* Sp. Strain E-3.
    H. Okuyama, D. Enari, T. Kusano and N. Morita.   84

Intracellular Distribution of Fatty Acid Desaturases in Cyanobacterial Cells and Higher Plant Chloroplasts.
    L. Mustardy, D.A. Los, Z. Gombos, N. Tsvetkova, I. Nishida and N. Murata.   87

Triacylglycerols Participate in the Eukaryotic Pathway of PUFAs Biosynthesis in the Red Microalga *Porphyridium cruentum*.
    I. Khozin, H.Z. Yu, D. Adlerstein, C. Bigogno and Z. Cohen.   90

Elucidation of the Biosynthesis of Eicosapentaenoic Acid (EPA) in the Microalga *Porphyridium cruentum*.
    I. Khozin, D. Adlerstein, C. Bigogno and Z. Cohen.   93

Immobilization of Hydroperoxide Lyase from Chlorella for the Production of C13 Oxo-Carboxylic Acid.
    A. Nuñez, T.A. Foglia and G.J. Piazza.   96

Allene Oxide Cyclase from Corn: Partial Purification and Characterization.
    J. Ziegler, M. Hamberg and O. Miersch.   99

SECTION 2: GLYCEROLIPID BIOSYNTHESIS

How is Sulpholipid Metabolised?
    C.E. Pugh, A.B. Roy, G.F. White and J.L. Harwood.   104

Biochemical Characterization of Cottonseed Microsomal N-Acylphosphatidylethanolamine Synthase.
    K.D. Chapman and R.S. McAndrew.   107

Phosphatidylcholine Biosynthesis in Soybeans: the Cloning and Characterization of Genes Encoding Enzymes of the Nucleotide Pathway.
    D.E. Monks, J.H. Goode, P.K. Dinsmore and R.E. Dewey.   110

Phospholipid metabolism by Castor Microsomes.
    J.T. Lin, C.L. Woodruff, O.J. Lagouche and T.A. McKeon.   113

Biosynthesis of Diacylglyceryl-N,N,N-Trimethylhomoserine (DGTS) in *Rhodobacter sphaeroides* and Evidence for Lipid-Linked N-Methylation.
    M. Hofmann and W. Eichenberger.   116

De Novo Biosynthesis of Eukaryotic Glycerolipids in Chloroplasts of Transgenic Tobacco Plants.
    D. Weier and M. Frentzen.   119

Fungal Oil Biosynthesis.
    F. Jackson, L. Michaelson, T. Fraser, G. Griffiths and K. Stobart.   122

Triton X-100 Solubilization of Diacylglycerol Acyltransferase from the Lipid Body Fraction in an Oleaginous Fungus.
    Y. Kamisaka.   125

Binding of Lipids on Lipid Transfer Proteins.
    F. Guerbette, A. Jolliot, J.-C. Kader and M. Grosbois.   128

Variations in UDGT Activity Imposed by Growth Conditions Reflect Internal Availability of Substrates.
    A. Livne and A. Sukenik.   131

The Role and Partial Purification of CTP: Ethanolaminephosphate Cytidylyltransferase in Postgermination Castor Bean Endosperm.
    F. Tang and T.S. Moore, Jr.   134

Identification of a Region in Stromal Glycerol 3-Phosphate Acyltransferase which Controls Acyl Chain Specificity.
    S.R. Ferri, O. Ishizaki-Nishizawa, M. Azuma and T. Toguri.   137

The Purification and Characterisation of Phosphatidate Phosphatase from Avocado.
    M. Pearce and A.R. Slabas.   140

Analysis of Plant Cerebrosides by C_{18} and C_6 HPLC.
    B.D. Whitaker.   143

SECTION 3: MEMBRANES

The Role of Membrane Lipids in the Arrangement of Complexes in Photosynthetic Membranes.
    P.J. Quinn.   148

Lipids in Cell Signalling: A Review.
    P. Mazliak.   151

Release of Lipid Catabolites from Membranes by Blebbing of Lipid-Protein Particles.
    J.E. Thompson, C.D. Froese, Y. Hong, K.A. Hudak and M.D. Smith.   154

Chloroplast Membrane Lipids as Possible Primary Targets for Nitric Oxide-Mediated Induction of Chloroplast Fluorescence in *Pisum sativum* (Argentum Mutant).
    Y.Y. Leshem, E. Haramaty, Z. Malik and Y. Sofer.   157

Glycosylphosphatidylinositol-Anchored Proteins in Plants.
    N. Morita, H. Nakazato, H. Okuyama, Y. Kim and G.A. Thompson, Jr.   160

Effects of Exogenous Free Oleic Acid on Membrane Fatty Acid Composition and Physiology of *Lemna minor* Fronds
    G. Grenier, A.M. Zimafuala and J.-P. Marier.   163

A New *Arabidopsis* Mutant with Reduced 16:3 Levels.
    M. Miquel and J. Browse.   166

Study on the Fatty Acids of the Lipids Bound to Peptides of Photosystem II of *Nicotiana tabacum*.
    A. Gasser, S. Raddatz, A. Radunz and G.H. Schmid.   169

The Cyanobacterium *Gloeobacter violaceus* Lacks Sulfoquinovosyl Diacylglycerol.
    E. Selstam and D. Campbell.   172

SECTION 4: ISOPRENOIDS AND STEROLS.

A Novel Mevalonate-Independent Pathway for the Biosynthesis of Carotenoids, Phytol and Prenyl Chain of Plastoquinone-9 in Green Algae and Higher Plants.
    H.K. Lichtenthaler, M. Rohmer, J. Schwender, A. Disch and M. Seemann.   177

Biosynthesis of Sterols in the Green Algae (*Scenedesmus, Chlorella*) According to a Novel, Mevalonate-Independent Pathway.
    J. Schwender, H.K. Lichtenthaler, A. Disch and M. Rohmer.   180

Plant Sterol Biosynthesis: Identification and Characterization of Δ^7-Sterol C5(6)-Desaturase.
    M. Taton and A. Rahier.   183

Phytosterol Synthesis: Identification, Enzymatic Characterization and Design of Potent Mechanism-Based Inhibitors of $\Delta^{5,7}$-Sterol-Δ^7-Reductase.
    A. Rahier and M. Taton.   186

The Occurrence and Biological Activity of Ferulate-Phytosterol Esters in Corn Fiber and Corn Fiber Oil.
    R.A. Moreau, M.J. Powell and K.B. Hicks.   189

The Occurrence of Long Chain Polyprenols in Leaves of Plants.
    E. Swiezewska, M. Szymanska, K. Skorupinska and T. Chojnacki.   192

Taxonomic Aspects of the Sterol and Δ^{11}-Hexadecenoic Acid (C16:1,Δ^{11}) Distribution in Arbuscular Mycorrhizal Spores.
A. Grandmougin-Ferjani, Y. Dalpé, M.-A. Hartmann, F. Laruelle, D. Couturier and M. Sancholle. 195

SECTION 5: ENVIRONMENTAL EFFECTS ON LIPIDS

Lack of Trienoic Fatty Acids in an *Arabidopsis* Mutant Increases Tolerance of Photosynthesis to High Temperature.
J.-M. Routaboul, P. Vijayan and J. Browse. 200

A Trienoic Fatty Acid Deficient Mutant of *Arabidopsis* is Defective in Recovery from Photoinhibition at Low Temperatures.
P. Vijayan, J.-M. Routaboul and J. Browse. 203

Growth Temperature and Irradiance Modulate Trans-Δ^3-Hexadecenoic Acid Content and Photosynthetic Light-Harvesting Complex Organization.
G.R. Gray, M. Krol, M.U. Khan, J.P. Williams and N.P.A. Huner. 206

Responses to Cold in Transgenic Russet Burbank and Bolivian Potato.
D. Guerra, K. Dziewanowska and J. Wallis. 209

Low-temperature Resistance of Higher Plants is Significantly Enhanced by a Non Specific Δ^9-Desaturase From Cyanobacteria.
O. Ishizaki-Nishizawa, T. Fujii, T. Ohtani and T. Toguri. 212

Biochemical Aspect on the Effect of Titavit Treatment on Carotenoids, Lipids and Antioxidants in Spice Red Pepper.
P.A. Biacs, H.G. Daood and Á. Keresztes. 215

Effect of Environmental Conditions on the Molecular Species Composition of Galactolipids in the Alga *Porphyridium cruentum*.
D. Adlerstein, I. Khozin, C. Bigogno and Z. Cohen. 218

Chilling Injury and Lipid Biosynthesis in Tomato Pericarp.
H. Yu and C. Willemot. 221

Cerebrosides in Seed-Plant Leaves: Composition of Fatty Acids and Sphingoid Bases.
H. Imai, M. Ohnishi, M. Kojima and S. Ito. 224

Changes in Phophatidylinositol Metabolism in Suspension Cultured Tobacco Cells Submitted to Stress.
P. Norberg, D. Chervin, M. Gawer, N. Guern, Z. Yaniv and P. Mazliak. 227

Inhibition of Polyunsaturated Fatty Acid Synthesis by Salicyclic Acid and Salicylhydroxamic Acid and Their Modes of Action.
A. Banas, G. Stenlid, M. Lenman, F. Sitbon and S. Stymne. 230

Idioblast as a Model System for the Study of Biosynthesis of Lipids with Antifungal
Properties in Avocado Fruits.
A.I. Leikin-Frenkel and D. Prusky. 233

Inhibition of Three Soybean Fungal Plant Pathogens by Lipid Derivatives and Natural
Compounds.
B.J. Barnes, H.A. Norman and N.L. Brooker. 236

Can Lipids from the Freshwater Alga *Selenastrum capricornutum* (CCAP 278/4)
Serve as an Indicator of Heavy Metal Pollution in Freshwater Environments?
C. Riches, C. Rolph, D. Greenway and P. Robinson. 239

SECTION 6: LIPID DEGRADATION.

Purification and Characterization of a Microsomal Phospholipase A2 From
Developing Elm Seeds.
U. Ståhl, B. Ek, A. Banas, M. Lenman, S. Sjödahl and S. Stymne. 244

Biogeneration of Green Odour Emitted by Green Leaves - On HPO Lyase and
Relationship of LOX-HPO Lyase Activities to Environmental Stimuli.
K. Matsui, T. Kajiwara and A. Hatanaka. 247

Do Lipoxygenases Initiate β-Oxidation?
I. Feussner, H. Kuhn and C. Wasternack. 250

Degradation of Acetylenic Triacylglycerols and the Inactivation of Membrane
Preparations from Moss Protonema Cells.
P. Beutelmann and K. Menzel. 253

Storage Lipid Mobilization During Nitrogen Assimilation in a Marine Diatom.
T. Larson and P.J. Harrison. 256

Biocatalytic Oxidation of Acylglycerol and Methyl Ester.
G.J. Piazza, T.A. Foglia and A. Nuñez. 259

Volatile Production by the Lipoxygenase Pathway in Olive Callus Cultures.
M. Williams and J.L. Harwood. 262

Immobilization of Lipoxygenase in Alginate-Silicate Sol-Gel Matrix: Formation of
Fatty Acid Hydroperoxides.
A.-F. Hsu, T.A. Foglia and G.J. Piazza. 265

Expression and Location of Lipoxygenases in Soybean Seeds and Seedlings.
C. Wang, K.P.C. Croft, D. Hapsoro and D.F. Hildebrand. 268

Relationship Between Lipoxygenase-Catalyzed Formation of Dihydroxylated
Unsaturated Fatty Acids and its Auto-Inactivation.
   M.R. Kim and D.-E. Sok.   271

Changes of Phospholipase D Activity During Rape Seed Development and Processing.
   O. Valentová, Z. Novotná, J.-C. Kader and J. Káš.   275

Lipase Activity in Germinating Oil Palm Seeds.
   R.D. Abigor and D.A. Okiy.   278

Effect of γ-Irradiation on Tomato Fruit Ripeness with Special Regard to
Lipoxygenase Activity, Carotenoid Content and Antioxidation Potency.
   A. Abushita, H.G. Daood, E. Hebshi and P.A. Biacs   281

Expression and Substrate Specificity of Lipoxygenase Isozymes from Embryos of
Germinating Barley.
   W.L. Holtman, G. van Duijn, J.R. van Mechelen, N.J.A. Sedee, A.C. Douma
   and N. Schmitt.   284

SECTION 7: OIL SEEDS AND FRUITS

Oleosins: Their Subcellular Targeting and Role in Oil-Body Ontogeny.
   D.J. Murphy, C. Sarmiento, J.H.E. Ross and E. Herman.   289

Evolution of Oleosins.
   A.H.C. Huang.   292

Oils-to-Oleosins Ratio Determines the Size and Shape of Oil Bodies in Maize
Kernels.
   J.T.L. Ting and A.H.C. Huang.   295

Development of Genetically Engineered Oilseeds: From Molecular Biology to
Agronomics.
   A.J. Kinney.   298

Regulation of Oil Composition in Oilseed Rape (*B. napus* L.): Effects of Abscisic
Acid and Temperature.
   J.A. Wilmer, J.P.F.G. Helsper and L.H.W. van der Plas.   301

Broad-Range and Binary-Range Acyl-ACP Thioesterases from Medium-Chain
Producing Seeds.
   T. Voelker, A. Jones, A. Cranmer, M. Davies and D. Knutzon.   304

Carbon Partitioning in Plastids During Development of *B. napus* Embryos.
   S. Rawsthorne, P.J. Eastmond, F. Kang, P.M.R. Da Silva, A.M. Smith, D.
   Hutchings and M.J. Emes.   307

Isolation and Purification of Membrane Fractions Involved in Triacylglycerol
Metabolism in Developing Seeds of *Brassica napus* L.
    D.J. Lacey, P. Pongdontri and M.J. Hills.   310

Altered Fatty Acid Composition of Membrane Lipids in Seeds and Seedling Tissues
of High-Saturate Canolas.
    G.A. Thompson and C. Li.   313

Mutants of *Brassica napus* with Altered Seed Lipid Fatty Acid Composition.
    B. Rücker and G. Röbbelen   316

Erucic Acid Distribution in *Brassica oleracea* Seed Oil Triglycerides.
    S.L. MacKenzie, E.M. Giblin, D.L. Barton, J.R. McFerson, D. Tenaschuk and
D.C. Taylor.   319

Fatty Acid Composition of Different Tissues During High Stearic or High Palmitic
Sunflower Mutants Germination.
    R. Álvarez-Ortega, S. Cantisán, E. Martínez-Force, M. Mancha and R.
Garcés.   322

Photosynthetic Carbon Metabolism of Olives.
    J. Sánchez and J.J. Salas.   325

Biogenesis of Alcohols Present in the Aroma of Virgin Olive Oil.
    J.J. Salas and J. Sánchez.   328

Dynamics of Triacylglycerol Composition in Developing Sea Buckthorn (*Hippophae
rhamnoides* L.) Fruits.
    A.G. Vereshchagin, O. Ozerinina and V.D. Tsydendambaev.   331

SECTION 8: MOLECULAR BIOLOGY AND BIOTECHNOLOGY

Molecular Biology of Genes Involved in Cuticular Wax Biosynthesis.
    J.D. Hansen, C. Dietrich, Y. Xia, X. Xu, T.-J. Wen, M. Delledonne, D.S.
Robertson, P.S. Schnable and B.J. Nikolau.   336

Expression of Genes Involved in Wax Biosynthesis in Leek.
    Y. Rhee, A. Hlousek-Radojcic, P.S. Jayakumar, D. Liu and D. Post-
Beittenmiller.   339

Expression of Castor and *L. fendleri* Oleate 12-Hydroxylases in Transgenic Plants:
Effects on Lipid Metabolism and Inferences on Structure-Function
Relationships in Fatty Acid Hydroxylases.
    P. Broun, N. Hawker and C.R. Somerville.   342

Characterization of Phospholipase D-Overexpressed and Suppressed Transgenic
   Tobacco and *Arabidopsis*.
   X. Wang, S. Zheng, K. Pappan and L. Zheng. 345

Purification and Molecular Cloning of Bell Pepper Fruit Fatty Acid Hydroperoxide
   Lyase.
   K. Matsui, M. Shibutan,i Y. Shibata and T. Kajiwara. 348

Production of γ-Linolenic Acid by Transgenic Plants Expressing Cyanobacterial or
   Plant Δ^6-Desaturase Genes.
   P.D. Beremand, A.N. Nunberg, A.S. Reddy, and T.L. Thomas. 351

cDNA Cloning of Cucumber Monogalactosyldiacylglycerol Synthase and the
   Expression of the Active Enzyme in *Escherichia coli*.
   H. Ohta, M. Shimojima, A. Iwamatsu, T. Masuda, F. Kitagawa, Y. Shioi and
   K.-I. Takamiya. 354

Differential Display of mRNA From Oil-Forming Cell Suspension Cultures of
   Brassica napus.
   R.J. Weselake, J.M. Davoren, S.D. Byers, A. Laroche, D.M. Hodges, M.K.
   Pomeroy and T.L. Furukawa-Stoffer. 357

Analysis of the *Arabidopsis* Enoyl-ACP Reductase Promoter in Transgenic Tobacco.
   G.-J. de Boer, T. Fawcett, A.R. Slabas, H.J.J. Nijkamp and A.R. Stuitje. 360

Molecular Biology of Biotin-Containing Enzymes Required in Lipid Metabolism.
   J.-K. Choi J. Ke, A.L McKean, L.M. Weaver, T.-N. Wen, J. Sun, T. Diez, F.
   Yu, X Guan, E.S. Wurtele and B.J. Nikolau. 363

Over-Expression, Purification and Characterisation of an Acyl-CoA Binding Protein
   from *Brassica napus* L.
   A.P. Brown, P.E. Johnson and M.J. Hills. 368

Sequence Analysis of CTP:Phosphocholine Cytidylyltransferase cDNA from
   Arabidopsis thaliana.
   S.H. Cho, S.-B. Choi J.C. Kim and K.-W. Lee. 371

Approaches to the Design of Acyl-ACP Desaturases with Altered Fatty Acid Chain
   Length and Double Bond Positional Specificities.
   E.B. Cahoon and J. Shanklin. 374

Effect of a Mammalian Δ^9-Desaturase on Specific Lipids in Transgenic Plant Tissues.
   H. Moon, M. Scowby, S. Avdiushko and D.F. Hildebrand. 377

Two Acyl-Lipid Δ^9-Desaturase Genes of the Cyanobacterium, *Synechococcus* sp.
   Strain PCC7002.
   T. Sakamoto, V.L. Stirewalt and D.A. Bryant. 380

Isolation and Characterisation of Two Different Microsomal ω-6 Desaturase Genes in Cotton (*Gossypium hirsutum* L.).
    Q. Liu, S.P. Singh, C.L. Brubaker, P.J. Sharp, A.G. Green and D.R. Marshall.   383

Can *E.coli* β-Hydroxydecanoyl-ACP Dehydrase and β-Keto-Acyl-ACP Synthase I Interact with *Brassica napus* Fatty Acid Synthase to Alter Oil Seed Composition?
    J.-A. Chuck, I. Verwoert, E. Verbree, M. Siggaard-Andersen, P. von Wettstein-Knowles and A.R. Stuitje.   386

Isolation of Cytochrome P-450 Genes from *Vernonia galamensis*.
    C. Seither, S. Avdiushko and D. Hildebrand.   389

Engineering Trierucin into Oilseed Rape by the Introduction of a 1-Acyl-sn-Glycerol-3-Phosphate Acyltransferase from *Limnanthes douglasii*..
    C.L. Brough, J. Coventry, W. Christie, J. Kroon, T. Barsby and A.R. Slabas.   392

Brassica napus CTP:Phosphocholine Cytidylyltransferase: Molecular Cloning and Subcellular Localization in Yeast, and Enzyme Characterization.
    I. Nishida, Y. Kitayama, R. Swinhoe, A.R. Slabas, N. Murata and A. Watanabe.   395

Cloning of a Putative Pea Cholinephosphate Cytidylyltransferase and Evidence for Multiple Genes.
    P.L. Jones, D.L. Willey, P. Gacesa and J.L. Harwood.   398

Immunological Identification of Sunflower (*Helianthus annuus* L.) Spherosomal Lipase.
    S. Bahri, R. Marrakchi, A. Landoulsi, J. Ben Hamida and J.-C. Kader.   401

Purification and Immunological Analysis of Phospholipase D from *Brassica napus* (Rape Seed).
    Z. Novotná, O. Valentová, J. Daussant and J. Káš.   404

Modification of Seed Oil Content and Acyl Composition in the Brassicaceae Utilizing a Yeast *sn*-2 Acyltransferase (SLC1-1) Gene.
    J.-T. Zou, V. Katavic, E.M. Giblin, D.L. Barton, S.L. MacKenzie, W.A. Keller and D.C. Taylor.   407

Author index   411
Subject index   415

PREFACE

The 12th International Symposium on Plant Lipids was held at the University of Toronto, Canada, from July 7th to 12th, 1996. The conference was attended by over 200 scientists from university, government and corporate laboratories from 24 different countries.

The topics covered in the symposium ranged from basic physiology, biochemistry and molecular biology of plant lipids to transformation and genetic engineering of crop plants. Oil seed plants were a particular focus of the symposium. There were 62 oral and 96 posters presentations. A special lecture in memory of the founder of this series of symposium, Terry Galliard, was presented by John Shanklin. This Proceedings Book has been dedicated to Grattan Roughan for his important contributions to our knowledge of plant lipid metabolism.

This volume contains manuscripts submitted from most of the presentations at the symposium. It provides a useful summary of the major fields of plant lipid studies and our present state of knowledge. The papers are arranged in eight sections covering the major areas in the field of plant physiology, biochemistry and molecular biology of plant lipids.

We would like to thank Valerie Imperial, Rajesh Khetarpal and Mary Williams for their invaluable help in organizing and running the meetings and excursions.

John P. Williams, Mobashsher U. Khan and Nora W. Lem

Toronto, Canada, October 1996

DEDICATION

This volume is dedicated to Grattan Roughan.

Grattan has contributed much to our knowledge in several key areas of the field of plant lipid metabolism. He first became interested in plant lipids in the 1960's when he undertook his Ph.D. on the subject of lipid turnover in plants at Massey University in New Zealand. At that time he discovered the unusual role of phosphatidylcholine in lipid metabolism in pumpkin leaves. This discovery eventually led to our present acceptance of an intermediate role for this cytosolic lipid in the biosynthesis of chloroplast lipids in most plants and algae. Following a break of 3-4 years, he returned to the field of plant lipids and their importance as sensors of chilling temperatures in susceptible plants at the Plant Physiology Division of DSIR, Palmerston North, NZ. At that time he began a long and productive partnership with Roger Slack on the metabolic pathways of unsaturated fatty acids in higher plants. This work nicely complemented the physical studies being carried out in Australia at the same time. The collaboration resulted in the theory of procaryotic and eucaryotic pathways of glycerolipids synthesis in plants. Since 1983 he has continued his research at DSIR, Auckland, NZ., and later at the Horticulture and Food Research Institute of New Zealand, Ltd. He is still very active in research and is presently studying the possibility that chloroplast fatty acid synthase channels acetate through acetyl-CoA synthetase and acetyl-carboxylase into oleate and palmitate.

Grattan has added not only to our knowledge of lipid metabolism but all some spice and added interest to the field. Without his continued participation we would not understand the metabolic pathways in leaves and seeds as well as we do today and the field would miss cheerful and colourful presence.

Section 1:

Fatty Acid Biosynthesis

IS THE CHLOROPLAST FATTY ACID SYNTHASE A MULTIENZYME COMPLEX?

GRATTAN ROUGHAN[1] AND JOHN OHLROGGE[2]

[1]*The Horticulture and Food Research Institute of New Zealand, Ltd., Private Bag 92169, Auckland, New Zealand, and* [2]*Department of Botany and Plant Pathology, Michigan State University, East Lansing, MI 48824, USA.*

1. Introduction

Prior to about 1982 it was often assumed that the enzymes of the chloroplast fatty acid synthase would be organized into some sort of multienzyme complex even though the absolute requirement of all known plant fatty acid synthase (FAS) preparations for added acyl carrier protein (ACP) precluded the type of complex observed in yeast and mammalian cells. Then in 1982 three laboratories [1-3] independently showed that the FAS consisted of individual proteins that were readily separable by conventional gel filtration techniques i.e. the plant synthase was of the prokaryotic type, similar to that in *E. coli*. After that, the concept of a multienzyme complex synthesizing fatty acids in chloroplasts was effectively in limbo.

However, there were some puzzling features of fatty acid synthesis from acetate by isolated chloroplasts that required an explanation. Firstly, it was known that light-dependent fatty acid synthesis by spinach chloroplasts proceeded almost as efficiently in hypotonic as in isotonic media even though bicarbonate-dependent oxygen evolving activity was totally lost [4]. The integrity of the envelope was presumed to be breached so that substrates and cofactors would be diluted into the reaction medium. Secondly, although acetate was efficiently incorporated into fatty acids by the "permeabilized" chloroplasts, acetyl-CoA was not [5] and nor did added acetyl-CoA interfere with acetate incorporation. Thirdly, the endogenous concentrations of CoA-SH, malonyl-CoA, and acetyl-CoA during fatty acid synthesis by spinach and pea chloroplasts [6] appeared to be to low to account for known rates of acetate incorporation. More recently, a significant accumulation of malonyl-CoA in assays optimized for measuring acetyl-CoA synthetase (ACS) activity in chloroplasts [7] suggested that ACS and acetyl-CoA carboxylase (ACC) might function cooperatively in situ to channel acetate into fatty acids. Channeling would permit high reaction rates at apparently low substrate concentrations, and might also explain the inability of chloroplasts to incorporate acetate into isoprenoids [8]. Therefore, it was decided to investigate more thoroughly fatty acid synthesis by chloroplasts in hypotonic media [9].

2. Results and Discussion

The rate of light-dependent acetate incorporation into fatty acids in both spinach and pea chloroplasts initially increased as the sorbitol concentration of the basal (unsupplemented) reaction mixture decreased (Figure 1) and was consistently higher in spinach chloroplasts incubated in 66 mM compared with 330 mM sorbitol. In pea chloroplasts the rate in 66 mM sorbitol was 80-90% of the rate in isotonic media, and in both chloroplasts the rate declined to near zero at the lowest sorbitol concentrations employed.

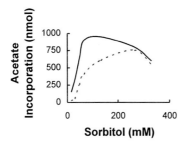
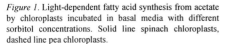
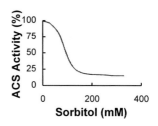

Figure 1. Light-dependent fatty acid synthesis from acetate by chloroplasts incubated in basal media with different sorbitol concentrations. Solid line spinach chloroplasts, dashed line pea chloroplasts.

Figure 2. Chloroplast acetyl-CoA synthetase activity (7) in media with different sorbitol concentrations. Reactions contained acetate, CoA and ATP.

Conversely, ACS activity [7] increased markedly at the lower sorbitol concentrations as the envelope became more permeable to added CoA and ATP (Figure 2). When spinach chloroplasts were illuminated in 66 mM sorbitol, rates of fatty acid synthesis were linear for at least 30 min and were directly proportional to chlorophyll concentration over the range 30 - 250 µg per ml of reaction, indicating that the synthase activity was stable for a reasonable period of illumination and was unaffected by the potential dilution of cofactors by up to 250 times. Acetate incorporation in the basal hypotonic medium was absolutely light-dependent but was not enhanced by adding ACP, ATP, CoA, NADPH and NADH. Although ADP, NADP, and CoA readily entered permeabilized chloroplasts and were metabolized (Table 1), acetyl-CoA and malonyl-CoA were neither incorporated into fatty acids nor able to inhibit acetate incorporation. Fatty acid synthesis from acetate by permeabilized chloroplasts was not inhibited by avidin [9], a result which, along with those for acyl-CoA, could be indicative of substrate channeling.

Products of acetate incorporation in isotonic and hypotonic media differed subtly but could be explained simply by loss of glycerol 3-phosphate from permeabilized chloroplasts. Therefore, when reaction media were supplemented with glycerol 3-phosphate, CoA and ATP, there was little or no difference in the products accumulated by intact and permeabilized chloroplasts Chloroplasts may combine all of the enzymes that process acetate into glycerides and oleoyl-CoA in some sort of complex which must also include the suite of cofactors required to allow the process to run unimpeded.

How permeable were the permeabilized chloroplasts? Chloroplast suspensions in 66 mM sorbitol were translucent, and the organelles packed to 2 - 2.5 time the volume of intact plastids. Transferring intact chloroplasts to 66 mM sorbitol resulted in the loss of 40 - 60% of soluble proteins and ACS activity to the medium within 5 min at 22 - 25° C. Although a small and highly soluble protein, ACP was presumable completely retained by permeabilized chloroplasts since fatty acid synthesis was not inhibited. The loss of some ACS activity into the medium was accompanied by a 2-3 fold increase in the Km for acetate in fatty acid synthesis (G. Roughan and J. Browse, unpublished result). Whereas intact chloroplasts did not metabolize NADP or CoA, the permeabilized chloroplasts efficiently reduced NADP, and acetylated CoA either in the presence of ADP or in the absence of any exogenous nucleotides (Table 1). There seems little doubt that the envelope of permeabilized chloroplasts was thoroughly breached.

Metabolite channeling is a function of multienzyme complexes and becomes probable when metabolite concentrations are very low. For one or more enzymes in a metabolic sequence, total metabolite concentrations are very low. For one or more enzymes in a metabolic sequence, total activity and Km for the substrate are used to calculate the maximum rate of substrate conversion from the known concentrations of substrates in the matrix during the operation of the process. In cases where the

TABLE 1. Ratios of reaction rates when permeabilized and intact chloroplasts were incubated with exogenous substrates.

Reaction		CoA → Ac-CoA	CoA + ADP → Ac-CoA	CoA + ATP →Ac-CoA	ADP → ATP	NADP → NADPH
Rate of Reaction	(Permeabilized/ Intact)	> 40	> 25	> 6	> 10	> 100

Intact chloroplasts were added to basal reaction media containing acetate in 330 mM (isotonic) or 66 mM (hypotonic) sorbitol. Following suspension in the latter, chloroplasts are considered permeabilized. Additions to the reaction media and products of the reactions were as shown. Ac = acetyl.

calculated rates for individual reactions fall far short of the rates measured for the whole process, substrate channeling may be assumed. Concentrations of CoA-SH, acetyl-CoA and malonyl-CoA were found to be < 0.1µM, 10-15 µM, and 0.1-3 µM respectively, during fatty acid synthesis from acetate by intact chloroplasts (G. Roughan, unpublished results). Using known Kms for and total activities of the enzymes, it may be calculated from these concentrations that maximum possible rates of the ACS and ACC reactions are just 5%, and of the malonyl-CoA:ACP transacylase reaction < 1%, of the rate of acetate incorporation into fatty acids. Hence, channeling of these substrates into fatty acid synthesis may be assumed.

Conclusion

Two different lines of evidence support the concept that fatty acid synthesis within chloroplasts is carried out by a multienzyme complex. Firstly, the reaction sequence proceeds as rapidly in permeabilized as in intact chloroplasts and is light dependent even in the presence of added cofactors. Therefore, the enzymes, substrates and cofactors of fatty acid synthesis in the residual stroma must remain integrated in some sort of complex that resists disruption and dilution by the hypotonic medium. The complex must also be linked to the energy generating reactions of thylakoids in some way to account for the light-dependency. Secondly, the concentrations of CoA-SH, acetyl-CoA, and malonyl-CoA measured within chloroplasts actively synthesizing fatty acids are clearly insufficient to drive acetyl-CoA synthetase, acetyl-CoA carboxylase, and malonyl-CoA:ACP transacylase at the rates required to account for rates of acetate incorporation into fatty acids. To allow the rates of fatty acid synthesis measured, those substrates may not be distributed evenly throughout the chloroplast stroma but must be concentrated in the vicinity of the active sites on the enzymes. That is, the substrates must be metabolized by being channeled through a multienzyme complex.

Acknowledgments

This work was supported in part by grants from the New Zealand Marsden Fund (GR) and from the U.S. National Science Foundation (JO).

References

1. Caughey, I., and Kekwick, R.G.O.: *Eur. J. Biochem.* **133** (1982), 553-561.
2. Hoj, P.B., and Mikkelsen, J.D.: *Carlsberg Res. Commun.* **47** (1982), 119-141.
3. Shimakata, T., and Stumpf, P.K., *Plant Physiol.* **69** (1982), 1257-1262.
4. Roughan, P.G., Kagawa, T. and Beevers, H. : *Plant Sci. Lett.* **18** (1980), 221-228.
5. Roughan, P.G., Holland, R., and Slack, C.R.: *Biochem. J.* **184** (1979), 565-569.
6. Beittenmiller, D., Roughan, P.G., and Ohlrogge, J.B.: *Plant Physiol.* **100** (1992), 923-930.
7. Roughan, P.G., and Ohlrogge, J.B.: *Anal. Biochem.* **216** (1993), 77-82.
8. Kleinig, H.: *Ann. Rev. Plant Physiol.* **40** (1989), 39-59.
9. Roughan, G., and Ohlrogge, J.B.: *Plant Physiol.* **110** (1996), 1239-1247.

STRUCTURE-FUNCTION STUDIES ON DESATURASES AND RELATED HYDROCARBON HYDROXYLASES

John Shanklin[1*], Edgar B. Cahoon[1], Edward Whittle[1],
Ylva Lindqvist[2], Weijun Huang[2], Gunter Schneider[2], and
Hermann Schmidt[3].
1, Biology, Brookhaven National Laboratory, Upton NY 11973
* e-mail shanklin@bnl.gov
2, Karolinska Institute, Stockholm, Sweden;
3, University of Hamburg, Germany.

All organisms require unsaturated fatty acids. This is because saturated fatty acids have high melting points and thus exist as solids at room temperature, and the correct functioning of biological membranes requires unsaturated fatty acids to provide membrane fluidity. Desaturation is a highly energy-demanding reaction because the carbon-hydrogen bond of a methylene group in an acyl lipid is one of the most stable bonds found in nature (98 kCal/mol), thus, desaturase enzymes have to be able to mediate high energy chemistry. Such high energy reactions typically involve a metal cofactor and molecular oxygen. Bloch's group was the first to demonstrate a requirement for both iron and oxygen in fatty acid desaturation.

Fatty acid desaturases fall into one of two distinct classes. The first is the soluble class found in the plastids of plants and in Euglena. Its archetype is the Δ^9 18:0-ACP desaturase, though several, less widespread, acyl-ACP desaturases have been identified that recognize substrates with shorter chain lengths and introduce double bonds at a variety of positions. The second class of desaturases are integral-membrane proteins. They are more widespread in nature, being found in some prokaryotes and almost all eukaryotes. The major impediment to studying these enzymes has been the difficulty in isolating sufficient quantities of protein from natural sources to perform biophysical characterizations. Overexpression of cDNAs encoding soluble desaturases in *E. coli* has facilitated the isolation of gram quantities of these enzymes, but to date, the membrane desaturases have not been overexpressed to such levels. However, we have successfully overexpressed the alkane hydroxylase from *Pseudomonas oleovorans* in *E. coli*, and we are currently using this enzyme as a model system to investigate the active sites of the membrane desaturases and hydroxylases. Recent advances in our understanding of each of these classes of enzymes will be described.

helices are part of a larger nine-helix bundle which is capped by a helix each at each end. The first interesting finding from the desaturase structure is that helices 1-8 and 10 have equivalents in both RB2 and MMO (see Fig. 1C for an overlay of the desaturase and MMO). What makes this remarkable is that aside from the diiron binding motif, there is almost no sequence homology between these enzymes. The conclusion we draw from this is that the fold of the protein and the diiron cluster ligands have been conserved, but that there has been no evolutionary conservation of amino acid side chains in the remainder of the enzymes, perhaps reflecting their diverse functions. There are several specific exceptions to this: T199, D223 and W62. T199 has an equivalent in MMO and cytochrome P450-camphor and lies in a hydrophobic cavity adjacent to the diiron center. It has been hypothesized to be involved in oxygen activation. D223 and W62 and one of the histidines that coordinate the iron center are conserved in RB2 and the desaturase. These residues have been proposed to mediate electron transfer into the iron center in RB2. We are in the process of evaluating the effects of site directed-mutants of these and other residues that may play a role in catalysis.

Identification of factors governing substrate binding and positional insertion of double bonds. A long hydrophobic cavity was identified in the crystal structure of the castor Δ^9 18:0-ACP desaturase leading from a depression in the surface to the diiron center and beyond (see Fig. 1D). The annulus surrounding the cavity is characteristic of those found at protein-protein interfaces and likely represents the ACP binding site. The cavity is long enough to accommodate a fully extended 18-carbon acyl chain. This is somewhat surprising as we had expected the acyl chain to enter the desaturase in some more energetically favored conformation such as a hairpin. However, there can be little doubt that this is the correct interpretation for several reasons. First, our modeling experiments suggest that the cavity is close to a perfect fit for oleic acid with its cis double bond. The second support for this interpretation comes from experiments designed to identify residues that determine chain length and regiospecific insertion of double bonds. For these experiments, the primary sequences of two *Thunbergia* enzymes representing the Δ^9 18:0-ACP and the Δ^6 16:0-ACP desaturases were compared, and chimeric desaturase cDNAs were constructed and evaluated for their chain length and positional specificity. Later, site-directed mutations were made in which specific residues from the Δ^9 18:0-ACP desaturase were substituted into the Δ^6 16:0-ACP desaturase. We were able to separate the two factors governing chain length and regiospecificity, and in one such construct we were able to convert a Δ^6 16:0-ACP desaturase into predominantly a Δ^9 18:0-ACP desaturase. The determinants of chain length specificity lie principally at the deepest part of the hydrophobic cleft, whereas the determinants of regiospecificity lie primarily between the annulus and the diiron cluster. It also appears that the determinants of chain length specificity exert direct effects because these residues lie at the bottom of the substrate pocket, whereas the determinants of regiospecificity lie back from the cavity and likely mediate more subtle interactions, perhaps by affecting the packing of the helices (see Fig. 1D).

THE SOLUBLE DESATURASES.

A combination of spectroscopies including Mössbauer and resonance Raman were used to identify the presence of a diiron-oxo cluster with properties similar to those identified in ribonucleotide reductase (RB2) and methane monooxygenase (MMO). These enzymes all share the ability to break unactivated carbon-hydrogen bonds with a nonheme diiron cluster cofactor. Fatty acid desaturation and methane oxidation require a two-electron reduction of the diiron cluster to initiate the oxygen activation reaction. Identification of a diiron cluster in the desaturase allowed us to propose a consensus diiron-oxo binding motif consisting of two repeats of (D/E)EXXH.

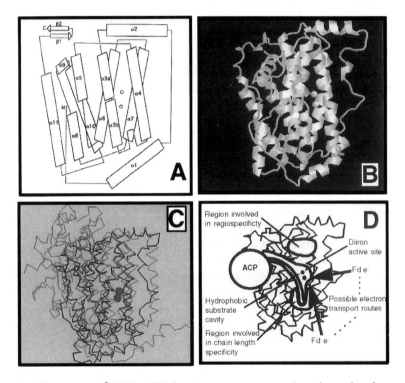

Figure 1. The castor Δ^9 18:0-ACP desaturase monomer. A, schematic of secondary structural elements; B, structural elements; C, overlay of the desaturase (darker) and MMO (lighter); D, cartoon of regions involved in the function of the desaturase.

Structure of the desaturase. Figures 1A and B show the arrangement of structural elements in the castor Δ^9 18:0-ACP desaturase monomer (Lindqvist et al. 1996). The molecule is almost entirely α-helical, with the exception of a small β-hairpin at the C-terminus. The core of the enzyme consists of a four helix bundle characteristic of all diiron proteins, which contains all the coordination ligands for the iron center. The consensus motif (D/E)EXXH is found in helices 4 and 7, and one additional carboxylate ligand is found in each of helices 3 and 6. These four

Active site and a proposed desaturase reaction cycle. Ligand arrangement for the iron ions in the oxidized desaturase, based on Mössbauer, resonance Raman spectroscopy, and crystallographic identification of the ligands, is shown in Fig. 2.

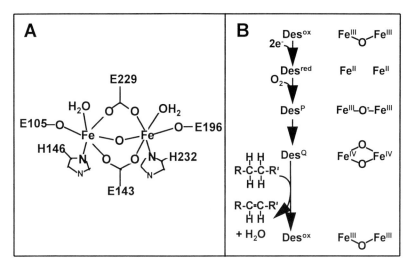

Figure 2. A, ligation sphere of the iron ions; B, proposed desaturase reaction cycle.

The active site in the crystal structure of the desaturase is the reduced form of the enzyme despite the fact that the oxidized protein was used for crystallization. The evidence for this is that the protein was likely photoreduced in the X-ray beam resulting in the long Fe-Fe distance (4.2 Å) in comparison with that determined by extended X-ray fine structure analysis, EXAFS, (3.1 Å) and is consistent with the lack of electron density that could be interpreted as an oxo bridge. This is consistent to that seen for RB2 in the reduced (FeII-FeII) form. Detailed investigations using a combination of Mössbauer, resonance Raman, EXAFS, and EPR spectroscopies has led to the identification of a common activated diiron-oxygen intermediate "Q" for RB2 and MMO (reviewed by Que and Dong 1996). Based on their common protein fold, diiron-cluster ligation sphere (Fig. 2A), and reaction chemistry, we propose the following activation cycle for the soluble desaturase (see Fig. 2B). Under this scheme, the oxidized form of the desaturase undergoes a two-electron reduction converting the iron ions from the III to the II form. This reduced cluster binds dioxygen to form Des^P, the peroxo intermediate. This isomerizes to form the diamond structure Des^Q equivalent to the activated form of both RB2 and MMO. Q is the high valent species that would have enough energy to abstract two hydrogens and oxidize them to water. Clearly, detailed spectroscopic studies of the desaturase are needed to test this hypothesized reaction scheme. In addition, it will be necessary to design experiments to determine if the reaction proceeds via radical or ionic intermediates, and to distinguish between sequential or concerted mechanisms.

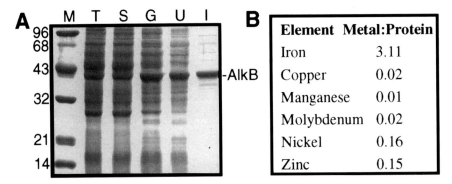

Figure 3. A, purification of AlkB; B, element analysis of AlkB by inductively coupled plasma emission spectroscopy (ICP) and quantitative amino acid analysis.

INTEGRAL-MEMBRANE CLASS OF DESATURASES AND HYDROXYLASES. The primary sequences of the integral-membrane protein class lack the soluble desaturase iron ligation motif $[(D/E)EXXH]_2$; however, they share a tripartite consensus sequence $HX_{(3or4)}H$, $HX_{(2\ or\ 3)}HH$, $HX_{(2or3)}HH$. Mutation of any of these histidines in either the rat Δ^9 18:0-CoA desaturase or the *Pseudomonas oleovorans* alkane hydroxylase AlkB results in a complete loss of activity, whereas mutation of adjacent histidines that are not part of the consensus motif has little effect on activity. The main objective for this class of enzyme is to obtain the 1-2 mM quantities of protein required for spectroscopic analysis, i.e., 50-100 mg/ml of purified protein. We have now achieved this for AlkB using a combination of differential ultracentrifugation and HPLC ion exchange chromatography (see Fig. 3). Surprisingly, we were able to purify this integral membrane protein to near homogeneity without detergent solubilization because it exists as a protein micelle consisting of 80+/-10 AlkB monomers each. The purified protein is typically 90+% pure as judged by Coomassie stained PAGE, has a specific activity of 4.5 U/mg, and contains 3.11+/- 0.2 moles of Fe per mole of AlkB (Fig. 3). Mössbauer spectroscopy confirms the presence of a diiron cluster, but the nature of the bridging oxygen species has yet to be defined. Many questions remain to be answered for the AlkB, including: what if any is the function of the extra mole of Fe, i.e., is there a second mono-Fe site, if so, is it needed for catalysis or to maintain structural integrity? Which histidines act as ligands to the diiron cluster? However, given the rarity of this motif in nature and the requirement of these histidines for catalysis, it seems likely that they are associated with the active site, and that the entire class of desaturases and hydroxylases will indeed contain diiron sites which mediate catalysis.

References:
Lindqvist Y., Huang W., Schneider G., and Shanklin, J. (1996) *EMBO J.* (in press)
Que, L. and Dong, Y. (1996) Acc. Chem. Res. 29:190-196.

ACETYL-COA CARBOXYLASE FROM *BRASSICA NAPUS*

Studies on the Type II multi-subunit form.

JONATHAN E. MARKHAM KIERAN M. ELBOROUGH AND
ANTONI R. SLABAS
University of Durham
Department of Biological Sciences, South Road, Durham DH1 3LE, U.K.

1. Introduction

Acetyl-CoA carboxylase (ACCase; E.C. 6.4.1.3) is a biotin containing enzyme found in all organisms and represents the first committed step towards fatty acid and flavanoid biosynthesis in higher plants. In the dicot *Brassica napus*, as in all other dicots studied to date, there exist two distinct forms of the enzyme [1, 2]. The Type I form has been well characterized in *Arabidopsis thalina* and *B. napus* and consists of a single polypeptide, some 220-250kDa in length. The Type II form has only recently been revealed in higher plants however. In contrast to the Type I form, Type II ACCase consists of four separate subunits, Biotin Carboxylase (BC), Biotin Carboxyl Carrier Protein (BCCP), and Carboxyl-Transferase α and β (CT α and β). This multi-subunit structure is consistent with the structure of acetyl-CoA carboxylase in prokaryotes and algae. The biotin containing protein of the Type II form has been shown to be located within purified chloroplasts from pea and oilseed rape [2]. The genetic organization of these subunits is also unique. The gene for CT β is encoded by a single gene located on the chloroplast genome whereas the other subunits are all nuclear encoded and are present as multiple copies in the *Brassica* genome. The genetic and multi-subunit structure of the Type II ACCase points to a complicated mechanism for co-regulation of expression of the three nuclear encoded genes, cross-talk with the chloroplast genome to regulate expression of the CT β subunit, and finally a molecular mechanism to arrange and organize the individual subunits into a complete and functional protein once translated into the chloroplast itself.
Biotin Carboxylase and Biotin Carboxyl Carrier Protein have been shown to co-purify and co-immunoprecipitate [3, 4] and biotin carboxylase will transfer CO_2 to BCCP or free biotin from HCO_3^-, in the absence of the carboxyl-transferase subunits [4]. Since the prokaryotic form of acetyl-CoA carboxylase is arranged in the same way as the plastid enzyme, we have attempted to reassemble two of the subunits of the *Brassica napus* Type II acetyl-CoA carboxylase by expression in *E. coli*, in order to demonstrate that our isolated genes have biological function.

2. Expression of BCCP and BC in *E. coli*

2.1 CONSTRUCTING THE EXPRESSION SYSTEM

We chose to use the pET system (Novagen) as this has previously produced large quantities of functional proteins with other genes in our hands. Three expression vectors were constructed;

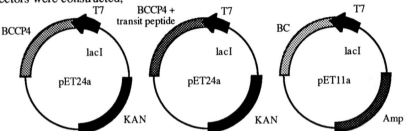

Fig 1: Plasmid maps of the three expression plasmids used to express Biotin Carboxyl Carrier Protein without transit peptide, BCCP with transit peptide and Biotin Carboxylase without transit peptide.

The plasmids were transformed into BL21(λDE3) in the following combinations; pETBC; pETBCCP4; pETBCCP4+transit peptide; pETBC and pETBCCP4; pETBC and pETBCCP+ transit peptide. Plasmids were either transformed singly, or where two plasmids were required, co-transformation was used with equal amounts of plasmid DNA. Transformants were allowed to grow overnight and transformed colonies were made into seed stocks for subsequent expression. A single seed stock was used to inoculate 100ml LB and grown to OD1.0 before inducing expression with 1.0mM IPTG. Cells were harvested after 3hrs growth and split into soluble and insoluble cell fractions by treatment with lysozyme and sonication. Protein expression was analyzed by SDS / PAGE and western blotting with anti-biotin antibodies.

2.2 SOLUBLE AND INSOLUBLE FRACTIONS FROM *E. COLI* EXPRESSING SUBUNITS OF *BRASSICA NAPUS* ACETYL -COA CARBOXYLASE

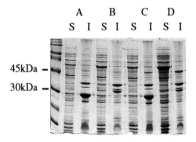

Fig. 2: 12 % polyacrylamide gel showing the soluble and insoluble fractions from *E.coli* overexpressing BCCP and BC. S, soluble fraction; I, insoluble fraction. Lanes A, pETBCCP4; B, pETBCCP4+transit peptide; C, pETBC and pETBCCP4; D, pETBC and pETBCCP4+transit peptide.

2.3 WESTERN BLOT OF SOLUBLE FRACTIONS USING ANTI-BIOTIN ANTIBODIES

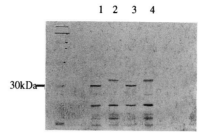

Fig. 3: Western analysis of the soluble fraction from *E. coli* over expressing BCCP and BC. The biotinylated proteins were detected with anti-biotin antibodies. 1, pETBCCP4; 2, pETBCCP4+transit peptide; 3, pETBC and pETBCCP4; 4, pETBC and pETBCCP4+transit peptide.

3. Discussion

We have overexpressed the *B. napus* biotin carboxyl carrier protein and biotin carboxylase. Whilst most of the protein is in an insoluble form, a fraction of the BCCP is present as soluble protein and this may or may not be complexed with the biotin carboxylase. Initial gel permeation experiments give inconclusive results, in that no substantial difference in flow through the column is seen if BC is expressed in the bacteria with BCCP or not. On the other hand, the initial elution fraction for the overexpressed BCCP equates to a molecular weight of some 60-70kDa, the same fraction as the *E. coli* BCCP elutes in. Clearly further experimentation is required to asses the degree of interaction and what proteins BCCP will interact with.
Similar experiments with other multi-subunit proteins have shown that it is often very difficult to overexpress a single subunit of a multi-subunit protein without it becoming insoluble. This may be because there are hydrophobic facets of the protein which would normally interact with the other subunits, but in the absence of other the other subunits they aggregate into an insoluble inclusion body.

4. References

1. Elborough, K.M., Swinhoe, R., Winz, R., Kroon, J.T.M., Farnsworth, L., Fawcett, T., Martinezrivas, J.M., and Slabas, A.R. (1994) Isolation of cDNAs from *Brassica-napus* encoding the biotin-binding and transcarboxylase domains of Acetyl-CoA Carboxylase - assignment of the domain-structure in a full-length *Arabidopsis-thaliana* genomic clone, *Biochemical Journal* **301**(2), 599-605
2. Elborough, K.M., Winz, R., Deka, R.K., Markham, J.E., White, A.J., Rawsthorne, S., Slabas, A.R. (1996) Biotin Carboxyl Carrier Protein and Carboxyltransferase of the multisubunit form of Acetyl-CoA carboxylase from *Brassica-napus* - cloning and analysis of expression during oilseed rape embryogenesis, *Biochemical Journal* **315**(1), 103-112
3. Alban, A., Jullien, J., Job, D., and Douce, R., (1995) Isolation and Characterisation of Biotin Carboxylase from Pea Chloroplasts, *Plant Physiology* **109**, 927-935.
4. Roesler, K.R., Savage, L.J., Shintani, D.K., Shorrosh, B.S., and Ohlrogge, J.B., (1996) Co-purification, co-immunoprecipitation, and coordinate expression of acetyl-coenzyme A carboxylase activity, biotin carboxylase, and biotin carboxyl carrier protein of higher plants, *Planta* **198**(4), 517-525

TRANSGENIC MODIFICATION OF ACETYL COA CARBOXYLASE AND ß-KETO REDUCTASE LEVELS IN *BRASSICA NAPUS* FUNCTIONAL AND REGULATORY ANALYSIS

Andrew J. White[1], Kieran M. Elborough[1], Helen Jones[2], Antoni R. Slabas[1]
1. Lipid Molecular Biology Group, University of Durham, DH1 3LE, UK
2. ZENECA Agrochemicals, Jealott s Hill Research Station, Bracknell, RG42 6EY, UK

Introduction

In plants *de novo* fatty acid biosynthesis occurs within the plastids. The first committed step in this process is the ATP dependant carboxylation of acetyl CoA to malonyl CoA. This is catalysed by acetyl CoA carboxylase (ACCase, EC 6.4.1.2), which is believed to be a key regulatory enzyme in this process (Post Beittenmiller *et al.* 1992). Two forms of ACCase have now been identified in dicots. A single large multifunctional polypeptide (ACCase I), and a multipeptide prokaryotic form (ACCase II). Subcellular localisation studies on ACCase II (Alban *et al.* 1995; Shorrosh *et al.* 1995; Elborough *et al.* 1996) indicate a chloroplast localisation. This is suggestive of a role in *de novo* fatty acid biosynthesis. The extrachloroplastic localisation (Alban *et al* 1994) of ACCase I suggests a different role. ACCase I is responsive to UV and fungal elicitors possibly indicating a role in cytosolic fatty acid elongation, and wax and flavanoid biosynthesis.
To further our understanding of lipid biosynthesis and its regulation in a major oil producing crop, we have generated *Brassica napus* plants containing antisense constructs to ACCase I . ß-Ketoacyl ACP reductase, is a central component of the fatty acid synthase complex, and in a complementary experiment we have also generated antisense plants to this. Each construct was placed under the control of both a constitutive CaMV 35S promoter and a seed specific promoter.

Results
CONSTRUCTION OF TRANSGENIC *B. napus* PLANTS
cDNA clones from *B. napus* were digested to release a 2.5 kb fragment containing the transcarboxylase domain of ACCase I and a 1.1kb fragment from ß-Ketoacyl ACP reductase. These were directly subcloned into the binary vetor pJRIi under the control of a CaMV promoter. For the ACP constructs DNA fragments containing coding regions of the genes were initially subcloned into the vector pAP1GUS from which the *gus* A gene had

been removed. The cassette containing the ACP promoter and antisense gene was isolated and ligated into pJRIi removing the CaMV promoter (Fig.1).

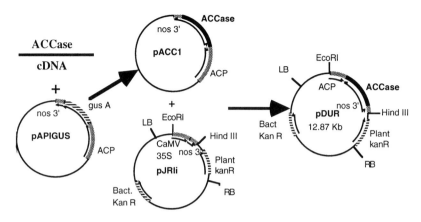

Fig.1. Construction of anti ACCase / ACP vector

ANALYSIS OF TRANSGENIC PLANTS.

Initial selection of F_1 plants carrying the transgene was carried out by PCR using primers specific to the *npt* II gene. Southern hybridisation of genomic DNA extracted from each transgenic plant confirmed these findings and has to date indicated 11 independant transformation events for the 4 constructs, each containing between 1 and 8 inserts.

Leaf and embryo tissue was harvested at equivalent developmental stages from both control and transgenic plants, during maximal expression of FAS genes. RNA was extracted using the TRIzOL procedure (Gibco BRL) according to the manufacturers instructions. Expression of the antisense constructs for both ACCase and ß-Ketoreductase has been confirmed in both leaf and embryo tissue via RT-PCR and northern hybridisation. A purified antibody produced from the overexpression of the rape ß-Ketoreductase was used to identify the affects of antisense expression on steady state levels of the protein. Crude protein extracts were obtained from the same samples as used for RNA extraction in order to provide a direct comparison. Equal amounts of protein extract, as measured by Bradfords reagent, were separated by SDS-PAGE and western blotted. Immunodetection has shown that the ß-Ketoreductase levels observed are reduced compared to those in the control plants (Fig 2.). In addition, the level of down regulation appears to vary between different transgenic lines. This would suggest that the antisense constructs prepared are effective in the down regulation of ß-Ketoreductase.

1 2 3 4 5 6 7 8

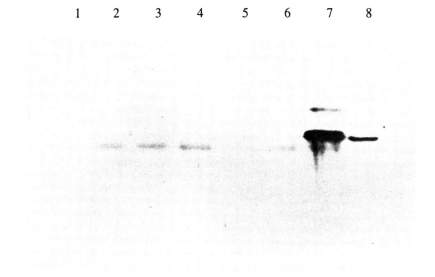

Fig..2. Immunodectection of ß-Ketoreductase in control and transgenic leaf extracts. Lane 1-6 Transgenic plant extracts from 7 day old leaves 25µg/ lane. Lane 7 & 8, 25µg of extract from Westar control plants. Lane 7, 3 day old leaf; Lane 8, 7 day old leaf.

Discussion

We have shown the controlled down regulation of ß-Ketoreductase levels and the expression of ACCase I antisense constructs in *B.napus*. At present their affect on the end products of fatty acid biosynthesis are being examined. Similar phenotypic alterations have been observed in some of these transgenic lines. The more pronouced alterations from the wildtype appearing to occur in those lines containing multiple inserts. These transgenics will aid studies into the regulation of other FAS components.

References

Alban C, Baldet P, Douce R (1994) Biochem J 300:557-565
Alban C, Jullien J, Job D, Douce R (1995) Plant Physiol 109: 927-935
Elborough KM, Winz R, Deka RK, Markham JE, White AJ, Rawsthorne S, Slabas AR (1996) Biochem J 315: 103-112
Post Beittenmiller D, Roughan G, Ohlrogge JB (1992) Plant Physiol 100:923-930
Shorrosh BS, Roesler KR, Shintani D, van de Loo FJ, Ohlrogge JB (1995) Plant Physiol 108: 805-812

ENZYME STUDIES ON ISOFORMS OF ACETYL-COA CARBOXYLASE

L.J. PRICE, D. HERBERT, C. ALBAN[+], D. JOB[+], D.J. COLE[*], K.E. PALLETT[*] AND J.L. HARWOOD
School of Molecular and Medical Biosciences, University of Wales, Cardiff, CF1 3US, U.K., [+]Rhone-Poulenc Agrochimie, 69263 Lyon, France and [*]Rhone-Poulenc Agriculture, Ongar, Essex, CM5 OHW, U.K.

Introduction

Acetyl-CoA carboxylase is recognised generally as being a key enzyme in acyl lipid formation. In plants it has been shown to be important as a regulatory enzyme in light-driven lipid synthesis [1] and has a high flux control coefficient under such conditions [2]. Recently, characterisation of different isoforms of acetyl-CoA carboxylase from plants has been made. In Poaceae such as maize there are two multifunctional proteins [3]. By contrast, the dicotyledon pea contains a multienzyme complex form of acetyl-CoA carboxylase in the mesophyll chloroplasts but a multifunctional protein isoform in the cytosol of epidermal cells [4].

Since acetyl-CoA carboxylase exerts strong flux control for leaf lipid synthesis, it is not surprising that it is a good target site for herbicides. What was unexpected is that such herbicides (the aryloxyphenoxypropionates (or fops) and the cyclohexanediones (or dims)) are selective for grasses. The reason for the selectivity has been shown to be substantially based on target-site differences [5]. In order to understand the reason(s) for different graminicide sensitivities in various acetyl-CoA carboxylases we have carried out some detailed enzymatic studies.

Experimental Procedures

Methods for the growth of plant material and assay of acetyl-CoA carboxylase (ACCase) or propionyl-CoA carboxylase activity are given in [6]. Purification of isoforms of maize ACCase and kinetic analysis is given in [7] and purification of Poa annua ACCase in [8].

Results

Two multifunctional forms of maize ACCase were separated by a three-step protocol [7]. The major isoform, termed ACCase 1, represented 80-85% of the total recovered activity and was sensitive to quizalofop, fluazifop or sethoxydim. It was localised in the chloroplast stroma of mesophyll cells. The minor isoform, ACCase 2, was relatively insensitive to graminicides and was shown by immunofluorescent microscopy to be concentrated in the cytosol of epidermal cells [9]. Both enzymes were multifunctional proteins of similar molecular mass which functioned as homodimers. The Km values of the three substrates (ATP, HCO_3^-, acetyl-CoA) for these two ACCase isoforms were similar (Table 1).

More detailed kinetic studies of the two maize isoforms were carried out. These showed that both their reaction mechanisms were Ter-Ter. Furthermore, the overall reaction was very close to an ordered Ter-Ter mechanism [7]. Thus, although maize ACCase 1 was considerably more sensitive to graminicides (2000-fold more sensitive to quizalofop, 150-fold more sensitive to fluazifop) than ACCase 2, both enzymes are very similar to each other in terms of their molecular form and mass, their kinetic parameters and reaction mechanism.

Poa annua (a naturally-resistant grass weed) also contained two multifunctional protein isoforms of ACCase. Partly purified preparations gave similar kinetic parameters to those for maize ACCases but showed a Ping-Pong reaction mechanism [8]. Whether this latter property relates to the insensitivity of the Poa ACCase towards graminicides is unclear but in pea the multienzyme complex form of ACCase is thought to be insensitive by virtue of its different molecular form [4].

Since the multifunctional forms of plant ACCase function as dimers, we examined the binding to and subsequent inhibition of the enzyme by graminicides. Curve fits for the binding of quizalofop or fluazifop were generated for an equation of simple hyperbolic inhibition or the Hill equation. Maize ACCase 1 exhibited little cooperativity in binding graminicides but ACCase 2 (of reduced sensitivity) showed

TABLE 1. Some characteristics of purified maize ACCases[6]

	ACCase 1	ACCase 2
Subunit mass (kDa)	227	219
Native structure	Dimer	Dimer
pH optimum	8.0	7.5
Km AcCoA (μM)	122±15	113±12
Km ATP (μM)	80±9	76±8
Km HCO_3^- (μM)	1091±242	397±60
I_{50} quizalofop (μM)	0.03	60
I_{50} fluazifop (μM)	10	1500
Mechanism	Ter-Ter	Ter-Ter

TABLE 2. Graminicide binding characteristics (apparent Hill coefficients) for three ACCase preparations

Herbicide used	Maize ACCase 1	Maize ACCase 2	Poa annua
Quizalofop	0.86±0.03	1.85±0.19	0.54±0.02
Fluazifop	1.16±0.09	1.59±0.09	-

strong positive cooperativity for herbicide binding (Table 2). Furthermore, the Poa annua ACCase preparation also showed cooperativity in binding (in this case negative). It is of note that, in a study of biotypes from Lolium multiflorum, the resistant biotype also showed strong negative cooperativity (n_{app} 0.62) while the susceptible biotype showed little cooperativity (n_{app} 1.2) [10].

Conclusions

We have reported detailed kinetic analyses of ACCases of different graminicide sensitivities [6-8]. The only consistant property which parallels herbicide insensitivity is cooperativity in herbicide binding to the two halves of the ACCase dimer. How this relates to the subtle changes in protein structure which confer resistance must await further research.

References

1. Post-Beittenmiller, D., Roughan, P.G. and Ohlrogge, J.B. (1992) *Plant Physiol.* **100**, 923-930.
2. Page, R.A., Okada, S. and Harwood, J.L. (1994) *Biochim. Biophys. Acta* **1210**, 369-372.
3. Herbert, D., Alban, C., Cole, D.J., Pallett, K.E. and Harwood, J.L. (1994) *Biochem. Soc. Trans.* **22**, 261.
4. Alban, C., Baldet, P. and Douce, R. (1994) *Biochem. J.* **300**, 557-565.
5. Harwood, J.L. (1991) In Target Sites for Herbicide Action (Kirkwood, R., ed.) pp. 57-94, Plenum Press, New York.
6. Herbert, D., Harwood, J.L., Cole D.J. and Pallett, K.E. (1995) *Brighton Crop Protec. Conf.* 387-392.
7. Herbert, D., Price, L.J., Alban, C., Dehaye, L., Job, D., Cole, D.J., Pallett, K.E. and Harwood, J.L. (1996) *Biochem. J.*, in press.
8. Herbert, D., Cole, D.J., Pallett, K.E. and Harwood, J.L. (1996) *Phytochemistry* submitted.
9. Robertson, E.J., Price, L.J., Herbert, D.J., Alban, C., Leech, R.M. and Harwood, J.L. (1995) *Plant Physiol.* submitted.
10. Evenson, K.J., Gronwald, J.W. and Wyse, D.L. (1994) *Plant Physiol.* **105**, 671-680.

EXPRESSION OF THE ACETYL-COA CARBOXYLASE MULTI-SUBUNIT ENZYME IN OILSEED RAPE

D.BURTIN AND S. RAWSTHORNE
Brassica and Oilseeds Research Department, John Innes Centre, Norwich UK.

Introduction

Acetyl-CoA carboxylase (ACCase) catalyses the ATP-dependant carboxylation of acetyl-CoA to form malonyl-CoA, thus providing the essential substrate for fatty acid biosynthesis. Dicotyledonous plants contain two forms of ACCase: a multifunctional enzyme (typeI) wich is presumed to be cytosolic, and a multi-subunit complex (typeII) located in the plastid wich is responsible for de novo fatty acid synthesis. In prokaryotes, the ACCase is a type II enzyme comprising biotin carboxylase (BC), biotin carboxyl carrier protein (BCCP) and a carboxyl transferase with two subunits (CTα and CTβ). The cDNA encoding the *B.napus* CTβ and BCCP have already been cloned (Elborough *et al.*,1995) as have the cDNAs encoding the BC and BCCP from tobacco and *Arabidopsis* respectively (Shorrosh *et al.*, 1995; Choi *et al.*, 1995).
To study the regulation of the two forms of acetyl-CoA carboxylase in relation to lipid metabolism in *B.napus* we undertook the isolation of cDNAs encoding the BC and the CTα subunits.
Oligonucleotide primers representing conserved amino acid motifs in the prokaryotic and eukaryotic BCs were used to amplify, by PCR, a single product from *B.napus* DNA. Sequencing of this product revealed that the amino acid sequence of the deduced open reading frame was similar to known BCs. This PCR fragment and a partial clone of the pea IEP96 gene (Hirsh and Soll, 1995 ; the predicted protein product of wich shows homology, in part, to the *E.coli* CTα) were used to screen a developing embryo or cDNA library.

Cloning of biotin carboxylase

Fifteen clones from the developing embryo cDNA library were isolated and partially sequenced. The longest cDNA insert contained 1667 bp excluding the poly(A) tail, with 67 bp of 5' untranslated leader sequence. The deduced amino acid sequence encoded by the open reading frame consists of 533 amino acids. It has 82% and 52% identity with the *N.tabacum* and *E.coli* biotin carboxylases respectively. The *B.napus* biotin carboxylase contains an extra 70 aa N-terminal fragment compared to the prokaryotic forms. This fragment is presumed to be the chloroplast transit peptide.

Cloning of carboxyl transferase CTα

Three clones with an N-terminal aa sequence showing 50% identity with the *E.coli* CTα were obtained by screening a developing embryo cDNA library with a partial clone of IEP96. A 5' end fragment of the longest *B. napus* partial clone was used to rescreen the cDNA library, which also produced partial clones. The N terminal deduced amino acid sequence encoded by these clones showed 50 and 76 % identity with the *E.coli* CTα and the pea IEP96 proteins respectively. The 5' end of the *B.napus* "CTα" was obtained by 5'RACE.
Out of the 760 aa of the deduced amino acid sequence 308 residues (92-400) align with the prokaryotic CTαs. An extra T, S, R rich 91 aa polypeptide (probably the chloroplast transit peptide) is present at the N-terminus part of the protein (1-91). The C-terminal aa sequence (360-760) shows homology only with the pea IEP96 and the *Glycine max* "CTα".

Expression of ACCase genes in *B.napus*

RNA isolated from a number of different tissues of *B.napus* cv Topas plants was analysed by northern blotting. The accumulation of BC and "CTα" transcripts was compared to that of ACCaseI mRNA (ACCase multifunctional).
The expression of the three genes showed the same pattern. The transcripts were detected in all the tissues tested with the strongest expression in developing embryos.
Developmental analysis of expression in the embryos showed that BC "CTα" and ACCaseI were expressed at highest levels in the early to mid cotyledon stages and then decline by the mid-late stage (Figure 1).

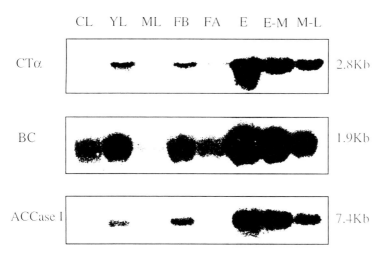

Figure 1. Nothern analyses of CTα, BC and ACCase I expression in B.napus. 15 mg of total RNA were loaded per tract. CL, cotyledon leaf; YL, young leaf; ML, mature leaf; FB, flower bud; FA, flower at anthesis; E, embryo early cot; E-M, embryo early-mid cot; ML, embryo mid-late cot.

Kang *et al.* (1994) showed that the rate of lipid accumulation/mg embryo fresh weight increased from the early-mid to the mid-late cotyledon stage whereas total ACCase activity/mg embryo fresh weight declined during the same period. The analysis of the expression of two subunits of ACCase II and of ACCase I correlated well with the ACCase activity described previously.

Conclusion

The genes encoding four subunits of the *B.napus* Accase II have now been cloned. However, the functional characterisation of the *B.napus* "CTα" has to be clearly established.
Expression analysis showed that ACCase I and ACCase II followed the same pattern of expression in leaves, flowers and during embryo development. The transcript accumulation of these genes correlated well with the total ACCase activity but not with lipid accumulation.

References

Choi,J.K. *et al.*, (1995) Molecular cloning and characterization of the cDNA coding for the biotin-containing subunit of the chloroplastic acetyl-CoA carboxylase. Plant Physiol. 109: 619-625

Elborough, K.M. *et al.*, (1995) Biotin carboxyl carrier protein and carboxyltransferase subunits of the multi subunit form of acetyl-CoA carboxylase from *Brassica napus* : Cloning and analysis of expression during oilseed rape embreyogenesis. Biochemical J. 315:103-112.

Hirsh, S. and Soll,J. (1995) Import of a new chloroplast inner envelope protein is greatly stimulated by potassium phosphate. Plant Mol. Biol. 27:1173-1181.

Kang *et al.*, (1994) The activity of acetyl-CoA carboxylase is not correlated with the rate of lipid synthesis during development of oilseed rape (*Brassica napus L.*) embryos. Planta 193: 320-325.

Shorrosh, B.S. *et al.*, (1995) Structural analysis, plastid localization, and expression of the biotin carboxylase subunit of acetyl-coenzyme-A carboxylase from tobacco. Plant Physiol.108: 805-812.

BIOTIN CARBOXYL CARRIER PROTEIN AND BIOTIN CARBOXYLASE SUBUNITS OF THE MULTI SUBUNIT FORM OF ACETYL COA CARBOXYLASE FROM *BRASSICA NAPUS*

KIERAN M. ELBOROUGH, JONATHAN E. MARKHAM, I. MARTA EVANS, *ROBERT WINZ, ANDREW J. WHITE, ANTONI R. SLABAS
Biological Sciences Department, University of Durham, South Road, Durham. DH1 3LE. U.K.
Nara Institute of Science & Technology, 8916-5 Takayama-cho, Ikoma-shi Nara-ken 630-01, JAPAN

1. Introduction

Plant acetyl CoA carboxylase [EC 6.4.1.3] is one of the pivotal enzymes of fatty acid biosynthesis in both seed and leaf tissue and is thought to be an important regulatory step of *de novo* fatty acid synthesis in chloroplasts (Post Beittenmiller *et al.*, 1992). Its central role reflects its importance as a target for commercial herbicides. Two forms of ACCase are present in dicot plants. The chloroplast is thought to be the site for *de novo* fatty acid synthesis in mesophyll cells and BCCP has been shown to reside within the chloroplast. It is therefore reasonable to suppose that a type II *Brassica napus* ACCase is mainly associated with *de novo* lipid synthesis. Specific herbicides differentiate between the two forms of acetyl CoA carboxylase (ACCase) found in dicotyledonous plants.

2. Results and Discussion

We have cloned the two BC and BCCP subunits of the *Brassica napus* multi-subunit form of ACCase. For the BCCP we isolated 6 different cDNA's (BCCP 1,2,3,4,6& 7)from an embryo library (Elborough *et al.*, 1996). The phylogenetic tree (fig 1.) shows that these clones could be grouped into two distinct classes as shown below.

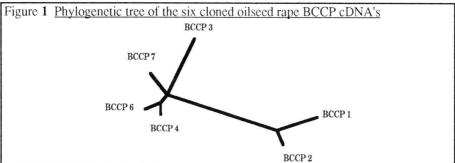

Figure 1 Phylogenetic tree of the six cloned oilseed rape BCCP cDNA's

The BC had previously been cloned from tobacco (Shorrosh *et al.*, 1995). We cloned the *Brassica napus* gene by a two step PCR protocol using high fidelity polymerase (figure 2). The 5' primer 1 was the forward sequencing primer whilst the internal primers (3&4) were derived from the available Arabidopsis EST data. The 3' primer 2 was the reverse sequencing oligonucleotide. A λ ZapII rape embryo library was used as the template. The two products obtained (one from primers 1&4 and one from 2&3) were fully sequenced and compared over their shared 400bp. The two sequences were identical in this region. The two products were stitched together using a the convenient *Nco*I restriction site. The full sequence has been submitted to the EMBL data base but is shown as AA sequence in figure 2B.

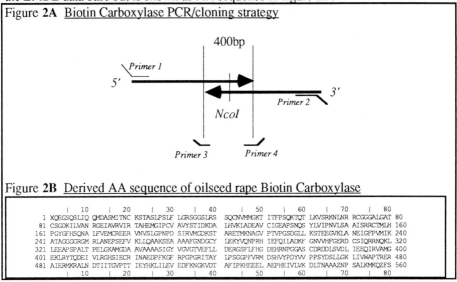

Figure 2A Biotin Carboxylase PCR/cloning strategy

Figure 2B Derived AA sequence of oilseed rape Biotin Carboxylase

```
        |   10     |   20     |   30     |   40     |   50     |   60     |   70     |   80
    1 XQEGSQSLIQ QMDASMITNC KSTASLPSIF LGRSGGSLRS SQCNVMMGKT ITFPSQKTQT LKVSRKNLNR RCGGGALGAT  80
   81 CSGDKILVAN RGEIAVRVIR TAHEMGIPCV AVYSTIDKDA LHVKLADEAV CIGEAPSNQS YLVIPNVLSA AISRRCTMLH 160
  161 PGYGFHSQNA LFVEMCREER VNVSLGPNPD SIRVMGDKST ARETMKNAGV PTVPGSDGLL KSTEEGVKLA NEIGFPVMIK 240
  241 ATAGGGGRGM RLANEPSEFV KLIQAAKSEA AAAFGNDGCY LEKYVQNPRH IEFQILADKF GNVVHFGERD CSIQRRNQKL 320
  321 LEEAPSPALT PELGKAMGDA AVAAAASIGY VGVGTVEFLL DERGSFLFHG DEHRNPGGAS CDRDDLSVDL IEEQIRVAMG 400
  401 EKLRYTQDEI VLRGHSIECR INAEDPFKGF RPGPGRITAY LPSGGPFVRM DSHVYPDYVV PPSYDSLLGK LIVWAPTRER 480
  481 AIERMKRALN DTIITGVPTT IEYHKLILEV EDFKNGKVDT AFIPKHEEEL AEPHEIVLVK DLTNAAAZNP SALKMKQZFS 560
        |   10     |   20     |   30     |   40     |   50     |   60     |   70     |   80
```

The resulting hybrid clone sequence showed very high homology with that of the E. coli and Tobacco BC at the amino acid level (figure 3).

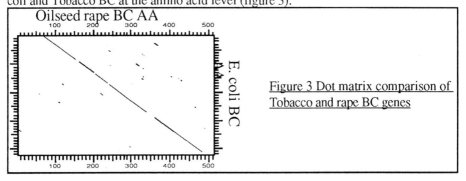

Figure 3 Dot matrix comparison of Tobacco and rape BC genes

In order to analyse the role of this form of ACCase we needed to look at both the relative expression of both subunits during oilseed rape embryogenesis. Figure 3 shows a Northern blot using both the BCCP and BC genes as probes against mRNA from five stages of rape embryogenesis. The peak of lipid synthesis occurred at stage 3 (data not shown). There were several notable features i) both BCCP and BC show very similar expression during oilseed rape embryogenesis, ii) The BC relative to BCCP was expressed at higher levels in leaf and root and iii) the full length transcript size shows that we had cloned both full length BCCP and BC clones.

Southern blot analysis using the BC clone showed that there are approximately four separate genes in oilseed rape encoding Biotin Carboxylase (data shown in poster).

Since the Northern data showed that a significant amount of BC was expressed in embryo, root and leaf it may be that the four genes code for temporal and tissue specific isoforms. Given the four BC cDNA's it may be possible to isolate the corresponding specific promoters. Since the BC gene is expressed at high levels in embryo's, with a peak during maximal lipid synthesis, its promoter could be used generally for use in antisense FAS constructs.

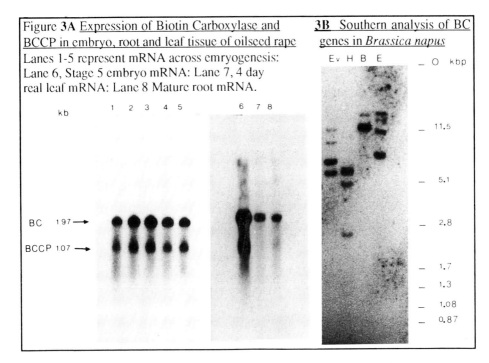

Figure 3A Expression of Biotin Carboxylase and BCCP in embryo, root and leaf tissue of oilseed rape
Lanes 1-5 represent mRNA across emryogenesis: Lane 6, Stage 5 embryo mRNA: Lane 7, 4 day real leaf mRNA: Lane 8 Mature root mRNA.

3B Southern analysis of BC genes in *Brassica napus*

In recent years it has become apparent that dicotyledonous plant species contain two distinctive forms of the enzyme acetyl CoA carboxylase, type I a single polypeptide of >220 kDa, and, type II a complex of smaller proteins BC, BCCP and CTα and β. To fully achieve the potential for the genetic manipulation of lipid products from crops a detailed understanding of fatty acid synthesis is required. This is dependent upon a full characterisation of the proteins involved, and more specifically on the pivotal enzyme ACCase.

We have cloned and analysed both the BCCP and the BC subunits of type II ACCase. With the isolation of the other subunits and the purification of expressed gene products the stage will be set for the full characterisation of type II ACCase from plants at the protein level. Since the subunits are thought to form an *in vivo* complex by protein-protein interactions it may be important to study the subunits from the same plant source. In isolating two of the three or four subunit types we have gone some way to achieving this in the agriculturally relevant crop *Brassica napus*.

Refs.
Elborough *et al.*, Biochem J. (1996) **315**, 103-112
Post Beittenmiller *et al.*, Plant Physiology (1992) **100**, 923-930
Shorrosh *et al.*, Plant Physiology (1995) **108**, 805-812

ACETYL-COA CARBOXYLASE ACTIVITY IN THE OIL PALM

Ravigadevi Sambanthamurthi[*] and John B. Ohlrogge[+]
[*] Plant Science and Biotechnology Unit, Palm Oil Research Institute of Malaysia, P.O. Box 10620, 50720 Kuala Lumpur, Malaysia. [+]Dept. of Botany and Plant Pathology, Michigan State University, East Lansing, Mi 48824 - 1321, USA.

Abstract

Immunoblot experiments confirmed the presence of a multisubunit form of acetyl-CoA carboxylase in the oil palm. ACCase activity was low in various oil palm tissues investigated presumably because of dissociation of the multi-subunit form. Higher activity was observed in callus cultures. Preliminary data suggest positive correlation between ACCase activity and embryogenesis in callus cultures. Amplification by PCR using degenerate primers designed to a conserved region of biotin carboxylase and genomic DNA from *Elaeis guineensis* and *Elaeis oleifera* as templates yielded ~ 850bp fragments.

Introduction

Acetyl-CoA carboxylase (ACCase) catalyzes the ATP dependent carboxylation of acetyl-CoA to form malonyl-CoA. This reaction is the first committed step in fatty acid biosynthesis and is a potentially rate-limiting step in the process. Two forms of ACCase have been observed - a prokaryotic and a eukaryotic form. In the prokaryotic form, the biotin carboxylase, biotin carboxyl carrier protein and carboxyl transferase components of ACCase are associated with different polypeptides in a multi-subunit (MS) complex (Samols *et al*. 1988). In contrast, the eukaryotic form comprises multimers of a single multifunctional (MF) polypeptide of ~ 200 kDa. Dicots have been reported to have an MS plastidial ACCase and an MF extra-plastidial ACCase while grasses and other monocots are believed to have the MF ACCase in both plastids and extra-plastidial sources (Sasaki *et al*. 1995).

We report here the presence of a multi-subunit form of ACCase in the oil palm.

Materials and Methods

Oil palm fruits were obtained from PORIM Research Station, UKM. Mesocarp and kernel tissue were removed from fruits at 20 and 12 weeks after anthesis respectively. Callus cultures (5 days after sub-culture) and young leaves of clonal ramets were obtained from Cik Rohani Othman (Applied Tissue Culture Group, PORIM). ACCase activity was measured by the incorporation of radioactivity from $NaH^{14}CO_3$ into an acid-stable product using a modification of the procedure of Nikolau et al., (1981). Results are expressed as dpm incorporated into malonyl-CoA. Crude extracts of oil palm mesocarp and kernel were probed with rabbit antibodies to castor biotin carboxylase. 100µg protein were loaded on the gels for both extracts. PCR amplification of genomic sequences was carried out in 100 µl reaction mixtures containing 150 ng genomic DNA from *Elaeis guineensis* or *Elaeis oleifera* and 25 pmol each of degenerate sense (AC1) and antisense (AC2)primers designed to a conserved region of biotin carboxylase. Taq DNA polymerase (Boehringer) and other components were used according to suppliers recommendation.

Results and Discussion

Table 1 shows the ACCase activity in various oil palm tissues. The activity was low in all tissues studied and especially so for the mesocarp and kernel. Leaf and callus material had comparatively higher ACCase activity. Preliminary data (not shown here) also suggested positive correlation between embryogenesis (of callus) and ACCase activity. During *in vitro* embryogenesis of oil palm, there is, increased synthesis of both storage and polar lipids (Turnham and Northcote, 1984) and this is preceded by an increase in ACCase activity. Lines that are limiting for ACCase activity thus may have a reduced capacity to proliferate into embryoids.

Western blots of palm mesocarp and kernel were probed with castor biotin carboxylase antibodies. Fig. 1 shows the presence of an ~ 50kDa protein in both kernel and mesocarp cross-reacting with the antibodies. The oil palm thus contains an MS form of ACCase. The signal in kernel tissue was much stronger compared to mesocarp although the same amount of protein was loaded for both tissues. This difference may be attributed to the stage of development of the tissues. Besides the signal at 50 kDa, a weaker signal was obtained at ~38 kDa for kernel. Most monocots have been reported to have the MF form of ACCase. Roesler (1996) however, reported the presence of the MS form in some monocots including *Iris* and onion. The oil palm is yet another example of a monocot with the MS form of ACCase. The low ACCase activity observed in oil palm tissues can thus be a result of dissociation of the multi-subunits of ACCase during the extraction procedure.

Table 1: ACCase activity in various oil palm tissues.

Tissue	ACCase activity dpm/mg protein/30 min
Mesocarp	945
Kernel	1255
Leaf	4195
Callus (line LPR 165-12)	5230

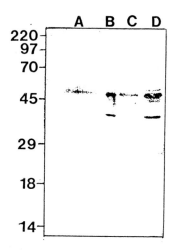

Fig. 1: Western blots of oil palm extracts probed with anti-biotin carboxylase. Lanes A and C: Crude mesocarp extract. Lanes B and D: Crude kernel extract.

References

Roesler, K.R., Savage, L.J., Shintani, D.K., Shorrosh, B.S. and Ohlrogge, J.B. (1996). Co-purification, co-immunoprecipitation and co-ordinated expression during oilseed development of acetyl-CoA carboxylase protein and biotin carboxyl carrier protein from higher plants, *Plant Physiol*.

Samols, D., Thornton, C.G., Murtif, V.L., Kumar, G.K., Haase, F.C. and Wood, G.H. (1988). Evolutionary conservation among biotin enzymes, *J. Biol. Chem.* **263**, 6461-6464.

Sasaki, Y., Konishi, T. and Nagano, Y. (1995). The compartmentation of acetyl-coenzyme carboxylase in plants, *Plant Physiol* 108, 445-449.

Turnham, E. and Northcote, D.H. (1984). The incorporation of [1-^{14}C] acetate into lipids during embryogenesis in oil palm tissue cultures. *Phytochemistry* **23**, 35-39.

SOYBEAN CHLOROPLAST ACETYL-COA CARBOXYLASE

S.V. REVERDATTO[1], V. BEILINSON[1], N.C. NIELSEN[1,2]
[1]*Department of Agronomy, Purdue University,* [2]*USDA-ARS, West Lafayette, IN 47907-1150, USA*

Acetyl-CoA carboxylase (ACCase, EC 6.4.1.2) catalyzes an ATP-dependent conversion of acetyl-CoA to malonyl-CoA, which is a regulated step in de novo synthesis of fatty acids that takes place in chloroplasts. In soybean, the chloroplast ACCase is apparently a multiprotein, prokaryotic type of complex [1], that is thought to consist of 4 subunits: biotin carboxylase (BC), biotin carboxyl carrier protein (BCCP), and the alpha and beta transcarboxylase subunits (α- and β-CT). The study of chloroplast ACCase by traditional methods of protein chemistry have proven difficult, hence we started with the isolation of ACCase genes and have initiated attempts to synthesize the subunits *in vitro* so that the interactions among the components of this enzyme can be studied more completely.

We recently reported the cloning of soy *accD* gene coding for the β-CT subunit as part of a bigger fragment of chloroplast DNA [2]. The screening of a soy cDNA library resulted in a full coding sequence for the BCCP subunit (gene accB-1) and several for α-CT. All α-CT clones have sequence differences which are very sparse in the coding part but are plentiful in 3'- and 5'- untranslated regions. Similar strategy with degenerated oligonucleotides allowed us to PCR a 871bp soy cDNA fragment highly homologous to tobacco and E. coli

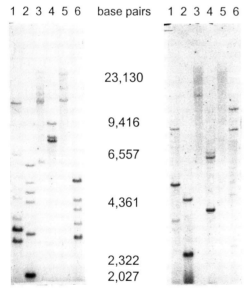

Figure 1. Southern blot analysis of soybean genomic DNA. Probes used are .αCT cDNA (A) and BCCP cDNA (B). Restriction enzymes used for digestions are (1) - Nde1; (2) - Ase1; (3) - Bsu36 1; (4) - EcoRV; (5) - Avr II; (6) - Afl II. All hybridizations were run at 60°C. Positions of lambda/HindIII markers are shown. Presented data suggest multiple gene patterns for both subunits.

BC genes. Some of the properties of the discussed genes are summarized in Table 1.

Southern blot analysis of soybean genomic DNA confirmed that BCCP and α-CT are encoded by multiple gene families (Fig. 1).

Deduced proteins have considerable similarity with corresponding E. coli subunits (67% for α-CT and 57% for BCCP), favoring their functional resemblance. Even more striking is the evident homology of the secondary structure patterns between E. coli and soy proteins (Fig. 3).

Being of nuclear origin, both α-CT1, α-CT2 and BCCP precursors are successfully imported into isolated pea

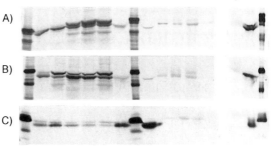

Figure 2. Import of the precursors of soybean ACCase into pea chloroplasts: A), α-CT1; B), α-CT2; C), BCCP. Translation reactions (Tr) products were imported at different concentrations of potassium phosphate (Kpi). A portion of import reactions was treated with thermolysine (Pr). Pea chloroplast were lysed and fractioned on sucrose step gradient into stroma (St), envelopes (En), fraction from 1M/1.2M sucrose interphase (In) and thylakoid membranes (Th) and lumen (Lm). Alternatively lysed chloroplasts were separated into stromal (St) and thylakoid (Ty) fractions by high speed centrifugation (RF). Proteins were separated by SDS-PAGE and radioautographed.

chloroplasts (Fig. 2) and are protected inside them from protease treatment. We did not observe any significant effect of K-phosphate on chloroplast import, except for α-CT2, which seems to be stimulated by its moderate (40-60mM) concentrations (Fig. 3). The distribution of imported proteins seems to be dependent on the fractionation scheme. Employing the protocol published by Bruce et al. [3], imported BCCP is found largely in stroma, but both α-CT proteins tend to stromal and chloroplast membrane fractions. In both cases no labelled product can be found in lumen (Fig. 3). The quick fractioning of lysed chloroplasts esults in almost 100% of the labelled product to accumulate in stroma. This might be due to the relative instability of the ACCase complex, and once released, individual subunits distributes according to their hydrophobic preferences, which probably guides some of the α-CT to the membrane fractions. This fits the recent hypothesis of Roughan and Ohlrogge [4] regarding the existence of a chain of multienzyme lipid-synthesizing complexes including ACCase, which is flexible enough to accommodate additions of several foreign proteins.

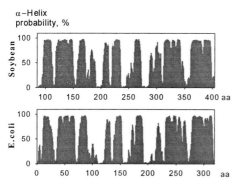

Figure 3. Predicted secondary structure alignment for α-CT subunits of E. coli and soybean chloroplastic ACCases. Probabilities are plotted against the amino acid sequences of homologous regions.

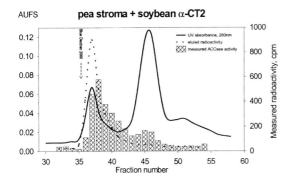

Figure 4 Coelution of the labeled soybean α-CT2, imported into isolated pea chloroplasts, with the pea ACCase activity during gel-filtration on Sephacryl S 300.

To test the functionality of the in vitro synthesized subunits we performed the fractionation of clarified pea stroma containing imported soy proteins by gel-filtration on Sephacryl S300. We found that the labelled soy proteins coelute with ACCase activity in high MW range (Fig.4), which we think is the evidence for their integration into pea ACCase. This supports our assumption that the discussed soy accA and accB genes are coding for the components of soybean chloroplastic ACCase complex, and that pea ACCase is structurally and functionally similar to the one from soybeans.

Sequencing of the labelled mature soy α-CT and BCCP, isolated from pea stroma after import of corresponding precursors, allowed us to determine the position of the transit peptide cleavage site, which is for α-CT2

.... -Thr^{45}-$\underline{Val}^{46}$-Ala^{47}-$\underline{Ala}^{48}$ ⇓ Lys^{49}-$\underline{Leu}^{50}$-Arg^{51}-Lys^{52}-$\underline{Val}^{53}$- ....

and for BCCP

.... -Arg^{44}-$\underline{Val}^{45}$-Lys^{46}-$\underline{Ala}^{47}$ ⇓ Gln^{48}-$\underline{Leu}^{49}$-Asn^{50}-Glu^{51}-$\underline{Val}^{52}$- ....

TABLE 1. Features of soybean chloroplastic ACCase genes and deduced proteins

Gene	Insert size	Coding region	Corresponding ACCase subunit	Deduced protein MW	Calculated pI
accD	5600bp	1296bp	β-CT	48.9kD	4.86
accB-1	1071bp	786bp	BCCP	27.7kD	10.22
accA-1	2839bp	2127bp	α-CT1	78.5kD	7.94
accA-2	2607bp	2070bp	α-CT2	76.9kD	8.96
accC(part)	871bp	867bp	BC(part)	30.3kD	-

Genes for soy chloroplastic ACCase subunits were expressed in E. coli, obtained material was used for generating antibodies which are currently being tested in our laboratory.

References

1. Sasaki Y, Konishi T, Nagano Y (1995) The compartmentation of acetyl-Coenzyme A carboxylase in plants. Plant Physiol 108:445-449.
2. Reverdatto S V, Beilinson V, Nielsen N C (1995) The rps16, accD, psaI, ORF 203, ORF 151, ORF 103, ORF 229 and petA gene cluster in the chloroplast genome of Soybean (Accession No. U26948)(PGR95-051). Plant Physiol 109(1): 338
3. Bruce B D, Perry S, Froehlich J, Keegstra K (1994) In vitro import of proteins into chloroplasts. In Gelvin S E, Shilperoort R A (eds.), Plant Molecular Biology Manual, Kluwer Academic Publishers, Dordrecht
4. Roughan P G, Ohlrogge J B (1996) Evidence that isolated chloroplasts contain an integrated lipid-synthesizing assembly that channels acetate into long-chain fatty acids. Plant Physiol 110:1239-1247

TWO FORMS OF ACETYL-COA CARBOXYLASE IN HIGHER PLANTS AND EFFECTS OF UV-B ON THE ENZYME LEVELS

Yukiko Sasaki and Tomokazu Konishi*
Laboratory of Plant Molecular Biology, School of Agricultural Sciences, Nagoya University, Nagoya, 464-01, Japan
*Laboratory of Molecular Genetics, Biotechnology Institute, Akita Prefectural College of Agriculture, 2-2 Minami. Ohgata, Akita, 010-04, Japan

Acetyl-CoA carboxylase (ACCase) catalyzes the first committed step in the fatty acid synthesis, namely, the formation of malonyl-CoA. Two forms of the enzyme, the prokaryotic and the eukaryotic forms, have been found in plants (1-4). The prokaryotic ACCase is composed of several subunits, one of which is encoded in plastid genome and named *acc*D, and exists in plastids. The eukaryotic ACCase is composed of a single multi-functional polypeptide and exists in cytosol. However, Gramineae do not have the prokaryotic ACCase in plastids, but have the eukaryotic ACCase because of the loss of *acc*D gene in plastids. The eukaryotic ACCase substitutes for the prokaryotic ACCase in Gramineae. It is unknown whether plants in other families loss the prokaryotic ACCase. To answer this question, we examined the prokaryotic ACCase in 28 plant families (5).

In plants, fatty acid synthase exists in plastids, so one function of the prokaryotic ACCase is to provide malonyl-CoA for *de novo* fatty acid synthesis. What is the role of the eukaryotic ACCase? Malonyl-CoA is used for flavonoid synthesis. We examined effect of UV-B on the two ACCases in pea seedlings because various enzymes involved in flavonoid synthesis are induced by UV-B.

1. ACCase substitution in plastids

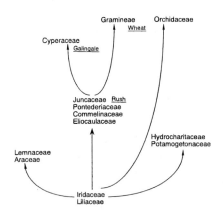

Fig. 1 The postulated origin of monocots

Prokaryotic ACCase is composed of four subunits, biotin carboxylase, biotin carboxyl carrier protein (BCCP), transcarboxylase-alpha and-beta subunits (*acc*D protein). To date *acc*D is found in all the plastid genomes sequenced except for Gramineae. The plastid genomes sequenced in monocots are limited to Gramineae, and we do not know whether the other monocots have *acc*D gene or not. It is postulated that Gramineae and Cyperaceae are evolved from Juncaceae as shown in Fig 1. So we examined whether Juncaceae and Cyperaceae plants have *acc*D gene using rush and galingale, respectively. Southern blot analysis showed that both plants have *acc*D gene.

Presumably, the loss of *acc*D gene is limited to Gramineae.

The prokaryotic ACCase is composed of four subunits. In these subunits, BCCP protein is biotinylated and easily detected by using streptavidin as a probe. Although higher plants contain several biotinylated polypeptides, these can be distinguished from BCCP, which yields a major band ca.35 kDa on gels after SDS-PAGE. We surveyed the biotinylated polypeptide of ca. 35 kDa in extracts from 28 plant families, namely, one Hepaticae, one Cycadopsida, one Coniferopsida, 8 Dicotyledoneae, and 17 Monocotyledoneae families. All the tested plants except for the plants belonging to Gramineae have a biotinylated polypeptide of ca. 35 kDa. About 20 kinds of Gramineae plants did not have such size of biotinylated polypeptide. These results suggest that the loss of BCCP protein is limited to Gramineae (5).

The above two observations suggest that the loss of prokaryotic ACCase is limited to Gramineae. Plastids of ancestral plants of Gramineae might have get a eukaryotic ACCase by accidental addition of transit peptide to this ACCase and the entered ACCase might have functioned efficiently. So an unnecessary prokaryotic ACCase might have lost. Usually *acc*D gene locates between *rbc*L and *psa* I in plastid genome. This genetic locus is known as a hot spot for mutation in Gramineae (6). Various mutations in the *acc*D locus suggest that reorganization of the plastid genome around this locus is still in progress in the Gramineae. Plastid genomes seem to have been under selective pressure to be compact and to lose genes. Many plastid proteins are encoded by nuclear genes that seem to have been located from the ancestral plastid genome. An alternative means of reducing the size of genomes is to replace a plastid-encoded protein by a nucleus-encoded one, as seems to be the case for a ribosomal protein (7). The deletion of *acc*D in the plastid genome and its replacement by a nuclear gene in Gramineae may be the first example of such replacement in the case of a gene for a non-ribosomal component.

2. Effect of UV-B on two ACCase levels

We examined effects of UV-B on the plastidic and cytosolic ACCase levels. Both ACCases have biotinylated polypeptides, and we measured the relative changes in the biotinylated polypeptides using the primary leaves of pea seedlings. Fig 2 indicates the changes in biotinylated poypeptides of the primary leaves during leaf growth. The ca. 35 kDa polypeptide is a subunit of the plastidic ACCase, the ca. 76 kDa polypeptide is 3-methylcrotonyl-CoA, and the ca. 220 kDa polypeptide is the cytosolic ACCase as previously identified (1, 2, 8). The band density of both ca. 35 kDa and ca. 220 kDa polypeptides decreased with age, reaching minima about 6 days after inbibition, and then the levels stayed almost constant. The primary leaves of 6-day-old seedlings were fully expanded. These results suggested that the levels of plastidic and cytosolic ACCases were abundant in early stage of leaf growth and decreased with age.

Effects of UV-B irradiation on ACCase levels were examined for the primary leaves of 9-day-old seedlings. Nine-day-old seedlings were exposed to UV-B for 24 hr and the irradiated seedlings were returned to the earlier light conditions. The total proteins in the primary leaves were extracted and the biotinylated polypeptides were measured (Fig. 3). The levels of the ca. 220 kDa polypeptide increased after the irradiation, reached a maximum at one day after the irradiation started, and then decreased to a low level at two days after the irradiation stopped. The level of the ca. 35 kDa polypeptide changed little. In pea leaves, flavonoids and chalcone synthase are found in epidermis (9). The cytosolic ACCase is also abundant in epidermis (4). These results suggest that one function of the cytosolic ACCase is to supply malonyl-CoA for flavonoid synthesis.

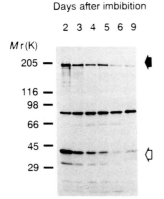

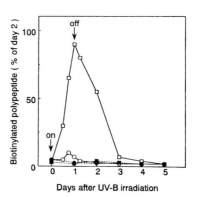

Fig. 2 Changes in the amounts of the two biotinylated polypeptides during leaf growth. Five milligrams of the total proteins was treated by SDS-PAGE (5-15% gradient gel). The separated proteins were probed with streptavidin conjugated to peroxidase. Open and closed arrows indicate the subunit of the plastidic and cytosolic ACCases, respectively.

Fig. 3. Effects of UV-B irradiation on the two biotinylated polypeptide levels in fully-expanded leaves. Nine-day-old seedlings were irradiated with UV-B for 24 hr (times on and off are shown). Relative levels of the ca. 35 kDa (O) and the ca. 220 kDa (□) polypeptides are shown, taking the value of the day 2 in Fig 2 as 100%. The value in the plants without UV-B irradiation was shown for the 35 kDa (●) and the 220 kDa (■) polypeptides.

Flavonoids as well as soluble phenolics absorb UV-B and may prevent the radiation from penetrating photosynthetic organ (10,11). The cytosolic isozyme can probably help to control the synthesis of protective compounds. Thus, the two compartmentalized ACCases seem to have different functions. Plant cells may survive exposure to sun light by using one compartmentalized ACCase for the synthesis of protective substances and the other for the synthesis of membrane constituents and energy stocks.

References
1. Sasaki Y, Hakamada K, Suama Y, Nagano Y, Furusawa I, Matsuno R (1993) J Biol Chem 268: 25118-25123
2. Konishi T and Sasaki Y (1994) Proc Natl Acad Sci U.S.A. 91: 3598-3601
3. Sasaki Y, Konishi T, Nagano Y (1995) Plant Physiol 108: 445-449
4. Alban C, Baldet P, Douce R (1994) Biochem J 300: 557-565
5. Konishi T, Shinohara K, Yamada K, Sasaki Y (1996) Plant Cell Physiol 37: 117-122
6. Clegg, M. T., Gaut, B. S., Learn, G. H. Jr. and Morton, B. R. (1994) Proc. Natl. Acad. Sci. USA 91: 6795-6801
7. Bubunenko, M. G., Schmidt, J. and Subramanian, A. R. (1994) J. Mol. Biol. 240: 28-41
8. Baldet P, Alban C, Axiotis S, Douce R (1992) Plant Physiol 99: 450-455
9. Beerhues L, Robenek H, Wiermann R (1988) Planta 173: 544-553
10. Lois R, Buchanan BB (1994) Planta 194: 504-509
11. Landry LG, Chapple CCS, Last RL (1995) Plant Physiol. 109: 1159-1166

CHARACTERIZATION AND CLONING OF BIOTIN HOLOCARBOXYLASE SYNTHETASE FROM HIGHER PLANTS

G. TISSOT, R. DOUCE and C. ALBAN
Laboratoire Mixte CNRS/Rhône-Poulenc (UMR41 associée au Centre National de la Recherche Scientifique), Rhône-Poulenc Agrochimie, 14-20 rue Pierre Baizet, 69263 Lyon cedex 9, France

Holocarboxylase synthetase (EC 6.3.4.10) (HCS or biotin ligase) catalyzes biotin incorporation into various carboxylases which are essential for housekeeping cellular metabolism, as for example ACCase that realizes the first committed step in fatty acid biosynthesis. These carboxylases are synthesized at first as inactive apoproteins which are then modified into active holocarboxylases by post-translational addition of D-biotin to a specific Lysine residue in the apoprotein. This covalent reaction occurs via an amide linkage between the biotin carboxyl group and the ε-amino group of a unique Lysine residue present in a specific sequence of all biotin-dependent carboxylases. This covalent attachment occurs in two steps (1) and (2) as follows:

(1) D-biotin + ATP —> D-biotinyl-5'-AMP + PPi
(2) D-biotinyl-5'-AMP + apocarboxylase —> holocarboxylase + AMP

Biotin ligase has been purified from *E. coli* and its gene cloned [1, 2]. This enzyme also called BirA is bifunctional since it also acts as a repressor of the biotin operon [3]. Its three dimensional structure has recently been determined at 2.3-Å resolution [4]. The corresponding enzyme from various mammalian species has been purified [5]. Recently, clones encoding *S. cerevisiae* HCS gene [6] and human HCS cDNAs have been obtained [7, 8]. In plants, biotin and biotinylated proteins play a central role in metabolism. Biotin-dependent carboxylases are present in different compartments of plant cells [9]. However, nothing is known concerning the mechanisms of biotinylation of these enzymes. Biotin-dependent carboxylases which are localized in chloroplasts or mitochondria, are encoded by nuclear genes and thus must be targeted in their respective compartment. Therefore, this raises the question at what stage the biotinylation occurs: prior or after translocation of biotin enzymes.

Recently, we have provided the first direct evidence for the existence of HCS activity in plant [10]. Indeed, we have partially purified and biochemically characterized HCS activity from pea leaves. The enzyme activity was assayed using D-[^{3}H]biotin and bacterial apo-BCCP (the biotinylated subunit of ACCase) from an *E. coli* mutant lacking *in vitro* HCS activity, as substrates. Several features of the reaction demonstrate that the plant HCS is responsible for the specific observed D-[^{3}H]biotin incorporation (Table 1). First, we have verified that the reaction catalyzed by the plant HCS was dependent upon the presence of ATP, MgCl$_2$ and apo-BCCP extract in the reaction medium (Table 1). Conditions for optimal catalytic activity and biochemical parameters of the plant enzyme were determined. HCS was active over a wide range of pH with an optimum between pH 7.5 and 8.5.

The influence of substrate concentrations on enzyme activity was assessed by determination of apparent K_m values. The apparent K_m values for D-biotin and ATP were 28 nM and about 1 mM, respectively.

TABLE 1. Characterization of the HCS reaction in a partially purified extract of pea leaves. These data are representative of an experiment repeated at least five times. 1 pmol of D-biotin corresponded to 95 000 dpm.

Reaction Medium	HCS Activity (dpm)
Complete	133 240
Minus ATP	560
Minus Mg^{2+}, Plus EDTA	4 950
Minus apo-BCCP	1 950
Minus enzyme	720

On the other hand, using fractionated pea leaf protoplasts and Percoll purified organelles, we have specifically localized HCS activity in three plant cell compartments [11]. Enzyme activity was mainly located in cytosol but significant activity was also identified in both chloroplast and mitochondrial soluble fractions (Table 2).

TABLE 2. Subcellular localization of HCS in pea leaves.

Distribution	Cytosol	Chloroplasts	Mitochondria
Total activity HCS (%)	90%	7%	3%

To determine whether HCS activity from pea leaves could be resolved into several forms by anion-exchange chromatography, extracts from pea leaves (crude leaf extract, chloroplast stroma, mitochondrial matrix and cytosol) were fractionated on a Mono Q HR 5/5 column [11]. We have observed that HCS activity from a crude extract could be resolved into two peaks. The minor peak, eluting at 50 mM NaCl represented approximately 10% of the total HCS activity and was detected in the chloroplast and mitochondrial fractions. The major peak (90% of the total activity), eluting at 140 mM NaCl was specific of the cytosolic fraction. These results strongly suggest the existence of two HCS isoforms, one localized in the cytosol and the other one in both chloroplasts and mitochondria. Thus, the occurrence of HCS isoforms in different cell compartments also suggest that the different biotin-dependent carboxylases present in plants are biotinylated in the cell compartment where these enzymes are active.

In order to obtain the cDNAs encoding these isoforms, we developed a method of functional complementation of an *E. coli birA* mutant with an *A. thaliana* cDNA expression library [11]. We have isolated and sequenced a full-length HCS cDNA. This clone encodes a protein of 367 residues (41 172 Da) and shows specific enclosed regions of homology with other known biotin ligases of bacterial, yeast and human origin. These homologies are restricted to the ATP binding domain and the biotin binding domain, respectively (Figure 1). These motifs, located in the central part of the protein are interconnected, thus reflecting the requirement for ATP and biotin to be spatially close to permit formation of biotinyl-5'-AMP. The N-terminal region of this *A. thaliana* HCS exhibits the characteristic features of a mitochondrial transit peptide. Also, the occurrence of two Methionine residues close together (Figure 1) may suggest the existence of cytosolic and "organelle targeted" HCS, synthesized from a single species

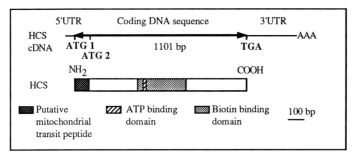

Figure 1. Representation of cDNA and predicted aminoacid sequences encoding *A. thaliana* HCS.

of mRNA by alternative translational initiation. Such possibility has been recently suggested to account for the synthesis of human mitochondrial and cytosolic HCS isoforms, together with an alternative splicing mechanism [8]. The obtention of two translation products of the expected sizes by *in vitro* transcription-translation experiments using the plant cDNA is consistent with this hypothesis. Further characterization of the HCS isozymes is needed to better understand the mechanisms of biotinylation of apocarboxylases in plants.

References

1. Eisenberg, M.A., Prahash, O. and Hsiung, S-C. (1982) Purification and properties of the biotin repressor, *J. Biol. Chem.* **257**, 15167-15173
2. Howard, P.K., Shaw, J. and Otsuka, A.J. (1985) Nucleotide sequence of the *birA* gene encoding the biotin operon repressor and biotin holoenzyme synthetase functions of *Escherichia coli*, *Gene* **35**, 321-331
3. Cronan, J.E., Jr. (1989) The *Escherichia coli bio* operon: transcriptional repression by an essential protein modification enzyme, *Cell* **58**, 427-429
4. Wilson, K.P., Shewchuk, L.M., Brennan, R.G., Otsuka, A.J. and Matthews, B.W. (1992) *Escherichia coli* biotin holoenzyme synthetase-*bio* repressor crystal structure delineates the biotin and DNA-binding domains, *Proc. Natl. Acad. Sci. USA* **89**, 9257-9261
5. Chiba, Y., Suzuki, Y., Aoki, Y., Ishida, Y. and Narisawa K. (1994) Purification and properties of bovine liver holocarboxylase synthetase, *Arch. Biochem. Biophys.* **313**, 8-14
6. Cronan, J.E., Jr. and Wallace, J.C. (1995) The gene encoding the biotin-apoprotein ligase of *Saccharomyces cerevisiae*, *FEMS.* **130**, 221-230
7. Suzuki, Y., Aoki, Y., Ishida, Y., Chiba, Y., Iwamatsu, A., Kishino, Y., Niikawa, N., Matsubara, Y. and Narisawa, K. (1994) Isolation and characterization of mutations in the human holocarboxylase synthetase cDNA, *Nat. Genet.* **8**, 122-128
8. León-Del-Rio, A., Leclerc, D., Akerman, B., Wakamatsu, N. and Gravel, R.A. (1995) Isolation of a cDNA encoding human holocarboxylase synthetase by functional complementation of a biotin auxotroph of *Escherichia coli.*, *Proc. Natl. Acad. USA* **92**, 4626-4630.
9. Alban, C., Baldet, P. and Douce, R. (1994) Localization and characterization of two structurally different forms of acetyl-CoA carboxylase in young pea leaves, of which one is sensitive to aryloxyphenoxypropionate herbicides, *Biochem. J.* **300**, 557-565.
10. Tissot, G., Job, D., Douce, R. and Alban, C. (1996) Protein biotinylation in higher plants: Characterization of biotin holocarboxylase synthetase activity from pea leaves, *Biochem. J.* **314**, 391-395.
11. Tissot, G., Douce, R. and Alban, C. (1996) Subcellular distribution of biotin holocarboxylase synthetase from *Pisum sativum* and characterization of an *Arabidopsis thaliana* cDNA encoding putative mitochondrial isoform, (submitted).

STRUCTURAL STUDIES ON PLANT AND BACTERIAL REDUCTASES INVOLVED IN LIPID BIOSYNTHESIS

[1] A.R.SLABAS, [2] J.RAFFERTY, [2] D.RICE, [2] C.BALDOCK, [1] J. W.SIMON, [1] N.THOMAS, [1] S.BITHELL & [3] A.R.STUITJE.
[1] Department of Biological Sciences, University of Durham, Science Laboratories, South Road, Durham, UK
[2] Krebs Institute for Biomolecular Research, Department of Molecular Biology and Biotechnology, University of Sheffield, Sheffield UK.
[3] Department of Genetics, Vrije Universiteit, De Boelelaan1087, 1081 HV Amsterdam, The Netherlands.

Introduction
The *de novo* biosynthesis of fatty acids is catalysed by a type 2, freely dissociable, fatty acid synthetase in both *E.coli* and the plastids of higher plants. This is distinct from the type 1, associated fatty acid synthetase, which is exemplified by both yeast - in which the activities are present on two polypeptide chains and animals - in which all the activities are present on one polypeptide. Understanding the nature of these enzymes and the way in which they function has moved a long way from the early days of enzyme isolation and cDNA cloning. With the ability to overproduce proteins in bacterial (Kater. M. M et al., 1991) yeast and baculovirus systems (Joshi.A. K & Smith. S., 1993)limitations on the availability of enzyme for X-ray crystallographic studies are largely overcome. Providing the protein will crystallise then a solution to the structure is possible. There are however other requirements if the structure is to be solved speedily and yield the required information. [1] The crystals have to be stable in the X-ray beam. [2] Diffraction has to occur to a high enough resolution to provide data for preparation of a sufficiently detailed model of the structure i.e. at least 3 Angstrom. [3] Suitable heavy metal derivatives have to be made to solve the phasing problem or a suitable highly structurally similar protein model needs to be known (a condition frequently impossible to know in advance of the new structure determination itself). [4] If the details of substrate/enzyme complexes are sought, ways of stabilizing the crystals when in complexes may be required or completely new conditions for crystallization may need to be found.
The space group to which the crystals belong, the unit cell dimensions and the number of molecules in the asymmetric unit will all influence the ease of solution. Additionally, even in models determined at high resolution there are often portions of the structure which are unobservable because of inherent local

disorder within the crystal lattice. These regions reflect underlying flexibility in the protein architecture as is seen with IgG where the carbohydrate region shows great flexibility and is ill defined.

For type 1 FAS the entire structure could be solved if all the above criteria were met although the size of the protein would make crystallization and subsequent structure determination a difficult undertaking. The structure of a type 1 FAS would however give insight into the oligomeric state of the molecule and its mechanism of action. With type 2 FAS the prospects are much better, with the opportunity for crystallising the individual components and solving their structures. However, even here there are at least 7 FAS components for structure determination plus possible co-crystallisations to obtain insight into the association between them. In this study we report on the structure of both bacterial and plant enoyl reductase and point to the reaction mechanism and interesting drug complexes with this enzyme. Some of this data has been reported elsewhere (Rafferty. J.B., et al., 1995) and we wish merely to highlight the central features.

Which reductive components of FAS have we crystallised?

Both *E.coli* and *B.napus* NADH specific enoyl ACP reductase [ENR] and the NADPH specific β-keto reductase [BKR] have been crystallised. The space groups and unit cell sizes are summarised in tabular form below.

ENZYME	SPACE GROUP	CELL DIMENSIONS
ENR (Oilseed rape)	$P4_22_12$	a=b=70.5Å c=117.8Å
ENR (*E. coli*)	$P2_1$	a=74Å b=81.2Å c=79Å
ENR (*E. coli*+diazoborine)	$P6_122$	a=b=80.9Å c=328.3Å
βKR (*E. coli*)	$P6_122$ or $P6_522$	a=b=67.8Å c=355.8Å

What have we learnt about substrate binding?

ENR binds two substrates; a nucleotide cofactor and an acylated ACP.

The NADH is bound in an extended conformation with its nicotinamide and adenine rings lying in pockets on the enzyme surface. The nicotinamide ring is in the syn conformation with the B face accessible for enoyl substrate reduction.

When NAD is bound to the enzyme the exact position of its nicotinamde ring is not observable because of local disorder but it can be unambiguously located when NAD is replaced with NADH by soaking crystals in buffered solutions containing NADH. In the NAD complex the sidechain of Tyr32 occupies a number of conformations, including that occupied in the NADH complex, but its major site overlaps with the position of the nicotinamide ring and thus it must move out of the way to permit NADH binding.

Cocrystallization studies using the substrate analogue crotonyl CoA have shown its adenine and associated ribose moiety to be capable of binding at the same site on ENR as its counterpart in NADH, which could explain why crotonyl CoA is a worse substrate than the crotonyl ACP derivative since the former would also act as a partial inhibitor.

Examination of the composition of the surface residues on ENR has revealed a patch of hydrophobic and aromatic amino acids near the nicotinamide ring of the NADH that may represent the binding site for the acyl substrate.

Searching the 3-dimensional structural data base of proteins reveals a common theme in the catalytic mechanism.

ENR has a strong overall 3-dimensional structural similarity to 3a,20b-hydroxysteroid dehydrogenase (HSD) from *Streptomyces hydrogenans*. This extends to the location of the putative active site residues in ENR, Tyr198 and Lys206, and the catalytically critical residues in HSD, Lys152 and Tyr156. Both enzymes are believed to donate protons from their catalytic tyrosine residues to partially reduced transition state intermediates which are stablized by the presence of the catalytic lysine residues. Thus despite catalyzing chemically distinct reactions the two enzymes appear to share common features in their mechanisms of action. The ENR active site residues are currently being confirmed by site directed mutagenesis.

Why are inhibitor studies of fundamental importance to other biological questions?

ENR from *M.tuberculosis* is the target site for tuberculosis treatment by the drug isoniazid. Unfortunately resistance to this treatment is building up in the causative agent. A way forward would be to design new inhibitors based on the information derived from structural studies on the enzyme inhibitor complex. This is currently not possible with the enzyme from *M.tuberculosis* as the active enzyme inhibitor is an uncharacterized metabolic product of isoniazid. There is however another approach in helping to design new antibiotics utilizing the enzyme from E.coli. The E.coli ENR is potently inhibited by diazaborine reagents and the structure of the enzyme/diazaborine complex has been solved. It reveals a new chemistry of inhibition of the enzyme involving covalent binding of the inhibitor to the cofactor. Using this model system it is hoped that new broad spectrum antibiotics might be designed in the future, including ones to target the *M.tuberculosis* enzyme.

Future perspective on structural studies on FAS and what it might hope to reveal.

Both the nature of the ACP binding domain on ENR and other FAS components remains to be elucidated. Earlier studies from our laboratory have clearly shown that ENR will bind to ACP as will a number of other components of FAS. The use of co-crystallisation studies and surface plasma resonance spectroscopy should enable us to define which units interact and the nature of the important interactions. This will be fundamental in understanding the complete molecular architecture of FAS and possible limitations on its assembly in vivo.

Alternative strategies in the solution of structure.

Structure solutions do not always emerge easily and so there are a few approaches that may help in overcoming some of the problems of which perhaps the most notable is the production of heavy atom derivatives for solving the phase problem.
[a] Selenomethionine can be introduced into the molecule by growing an

overproducing methionine mutant strain of a bacteria on selenomethionine .In such a way a heavy metal derivative is made in vivo.

[b] Cysteine residues can be engineered into an enzyme by recombinant means. Often this approach involves conversion of a serine to a cysteine and generates a potential site for the introduction of a heavy metal usually by covalent modification using mercury compounds.

[c] Modelling from existing structures of similar function can speed up the process of structural elucidation.

[d] Alternative crystal forms. Sometimes the first crystals that one obtains are not suited to rapid structural determination. They can either be of too large a unit cell size or be unstable. Alternative crystal forms might be found which overcome these problems.

Luck, protein purity and source of the protein.

There is no doubt that as with all science there will be an element of luck involved. In this case part of it was the chance meeting of individuals who had a common interest! This was followed by a bit of careful discussion and extra experiments. Some proteins will crystallise out from crude solutions, urease is a classic example, but there is no doubt that high protein purity and careful handling, especially in respect of potential sources of proteolysis, definitely improve the chances of successful crystallization. If however you want the structure badly enough then you should also be prepared to obtain the enzyme from several different sources as there is no predicting which source will turn up trumps!

Dreams.

One day there will be an understanding of the rules of protein folding that may allow the prediction of 3-dimensional structure from primary sequence. However the interaction of one protein with another and the role of molecular chaperones in protein folding may make this a pious hope. Understanding the structure of an enzyme is important for the functional insights it gives and in its own right for the personal satisfaction. It also allows one to see new things which were perhaps only previously imagined. It is somewhat like the unicorn meeting the lion in "Through the Loooking Glass" (Carrol. L. 1942).

"I never saw one alive before," said the lion.

"Well, now that we have seen each other," said the unicorn.

"If you believe in me, I'll believe in you. Is that a bargain?"

So once we have seen the structure all sorts of things can emerge from it!

References

Carrol .L (1942) Through the looking-glass and what Alice found there.. Macmillan and Co Ltd, London.

Joshi. A. K. & Smith. S., (1993) Construction of a cDNA encoding the multifunctional animal fatty acid synthase and expression in *Spodoptera frugiperrda* cells using baculovirus vectors. *Biochem. J.* 296: 143-149.

Kater . M .M., et al., (1991) cDNA cloning and expression *of Brassica napus* enoyl-acyl carrier protein reductase in *Escherichia coli.. Plant Mol. Biol.* **17**: 895-909.

Rafferty. J. B., et al., (1995) Common themes in redox chemistry emerge from the X-ray structure of oil seed rape (Brassica napus) enoyl acyl carrier protein reductase. *Structure* **3**: 927-38.

A NOVEL PATHWAY FOR THE BIOSYNTHESIS OF STRAIGHT AND BRANCHED, ODD- AND EVEN-LENGTH MEDIUM CHAIN FATTY ACIDS IN PLANTS

G. J. WAGNER AND A. B. KROUMOVA
Plant Physiology/Biochemistry/Molecular Biology
Department of Agronomy
Room N-212, ASCN Bldg.
University of Kentucky, Lexington, KY 40546-0091 U.S.A.

Much effort is currently focused on modifying plants and microbes to over-produce medium-chain fatty acids (mcFAs) for commercial and dietary use [1]. The value of biologically derived mcFAs and their conjugates lies in their favorable properties for use as renewable and biodegradable components of surfactants, adhesives, emulsifiers, edible oils, fat substitutes, flavorants, fragrances, and natural product pesticides. mcFAs are found in microorganisms; in mammalian milk, blood, and other tissues; and in certain vegetable oils and, together with short-chain fatty acids (scFAs), are components of a large variety of volatile and nonvolatile esters produced by plants, including sugar esters (SE) secreted by trichomes of *Solanaceae* species. Esters containing mcFAs and scFAs are often key factors in plant-pollinator and plant-pest interactions [2,3].

Mechanisms for biosynthesis of long-chain fatty acids (lcFAs) in plants are well studied [1]. In contrast, metabolism leading to the biosynthesis of mcFAs is not as well understood [4,5], particularly that leading to the formation of branched and odd-length mcFAs. Recent studies show that straight-chain mcFAs of many oil seed triacylglycerols are formed by action of chain-length-specific ACP thioesterases that terminate chain elongation before the 7 cycles of the fatty acid synthase (FAS) required for formation of palmitic acid [5].

We showed previously that straight, iso-, and anteiso-branched, odd- and even-length scFAs are synthesized via modified branched-chain amino acid (bcAA) metabolism in trichome glands producing these components as constituents esterified in sugar esters [6,7]. Similar observations relating to the involvement of bcAA metabolism in donating carbon to mcFA esterified in tomato glucose esters and tobacco wax esters have been made, but the mechanism of elongation was not defined or was incorrectly assumed [see 8]. More recently, our studies were extended to plant types producing sugar esters containing mcFA acyl groups [8,9]. These studies have led to the novel elongation concept of α-KAE as described in Figure 1. The net result of one cycle of α-KAE is elongation by one carbon, in contrast to elongation by 2 carbons per cycle in the FAS reaction [8].

Glandular secreting trichomes of several members of the plant family *Solanaceae* produce a variety of acyl scFA and mcFA esterified in secreted SE. These vary in chain length from C3 - C12 and, depending on the species, can be straight-, iso- or anteiso- in conformation.

We are currently studying several aspects of α-KAE metabolism. First we have surveyed various plants to determine if α-KAE is restricted to highly specialized trichome glands, or if it also participates in formation of storage TAG (however unlikely), or acid moieties (particularly branched) of wax esters, or in synthesis of volatile aliphatic components (particularly branched) produced by plants. We recently obtained evidence that mcFAs (8:0 to 14:0) as well as lcFAs (16:0, 18:0, 18:1, 18:2) of mature-imbibing, and developing seeds of *Cuphea*, endosperm of coconut, and developing seeds of soybean synthesize these components of TAG directly via FAS/ACP thioesterase mediated

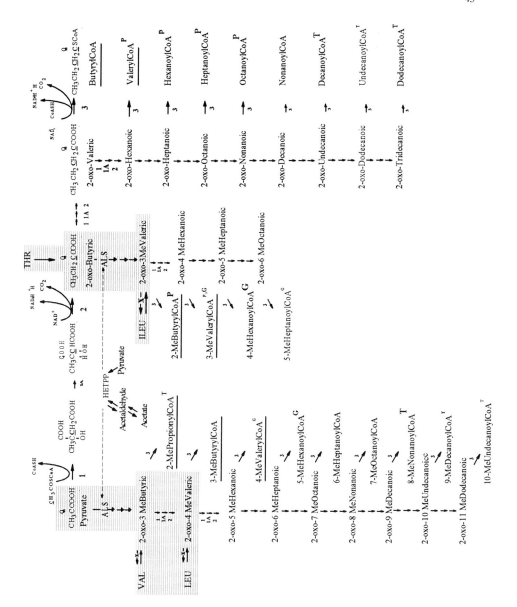

Figure 1. Reactions in the biosynthesis of bcAAs (shaded areas) and the proposed α-ketoacid elongation (α-KAE) pathway for biosynthesis of branched and straight mcFAs and scFAs in trichome glands. The conversion of pyruvate to 2-oxobutyrate and then to 2-oxovalerate via cycling of ketoacid products in α-KAE can lead, minimally, to synthesis of straight-chain fatty acid derivatives up to C12 and anteiso derivatives up to C8. Similar cycling from 2-oxo-3-methylbutyric acid leads to iso-branched products up to C12. Lines under atoms of carbon skeletons at top indicate the position of label derived from [2-^{14}C]acetate. Superscripted letters represent principal sugar ester alkyl groups formed (and labeled from [2-^{14}C]acetate) in trichome glands of tomato (T), petunia (P), and *Nicotiana glutinosa* (G). The relative labeling from [2-^{14}C]acetate of extended acyl moieties in sucrose esters formed in these plant systems is denoted by the relative size of T, P, and G after products. ALS, position of acetolactate synthase.

reactions. As expected, no evidence was found for the involvement of α-KAE in their synthesis. Analysis of seeds and fruits of *Arabidopsis* provided different and interesting results. Our observation with *Arabidopsis* seed, but not bracts and stems, suggest formation of 16:0, 18:0, 18:1, 18:2 and 18:3 via FAS, but with a substantial amount of carbon (exogenously supplied acetate) being derived via mitochondrial malate. We must emphasize here that these results were obtained using tracer experiments. This aspect is described in detail in the poster presented at this meeting by Kroumova and Wagner. We are currently investigating the role of α-KAE in wax ester and volatile aliphatic compound production. We have shown directly that SE formation is restricted to glands of trichomes (unpublished).

A second question of current interest in our laboratory is the basis of chain-type-specificity (straight versus iso-branched versus anteiso-branched) in SE acyl group formation. As shown in Figure 1, a given species tends to form primarily one acyl group chain type (straight or iso or anteiso). We have speculated that chain type may be dictated by species-specific isozymes of isopropylmalate synthase (reaction 1, Figure 1). This enzyme is the key regulated enzyme of leucine biosynthesis and is therefore a strong candidate for regulation of α-KAE. We have found that *in vivo* this activity from a tobacco type primarily producing anteiso-branched SE acyl moieties prefers 2-oxo-3 Me-valeric acid (anteiso precursor) over 2-oxo-4 methylvaleric acid (iso precursor). The specific activities with the form and latter were 0.63 and 0.35, respectively. The corresponding distribution between anteiso versus iso branched type groups found in this tobacco are 73 and 25, respectively [11]. While preliminary, these results are consistent with the above-stated isozyme working-hypothesis. Further study is underway.

A third question of current interest is whether we can modify α-KAE in trichome glands to over-produce a SE type having acyl group composition that confers useful properties. Targets might be SE having enhanced pest resistance or those having commercial value. In general, longer chain acyl groups appear to confer higher insect resistance, and certain plants producing SE have high secretion potential and high biomass, making them potential "factory" plants [3].

We have shown that α-KAE can be manipulated to redirect carbon flow from branched to straight chain acyl group formation. Inhibitors of acetolactate synthase (ALS in Figure 1) were shown to inhibit formation of branched scFA acyl constituents and enhance synthesis of straight chain groups [11]. Thus, carbon flow was diverted from primarily anteiso-branched to straight chain group formation. These results suggest possibilities for manipulation of α-KAE.

In conclusion, we have discovered a novel pathway for biosynthesis of scFA and mcFA having straight, iso or anteiso configuration. Thus, far this pathway is only shown to occur in specialized glands of trichomes on plants in the family *Solanaceae*. The potential for manipulation of this metabolism to enhance natural pest resistance and to exploit trichome secretion to produce commercially valuable biochemicals is being explored.

References

1. Ohlrogge, J.B. *Plant Physiol.* **104**, (1994) 821-826.
2. Severson, R.F., Jackson, D.M., Johnson, A.W., Sisson, V.A. and Stephenson, M.G. in Hedin, P.A. (ed.), *Naturally Occurring Pest Bioregulators*, American Chemical Society Symposium Series, No. 449, pp. 264-279, 1991.
3. Wagner, G.J. *Plant Physiol.* **96**, (1991) 675-679.
4. Dormann, P., Spener, F. and Ohlrogge, F.B. *Planta* **189**, (1993) 425-432.
5. Jones, A., Davies, H.M. and Voelker, T.A. *Plant Cell* **7**, (1995) 359-371.
6. Kandra, L., Severson, R. and Wagner, G.J. *Eur. J. Biochem.* **188**, (1990) 385-391.
7. Kandra, L. and Wagner, G.J. *Plant Physiol.* **94**, (1990) 906-912.
8. Kroumova, A.B., Xie, Z. and Wagner, G.J. *Proc. Natl. Acad. Sci.* **91**, (1994) 11437-11441.
9. Kroumova, A.B. and Wagner, G.J. *Analyt. Biochem.* **225**, (1995) 270-276.
10. Wagner, G.J. *Rec. Adv. in Tob. Sci.* 46th Tobacco Chemists' Research Conference, Vol. 18, pp. 3-4, 1992.
11. Kandra, L. and Wagner, G.J. *Plant Physiol.* **94**, (1990) 906-912.

DOES ENDOGENOUS ACYL-COA SYNTHETASE ACTIVITY GENERATE THE PRIMER IN BIOSYNTHESIS OF VERY LONG CHAIN FATTY ACIDS?

A. HLOUSEK-RADOJCIC[1], K. J. EVENSON[2], J. G. JAWORSKI[3] AND
D. POST-BEITTENMILLER[1]
[1]Plant Biology Division, S.R. Noble Foundation, P.O.Box 2180, Ardmore, OK 73402
[2]Department of Biology, Winona State University, Winona, MN 55987
[3]Chemistry Department, Miami University, Oxford, OH 45056

Introduction

The group of enzymes that catalyze elongation of long chain fatty acids to very long chain fatty acids (VLCFA) are generally referred to as elongases. The complete elongation cycle includes four enzymatic reactions, initiating with a condensation reaction of an acyl primer with malonyl-CoA and continuing with reduction, dehydration and reduction reactions. In this respect, biosynthesis of very long chain fatty acids is similar to *de novo* fatty acid biosynthesis. However, contrary to the enzymes involved in *de novo* fatty acid biosynthesis, elongases are membrane bound and are found in microsomal and, in some cases, oil body fractions.

The acyl primer in biosynthesis of VLCFA has been assumed to be an acyl-CoA, mainly based on the following evidence: i) the intermediates of the elongation reaction were found to be linked via a thioester bond to CoA [1], and ii) exogenously added acyl-CoAs are readily elongated [2,3]. However, during our initial characterization of elongases from developing seeds of high erucic acid variety of *Brassica napus* var. Reston and from epidermal tissues of *Allium porrum* L. leaves, we found evidence suggesting that acyl-CoAs may not be the direct substrate [4,6]. First, ATP is required for maximum elongation activities; second, oil bodies of *B. napus* developing seeds have very small endogenous levels of acyl-CoA; third, elongation activity was higher with [^{14}C]malonyl-CoA than with [^{14}C]18:1-CoA; and fourth, the specific activities of the elongation products ([^{14}C]20:1-CoA) were lower than specific activities of the [^{14}C]18:1-CoA, a putative primer. While the specific activity of [^{14}C]18:1-CoA decreased from 88% ^{14}C to 48% in a thirty minute reaction, the specific activity of 20:1-CoA did not change significantly. These data indicate that exogenously added [^{14}C]acyl-CoA entered an endogenous, unlabeled intermediate pool [4].

To address the question of what is the direct substrate for elongase, as well as to tackle the intriguing issue of the dependence of elongase activity on ATP, we have examined the involvement of acyl-CoA synthetase as a possible endogenous source of a substrate for elongases. All acyl-CoA synthetase (ACS) require ATP, Mg^{2+}, CoASH and free fatty acid [5]. In addition, their activities have been detected in various cellular compartments including microsomal pellets from maturing safflower seeds [5] and from leek epidermis [3]. It has been proposed that the ATP requirement for maximal elongation rates is due to ACS activity and maintenance of an acyl-CoA pool [3].

Materials and Methods

Isolation of oil bodies and microsomes from developing seeds of *Brassica napus* var. Reston and microsomes from epidermal cells of *Allium porrum* L. leaves was carried out as described previously [4,6]. Elongase and acyl-CoA synthetase assays were done as outlined before using either [^{14}C]malonyl-CoA (58μCi/μmol), [^{14}C]18:1 (56 μCi/μmol), [^{14}C]18:0 (50 μCi/μmol), [^{14}C]18:1-CoA (56 μCi/μmol),

[^{14}C]18:0-CoA (50 µCi/µmol). The products of elongase and acyl-CoA synthetase reactions were processed for further analyses by thin layer chromatography as in [4,6].

Results and Discussion

Analyses of elongation activities in the presence or absence of various cofactors indicates that ATP is essential for maximum elongation rates in *B.napus* oil bodies [4]. In leek, ATP dependent activity is localized in young, rapidly expanding areas of leaf epidermis, whereas ATP independent elongase activity is found throughout the epidermis. Furthermore, elongation of free fatty acids in leek microsomes is completely dependent on ATP, whereas elongation of stearoyl-CoA is only partially dependent on ATP (Fig.1). As it was observed in *B.napus* oil body preparations [4], elongation activity in leek microsomes was higher if radiolabel was on [^{14}C]malonyl-CoA (Fig.1 lanes 4 & 5) rather than on free fatty acid or acyl-CoA (Fig.1, lanes 1 & 2).

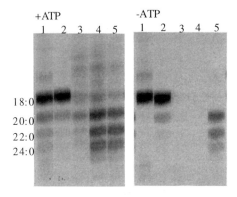
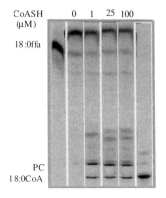

Figure 1. Effect of ATP on elongation in leek microsomes. CoA[^{14}C]18:0, lane 1, [^{14}C]18:0-CoA, lane 2, no primer, lane 3, 18:0, lane 4, 18:0-CoA,. lane 5. In samples (1) and (2) malonyl-CoA was added, whereas in samples (3)-(5) [^{14}C]malonyl-CoA was added. NADPH, DTT, Mg^{+2}, HEPES were added to all samples as described previously [6].

Figure 2. Effect of CoASH on acyl-synthetase activity in leek microsomes. Acyl- CoA synthetase was measured in the presence of [^{14}C]18:0, ATP, Mg^{+2}, DTT and indicated concentrations of CoASH as outlined before [6].

Because ATP involvement may be required for the endogenous synthesis of an acyl primer, such as acyl-CoA, we examined the effects of preincubation of leek or *B.napus* microsomes with [^{14}C]free fatty acids on elongation rates in the presence or absence of ATP. Preincubation did not have any significant effect on the elongation rate, and we did not observe accumulation of acyl-CoA during preincubation and before the elongation reaction was initiated by the addition of malonyl-CoA and NADPH (data not shown).

Acyl-CoA synthetase activity is dependent on the presence of ATP, Mg^{+2}, free fatty acid and CoASH. Therefore, we examined the effect of CoASH on elongase and acyl-CoA synthetase activities. In leek microsomes, acyl-CoA synthesis was stimulated by increasing concentrations of CoASH (Fig.2). In the absence of exogenously added CoASH, no acyl-CoA was detected (Fig.2). However, elongation of 18 carbon acyl chain was higher in the absence of CoASH than in the presence of CoASH, regardless of concentration (Fig.3). Elongation of 20 carbon acyl chain was stimulated with CoASH (Fig.3).

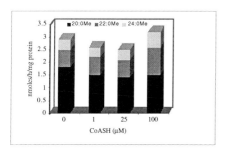

Figure 3. Effect of CoASH on elongation activity in leek microsomes. Elongase activity was measured in the presence of [^{14}C]18:0, malonyl-CoA, ATP, Mg^{+2}, DTT and indicated concentrations of CoASH, as outlined before [6].

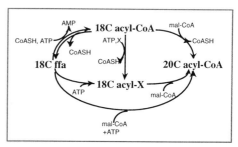

Figure 4. Possible reactions that may lead to the elongation of 18 carbon acyl chain into 20 carbon acyl chain.

Based on these data it appears that endogenous acyl-CoA synthetase activity does not support elongation by synthesizing acyl-CoA primers *in situ*. It is possible that ATP is involved in the synthesis of an elongase primer other than acyl-CoA or that ATP activates acyl chain directly to the enzyme (Fig.4).

References

1. Fehling, E. and Mukherjee, K.D. (1991) Acyl-CoA elongase from a higher plant (*Lunaria annua*): metabolic intermediates of very-long-chain acyl-CoA products and substrate specificity, *Biochim.Biophys.Acta* **1082**, 239-246.
2. Agrawal, V.P. and Stumpf, P.K. (1985) Characterization and solubilization of an acyl chain elongation system in microsomes of leek epidermal cells, *Arch.Biochem.Biophys.* **240**, 154-165.
3. Lessire, R. and Cassagne, C. (1979) Long chain fatty acid CoA-activation by microsomes from *Allium porrum* epidermal cell. *Plant Sci.Lett.* **16**, 31-39.
4. Hlousek-Radojcic, A., Imai, H. and Jaworski, J.G. (1995) Oleoyl-CoA is not an immediate substrate for fatty acid elongation in developing seeds of *Brassica napus*, *Plant J.* **8**, 803-809.
5. Ichihara, K., Nakagawa, M. and Tanaka, K. (1993) Acyl-CoA synthetase in maturing safflower seeds, *Plant Cell Physiol.* **34**, 557-566
6. Evenson, K.J. and Post-Beittenmiller, D. (1995) Fatty acid-elongating activity in rapidly expanding leek epidermis, *Plant Physiol.* **109**, 707-716.

MODIFICATION OF PLANT EPICULAR WAXES

R.K. DEKA and D. POST-BEITTENMILLER
The Samuel Roberts Noble Foundation
P.O. Box 2180, Ardmore, OK 73402

Introduction

Plants have defense mechanisms against biotic and abiotic stresses. All land plants have a waxy protective barrier, on their aerial surfaces, with which insects first come in contact. Many herbivorous insects may to select their host plants on the basis of the chemical and structural characteristics of the plant surface (Eigenbrode and Espelie, 1995). Certain plant epicuticular lipids can contribute to insect resistance by affecting insect behavior (Eigenbrode *et al.*, 1991). The molecular and biochemical basis of epicuticular wax biosynthesis and the composition that is involved in plant defense is poorly understood. In this study our purpose is to see whether the introduction and expression of an heterologous elongase condensing enzyme, with altered substrate specificities, could result in a change in epicuticular waxes, and if so, to determine if the change affects wax function.

In plants, very long chain fatty acids (VLCFAs, >C18) are synthesized by the action of elongases. VLCFs are the precursors for the long chain aldehydes, alcohols, alkanes, ketones and esters which comprise the wax layer (Post-Beittenmiller, 1996). The condensing enzyme component of an jojoba seed elongase, elongates both very long chain monounsaturated and saturated substrates (e.g., C18:1 and C20:1 to C24:1; C18:0 to C20:0) *in vitro*, but *in vivo*, the monounsaturates are elongated prefentially (Lassner *et al.*, 1996). This is in contrast to the epidermal elongases which utilize saturated primers both *in vitro* and *in vivo* (e.g., C18:0 to C20:0) for wax production. In order to test whether or not the jojoba seed elongase condensing enzyme can function in a heterologous epidermal elongation system, we developed transgenic alfalfa plants.

Materials and methods

Construct in binary vector: The *Arabidopsis* chalcone synthase (CHS) promoter was used in this study (Feinbaum and Ausubel, 1988). This promoter is highly expressed in epidermal cells and is light regulated. Therefore, the expression level can be modified by changing light intensity.

The jojoba elongase (JEL) under the control of CHS promoter and NOS terminator was subcloned into pBluescript (pBS-CHS-JEL-NOS), and the cassette was amplified by PCR with primers RKD1 and RKD2 containing *Xba* I restriction sites. The PCR product was digested with *Xba* I and ligated into *Xba* I-digested calf-intestinal alkaline phosphatase (CIAP) treated pBIN19, yielding pBin-CHS-JEL-NOS. pBIN19 was used as a control. *Agrobacterium* strain (LBA4404) was transformed with the construct by electroporation.

Plant transformation: Alfalfa plants were tansformed by transfecting leaf with a *Agrobacterium* strain containing the binary vector (pBin-CHS-JEL-NOS and pBin19) and selected with kanamycin. The transformed plants used in this study were cultured under high light for 16 h photoperiod. Regenerated plants were then grown in a greenhouse (25 ± 2^0 C) under 16 h of supplemental lighting.

DNA/RNA isolation, RNA Blotting and Hybridisation: Alfalfa trifoliates were frozen in liquid nitrogen and stored at -80°C for processing. After grinding tissues to a fine powder in liquid nitrogen, genomic DNA was isolated from 150 mg material, using a Easy-DNA kit (Invitrogen).

Total RNA was extracted from 150 mg material, using a Qiagen RNeasy extraction kit (Qiagen). Fifteen micrograms of total RNA/lane were loaded onto a 1.2% agarose/ formaldehyde gel and transferred to Gene Screen Plus membrane according to the manufacturer's recommendation (Dupont). The probe used for RNA blot analyses was a full length cDNA clone of an jojoba elongase condensing enzyme (Lassner *et al.*, 1996), labeled with [a-^{32}P]dCTP, by a random priming method (Stratagene). Prehybridization and hybridization were performed according to the Gene Screen Plus protocol (Dupont).

PCR analyses: The reactions were performed in a GeneAmp PCR System 2400 (Perkin Elmer) using 50 ml reaction volumes with the following program: initial denaturing, 3 min at 95°C followed by 35 cycles of 30 sec denaturing at 94°C, 1 min annealing at 53°C, 1 min extension at 72°C. Extension at 72°C was allowed to continue for 10 min at the end of the program. 10% reaction products were examined on a 1.5% agarose gel.

Results

Phenotypes of plants: All plants showed normal habit except the leaf appearance of plants 4668 and 4671 (not shown). Plant 4668 exhibited slower growth than the other plants.

Stable integration of elongase gene in alfalfa: Putative, transformed plants were tested for expression of the neomycin phosphotransferase (*nptII*) gene by ELISA-based estimates of NPTII protein (Kit supplied by 5 prime -> 3 prime, Inc). Plants possessing the *nptII* gene showed high level of protein compared to the population lacking the gene. This served as a selectable marker in transformed tissue (results not shown).

Confirmation of stable integration of the elongase gene by PCR: Primer ELON1 (5'-GAGATCCAAGTCTCCACGACCAT-3', position 33 to 56 of the jojoba elongase) and primer ELON2 (5'-CAGCGACGGCGTTGGGTTAAACA-3', position 646 to 669 of the jojoba elongase) were used in PCR to amplify the enclosed region (636 bp) using 100ng of genomic DNA from plants transformed with the different constructs as template. As expected, plants transformed with pBin-CHS-JEL-NOS yielded an amplification product of 636bp. Alfalfa genomic DNA from nontransformed and vector transformed (pBin19) plants did not produce an amplification product with the same set of primers (Fig. 1).

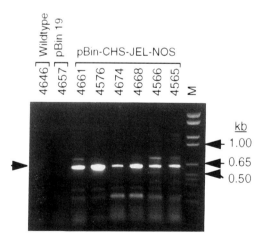

Figure 1. Confirmation of stable integration of jojoba elongase condensing enzyme gene in the alfalfa genome using PCR. Genomic DNAs of both transformed and nontransformed plants were used to amplify an intervening region (indicated by arrow).

Accumulation of jojoba elongase transcript in alfalfa leaf: To see the expression level of jojoba elongase transcript, a RNA gel blot experiment was performed. As shown in Fig. 2, the probe hybridized to the plant that are transformed with jojoba elongase cDNA. Plant 4661 which shows a band of expected size on PCR analysis (Fig. 1) do not show any accumulation of transcript.

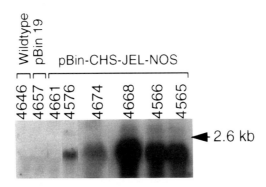

Figure 2. RNA blot analysis of jojoba elongase condensing enzyme gene expression in transgenic alfalfa. Total RNAs (~15 µg) were separated on a formaldehyde gel, transferred onto a nylon membrane and hybridized with ^{32}P-labeled jojoba condensing enzyme cDNA. RNA from vector transformed and nontransformed tissue culture-regenerated plants were used as control.

Discussion

We have generated transgenic alfalfa plants containing jojoba elongase gene for wax modification. Integration and expresssion of the elongase gene in the transgenic plants was confirmed by PCR (Fig. 1) and RNA blot (Fig. 2) analyses. The plant 4668, which shows some difference in their leaf appearance, has also very high level expression of elongase transcript. Crosses are in progress to determine if the two phenotypes are linked.

The epicuticular wax of transformed and nontransformed plants were analyzed by gas chromatography. There was no quantitative or qualitative change of lipid profiles of these plants (results not shown). From this data it appears that the expression of a heterologous elongase gene may not be enough to see the presence of an elongated monounsaturated product in the epicuticular wax, if the *in vivo* pool of monounsaturated fatty acids is limiting for the heterologous elongase.

We attempted to determine whether the expressed elongase gene product is enzymatically active by assaying elongase activity in the microsomal preparation of alfalfa leaf using the protocol described for leek elongase assay (Evenson and Post-Beittenmiller, 1995). There was low activity for [^{14}C] mal-CoA incorporation. Addition of primers (18:0 and 18:1) did not affect elongation rates in wild type or transgenic plants (data not shown). There are two possible explanations for this: either the assay conditions developed for leek may not work for alfalfa. Alternatively, the microsomes prepared from total leaf extract of alfalfa has very little elongase activity. This may be due to the presence of some inhibitory substance in the crude extract. Currently, we are using *in vivo* enzyme assay to determine any elongated product using ^{14}C labeled 18:0 and 18:1-primers. Also we are trying to express the jojoba elongase condensing enzyme in a baculovirus system for antibody production. This will help us in immunoblot analyses of the transgenic plants.

References

Eigenbrode, S.D. and Espelie, K.E. (1995) *Ann. Rev. Entomol.* **40:** 171-194.
Eigenbrode, S.D., Espelie, K.E. and Shelton, A.M. (1991) *J. Chem. Ecol.* **17:** 1691-1704.
Evenson, K.J. and Post-Beittenmiller, D. (1995) *Plant Physiol.* **109:** 707-716.
Feinbaum, R.L. and Ausubel, F.M. (1988) *Mol. Cell. Biol.* **8:** 1985-1992.
Lassner, M.W., Laradizabal, K. and Metz, J.G. (1996) *Plant Cell* **8:** 281-292.
Post-Beittenmiller, D. (1996) *Annu. Rev. Plant Physiol. Plant Mol. Biol.* **47:** 405-430.

BIOCHEMISTRY OF SHORT-CHAIN ALKANES: EVIDENCE FOR AN ELONGATION/REDUCTION/C1-ELIMINATION PATHWAY

T.J. SAVAGE, M.K. HRISTOVA, and R. CROTEAU

Institute of Biological Chemistry
Washington State University
Pullman, WA 99164-6340 USA

1. Introduction

C_7-C_{11} alkanes accumulate in the oleoresins of several *Pinus* species native to western North America, most notably in the xylem oleoresin of Jeffrey pine, *Pinus jeffreyi* Grev. & Balf. (Savage et al., 1990a). Biological production of light hydrocarbons is of interest because they possess the same excellent combustion properties as petrochemical hydrocarbons in gasoline. Whereas production levels of short-chain alkanes in plants are insufficient to provide an economically viable fuel source, the genes encoding the alkane biosynthetic pathway may provide a biotechnological resource for engineering fermentation organisms with the capability to convert biomass to an alkane-based fuel. However, the feasibility of transgenic alkane biosynthesis depends upon the complexity of the alkane biosynthetic pathway.

Aliphatic hydrocarbons in other organisms appear to arise via acyl lipid intermediates. Long-chain waxy hydrocarbons in plants, animals and microalgae apparently are synthesized by elongation of a C_{16} or C_{18} acyl CoA to generate an acyl thioester one carbon longer than the hydrocarbon product (Kolattukudy, 1987; Vaz et al., 1988). The long-chain acyl CoA is then reduced to an aldehyde and free CoA, and the former undergoes either decarbonylation to form a hydrocarbon and carbon monoxide (Cheesbrough and Kolattukudy, 1984, 1988; Dennis and Kolattukudy, 1991, 1992), or decarboxylation to form a hydrocarbon and carbon dioxide, as has been recently demonstrated for hydrocarbon biosynthesis in insects (Reed et al., 1994; Mpuru et al., 1996). *n*-Pentane found in developing peanuts also appears to arise from an acyl lipid, albeit directly from the lipoxygenase-catalyzed degradation of linoleic acid without the involvement of an aldehyde intermediate (Pattee et al., 1970, 1974). In an early study of short-chain alkane biosynthesis in *P. jeffreyi* (Sandermann et al., 1960), the labeling pattern of *n*-[^{14}C]heptane generated by incubating tissue slices with [2-^{14}C]acetate was consistent with chain elongation via a fatty acid synthase-type polymerization of acetate units. However, this study did not determine whether the aliphatic carbon chain arises by decarbon(x)ylation of octanal (generated from the reduction of a C_8 thioester), or by degradation of a longer-chain fatty acyl group without involvement of an aldehyde intermediate.

We recently developed an *in vivo* experimental system using *P. jeffreyi* xylem sections that rapidly incorporate radiolabel from [^{14}C]acetate into *n*-heptane (Savage et al., 1996a). Here, we examine the effect of an aldehyde trapping reagent (hydroxylamine) and a thiol reagent (β-mercaptoethanol) on the incorporation of radiolabel into *n*-heptane and putative biosynthetic intermediates. These results, along with observation of the direct *in vivo* conversion of [^{14}C]octanal to *n*-[^{14}C]heptane, suggest that the pathway to the short-chain alkane involves the decarbonylation or decarboxylation of octanal to generate *n*-heptane.

2. Results

Incubation of xylem sections with [^{14}C]acetate not only resulted in radiolabel incorporation into n-heptane, but also into octanal and 1-octanol. Incorporation of radiolabel into the C_8 alcohol was unexpected because there are no reports of alcohol intermediates in long-chain hydrocarbon biosynthesis, and 1-octanol has not been reported to accumulate in P. jeffreyi xylem or xylem oleoresin (Savage et al., 1996a). Thus, potential roles of both octanal and 1-octanol in the biosynthetic pathway to n-heptane were further investigated.

Early evidence for long-chain aldehydes as precursors of long-chain alkanes was provided by [^{14}C]acetate feeding studies of wax biosynthesis in garden peas (Pisum sativum), where β-mercaptoethanol inhibited radiolabel incorporation into alkanes and stimulated radiolabel incorporation into long-chain aldehydes (Buckner and Kolattukudy, 1973). Co-incubation of P. jeffreyi xylem sections with 5 μCi [^{14}C]acetate and 1 to 50 mM β-mercaptoethanol similarly resulted in inhibition of n-heptane biosynthesis relative to incubations with [^{14}C]acetate alone, although radiolabel accumulated into both octanal and 1-octanol in the presence of β-mercaptoethanol.

Hydroxylamine serves as a useful probe for biosynthetic pathways involving aldehyde intermediates by reacting with free aldehydes to generate metabolically inactive oximes. To determine whether octanal is a precursor to both n-heptane and 1-octanol, xylem sections were co-incubated with 5 μCi [^{14}C]acetate and 0.1 to 1 M hydroxylamine. Radiolabel was incorporated into octyl oximes in co-incubations with hydroxylamine, and increasing concentrations of the reagent inhibited radiolabel incorporation into n-heptane and 1-octanol, with a corresponding accumulation of radiolabel into octyl oximes.

To more directly assess the role of octanal and 1-octanol in n-heptane biosynthesis, xylem sections were incubated with [^{14}C]1-octanol and [^{14}C]octanal. Incubation of xylem sections with 10^5 dpm [^{14}C]octanal resulted in the incorporation of ^{14}C into n-heptane, although the majority of radiolabel was recovered as 1-octanol. Incubation of xylem sections with 10^5 dpm [^{14}C]1-octanol also resulted in incorporation of radiolabel into n-heptane, and a small amount of radiolabel was incorporated into octanal. To determine whether 1-octanol is first oxidized to the aldehyde before conversion to n-heptane, xylem sections were co-incubated with 10^5 dpm [^{14}C]1-octanol and 500 mM hydroxylamine. No radiolabel was incorporated into n-heptane, but ^{14}C was incorporated into octyl oximes.

3. Conclusions

Taken together, these results support a pathway for the biosynthesis of n-heptane whereby acetate is polymerized via a typical fatty acid synthase reaction sequence to yield a C_8 thioester, which subsequently undergoes a two-electron reduction to generate a free thiol and octanal, the latter of which alternately undergoes an additional, reversible, reduction to form 1-octanol or direct loss of C1 to generate n-heptane. Evidence supporting the role of acyl lipids in this pathway include the labeling pattern of n-[^{14}C]heptane generated from [2-^{14}C]acetate (alternately labeled carbons beginning with C1) previously reported (Sandermann et al., 1960), and the rapid incorporation of radiolabel into n-heptane from [^{14}C]acetate reported here. Evidence for the central role of octanal in n-heptane and 1-octanol biosynthesis is provided by the inhibition of radiolabel incorporation into n-heptane by β-mercaptoethanol, with corresponding accumulation of label in both the C_8 aldehyde and alcohol. Hydroxylamine inhibition of radiolabel incorporation into both n-heptane and 1-octanol, with corresponding accumulation of radiolabel into octyl oximes, provides additional evidence that octanal serves as a precursor of both the alkane and the alcohol. Incorporation of radiolabel into n-heptane and 1-octanol from [^{14}C]octanal, and the hydroxylamine-sensitive incorporation of radiolabel into n-heptane from [^{14}C]1-octanol, confirm the direct relationship of the aldehyde to the alkane and the reversibility of the aldehyde to alcohol reduction. Further evidence supporting this pathway, including the effect of the β-keto acyl acyl carrier protein synthase inhibitor cerulenin on n-heptane biosynthesis, has recently been published (Savage et al., 1996b).

Several details of the biosynthetic scheme proposed remain unresolved. For example, there is no evidence for the identity of the malonyl thioester substrate of the condensation reaction involved in chain elongation leading to n-heptane biosynthesis. Plant fatty acid synthases employ malonyl ACP as the two carbon donor, yielding elongated ACP thioesters, whereas chain elongation reactions producing long-chain thioesters in wax biosynthesis employ malonyl CoA as the two carbon donor, yielding elongated CoA thioesters (Agrawal et al, 1984). Ambiguity in the identity of the specific thioester product of chain elongation also leads to uncertainty as to the nature of the reduction to generate the aldehyde. Octanal could arise directly from reduction of either octanoyl CoA or octanoyl ACP, or alternatively from the reduction of free octanoic acid generated from the thioesterase-catalyzed cleavage of either thioester. Finally, there is no evidence to indicate whether n-heptane biosynthesis arises via decarbonylation or decarboxylation of the octanal precursor. These questions are currently being addressed in experiments employing cell-free extracts of *P. jeffreyi* xylem.

4. References

Agrawal VP, Lessire R, Stumpf PK (1984) Biosynthesis of very long chain fatty acids in microsomes from epidermal cells of Allium Porrum L. Arch. Biochem Biophys **230**:580-589

Buckner JS, Kolattukudy PE (1973) Specific inhibition of alkane synthesis with accumulation of very long chain compounds by dithioerythritol, dithiothreitol, and mercaptoethanol in Pisum sativum. Arch Biochem Biophys **156**:34-45

Cheesbrough TM, Kolattukudy PE (1984) Alkane biosynthesis by decarbonylation of aldehydes catalyzed by a particulate preparation from Pisum sativum. Proc Natl Acad Sci **81**:6613-6617

Cheesbrough TM, Kolattukudy PE (1988) Microsomal preparation from an animal tissue catalyzes release of carbon monoxide from a fatty aldehyde to generate an alkane. J Biol Chem **263**:2738-2743

Dennis MW, Kolattukudy PE (1991) Alkane biosynthesis by decarbonylation of aldehyde catalyzed by a microsomal preparation from Botryococcus braunii. Arch Biochem Biophys **287**:268-275

Dennis MW, Kolattukudy PE (1992) A cobalt-porphyrin enzyme converts a fatty aldehyde to a hydrocarbon and CO. Proc Natl Acad Sci **89**:5306-5310

Kolattukudy PE (1987) Lipid-derived defensive polymers and waxes and their role in plant-microbe interaction. In PK Stumpf, ed, The Biochemistry of Plants: A Comprehensive Treatise, Vol.9, Lipids: Structure and Function. Academic Press, Orlando, pp 291-314

Mpuru S, Reed JR, Reitz RC, Blomquist GJ (1996) Mechanism of hydrocarbon biosynthesis from aldehyde in selected insect species: requirement for O_2 and NADPH and carbonyl group released as CO_2. Insect Biochem Molec Biol **26**:203-208

Pattee HE, Singleton JA, Johns EB, Mullin BC (1970) Changes in the volatile profile of peanuts and their relationship to enzyme activity levels during maturation. J Agr Food Chem **18**:353-356

Pattee HE, Singleton JA, Johns EB (1974) Pentane production by peanut lipoxygenase. Lipids **9**:302-306

Reed JR, Vanderwel D, Choi S, Pomonis JG, Reitz RC, Blomquist GJ (1994) Unusual mechanism of hydrocarbon formation in the housefly: cytochrome P450 converts aldehyde to the sex pheromone component (Z)-9-tricosene and CO_2. Proc Natl Acad Sci **91**:10000-10004

Sandermann W, Schweers W, Beinhoff O (1960) Uber die biogenese von n-heptan in Pinus jeffreyi Murr. Chem Ber **93**:2266-2271

Savage TJ, Hamilton BS, Croteau R (1996a) Biochemistry of short-chain alkanes: tissue-specific biosynthesis of n-heptane in Pinus jeffreyi. Plant Physiol **110**:179-186.

Savage TJ, Hristova MK, Croteau R (1996b) Evidence for an elongation/reduction/C1-elimination pathway in the biosynthesis of n-heptane in Jeffrey pine (Pinus jeffreyi Grev. & Balf.). Plant Physiol, in press.

Vaz AH, Blomquist GJ, Reitz RC (1988) Characterization of the fatty acyl elongation reactions involved in hydrocarbon biosynthesis in the housefly, Musca domestica L. Insect Biochem **18**:177-184

A RE-EXAMINATION OF PATHWAYS FOR THE BIOSYNTHESIS OF MEDIUM AND LONG CHAIN FATTY ACIDS IN SEEDS OF SEVERAL PLANTS

A.B. KROUMOVA and G.J. WAGNER
*Plant Physiology/Biochemistry/Molecular Biology Program,
University of Kentucky, Lexington, Kentucky 40546-0091*

1. Introduction and Methods

The discovery of a novel pathway for biosynthesis of medium and short chain fatty acids in plants (α-keto acid elongation pathway, 1) raises the possibility (however unlikely) that medium-chain fatty acids (mcFAs) of certain oil seeds producing them may be derived by this pathway. Alternatively, these may be formed after release of elongating fatty acid chains from fatty acid synthase mediated biosynthesis (FAS) by specific medium chain thioesterases [2, 3, 4]. Thus far the αKAE pathway is only known to occur in trichome glands of plants in the family *Solanaceae*. In the αKAE pathway, iso-, anteiso- or straight-chain keto acid products of branched-chain amino acid metabolism are elongated by one carbon (via acetate) per cycle. The final step is predicted to be oxidative decarboxylation to yield CoA activated acids. The mechanism that determines the chain length of αKAE products is not understood [1].

The αKAE and FAS pathways can be distinguished by incubation of tissue with specifically labeled precursors, followed by microscale chemical degradation of radiolabeled products to determine carboxyl-carbon labeling. If the precursor is [2-^{14}C]acetate, products of this pathway are carboxyl labeled while products of FAS-mediated synthesis are not. Here we report on chemical degradation studies of [2-^{14}C]acetate-labeled medium and long chain FAs of various oil seeds. After labeling, tissues were saponified directly, or lipids were extracted and then saponified to release fatty acids [5, 6]. Aliquots of acids were separated by HPLC as free acids [7], then subjected to Schmidt degradation to recover the carboxyl carbon [7, 8]. Fatty acids from soybean seeds and *Arabidopsis* fruits were degraded by chemical α-oxidation [9], products were separated by HPLC and these were analyzed individually by Schmidt degradation. In this way labeling in carbon atoms 10-to-16 or 10-to-18 of 16:0 or 18:0, respectively, were independently characterized.

2. Results and Discussion

Lack of labeling in carboxyl carbons obtained after Schmidt degradation of medium (10:0 to 14:0) and long chain (16:0, 18:0, 18:1, 18:2, 18:3) fatty acids of *Cuphea* and coconut seeds was consistent with direct utilization of [2-^{14}C]acetate by FAS (Table 1).Similar results were obtained for soybean seeds (not shown). Long chain fatty acids of *Arabidopsis* bracts and stems also appeared to be formed via direct utilization of acetate by FAS (not shown). In contrast, labeling patterns of long-chain fatty acids (lcFAs) of *Arabidopsis* fruits were inconsistent with direct utilization of [2-^{14}C]acetate by FAS and also inconsistent with αKAE origin (Table 2, results from decarboxilation of 16:0 and 18:0). Degradation (chemical α-oxidation) of 16:0 and combined C18 products of developing soybean seeds, followed by separation of acids and subsequent Schmidt degradation of these showed labeling patterns that were consistent with their formation via FAS (no labeling in carboxyl groups of even carbon FAs and substaintial labeling in that of odd carbon FAs). In contrast, similar degradation of these acids from *Arabidopsis* fruits resulted in labeling patters in carbon atoms 9-to-16 and 10-to-18, respectively, inconsistent with direct utilization of [2-^{14}C]acetate by FAS and also inconsistent with αKAE origin (Table 2).

Labeling patterns found for *Arabidipsis* fruits might be explained by both direct incorporation of tracer [2-^{14}C]acetate into lcFAs in the plastids, and utilization by plastids of mitochondria-derived [1,2-^{14}C]acetate formed from the tracer.

TABLE 1. Analysis by Schmidt degradation to predict biosynthetic pathways for fatty acid biosynthesis from [2-^{14}C]acetate in various oil seeds.

FA	C. procumbens	C. viscosissima		Coconut	Predicted Radioioactivity in carboxyl carbon	
	Mature-Imbibing Seeds	Mature-Imbibing Seeds	Developing Seeds	Endosperm	FAS	αKAE
	% Total label in carboxyl carbon					
8:0	2.5 (1, FAS)	6.1 & 5.7 (2, FAS or ?)	1.8 ± 0.6 (3, FAS)		0	16.6
10:0	1.9 ± 0.8 (4, FAS)	5.5 & 2.7 (2, FAS, or ?)	1.8 ± 0.3 (3, FAS)		0	12.5
12:0	3.4 (1, FAS)		1.9 (1, FAS)	3.9 & 1.0 (2, FAS)	0	10.0
14:0			0.9 (3, FAS)	0.4 ± 0.2 (3, FAS)	0	8.3
16:0	4.9 ± 2.6 (3, FAS)	0.7 (1, FAS)	3.3 ± 2.9 (6, FAS)	3.9 & 0.2 (2, FAS)	0	7.1
18:0		1.2 (1, FAS)	2.3 ± 1.4 (3, FAS)		0	6.2
18:1	2.1 ± 1.2 (3, FAS)	1.5 (1, FAS)	1.9 ± 1.1 (5, FAS)	1.1 & 0.5 (2, FAS)	0	6.2
18:2	2.9 ± 0.7 (3, FAS)		0.3 & 1.0 (2, FAS)		0	6.2

Values represent the percent radioactivity (of total label in fatty acid) observed in the carboxyl-carbon. In brackets are the number of experiments form which data are summarized followed by the pathway of synthesis (FAS, direct utilization of acetate in FAS; or synthesis via αKAE) concluded from the data. The two right hand columns represent percentages predicted if fatty acids were derived from direct utilization of acetate by FAS or αKAE. When label of a particular acid was low, repeated collections form HPLC were pooled to obtain sufficient radioactivity for degradation. The method used for Schmidt degradation was reproducible to about ± 3% [7].

TABLE 2. Analysis by combined α-oxidation and Schmidt degradation to predict biosynthetic pathways for 16:0 and 18 carbon FAs biosynthesis from [2-^{14}C]acetate in *Arabidopsis* fruits and soybean seeds.

FA	*Arabidopsis* fruits		Soybean seeds		Predicted Radioactivity in carboxyl carbon		
	After α-oxidation of 18:0	After α-oxidation of 16:0	After α-oxidation of 18:0	After α-oxidation of 16:0	FAS	FAS*	αKAE
	% of Total label in carboxyl carbon						
9:0		13.9 (1, FAS*)			20.0	12.2	14.3
10:0	11.5 (1, FAS*)	13.7 (1, FAS*)	2.8 (1, FAS)	5.9 (1, FAS)	0	8.8	12.5
11:0	12.5 & 12.3 (2, FAS*)	16.0 (1, FAS*)	16.8 & 22.3 (2, FAS)	22.0 (1, FAS)	16.7	10.0	11.1
12:0	10.5 & 8.7 (2, FAS*)	9.8 & 14.0 (2, FAS*)	0.0 & 3.5 (2, FAS)	3.7 & 4.1 (2, FAS)	0	7.4	10.0
13:0	10.6 & 11.8 (2, FAS*)	12.6 & 12.8 (2, FAS*)	24.1 & 15.2 (2, FAS)	17.8 & 20.2 (2, FAS)	14.3	8.5	9.1
14:0	8.6 & 11.3 (2, FAS*)	9.9 & 8.0 (2, FAS*)	1.8 (1, FAS)	0.8 & 1.1 (2, FAS)	0	6.3	8.3
15:0	11.9 & 19.2 (2, FAS*)	11.7 & 10.6 (2, FAS*)	15.9 & 16.1 (2, FAS)	18.8 & 14.1 (2, FAS)	12.5	7.3	7.7
16:0	6.0 & 11.4 (2, FAS*)	7.8 & 6.9 (2, FAS*)	0.9 & 3.3 (2, FAS)	0.3 & 1.0 (2, FAS)	0	5.4	7.1
17:0	9.1 & 10.4 (2, FAS*)		22.6 & 18.6 (2, FAS)		11.1	6.5	6.7
18:0	7.2 & 4.6 (2, FAS*)		0.2 & 0.0 (2, FAS)		0	4.9	6.2

Values represent the percent radioactivity (of total label in fatty acid) observed in the carboxyl-carbon. In brackets are the number of experiments form which data are summarized followed by the pathway of synthesis (FAS, direct utilization of acetate in FAS; or synthesis via αKAE) concluded from the data. The three right hand columns represent percentages predicted if fatty acids were derived from direct utilization of acetate by FAS or αKAE, or from combined direct and indirect utilization of acetate by FAS (FAS*). When label of a particular acid was low, repeated collections form HPLC were pooled to obtain sufficient radioactivity for degradation. The method used for Schmidt degradation was reproducible to about ±3% [7].

References

1. Kroumova, A.B., Xie, Z., and Wagner, G.J.: A pathway for the biosynthesis of straight and branched, odd- and even-length, medium-chain fatty acids in plants, *Proc. Natl. Acad. Sci* **91** (1994), 11437-11441.
2. Voelker, T.A., Worrell, A.C., Anderson, L., Bleibaum, J., Fan, C., Hawkins, D.J., Radke, S.E., and Davies, H.M.: Fatty acid biosynthesis redirected to medium chains in transgenic oilseed plants, *Science* **257** (1992), 72-74.
3. Dormann, P., Spener, F., and Ohlrogge, B.: Characterization of two acyl-acyl carrier protein thioesterases from developing *Cuphea* seeds specific for medium-chain- and oleoyl-acyl carrier protein, *Planta* **189** (1993), 425-432.
4. Davies, H.M., and Voelker, T.A.: Mechanisms of chain length determination and medium-chain fatty acid biosynthesis, in N. Murata, C.R. Somerville (eds.), *Biochemistry and Molecular Biology of Membrane and Storage Lipids of Plants*, The American Society of Plant Physiologists, (1993) 1-5.
5. Slabas, A.R., Roberts, P.A., Ormesher, J., and Hammond, E.W.: *Cuphea procumbens* -A model system for studying the mechanism of medium-chain fatty acid biosynthesis in plants, *Biochim. Biophys. Acta* **711** (1982), 411-420.
6. Oo, K.C., and Stumpf, P.K.: Fatty acid bisynthesis in the developing endosperm of *Cocos nucifera*, *Lipids* **14** (1979), 132-143.
7. Kroumova, A.B., and Wagner, G.J.: Methods for separation of free, short, medium, and long chain fatty acids and for their decarboxylation, *Anal. Biochem.* **225** (1995), 270-276.
8. Brady, R.O., Bradley, R.M., and Trams, E.G.: Biosynthesis of fatty acids I. Studies with enzymes obtained from liver, *J. Biol. Chem.* **235** (1960), 3093-3098.
9. Harris, R.V., Harris, P., and James, A.T.: The fatty acid metabolism of *Chlorella vulgaris*, *Biochim. Biophys. Acta* **106** (1965), 465-473.

BIOSYNTHESIS OF AN ACETYLENIC FATTY ACID IN MICROSOMAL PREPARATIONS FROM DEVELOPING SEEDS OF *CREPIS ALPINA*

A. BANAS[1], M. BAFOR[1], E. WIBERG[1], M. LENMAN[2], U. STÅHL[1] AND S. STYMNE[2]

[1]*Department of Plant Physiology, Swedish University of Agricultural Sciences, Box 7047, S-750 07 Uppsala, Sweden* [2]*Department of Plant Breeding Research, Swedish University of Agricultural Sciences, Herman Ehles v. 2-4, S-268 31 Svalöv, Sweden*

1. Introduction

Over 600 naturally occuring compounds with acetylenic bonds (triple bonds) have been characterized (Bohlmann *et al.*, 1973). Previous *in-vivo* studies, from mosses accumulating acetylenic fatty acids, indicate that the acetylenic bond is formed by the substraction of two hydrogen atoms from a double bond (Kohn *et al.*, 1994).

Crepis alpina seeds have oils (triacylglycerols) where 65-80% of the fatty acids consist of crepenynic (9-octadecen-12-yonic, 18:1A) acid. We have studied the formation of 18:1A in developing seeds from *Crepis alpina* and can, for the first time, present evidence for the *in-vitro* activity of an acetylenic forming enzyme (acetylenase) and the partial characterization of this enzyme and its substrate.

2. Materials and Methods

Crepis alpina plants were grown in green houses and seeds harvested at the early to mid-stage of development (17-20 days after flowering). Cotyledons were squeezed out from their seed coats and homogenized with mortar and pestle in 0.1M phosphate buffer, pH 7.2 containing 0.33M sucrose, 4 mM NADH, 2 mM CoASH, 1 mg of bovine serum albumin/ml and 4000 units of catalase/ml. The homogenate was centrifuged for 10 min at 18 000 x g and the resulting supernatant centrifuged for 60 min at 150 000 x g to obtain a microsomal pellet. Standard desaturase or acetylenase assays were performed at 25°C with microsomal preparations equivalent to 0.2 mg of microsomal protein resuspended in fresh homogenisation buffer and 10 nmol of either [1-^{14}C]18:1-CoA (spec. activity 50 000 d.p.m./nmol) or [^{14}C]18:2-CoA (speci. activity 85 000 d.p.m./nmol) in a total volume of 360 µl.

3. Results and Discussion

The ammonium salt of [^{14}C]linoleic (9,12 octadecadienoic, 18:2) acid was converted into radioactive crepenynic acid (18:1A) by detached developing cotyledons of *Crepis alpina*, thus demonstrating that the formation of the acetylenic bond occurs via substraction of two hydrogen atoms from a Δ12 double bond.

Since the precursor fatty acid (18:2) for 18:1A formation is likely to be formed from oleoyl-phosphatidylcholine in the E.R. (Stymne and Appelqvist, 1978), it is reasonable to assume that the acetylenase is localized outside of the plastids, and perhaps in the E.R. itself. Therefore, microsomal preparations from developing cotyledons of *Crepis alpina* seeds were tested for their ability to convert [^{14}C]18:2 into [^{14}C]18:1A. Microsomes from homogenate prepared in the presence of high amount of NADH were active in converting [14C]18:2-CoA into [14C]18:1-A.

The *C. alpina* microsomal Δ12 acetylenase and Δ12 desaturase activities were characterized with resepect to cofactor requirement and inhibitors (Table 1). Both enzymes showed similar cofactor requirements and responded similarly to the different inhibitors indicating that they may both belong to the same type of monooxygenases. It is pertinent to note that the acetylenase reaction is a desaturation.

TABLE 1. Cofactor requirement and effects of carbon monoxide, anti-cytochrome-b_5 antibodies and KCN on Δ12 desaturase and Δ12 acetylenase activities in microsomal preparations from developing cotyledons of *Crepis alpina*

	Δ12 desaturase activity[a]	Δ12 acetylenase activity[a]
Control[b]	100	100
-NADH, +NADPH	95	73
Minus coreductant	14	0
Carbon monoxide	85	84
Anti-b5 antibodies	66	7
KCN	16	0

[a]Activity expressed as percentage of activity in control incubations (control Δ12 desaturase activity =5.0 nmol [^{14}C]18:1 desaturated in 30 min; control Δ12 acetylenase activity=0.76 nmol [^{14}C]linoleate desaturated in 180 min).
[b]Control incubations performed as described in Materials and Methods with 10 nmol of either [^{14}C]18:1- CoA or [^{14}C]18:2-CoA.

Since we previously showed that both fatty acid hydroxylation and epoxygenation can occur while substrate acyl groups are esterified to phosphatidylcholine (PC) (Bafor *et al.*, 1991, 1993), we investigated if this was also the case for the acetylenic fatty acid formation. Lipids in microsomal preparations from *C. alpina* seeds were first pre-labelled with [^{14}C]18:2 from [^{14}C]18:2-CoA afterwhich remaining acyl-CoA was removed by re-pelleting the microsomes. All of the radioactivity in the labelled microsomes was present in 18:2; of which 62% resided in PC, 26% in diacylglycerols (DAG) and triacylglycerols (TAG) and only small amounts in free fatty acids (3.4%) and other lipids (8.6%). During a 2 hour incubation of the *in-situ* labelled microsomes with NADH, 6% of the [^{14}C]18:2 was converted to [^{14}C]18:1A (Fig. 1), which was equivalent to the conversion seen when incubating with a corresponding amount of [^{14}C]18:2-CoA added directly to the microsomes. The [^{14}C]crepenynic acid was found exclusively as free fatty acid (see Fig.1). Very little 18:1A was formed in incubations without NADH. The amount of radioactivity in PC decreased during the incubation whereas free fatty acids and other neutral lipids gained activity, regardless wether NADH was present or not. However, the decrease in radioactivity in PC when incubated with NADH was greater than without coreductant and the lower radioactivity in PC was compensated by a corresponding increase in free [^{14}C]18:1A. This suggests that the [^{14}C]18:2 substrate used by the acetylenase was derived from PC. We have shown that the removal of ricinoleic acid and vernolic acid from PC in oil seeds are catalyzed by specific phospholipases (Ståhl et al., 1995). However, we failed to demonstrate a 18:1A specific phospholipase in incubations of C. alpina microsomes with exogenous 18:1A-PC when using similar conditions as for assaying hydroxy and epoxy specific phospholipases in other oil seeds (Ståhl et al., 1995).

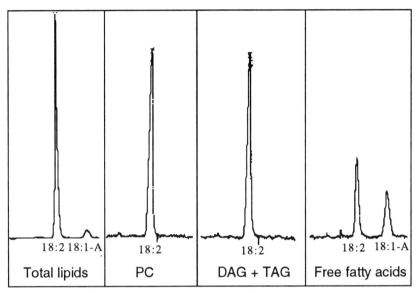

Figure 1. Radio-GLC chromatogrammes showing the distribution of radioactivity among acyl groups in different microsomal lipids after incubation (2 h) of microsomal preparations of developing C. alpina seeds with [^{14}C]18:2 in situ labelled lipids and NADH. Microsomal lipids were extracted after incubation and the different ipid classes were separated on TLC and methylated before injection into radio-GLC. Abbreviations: 18:2, linoleic acid; 18:1-A, crepenynic acid; PC, phosphatitdylcholine; DAG, diacylglycerol; TAG, triacylglycerol.

4. References

Bafor, M., Smith, M., Jonsson, L., Stobart, K. and Stymne, S. (1991) Ricinoleic acid biosynthesis and triacylglycerol assembly in microsomal preparations from developing castor bean endosperm, *Biochem. J.* **280**, 507-514.

Bafor, M., Smith, M.A., Jonsson, L., Stobart, K. and Stymne, S. (1993) Biosynthesis of vernoleate (cis-12-epoxyoctadeca-cis-9-enoate) in microsomal preparations from developing endosperm of *Euphorbia lagascae*, *Arch. Biochem. Biophys.* **303**, 145-151.

Bohlmann, F., Burkhardt, T., and Zdero, C. (1973) *Naturally occuring acetylenes*, Academic Press, New York.

Kohn, G., Hartmann, E., Stymne, S. and Beutelmann, P. (1994) Biosynthesis of acetylenic fatty acids in the moss *Ceratodon purpureus* (Hedw.) Brid., *J. Plant Physiol.* **144**, 265-271.

Stymne, S. and Appelqvist, L-A. (1978) The biosynthesis of linoleate from oleoyl-CoA via oleoyl-phosphatidylcholine in microsomes of developing safflower seeds, *Eur. J. Biochem.* **90**, 223-229.

Ståhl, U., Banas, A., and Stymne, S. (1995) Plant microsomal phospholipid acyl hydrolases have selectivities for uncommon fatty acids, *Plant Physiol.* **107**, 953-962.

METABOLISM OF PALMITATE DIFFERS IN *NEUROSPORA CRASSA* MUTANTS WITH IMPAIRED FATTY ACID SYNTHASE

M. GOODRICH-TANRIKULU, A.E. STAFFORD, and T.A. McKEON
Western Regional Research Center, Agricultural Research Service, U.S. Department of Agriculture, 800 Buchanan St., Albany, CA 94710 USA

1. Abstract

Fatty acid metabolism in the fungus Neurospora crassa resembles that in higher plants. Neurospora additionally has the advantages of microbial systems, allowing biochemical and genetic studies of the metabolism of both common and unusual fatty acids. The Neurospora cel mutant is impaired in fatty acid synthase, and requires supplementation with palmitate (16:0) for growth. By following metabolism of exogenous [2H]16:0 using GC/MS, we have observed that wild-type Neurospora only poorly converts exogenous 16:0 to unsaturated fatty acids. In contrast, cel readily converts 16:0 to the usual major fatty acids formed by Neurospora, 18:0, 18:1Δ9, 18:2Δ9,12 and 18:3Δ9,12,15. It also forms high levels of two other fatty acids not normally observed in Neurospora. Addition of the cyclopropane fatty acid dihydrosterculate suppresses formation of these fatty acids in cel. This altered metabolism of exogenous 16:0 may be directly due to impaired flux through the endogenous biosynthetic pathway, or may result from altered regulation of the synthesis of unsaturates in the mutant.

2. Introduction

The major fatty acids synthesized by wild-type Neurospora crassa are 16:0, 18:0, 18:1Δ9, 18:2Δ9,12 and 18:3Δ9,12,15. Several mutants in the fatty acid biosynthetic pathway exist. One such mutant is cel, which requires fatty acid supplementation for growth. 100 μM 16:0 is sufficient to restore wild-type growth rates [1]. The defect in cel is in the cytosolic fatty acid synthase. The enzyme complex has only 2% of the wild-type level of phosphopantetheine [2]. The cel mutant was used to show that some fatty acids are synthesized in mitochondria [3].

In metabolic tracer studies using [2H]16:0, we found that wild-type N. crassa poorly converts exogenous 16:0 to unsaturated fatty acids. Since cel grows well on 16:0 alone, and an earlier study suggested no impairment in conversion of [14C]16:0 to 18:2 and 18:3 [4], we tested whether it differs from wild type in the extent of conversion of 16:0 to unsaturates.

In addition, we compared the extent of incorporation of a fatty acid not synthesized by N. crassa, the cyclopropane fatty acid dihydrosterculate (DHS), in wild type and cel.

3. Materials and Methods

N. crassa wild type and cel strains were obtained from the Fungal Genetics Stock Center, Kansas City, KS. Conidia (100,000) were incubated at 34°C in Vogel's medium N for 2 days. 30 µM 7,7,8,8-[2H4]16:0 (98%, Cambridge Istotope Laboratories, Andover, MA) was added after 1 day of incubation (DHS when present, and, for cel, 16:0, were added at inoculation). Lipids were extracted from filtration-harvested mycelia with chloroform:methanol:water (5:10:4; 0.25% HCl). Fatty acid methyl esters were prepared, and determination of 2H incorporation was by GC/MS.

4. Results

N. crassa cel cultures differed from wild type in fatty acid composition (compare Fig. 1 A, B), accumulating two longer-chain polyunsaturated fatty acids, 20:2Δ11,14 and

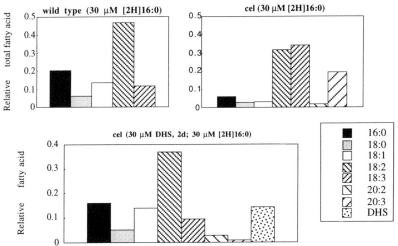

Fig. 1. Fatty acid composition of N. crassa. A (upper left), wild type; B (upper right), cel; C (lower panel), cel with DHS

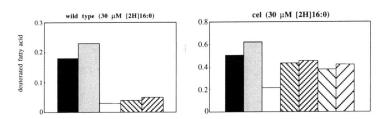

Fig 2. 2H incorporation into N. crassa. A (left panel), wild type; B (right panel), cel

20:3Δ11,14,17, found only in trace amounts in wild type. The fatty acid composition of cel more closely resembled wild type at higher levels of 16:0 (not shown), or in the presence of 30 μM DHS (Fig. 1C). DHS-grown cel accumulated more DHS than wild type, and had a higher mass than cel grown without DHS.

Incorporation of deuterated fatty acids from [2H]16:0 differed in wild type and cel. Wild type incorporated 2H poorly into unsaturates, whereas cel incorporated 2H readily into unsaturates, including longer-chain polyunsaturates (compare Fig. 2A, B).

5. Discussion

The cel mutant accumulates higher levels of 18:3 than does wild type. In addition, it forms significant amounts of two fatty acids made only in trace amounts by wild type. These fatty acids, 20:2Δ11,14 and 20:3Δ11,14,17, are elongation products of 18:2Δ9,12 and 18:3Δ9,12,15.

The high level of polyunsaturates in cel may result from its increased ability to convert exogenous 16:0 to unsaturates, compared to wild type. When grown from the time of inoculation in the presence of [2H]16:0, on average more than 70% of its fatty acids are derived from the supplement. The remaining fatty acids are presumably synthesized by a combination of residual activity of cytosolic fatty acid synthase, and mitochondrial fatty acid synthase.

The more extensive formation of unsaturates (including longer-chain polyunsaturates) observed in cel could result from several factors. Substrates derived from the supplement may be more available for elongation and desaturation due to the limited flux of substrates formed de novo from the impaired fatty acid synthase. Alternatively, the regulation of 18:0 desaturation may differ in the cel mutant, to compensate for the normally more restricted conversion of exogenous 16:0 to 18:1Δ9.

As expected, since cel obtains the majority of its fatty acid from the supplement, cel incorporated a greater proportion of exogenous DHS into its lipids than did wild type. In addition, DHS treatment inhibited the formation of the longer-chain polyunsaturates. When added at the time of inoculation, DHS also promoted growth and de novo fatty acid biosynthesis by cel to near wild type. We are currently investigating the mechanism of this effect.

6. References

1. Henry, S.A. and Keith, A.D. (1971) Saturated fatty acid requirer of Neurospora crassa, J. Bacteriology **106**:174-182
2. Elovson, J. (1975) Purification and properties of the fatty acid synthetase complex from Neurospora crassa, and the nature of the fas- mutation. J. Bacteriology **124**:524-533
3. Mikolajczyk, S. and Brody, S. (1990) De novo fatty acid synthesis mediated by acyl-carrier protein in Neurospora crassa mitochondria. Eur. J. Biochem. **187**:431-437
4. Coté, G.G. (1986) Ph.D. Thesis, University of California, San Diego

HYDROXY FATTY ACID BIOSYNTHESIS AND GENETIC TRANSFORMATION IN THE GENUS *LESQUERELLA* (*BRASSICACEAE*)

DARWIN W. REED[1], JOE K. HAMMERLINDL[1], C.E. PALMER[2], DAVID C. TAYLOR[1], WILFRED KELLER[1], PATRICK S. COVELLO[1]
[1]*NRC Plant Biotechnology Institute, 110 Gymnasium Place, Saskatoon, Saskatchewan, Canada S7N 0W9*
[2]*Department of Plant Science, University of Manitoba, Winnipeg, Manitoba, Canada R3T 2N2*

1. Introduction

Members of the genus *Lesquerella* produce seed oils containing a high proportion of hydroxy fatty acids. There are three types of *Lesquerella* which are distinguished by the most abundant seed oil fatty acid - lesquerolic acid (20:1OH), densipolic acid (18:2OH) or auricolic acid (20:2OH), as represented by *L. fendleri*, *L. kathryn* and *L. auriculata*, respectively. We have investigated the conversion of radiolabelled intermediates of hydroxy fatty acid biosynthesis in the above three species.

There is a growing interest in the use of transformable specialty crops for biotechnological applications. To explore the usefulness of *Lesquerella* in this regard, experiments were conducted on plant regeneration and *Agrobacterium*-mediated transformation of explants of *L. fendleri*.

2. Results and Discussion

2.1. METABOLISM OF ^{14}C-LABELED FATTY ACIDS BY DEVELOPING EMBRYOS

Six stages of development were assigned on the basis of seed and pod morphology. The stages which were the first to show substantial (>5%) hydroxy fatty acid content were used in biochemical experiments. The *in vivo* conversion of various ^{14}C-labeled fatty acids was investigated by incubating approximately 50 Stage III and IV embryos (seed coat removed) overnight in their presence. Lipids from incubated, washed embryos were saponified, converted to methyl esters and analyzed by radio-HPLC. The results of the labeled fatty acid feedings are shown in Table 1. Figure 1 summarizes the results pictorially. Some conclusions that can be drawn from the data are:

Table 1. Fatty acids detected by radio-HPLC upon embryo feeding with ^{14}C-labeled fatty acids. Percentage of total peak area and standard errors of the means of 3 or more replicates (in parentheses) are shown.

Species	^{14}C-Labelled Fatty Aid	18:1	18:2 + 18:3	20:1	18:1-OH	18:2-OH	20:1-OH	20:2-OH
L. fendleri	18:1	49.0 (0.6)	11.6 (1.6)	0.7 (0.02)	1.3 (0.8)		25.4 (0.6)	0.7 (0.7)
L. kathryn	18:1	60.5 (12.1)	5.9 (0.7)		5.3 (1.2)	16.4 (6.6)		
L. auriculata	18:1	59.1 (1.0)	10.6 (0.9)	1.6 (0.2)	11.0 (0.07)	0.6 (0.3)	14.8 (0.3)	
L. fendleri	18:1-OH				52.2 (19.9)	4.7 (2.5)	36.5 (20.1)	0.03 (0.03)
L. kathryn	18:1-OH				73.0 (2.2)	20.9 (2.4)		
L. auriculata	18:1-OH				53.4 (7.3)	10.3 (5.2)	23.3 (7.4)	0.1 (0.00)
L. fendleri	18:2-OH					82.7 (8.7)		
L. kathryn	18:2-OH					91.3		
L. auriculata	18:2-OH					84.7 (0.9)		1.4 (0.4)
L. fendleri	20:1-OH						97.1	0.9
L. kathryn	20:1-OH						75.6 (2.3)	18.8 (3.3)
L. auriculata	20:1-OH						81.0 (3.3)	12.8 (2.6)

- 18:1OH is produced by hydroxylation of 18:1
- 18:1OH is elongated to 20:1OH in *L. fendleri* and *L. auriculata*
- *L. kathryn* can desaturate both 18:1OH and 20:1OH
- 20:2OH is synthesized mainly by desaturation of 20:1OH
- The results are explicable by differences in levels of elongase and desaturase activities in different species

2.2 TISSUE CULTURE AND TRANSFORMATION

Leaf and root explants from axenically grown seedlings of *L. fendleri* regenerate shoots profusely on MS medium with 3% sucrose and 1 mg/L each of benzyl adenine and zeatin. These explants were cocultivated with *Agrobacterium tumefaciens* harbouring a binary transformation vector whose T-DNA carries genes conferring resistance to either the herbicide L-phosphinothricin (L-PPT) or to kanamycin.

When leaf explants were used, the transformation frequency was in the range of 3-5% whether selection was done on kanamycin or L-PPT. Root explants showed less tendency to be damaged by exposure to the bacterium and gave somewhat higher transformation frequencies (12-16%). Both kanamycin and L-PPT gave similar transformation frequencies with equivalent explants, but selection was cleaner using the herbicide, giving rise to fewer non-transformed escapes. Transformed shoots have been successfully rooted and transferred to soil. Southern analysis confirmed stable transformation of the plants.

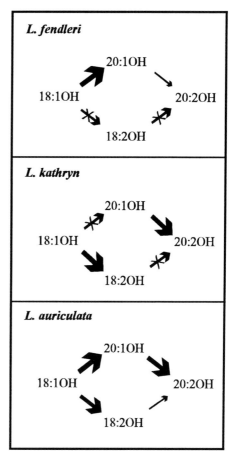

Figure 1. Proposed scheme for the biosynthesis of hydroxy fatty acids in *Lesquerella* species.

3. Reference

1. Engeseth,N. and Stymne,S. (1996) Desaturation of oxygenated fatty acids in Lesquerella and other oil seeds, *Planta* **198**, 238-245

DEVELOPMENTAL CHANGES IN SUBSTRATE UTILIZATION FOR FATTY ACID SYNTHESIS BY PLASTIDS ISOLATED FROM OILSEED RAPE EMBRYOS

PETER J. EASTMOND AND STEPHEN RAWSTHORNE
Brassica and Oilseeds Research Department,
John Innes Centre,
Norwich, NR4 7UH, UK

Introduction

Within the developing seeds of *Brassica napus* (L.) carbon in the form of sucrose is allocated towards the biosynthesis of storage lipids. The pathway of fatty acid synthesis is localized within the plastid. Carbon precursors and ATP required to support the pathway are imported from the cytosol. Fatty acid synthesis also requires an endogenous supply of NAD(P)H generated by plastidial metabolism. Plastids isolated from *B. napus* embryos during the early to mid stages of lipid deposition, contain a complete set of glycolytic enzymes and pyruvate dehydrogenase and incorporate carbon from a number of different metabolites into fatty acids in an ATP- and intactness-dependent manner [1]. Of the substrates tested glucose 6-phosphate (Glc6P) and pyruvate were the strongest precursors for fatty acid synthesis [1,2].

The aim of this study was to use isolated plastids to examine how carbon partitioning towards acetyl-CoA and therefore fatty acids might be regulated within the plastid during development. The rates of incorporation of carbon from exogenous metabolites into fatty acids were measured in plastids from embryos at different stages in development and compared to the *in vivo* rates of lipid accumulation. Changes in carbon flux were subsequently compared to enzyme and membrane transporter activities within the plastid in order to identify potential sites of regulation.

Materials and methods

Intact plastids were isolated from early, early to mid and mid to late cotyledon embryos (1.5, 2.5 and 3.5 mg embryo fresh weight) of *B. napus* cv. Topas (according to [1]). Isolated plastids were incubated with ^{14}C-labelled metabolites and the incorporation of ^{14}C into fatty acids determined [1]. The activities of enzymes were measured using standard assays [1] optimized for plastid extracts. The proportions of glycolytic enzyme activities attributable to the plastid were determined by adding increasing concentrations of cytosol (supernatant) back to a plastid pellet. The plastidial activity of each enzyme was then calculated from the relationship between its total activity and that of a cytosolic and plastidial marker enzyme essentially as previously described [3]. Short term assays for the rate at which isolated plastids take up ^{14}C-labelled metabolites were performed using silicone oil filtration [4].

Results and discussion

In order to study how the metabolism of the plastid changes over the course of embryo development intact plastids were isolated and characterized from early, mid and mid-late cotyledon embryos (1.5, 2.5 and 3.5

mg fresh weight). These stages, refered to as A, B and C, represent a low, an intermediate and a maximal rate of lipid accumulation *in vivo* (11, 17 and 23 nmol acetyl-CoA mgFW^{-1} h^{-1}) [5]. The intactness of the plastids, estimated from the latency of NADP-glyceraldehyde 3-phosphate dehydrogenase (GAPDH) activity [1], was between 65 and 75 % at all three stages of development. The level of contamination of preparations by the cytosol and other sub-cellular compartments was comparable to that previously reported [1].

The rates of incorporation of carbon from Glc6P into fatty acids by plastids isolated from embryos at stages A and B were approximately 450 nmol acetyl-CoA unit^{-1} GAPDH h^{-1} (Fig. 1). This rate was equivalent to around 80 % of the flux necessary to support the *in vivo* rate of lipid synthesis at stage A [5] (calculated from the latency and activity per embryo of NADP-GAPDH [1]). By stage C there was a two-fold drop in the rate of incorporation of carbon from Glc6P into fatty acids *in vitro* (Fig. 1). In contrast, the rate of incorporation of carbon from pyruvate into fatty acids increased more than 20-fold over the course of the three developmental stages (Fig. 1). By stage C the rate of incorporation of carbon from pyruvate into fatty acids (1200 nmol acetyl-CoA unit^{-1} GAPDH h^{-1}) could account for nearly 80 % of the flux required for *in vivo* lipid synthesis [5]. The rate of incorporation of carbon from pyruvate was more than four-fold greater than that from Glc6P, dihydroxyacetone phosphate, malate or acetate at stage C.

In order to investigate why there was a decline in the rate of incorporation of carbon from Glc6P into fatty acids *in vitro* between stages B and C, despite an increase in the rate of incorporation from pyruvate, the capacity of plastids to import Glc6P and metabolize it to pyruvate was examined. The movement of Glc6P across plastid envelopes has previously been reported to be mediated by a phosphate-exchange translocator [6]. In *B. napus* embryos at all three stages of development Glc6P uptake by isolated plastids displayed saturation kinetics consistent with protein mediated transport. The Km was 0.13mM at each of the three stages. The maximum rate of transport was similar at stages A and B (1200 nmol Glc6P unit^{-1} GAPDH h^{-1}) but fell by 30 % at stage C. Once in the plastid, Glc6P is metabolized to pyruvate via glycolysis. The plastidial activities of the glycolytic enzymes either remained constant or declined during development. Of the enzymes with the lowest activities aldolase, phosphoglycerate mutase and enolase displayed the largest decline in activity (more than 3-fold) between stages B and C (Table 1).

The rate of incorporation of carbon from pyruvate into fatty acids by isolated plastids increased greatly during embryo development. The low rate of incorporation of carbon from pyruvate at stage A is unlikely to be a result of limitations in the committed pathway of fatty acid synthesis or pyruvate dehydrogenase since Glc6P shares a common pathway and is incorporated at a higher rate (Fig. 1). The only apparent barrier to pyruvate metabolism is the uptake of pyruvate across the plastid envelope. Evidence for chloroplastic pyruvate transporters from C_4 and C_3 plants has previously been provided [7,8]. However, no pyruvate transporter has so far been reported in embryo plastids. In plastids isolated from stage A embryos, the rate of pyruvate uptake increased linearly with increasing concentration suggesting that uptake was by passive diffusion. In contrast, at stages B and C pyruvate uptake showed a hyperbolic response to concentration indicative of protein mediated transport. The Km was 0.23mM at both stages. The maximum rate of pyruvate transport increased coincidently with the rate of incorporation of carbon from pyruvate into fatty acid (Fig. 1). At each stage of development the rate of uptake was less than 1.4-fold greater than the corresponding rate of incorporation into fatty acids (Fig. 1).

In conclusion we have shown that changes in plastidial metabolism could account for the pattern of lipid accumulation in developing *B. napus* embryos. This implies that plastidial metabolism is likely to be important in the regulation of lipid synthesis, although it does not preclude regulation outside the plastid. During development the substrate used preferentially for fatty acid synthesis by isolated plastids changed from Glc6P to pyruvate. The decline in carbon flux from Glc6P into fatty acids may be explained by reductions in the capacity of a Glc6P translocator and enzymes in plastidial glycolysis. The dramatic increase in carbon flux from pyruvate into fatty acids may be dependent upon the induction of the activity of a pyruvate transporter. The activity of this transporter correlated positively with the rate at which the isolated plastids synthesised fatty acids from pyruvate during embryo development. Further studies relating our *in vitro* findings to lipid accumulation *in vivo* are underway.

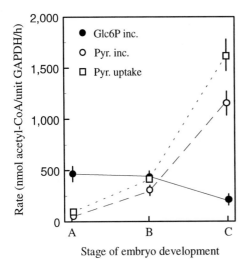

Figure 1. The rate of incorporation of carbon from Glc6P and pyruvate (1mM) into fatty acids and the rate of pyruvate (1mM) uptake by plastids isolated from embryos at three stages in development (A: 1.5 mgFW, B: 2.5 mgFW and C: 3.5 mgFW). A unit of GAPDH (NAPH-dependent glyceraldehyde 3-phosphate dehydrogenase) activity is one μmol min^{-1}. Each value is the mean $\pm$SE of measurements from three separate preparations.

Enzyme	Stage of embryo development		
	A	B	C
Aldolase	1.21 ±0.17	0.95 ±0.15	0.21 ±0.14
PGlyM	2.84 ±0.76	2.12 ±0.15	0.38 ±0.11
Enolase	2.56 ±0.33	2.04 ±0.38	0.48 ±0.12

Table 1. The plastidial activities of aldolase, phosphoglycerate mutase and enolase at three stages in embryo development (A: 1.5 mgFW, B: 2.5 mgFW and C: 3.5 mgFW). Activity is expressed as nmol mg^{-1} embryo FW min^{-1}. Each value is the mean $\pm$SE of measurements from three separate preparations.

References

1, Kang F and Rawsthorne S (1994). Starch and fatty acid synthesis in plastids from developing embryos of oilseed rape (*Brassica napus* L.). Plant J. **6**: 795-805.
2, Kang F and Rawsthorne S (1996). Metabolism of glucose 6-phosphate and utilization of multiple metabolites for fatty acid synthesis by plastids from developing oilseed rape embryos. Planta **199**: 321-327.
3, Denyer K and Smith AM (1988). The capacity of plastids from developing pea cotyledons to synthesise acetyl-CoA. Planta **173**: 172-182.
4, Heldt HW and Sauer F (1971). The inner membrane of the chloroplast envelope as the site of specific metabolite transport. Biochim. Biophys. Acta **234**: 83-91.
5, Kang F, Ridout CJ, Morgan CL and Rawsthorne S (1994). The activity of acetyl-CoA carboxylase is not correlated with the rate of lipid synthesis during development of oilseed rape (*Brassica napus* L.). Planta **193**: 320-325.
6, Borchert S, Grosse H and Heldt HW (1989). Specific transport of inorganic phosphate, glucose 6-phosphate, dihydroxyacetone phosphate and 3-phosphoglycerate into amyloplasts. FEBS Letters **253**: 183-186.
7, Huber SC and Edwards GE (1977). Transport in C_4-mesophyll chloroplasts. Characterization of a pyruvate carrier. Biochim. Biophys. Acta **462**: 583-602.
8, Proudlove MO and Thurman DA (1981). The uptake of 2-oxoglutarate and pyruvate by isolated pea chloroplasts. New Phytol. **88**: 255-264.

ß-KETOACYL-ACYL CARRIER PROTEIN [ACP] SYNTHASE II IN THE OIL PALM (*ELAEIS GUINEENSIS JACQ.*) MESOCARP

UMI SALAMAH RAMLI AND RAVIGADEVI SAMBANTHAMURTHI.
Biology Division, Palm Oil Research Institute of Malaysia
(PORIM), P.O. Box 10620, 50720 Kuala Lumpur (Malaysia).

1. INTRODUCTION

Very little is known about the regulation of fatty acid biosynthesis in the oil palm mesocarp which comprises 44% palmitic acid (C16:0), 39% oleic acid (C18:1) and less than 5% stearic acid (C18:0) in the commercial variety. Based on the fatty acid composition, it was postulated that relatively low activity of β-ketoacyl ACP synthase (KAS) II and high activity of palmitoyl ACP thioesterase may account for the high level of palmitic acid. Screening of oil palm varieties having a range of fatty acid compositions showed strong positive correlation between KAS II activity and unsaturated fatty acid (C18:1 + C18:2 + C18:3) but negative correlation with palmitic acid thus confirming the postulation. The enzyme was purified 10000-fold by acetone extraction followed by CM-Sepharose, HR-DEAE cellulose, hydroxylapatite and ACP-Sepharose chromatography.

2. MATERIALS AND METHODS

KAS II activity was assayed as described by MacKintosh *et. al.* [1]. One unit KAS II activity was defined as 1 nmol of [^{14}C]-malonyl CoA formed per minute derived from the reverse reaction. Determination of iodine value of oil palm mesocarp was carried out according to PORIM Test Methods [2].

3. RESULTS AND DISCUSSION

KAS II activity showed strong correlation with iodine value which is a measure of unsaturation level in the oil. Comparison of C16:0 and C18:1 indicated negative correlation between these fatty acids. It is also interesting to note that KAS II activity is positively correlated with the level of unsaturated fatty acids (C18:1 + C18:2 + C18:3). This finding suggests that the accumulation of palmitic acid in the oil palm is a consequence of limiting KAS II activity.

TABLE 1. Determination of KAS II activity, iodine value and fatty acid composition.

KAS II activity (nmol/min)	Iodine value	C16:0	C18:0	C18:1	C18:2	C18:3	C18:1+C18:2+ C18:3
5.60	84.30	16.40	0.90	61.80	18.70	0.97	81.50
5.37	83.40	15.00	0.87	59.90	18.70	1.30	79.90
3.39	68.80	22.20	2.60	69.20	5.90	0.40	75.50
2.25	67.10	26.90	2.30	63.90	5.40	0.61	69.91
2.15	60.80	30.80	6.60	49.00	11.70	0.50	61.20
1.85	57.30	36.50	4.50	50.30	8.00	0.21	58.51

KAS II was purified from the oil palm mesocarp using a combination of steps, with results being shown in Table 2. Attempts to fractionate the crude extract by ammonium sulphate precipitation resulted in loss of activity suggesting that oil palm KAS II is sensitive to or not stable in the presence of ammonium sulphate. Acetone extraction did not result in major loss of KAS II activity. This method was also useful in removing large amounts of lipid from the crude extract. The supernatant in Table 2 refers to acetone powder that was resuspended in 0.1 M phosphate buffer (pH 7.0) containing 2 mM dithiothreitol, 1 µM leupeptine, 1µM pepstatine A and 0.1 mM PMSF.

Table 2. Purification table of KAS II from oil palm mesocarp.

FRACTION	PROTEIN TOTAL (mg)	ACTIVITY (U)	SPECIFIC ACTIVITY (U/mg)	PURIFICATION FACTOR	RECOVERY (%)
Supernatant	800	4.05	0.005	1	100
CM-Sepharose	154	3.92	0.025	5.08	96.7
HR-DEAE	15	2.63	0.175	35.0	64.9
Hydroxylapatite	3	1.75	0.586	117.2	43.5
ACP-Sepharose	0.02	1.03	51.5	10300	25.4

The final and most important step in the purification involved ACP-Sepharose chromatography. Commercial ACP was not suitable and it was important that the ACP be highly purified for KAS II to bind. Under standard incubation conditions, KAS II activity was linear for 10 minutes. Enzyme activity favoured acidic conditions with optimum activity between pH 4.5 to 5.0 using 50 mM acetate buffer. With subsequent increase of pH in the assay, the activity decreased gradually. KAS II activity was optimum at 30°C. Sulfhydryl groups at 10 mM concentration efficiently increased enzyme activity with β-mercaptobenzothiozol giving the highest stimulation. The influence of divalent cations on KAS II activity was determined in the presence of various cations at 1 mM and 10 mM concentrations. KAS II activity increased 3 to 10-fold in the presence of Zn^{2+}, Mg^{2+} and Mn^{2+} at both concentrations. The highest activity was obtained with Mn^{2+} at 10 mM concentration. On the other hand neither Cu^{2+} nor Ca^{2+} at both concentrations influenced KAS II activity significantly. EDTA decreased KAS II activity by approximately 50%. However prior incubation of KAS II with divalent cation (Mn^{2+} or Mg^{2+}) prevented the inhibitory action of this chelating agent.

4. CONCLUSION

Our studies indicated that KAS II may play an important role in the regulation of fatty acid biosynthesis in the oil palm mesocarp. This study established that limiting KAS II activity is one of the factors that contributed to the accumulation of palmitic acid in the oil palm mesocarp.

5. REFERENCES

1. MacKintosh, R.W., Hardie, D.G. and Slabas, A.R.: A new assay procedure to study the induction of β-ketoacyl ACP synthase I and II and the complete purification of β-ketoacyl ACP synthase I from developing seeds of oilseed rape (*Brassicca napus*), *Biochim. Biophys. Acta.* 1002 (1989), 114-124.
2. PORIM test methods for oil palm and palm oil products (1986).

THE PRODUCTS OF THE MICROSOMAL FATTY ACID ELONGASE ARE DETERMINED BY THE EXPRESSION AND SPECIFICITY OF THE CONDENSING ENZYME

ANTHONY MILLAR AND LJERKA KUNST
The University of British Columbia, Department of Botany
6270 University Boulevard Vancouver, B.C. Canada, V6T 1Z4

Introduction.
Very long chain fatty acids (VLCFAs) have chain lengths greater than 18 carbons. They are found in waxes, suberin and in the seed triacylglyerides (TAGs) of some plant species. They are synthesised by a microsomal fatty acid elongation (FAE) system by sequential additions of C2 moieties to C18 fatty acids derived from the *de novo* fatty acid synthesis (FAS) pathway of the plastid. Analogous to *de novo* FAS it is thought that each cycle of FAE involves four enzymatic reactions; (1) condensation of malonyl-CoA with a long chain acyl-CoA, (2) reduction to b-hydroxyacyl-CoA, (3) dehydration to an enoyl-CoA and (4) reduction resulting in the elongated acyl-CoA. The Arabidopsis *FAE1* gene encodes a seed specific b-ketoacyl-coenzyme A synthase (KCS)(James et al 1995), the condensing enzyme that carries out the first reaction of the microsomal fatty acid elongation system. Despite the fact that four enzymatic reactions are required for elongation of C18, the *FAE1* gene was the only one found by mutational analysis of Arabidopsis that resulted in almost complete abolishment of VLCFA synthesis in the seed. There are several possibilities as to why mutations in gene/s encoding the reductases and the dehydrase were not found. For example, these enzymes may be encoded by members of gene families. Alternatively, they may not be seed specific, but ubiquitously present throughout the plant and shared with other FAE systems involved in VLCFA formation. If any of these VLCFA products are essential for the plant's viability, a mutation in anyone of the reductase or dehydrase genes would be lethal.
To examine the latter hypothesis and determine whether the expression of *FAE1* alone can result in the formation of VLCFAs, we have expressed this enzyme in cells which contain no detectable amounts of VLCFAs; vegetative and floral tissues of Arabidopsis, tobacco seeds and yeast. In addition we present data demonstrating that FAE1 is the limiting factor of VLCFA accumulation in seed TAGs.

Results

(1) A 35S-*FAE1* transgene is able to direct the synthesis of VLCFAs in vegetative and floral tissues in transgenic Arabidopsis. The coding region of the *FAE1* gene was placed behind the constitutive 35S promoter. Transgenic 35S-*FAE1* plants were able to synthesize 20:0, 20:1, 20:2, 20:3, 22:0, 22:1, 22:2, 22:3 and 24:0 (as confirmed by GC-MS) in all tissues examined (Table 1). VLCFAs accumulated to very high levels (> 30%).

Table 1. VLCFA composition of total lipids from 7 week old transgenic (35S-*FAE1*) and wild-type Arabidopsis. Values are the mol % of total fatty acids.

	20:0	20:1	20:2	22:0	22:1	TOTAL
35S-*FAE1*						
Leaves	4.02	2.80	1.27	2.01	2.87	18.92
Roots	9.20	4.59	2.28	14.80	3.17	32.62
Stems	3.38	1.61	2.78	1.78	1.45	12.34
Flowers	6.23	3.56	4.55	3.67	4.95	24.87
Wild-type						
Leaves	0	0	0	0	0	0
Roots	4.15	0	0	14.31	0	20.11
Stems	0	0	0	0	0	0
Flowers	0	0	0	0	0	0

(2) A Napin-*FAE1* transgene is able to direct the synthesis of VLCFAs in the seeds of tobacco. The *FAE1* gene was subcloned behind the seed-specific napin promoter and transgenic plants were generated. Analysis of seed showed that *FAE1* was able to synthesize VLCFAs (Table 2).

Table 2. VLCFA Composition of total lipids from seed of wild-type and transgenic (Napin-*FAE1*) tobacco. Values are the mol% of total fatty acids.

	20:0	20:1	20:2	Total
Wild-type	0.15	0.08	0.08	0.38
Napin-*FAE1* (2)	0.31	1.43	0.69	2.48
Napin-*FAE1* (2-F1)	0.40	1.97	0.85	3.22

(3) The Arabidopsis *FAE1* gene is able to direct the synthesis of VLCFAs in yeast. The *FAE1* coding region was subcloned into a yeast expression vector (pYES2). Transformed yeast cells with the pYES-*FAE1* plasmid accumulated VLCFAs (Table 3).

Table 3. VLCFA composition of transgenic (pYES-*FAE1*) yeast. Values are the mol% of total fatty acids.

	16:0	16:1	18:0	18:1	20:1	22:1
pYES-no insert	10.12	54.97	3.13	28.15	0	0
pYES-*FAE1*	8.52	60.14	2.13	19.52	4.38	0.45

(4) Introduction of the Napin-*FAE1* transgene resulted in the synthesis of increased amounts of VLCFAs in Arabidopsis seed. The mutant *fae1* alleles are semidominant suggesting that FAE1 activity is the limiting factor in the synthesis of VLCFAs. To determine whether the proportion of VLCFAs can be increased by increasing the amount of *FAE1* gene product, we overexpressed the *FAE1* gene under the control of the seed specific napin promoter. The content of VLCFAs in several lines had increased from approximately 28% (wt/wt) to over 40% (wt/wt) of the total. In contrast to a Napin-jojoba KCS transgene in Arabidopsis (Lassner et al 1996), insignificant amounts of 24C were made suggesting different specificities of the two condensing enzymes (Table 4).

Table 4. VLCFA composition of seed from wild-type Arabidopsis, and transgenic Napin-*FAE1* and transgenic Napin-Jojoba Ketoacyl-CoA Synthase Arabidopsis. *data from Lassner et al (1996). Values are the mol% of total fatty acids.

	20:0	20:1	20:2	22:0	22:1	24:0	24:1	Total
Wild-type	1.89	20.4	1.91	0.27	1.71	0.16	0.18	26.9
Napin-*FAE1*	3.76	26.1	3.79	0.42	6.75	0.14	0.20	42.5
Napin-Jojoba-KCS*	N/A	15.7	1.8	1.3	5.7	1.1	1.6	27.8

Conclusions
(1) The expression of *FAE1* alone results in the formation of VLCFAs
(2) If four enzyme activities are necessary for each elongation step, and *FAE1* only encodes one of these activities, it would appear that in Arabidopsis the other three enzyme activities are expressed in all tissues, for wherever *FAE1* is expressed VLCFAs are formed.
(3) Expression of *FAE1* in tobacco seed results in formation of VLCFAs, suggesting that the expression of *FAE1* is the determining factor of whether the oil of a plant's seed contains VLCFAs or not.
(4) FAE1 activity is the limiting factor in VLCFA synthesis of the wild-type Arabidopsis seeds and by increasing FAE1 activity a higher proportion of VLCFA in the seed oil was obtained.
(5) Similar to *de novo* FAS, where the condensing enzyme exhibits a certain acyl chain length specificity, the chain length specificity of the condensing enzyme seems to be a determining factor in the length of VLCFAs made. For example, where the Napin-Jojoba KCS transgene resulted in the production of significant amounts (9.7 mol% of total VLCFAs) of 24C in Arabidopsis seed (Lassner et al 1996), the increased expression of *FAE1* was unable to achieve this (0.8 mol% of total VLCFAs).

Summary
Our results suggest a general process for the regulation of both the amount of VLCFAs synthesized and their chain length. Whether the VLCFAs that are synthesised in a particular cell and the type of VLCFAs made will depend mainly on the expression and specificity of the condensing enzyme. Thus, to determine the amount/length of VLCFAs made the plant only has to regulate the expression of the condensing enzyme/s, instead of three or four enzymes.
The control of acyl chain length by the condensing enzyme may not just be limited to plants. Because FAE1 activity in yeast results in VLCFA formation, this may be a general eukaryotic mechanism.

References.
James, D. W., Lim, E., Keller, J., Plooy I., Ralston E. and Dooner H. K. (1995) Directed tagging of the Arabidopsis *FATTY ACID ELONGATION1 (FAE1)* gene with the maize transposon activator. Plant Cell 7, 309-319.
Lassner M. W. Lardizabal, K., and Metz J.G. (1996). A Jojoba b-ketoacyl-CoA synthase cDNA complements the Canola fatty acid Elongation Mutation in Transgenic plants. Plant Cell 8, 281-292.

CHLOROPLASTIC CARNITINE ACETYLTRANSFERASE

C. WOOD[1], C. MASTERSON[1] & J.A. MIERNYK[2].
[1]*Biological and Nutritional Sciences, University of Newcastle upon Tyne, UK.* [2]*USDA/NCAUR, Peoria, Illinois, USA.*

1. Introduction.

Carnitine acetyltransferase (CAT) is known to exist in plant mitochondria (see review [2]). CAT catalyses the reversible reaction
 Carnitine + short-chain acyl CoA <--->
  short-chain acylcarnitine + CoASH
and is involved in the transport of activated acetyl moieties from the mitochondrion.

The presence of CAT has been reported in chloroplasts of barley [7] and pea [5] leaves and its possible function was described in detail by Masterson et al. [3,4]. The experimental procedures and conclusions of these reports were strongly criticized by Roughan et al. [6] who could not detect any CAT activity associated with chloroplasts or indeed leaf homogenates, which also implies that CAT is not found in mitochondria, despite several reports to the contrary [2]. This work describes the presence, location and properties of chloroplastic CAT.

2. Results and Discussion.

Isolated pea leaf chloroplasts were devoid of mitochondrial contamination as evaluated by Western blotting using specific anti-marker protein antibodies. CAT activity, detected when chloroplasts were ruptured but not with intact chloroplasts, was found to be exclusively stromal (Table 1) and was stabilized by precipitation with 75% saturation $(NH_4)_2SO_4$. When assayed for acetyl CoA formation from acetylcarnitine, abundant CAT activity was detected, with negligible activity in the various controls (Figure 1). This activity corresponded to 300-400 nmol min^{-1} mg^{-1} chl., which is more than enough to account for

TABLE 1. Location of chloroplastic CAT.

SAMPLE	TOTAL UNITS (nmol min^{-1})	TOTAL PROTEIN (mg)	S.A. (nmol min^{-1} mg^{-1} protein)
Thawed chloroplasts	1558.7	137	11.4
Supernatant after centrifugation at 144000xg for 30min (soluble proteins)	1558.7	58	26.9
Pellet after centrifugation at 144000xg for 30min (membrane protein)	0	79	-

the rates of fatty acid synthesis from acetylcarnitine reported by Masterson et al. [3].

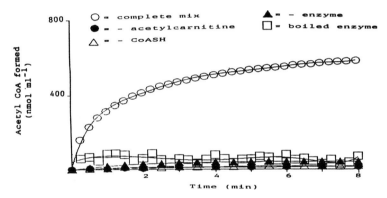

Figure 1. Activity of chloroplastic carnitine acetyltransferase.

Chloroplastic CAT was specific for C2 and C4 substrates (100% and 3.4% activity respectively) and had a pH optimum of 8.5 in contrast to CAT from pea cotyledon mitochondria which had an optimum activity at pH 7.0.

The presence of chloroplastic CAT was confirmed by Western blotting and radio-labelling with ^{14}C-carnitine. An anti-rat liver peroxisomal CAT polyclonal antibody gave a signal with both stromal CAT preparations and pigeon breast CAT from a commercial source. The recognition of chloroplastic CAT by a mammalian CAT antibody suggests that the plant protein has a certain degree of homology with the mammalian enzyme. The protein has a *Mr* 50-53k similar to that of the mammalian enzyme. ^{14}C-carnitine labelled a protein at this same position on a fluorograph of an SDS gel.

Roughan et al. [6] were unable to detect CAT activity in leaf homogenates of spinach, amaranthus and pea, or in the purified chloroplasts of spinach and pea. The fact that they found it necessary to add CoASH to the medium for fatty acid synthesis would suggest that their chloroplasts were damaged. Intact chloroplasts should be self-sufficient in CoASH as it cannot cross the chloroplast membrane barrier [1]. Addition of an external, non-penetrating CoASH should have no discernible effect upon fatty acid synthesis. If the chloroplasts were damaged then the soluble stromal CAT would have been lost during isolation.

It is also surprising that Roughan et al. [6] found no CAT activity in homogenates of pea, spinach and amaranthus as it is well established that mitochondria possess CAT [2] and this at least should have been detected. However, they homogenized their tissue in a medium containing mercaptoethanol and desalted the extracts with a medium containing dithiothreitol, both of which are thiol group protectors. It has been our experience since our carnitine work began in the late 1960s that the presence of these compounds was detrimental to the carnitine system and their inclusion has always been avoided.

3. References.

1. Brooks, J.L. and Stumpf, P.K. (1965) A soluble fatty acid synthesizing system from lettuce chloroplasts. Biochim. Biophys. Acta, 98, 213-216.
2. Masterson, C. Wood, C. and Miernyk, J.A. (1995) Inter-organellar transport of acyl-groups in plant cells, in S.S. Purohit and A. Kumar (eds) Agro's Annu. Rev. Plant Physiol., 2, Agro Botanical Publishers, India, pp. 65-124.
3. Masterson, C. Wood, C. and Thomas, D.R. (1990a) L-acetylcarnitine, a substrate for chloroplast fatty acid biosynthesis. Plant Cell & Envir., 13, 755-765.
4. Masterson, C. Wood, C. and Thomas, D.R. (1990b) Inhibition studies on acetyl group incorporation into pea chloroplast fatty acids. Plant Cell & Envir., 13, 767-771.
5. McLaren, I. Wood, C. Noh Hj Jalil, M. Yong, B.C.S and Thomas, D.R. (1985) Carnitine acyltransferases in pea chloroplasts. Planta, 163, 197-200.
6. Roughan, G. Post-Beittenmiller, D. Ohlrogge, J. and Browse, J. (1993) Is acetylcarnitine a substrate for fatty acid synthesis in plants? Plant Physiol., 101, 1157-62.
7. Thomas, D.R. Noh Hj Jalil, M. Ariffin, A. Cooke, R.J. McLaren, I. Yong, B.C.S and Wood, C. (1983) The synthesis of short- and long-chain acylcarnitine by etio-chloroplasts of greening barley leaves. Planta, 158, 259-263.

4. Acknowledgements.

C.M. thanks BBSRC, The Royal Society and CORF of the University of Newcastle upon Tyne for funding. J.A.M. thanks the University of Newcastle upon Tyne Research Committee for funding his Senior Visiting Research Fellowship.

FUNCTION OF CHLOROPLASTIC CARNITINE PALMITOYLTRANSFERASE

C. MASTERSON[1], C. WOOD[1] & J.A. MIERNYK[2]
[1]*Biological and Nutritional Sciences, University of Newcastle upon Tyne, UK.* [2]*USDA/NCAUR Peoria, Illinois, USA.*

1. Introduction.

Carnitine palmitoyltransferase (CPT) catalyses the reversible exchange of activated long-chain acyl groups between CoASH and L-carnitine.

Long-chain acyl CoA + L-carnitine <---> Long-chain acylcarnitine + CoASH

The presence of carnitine palmitoyltransferase (CPT) has been confirmed in pea leaf chloroplasts. It has been partially purified and some of its properties elucidated as previously published [3]. In brief, there is both an overt CPT activity (CPTo) bound to the outside (cytosol side) of the inner chloroplast envelope, and a latent CPT activity (CPTi), bound to the inside (stromal side) of the inner chloroplast envelope. This report concerns the possible function of these chloroplastic CPTs and their regulation *in vivo.*

2. Results and Discussion.

The CPTi and CPTo of the pea chloroplast inner envelope are ideally placed to facilitate the export/import of fatty acids in the eukaryotic pathway of fatty acid desaturation and provide an additional, alternative pathway to the one currently proposed in the literature [1].

The substrate specificities of CPTo and CPTi (Figure 1) show that the enzymes exhibit increased activity with desaturated long-chain fatty acids, exhibiting a preference for linolenic acid. This may indicate that the primary function of the CPTs is the import of desaturated fatty acids into the chloroplast. This would fill a gap in current knowledge of how desaturated fatty acids enter chloroplasts as the previously proposed mechanism of lipid transfer as DAG between the ER and the chloroplast is not known. Lipid transfer proteins have been postulated to transfer lipids between different cellular membranes, but their involvement in the transfer from ER to chloroplast is uncertain owing to the vacuolar and cell wall location of these proteins [1].

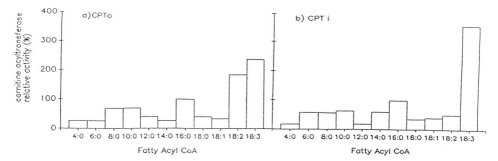

FIGURE 1. Substrate specificity of CPTo and CPTi.

Two peaks of CPT activity showing differing responses to malonyl CoA are separable by ion-exchange chromatography. The CPT activity which binds to a Q-Sepharose column was inhibited 70% by 150μM malonyl CoA whilst a peak of CPT activity that does not bind exhibited a 50% increase in activity with the same concentration of malonyl CoA. These two forms of chloroplastic CPT can be physically separated from the original inner envelope membrane by differential detergent treatments. Mild detergent treatment with 2% (v/v) Tween 20 solubilized the CPT which was stimulated by malonyl CoA. The membrane pellet after Tween 20 treatment still possessed CPT activity which was inhibited by malonyl CoA. This latter CPT protein was solubilized from the membrane pellet with 2% (v/v) Triton X-100. The supernatant here retained CPT activity which was again inhibited by malonyl CoA. The pelleted membrane after Tween 20 and Triton X-100 treatment was devoid of CPT activity. Studies with intact and broken chloroplasts indicate that these two forms of CPT correspond to CPTo and CPTi (Table 1). By assaying CPT in intact chloroplasts (which exhibit CPTo activity) and broken chloroplasts (which exhibit CPTo + CPTi activity) it can be shown that the malonyl CoA-inhibited CPT is the outer enzyme (CPTo) which is firmly bound to the membrane (requiring Triton X-100 solubilization) and the malonyl CoA-stimulated CPTi is less firmly bound, being solubilized by Tween 20. The inhibition of CPTo by malonyl CoA is less pronounced when the enzyme is bound to the membrane compared with the solubilized enzyme.

TABLE 1. Chloroplastic CPT activity in the presence and absence of malonyl CoA. (pmol of palmitoylcarnitine formed min^{-1} ml^{-1})

Chloroplast state	150µM Malonyl CoA −	+	% of Control
Intact (CPTo)	77.1	75.2	97.5
Broken (CPTo & CPTi)	144.1	179.1	124.3
Broken − intact (CPTi)	67.0	103.9	155.1

The rate limiting step in fatty acid synthesis in leaves is that catalysed by acetyl CoA carboxylase [4], which synthesizes malonyl CoA from acetyl CoA and CO_2. Malonyl CoA is thus a logical regulator of CPT activity. When malonyl CoA levels are high and fatty acid synthesis is proceeding, CPTi is stimulated to increase export of fatty acids for desaturation. The stimulation of CPTi by malonyl CoA means that CPT is responsive to the composition of the stromal acyl CoA pool. Acetyl CoA carboxylase is also found in the cytosol [2] where it synthesizes malonyl CoA for use in acyl CoA elongation reactions and/or aromatic ring biosyntheses. The CPTo may be regulated by this cytosolic pool of malonyl CoA, although the inhibition by malonyl CoA observed on the solubilized enzyme is much less marked when the enzyme is membrane bound (Table 1).

3. References.

1. Browse, J. and Somerville, C. (1991) Glycerolipid synthesis: biochemistry and regulation. *Annu. Rev. Plant Physiol. and Plant Mol. Biol.*, 42, 467-506.
2. Kannangara, C.G. and Stumpf, P.K. (1973) Fat metabolism in higher plants LVI, Distribution and nature of biotin in chloroplasts of different plant species. *Arch. Biochem. Biophys.*, 155, 391-399.
3. Masterson, C. Wood, C. and Miernyk, J.A. (1995), Inter-organellar transport of acyl-groups in plant cells, in S.S. Purohit and A. Kumar, (eds) *Agro's Annu. Rev. of Plant Physiol.*, 2, Agro Botanical Publishers, India, pp. 65-124.
4. Post-Beitenmiller, M.A. Jaworski, J. and Ohlrogge, J.B. (1991) *In vivo* pools of free and acylated acyl carrier protein in spinach: evidence for sites of regulation of fatty acid biosynthesis. *J. Biol. Chem.*, 266, 1858-1865.

4. Acknowledgements.

C.M. thanks BBSRC, The Royal Society and CORF of the University of Newcastle upon Tyne for funding. J.A.M. thanks the University of Newcastle upon Tyne Research Committee for funding his Senior Visiting Research Fellowship.

KINETIC ANALYSIS OF THE MECHANISM OF ENOYL-ACP REDUCTASE

Use of substrate analogues to determine kinetic mechanism.

TONY FAWCETT AND CATHERINE OVEREND
*Lipid Molecular Biology Group
Department of Biological Sciences, University of Durham, South Road, Durham, DH1 3LE, UK.*

1. Introduction

Enoyl-ACP reductase(ER) catalyses the second reductive step during the cyclical reactions of fatty acid synthesis. *Brassica napus* NADH-dependent ER has been purified in milligram quantities from *Escherichia coli* harbouring an expression plasmid containing the plant cDNA. The availability of large quantities of protein has enabled successful crystallization[1] and elucidation of the complete structure of this enzyme[2]. In addition the supply of recombinant protein has allowed detailed kinetic investigations to be carried out. This was not possible with the small quantities of protein available from rape seed material[3].
The natural substrate for *B. napus* ER is enoyl-ACP; the enzyme will also use enoyl-CoA and enoyl-cysteamine derivatives as substrates but with higher K_M values. Crotonyl-CoA has most often been used in studies with this enzyme as it is readily available commercially.

2. Results and Discussion.

2.1 CHARACTERIZATION OF RECOMBINANT PLANT ENOYL REDUCTASE.

During cloning procedures used to introduce the plant ER cDNA into an *E. coli* expression vector the N-terminal amino acid sequence was changed from SESSES to MAESSES[4]. Protein sequence analysis of the expressed recombinant protein revealed the sequence AESSES, indicating removal of the initiating formyl-methionine residue. The recombinant protein has similar subunit and native molecular weights and Michaelis constants for NADH and crotonyl-CoA as the native plant-derived enzyme. These criteria indicate that the cDNA is correctly expressed, and the protein is also processed and assembled into a fully functional homotetrameric enzyme.

2.2 STABILIZATION OF CATALYTIC POWER OF DILUTED ENZYME

It was necessary to carry out enzyme assays over an extended period of time, using the same dilution of enzyme, therefore conditions were sought to minimize time-dependent changes in catalytic power. Dilution of the enzyme in assay buffer alone(10 mM sodium phosphate, pH 6.2) caused approximately 20% decrease in catalytic activity after incubation at 4^{0}C for 4 h. This loss was ameliorated by a combination of raising the ionic strength of the enzyme dilution buffer to 37.5 mM and including Bovine Serum Albumin(BSA; 1 mg/ml); DTT(1 mM); EDTA(1 mM) and glycerol(20% w/v). When this solution was used to dilute the enzyme preparation, full stability was retained over a 5 h period and 98% of the activity, present immediately after dilution, was observed more than 24 h later.

Each of the components of the mixture was shown to improve the stability of the diluted enzyme; the addition of BSA did not reduce the concentration of crotonyl-CoA in the assay mixture, as determined by initial velocity measurements of freshly prepared enzyme dilution in the presence and absence of BSA.

2.3 MECHANISTIC DETERMINATIONS: SEQUENTIAL OR SUBSTITUTION?

We carried out an investigation to differentiate between sequential and substitution reaction mechanisms, using CoA substrate analogues. In a series of assays using fixed, non-saturating levels of NADH and varying concentrations of crotonyl-CoA,

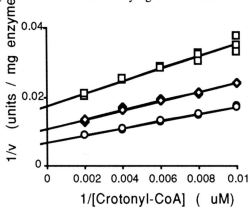

Figure 1: Double reciprocal plots achieved using varying crotonyl-CoA concentrations at fixed, non-saturating concentrations of NADH(3.44 µM[□], 9.09 µM[◇] and 14.20 µM[○]). Assays were performed in triplicate and data analyzed by linear regression.

the family of lines produced on a double reciprocal plot converged(Fig 1). A similar converging pattern of lines were seen when fixed, non-saturating levels of crotonyl-CoA and varying NADH concentrations were employed. This pattern is characteristic of a sequential or ternary complex reaction mechanism

2.4 PRODUCT INHIBITION STUDIES.

The products of the ER reaction, butyryl-CoA and NAD^+, were utilized as inhibitors in an attempt to determine whether the reaction proceeds via a random or compulsory order ternary-complex mechanism using the rules of Cleland[5]. Assays were carried out in triplicate. The data was analyzed by linear regression and statistical t-tests were employed to determine if lines intersected at the same or significantly different point on the vertical axis.

Analysis of the results in the form of double reciprocal plots indicated competitive inhibition when NADH was varied in the presence of NAD^+ (at fixed, non-saturating concentrations of crotonyl-CoA). When NADH was varied in the presence of butyryl-CoA and when crotonyl-CoA was varied in the presence of NAD^+, mixed patterns of inhibition were observed.

When crotonyl-Co was varied in the presence of butyryl-CoA the lines appeared to intersect to the left of the vertical axis. However, statistical analysis of the y-axis intercepts showed that they in fact intercept the vertical axis at points which are not significantly different from each other. That is, they intercept at a common point and therefore show a competitive pattern of inhibition. This combination of inhibition patterns is indicative of a random order ternary complex mechanism.

Crystallographic studies have shown that crotonyl-CoA is a competitive inhibitor of the nucleotide cofator, due to their structural similarities[2]. As the *in vivo* substrates for the reaction are enoyl-ACP's, care must be taken in interpretation of kinetic data using CoA substrate and product analogues. Further experiments are required to validate the proposed mechanism.

3. Conclusion.

Plant ER has been expressed in *E. coli*. The recombinant protein has a single amino acid change at the amino terminus, but this does not affect assembly into the homotetrameric state or correct functioning of the enzyme activity as judged by kinetic constants for the two substrates.

Analysis of the reaction mechanism, using CoA substrate analogues indicates that the reaction mechanism proceeds sequentially through a ternary complex, probably by a random order of substrate binding. This proposal requires verification using the *in vivo*, acyl-ACP, substrates.

4. References.

[1] Rafferty, JB. *et al.*, J. Mol. Biol.(1994) 237:240-242.
[2] Rafferty, JB *et al.*, Structure(1995) 3:927-938.
[3] Slabas, AR. *et al.*, Biochim. Biophys. Acta(1986) 877:271-280.
[4] Kater, MM *et al.*, Plant Mol. Biol.(1991) 17:895-909.
[5] Cleland, WW. Biochim. Biophys. Acta(1963) 67:188-192.

IDENTIFICATION AND CHARACTERIZATION OF 9-*CIS*-HEXADECENOIC ACID *CIS-TRANS* ISOMERASE OF *PSEUDOMONAS* SP. STRAIN E-3

H. OKUYAMA[1], D. ENARI[1], T. KUSANO[1] AND N. MORITA[2]
[1]*Laboratory of Environmental Molecular Biology, Graduate School of Environmental Earth Science, Hokkaido University, Kita-ku, Sapporo 060, Japan, and* [2]*Laboratory of Biochemistry, Hokkaido National Industrial Research Institute, AIST, Toyohira-ku, Sapporo 062, Japan.*

1. Introduction

Some bacterial cells contain mono-unsaturated fatty acids with a *trans*-confuguration as acyl components of membrane phospholipids. We found the temperature-dependent alteration in levels of a *trans*-mono-unsaturated fatty acid, 16:1(9t), in a psychrophilic bacterium, *Vibrio* sp. strain ABE-1 [1]. Morita *et al.* [2] presented the evidence for the direct *in vivo* isomerization between 16:1(9c) and 16:1(9t) in cells of this bacterium using stable isotope tracers, namely, $[2,2-^2H_2]16:1(9c)$ and $[2,2-^2H_2]16:1(9t)$. This isomerization is quite unique because of configurational isomerization of a double bond without shifting of the double bond position. In this article, we decribe purification and characterization of *cis-trans* isomerase from a psychrotrophic bacterium, *Pseudomonas* sp. strain E-3 (*Pseudomonas* E-3), another organism that contains 16:1(9t) [1], and the possible physiological roles of *trans*-mono-unsaturated fatty acid in this bacterium.

2. Purification of *cis-trans* isomerase

The *cis-trans* isomerase of 16:1(9c) (9-Iase), which locates in the cytosolic fraction [3], was purified by ammonium sulfate precipitaion and column chromatography. At the final step the 9-Iase was 5340 times purified and its recovery of activity was 5%. The purified enzyme preparation exhibited the sole band with a molecular mass of 80 kDa on incompletely-denaturing and denaturing polyacrylamide gel electrophoresis systems, indicating that the 9-Iase is a monomeric protein with a molecular mass of 80 kDa.

3. Characterization of *cis-trans* isomerase

The most effective substrate for 9-Iase was 16:1(9c). As shown in TABLE 1 the 9-Iase utilized 14:1(9c), 15:1(9c), 17:1(9c) and iso-17:1(9c) but did not utilize 14:1(7c), 15:1(7c), and 16:1(7c) as substrates, however, 15:1(10c), 16:1(11c) and 17:1(10c) could also serve as rather poor substrates. These results suggest that the specificity of the enzyme with respect to the position of the double bond is a little ambiguous and that a

---

The abbreviations used are: X:1(Yc) or X:1(Yt), fatty acid containing X carbon atoms with one double bond at position Y counted from the carboxyl terminus in the *cis* (c) or *trans* (t) configuration; PC, phosphatidylcholine; PE, phosphatidylethanolamine; 16:1(9c)/16:1(9c)-PE, PE esterified with 16:1(9c) at *sn*-positions 1 and 2.

TABLE 1. Substrate specificity of 9-Iase

Substsrate	Relative activity (%)
14:1(7c)	0
14:1(9c)	54
15:1(7c)	0
15:1(9c)	52
15:1(10c)	35
16:1(7c)	0
16:1(9c)	100
16:1(11c)	69
17:1(9c)	23
17:1(10c)	10
iso-17:1(9c)	46
18:1(9c)	0
18:1(11c)	0
16:1(9c)-CoA*	0
16:1(9c)-ME**	0
16:1(9c)-Na***	5
16:1(9c)/16:1(9c)-PC	0
16:1(9c)/16:1(9c)-PE	0
16:1(9c)/16:1(9c)-PC****	0
16:1(9c)/16:1(9c)-PE****	48

*CoA ester of 16:1(9c).
** Methyl ester of 16:1(9c).
*** Sodium salt of 16:1(9c).
**** The reaction was carried out in the presence of membrane fraction (5 µg of protein) prepared from *Pseudomonas* E-3

double bond between the ω-terminus and the 9 position in fatty acids is more susceptible than others to isomerization.

The 9-Iase utilized not only C16 fatty acids but also C15 and C14 fatty acids as substrates. However, 18:1(11c) and 18:1(9c) were not utilized at all. These results suggest that the enzyme has very strict specificity with respect to the chain length of the fatty acid and does not recognize chains of more than 17 carbons. The CoA and methyl esters of 16:1(9c), as well as 16:1(9c) esterified to PE and PC, could not be used as substrates and the sodium salt of 16:1(9c) was a very poor substrate. Thus, the enzyme isomerizes unsaturated fatty acids in the free acid form. Although the membranous factor(s) has not been identified, the combination of the 9-Iase and the membrane fraction caused the isomerization of 16:1(9c) esterifed to PE but not that of 16:1(9c) esterified to PC, an indication that the phospholipid-isomerizing activity of this bacterium exhibits specificity with respect to the polar head group of the phospholipid.

4. Effect of temperature on fatty acid compositions

When *Pseudomonas* E-3 cells were grown at 30°C, level of 16:1(9t) was from 2% to 3% of which value was almost the same with that of cells grown at 4°C (TABLE 2). However, when the culture of *Pseudomonas* E-3 grown at 4°C for 24 h was warmed up to 30°C at the high rate (about 20°C/min), the level of 16:1(9t) significantly increased from 2% to 14% at the expense of 16:1(9c) (from 46% to 30%, see TABLE 2) in 2 h after the

TABLE 2. Effect of growth conditions on fatty acid composition of *Pseudomonas* E-3

Growth conditions	16:0	16:1(9c)	16:1(9t)	18:1(11c)	Others
			% of total		
4°C, 24 h	24	46	2	26	2
30°C, 24 h	57	18	2	13	10
30°C, 48 h	56	16	3	11	14
2 h after shift to 30°C*	25	30	14	21	10
24 h after shift to 30°C*	43	24	5	15	13

*A culture of *Pseudomonas* E-3 grown at 4 °C was shifted to 30°C.

transfer. The abrupt changes in growth temperature seem to induce the activation of the 9-Iase. We suggest that 16:1(9t) might be urgently synthesized so as to adjust the hyper-fluid conditions of the cytoplasmic membrane, which was suddenly brought in this bacterium.

References

1. Okuyanma, H., Sasaki, S., Higashi, S., and Murata, N. (1990) A *trans*-unsaturated fatty acid in a psychrophilic bacterium *Vibrio* sp. strain ABE-1. *J. Bacteriol.*, **172**, 3515-3518.
2. Morita, N., Shibahara, A., Yamamoto, K., Shinkai, K., Kajimoto, G., and Okuyama, H. (1993) Evidence for *cis-trans* isomerization of a double bond in the fatty acids of the psychrophilic bacterium *Vibrio* sp. strain ABE-1. *J. Bacteriol.*, **175**, 916-918.
3. Okuyama, H., Enali, D., Shibahara, A., Yamamoto, K., and Morita, N. (1996) Identification of activities that catalyze the *cis-trans* isomerization of the double bond of a mono-unsaturated fatty acid in *Pseudononas* sp. strain E-3. *Arch. Microbiol.*, in press.

INTRACELLULAR DISTRIBUTION OF FATTY ACID DESATURASES IN CYANOBACTERIAL CELLS AND HIGHER-PLANT CHLOROPLASTS

L. MUSTARDY, D.A. LOS, Z. GOMBOS, N. TSVETKOVA,
I. NISHIDA AND N. MURATA

Department of Regulation Biology, National Institute for Basic Biology Myodaiji, Okazaki, 444 Japan

1. INTRODUCTION

According to current schemes of the glycerolipid synthesis, fatty acids are synthesized and then desaturated in plastids (1). Since the activity of UDP-galactose:1,2-diacylglycerol galactosyltransferase, which is involved in the biosynthesis of galactolipids, was first found in envelope membranes isolated from chloroplasts (2), it has been assumed that the sites of synthesis of glycerolipids and of the desaturation of fatty acids in plastids are confined to the envelope membranes and are not present in the thylakoid membranes (3-5). However, several lines of evidence suggest that the thylakoid membranes might also be a site of glycerolipid synthesis and desaturation (6-8).

All desaturases in cyanobacteria and higher plants, except $\Delta 9$ acyl-ACP desaturase in higher plants, are of the acyl-lipid and membrane-bound type (9). *Synechocystis* sp. PCC 6803 contains four desaturases which introduce double bonds at their respective positions, namely at the $\Delta 6$, $\Delta 9$, $\Delta 12$, and $\omega 3$ positions of C_{18} fatty acids. They are known as $\Delta 6$, $\Delta 9$, $\Delta 12$ and $\omega 3$ acyl-lipid desaturases, respectively (9). *Arabidopsis thaliana* contains $\Delta 12$ and $\omega 3$ acyl-lipid desaturases in chloroplasts. The objective of the present study was to localize these desaturases in cyanobacterial cells and higher-plant chloroplasts by immunocytochemistry to identify the sites of desaturation of fatty acids.

2. MATERIAL AND METHODS

Cells of the wild-type strain and the Fad12 mutant strain (10) of *Synechocystis* were grown photoautotrophically under continuous light. The temperature for growth was 34°C. For acclimation to low temperature, the cells were exposed to 24°C for 3 hours. Thylakoid and cytoplasmic membranes were isolated from these cells. *A. thaliana* was grown for 30 days at 22°C with a 16 hour light period. Chloroplasts were isolated from these plants.

Antibodies were raised in rabbits against synthetic oligopeptides that corresponded to the 15 amino acid residues at the carboxyl termini of the four desaturases of *Synechocystis* (9, 11) and the $\omega 3$ desaturase of *A. thaliana*. (12). These carboxyl-terminal regions were chosen because they are specific to the respective desaturases. For ImmunoGold labeling, cells of *Synechocystis* and isolated thylakoid membranes were pelleted and fixed for in 1% glutaraldehyde. Thin sections were incubated with the IgGs raised against the desaturases for 18 hours under which conditions almost all the epitopes were labeled. Then they were incubated with second antibodies raised in goat against rabbit IgG, that had been coupled to 10- nm colloidal gold particles.

3. RESULTS AND DISCUSSION

3. 1. CYANOBACTERIAL CELL

When the IgGs against Δ12, Δ6 and Δ9 desaturases were applied to sections of wild-type cells of *Synechocystis*, ImmunoGold particles were found in the regions of both the thylakoid membranes and the cell envelopes but not in the cytoplasmic region (Fig.1). In the section of an Fad12 mutant cell, which lacked the Δ12 desaturase, no labeling specific for this desaturase was visible. When the IgG against the ω3 desaturase was applied to sections of wild-type cells, gold particles were practically absent in the case of cells grown at 34°C but they were found in the regions of the thylakoid membranes and the cell envelope in cells that had been acclimated to 24°C for 3 hours (Fig.1).

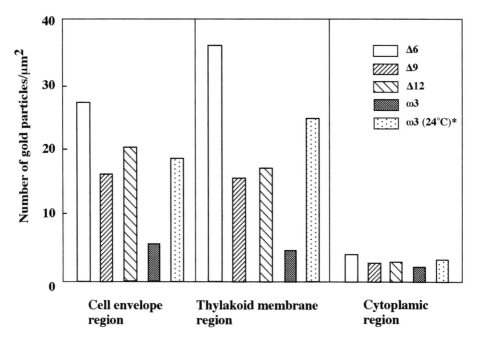

Fig. 1. Density of gold particles in the cell envelope, thylakoid membrane, and cytoplasmic regions of wild-type cells of *Synechocystis* sp. PCC 6803 grown at 34°C. The IgGs against the carboxyl-terminal peptides of the individual desaturases were used for the immunogold labeling. ω 3 (24°C) * corresponds to the ω3 desaturase in cells which had been acclimated to 24°C for 3 hours.

Western blot analysis of the Δ12 desaturase in thylakoid membranes that had been isolated from the wild-type cells revealed one band at the position of a protein of about 42 kDa, which corresponded to the molecular mass of the Δ12 desaturase. In the case of the Fad12 mutant, this band was absent, suggesting that the Δ12 desaturase was absent from the thylakoid membranes of these cells.

3. 2. CHLOROPLASTS OF *A. thaliana*

Labeling of leaf sections of *A. thaliana* with the IgGs against the presence of the chloroplastic ω3 desaturase revealed differences in the level of labeling among different cells. In old cells located toward the center of the leaf, the labeling was very low. In contrast young cells with developing chloroplasts at the margin of the leaf showed a high density of labeling. It is highly significant that in either chloroplasts in leaf or isolated chloroplasts the gold particles are located at the thylakoid membranes and that there is no distinct labeling at the envelope membranes. This observation contradicts current schemes of desaturation which propose the envelope membrane and not the thylakoid membrane as the site of fatty acid desaturation in the synthesis in membrane lipids in chloroplasts

4. CONCLUSION

All the data of the present study indicate that the four fatty-acid desaturases are localized on both the thylakoid membranes and the cytoplasmic membranes in the cyanobacterial cell and on both the thylakoid membrane and the envelope membrane in the higher-plant chloroplasts. These findings lead us to conclude that the sites of desaturation of fatty acids are not only the cytoplasmic membranes or the envelope membranes, but also the thylakoid membranes.

REFERENCES

1. Stumpf PK (1987) in Stumpf PK (ed.), *The Biochemistry of Plants* Vol. 9, Academic Press, Orlando, FL, pp. 121-136.
2. Douce R (1974) Science 183: 852-853.
3. Browse J, Kunst L, Anderson S, Hugly S and Somerville CR (1989) Plant Physiol. 90: 522-529.
4. Kunst L, Browse J and Somerville CR (1989) Plant Physiol. 91: 401-408.
5. Schmidt H and Heinz E (1990) Proc. Natl. Acad. Sci. U.S.A. 87: 9477-9480.
6. Sandelius AS and Selstam E (1984) Plant Physiol. 76: 1041-1046.
7. Slabas AR and Smith C G (1988) Planta 175: 145-152.
8. Omata T and Murata N (1986) Plant Cell Physiol. 27: 485-490.
9. Murata N and Wada H (1995) Biochem J. 308:1-8.
10. Wada H and Murata N (1989) Plant Cell Phyiol. 30: 971-978.
11. Mustardy L, Los D, Gombos Z and Murata N (1996) Proc. Natl. Acad. Sci. U.S.A., in press
12. Iba K, Gibson S, Nishiuchi T, Fuse T, Nishimura M, Arondel V, Hugly S and Somerville C (1993) J. Biol. Chem. 268: 24099-24105.

TRIACYLGLYCEROLS PARTICIPATE IN THE EUKARYOTIC PATHWAY OF PUFAS BIOSYNTHESIS IN THE RED MICROALGA *PORPHYRIDIUM CRUENTUM*

I. Khozin, H. Zheng Yu*, D. Adlerstein, C. Bigogno, and Z. Cohen

*The Laboratory for Microalgal Biotechnology, Jacob Blaustein Institute for Desert Research, Ben-Gurion University of the Negev, Sde-Boker, Israel; *Inst. of Hydrobiology, Academica Sinica, Wuhan Hubei, P.R. China*

Wada and Murata (1) selected desaturase-deficient mutants of the cyanobacterium *Synechocystis* based on their chill sensitivity. Wada et al. (2) have further shown that PUFAs are necessary for growth and tolerance to photoinhibition at low temperatures. In the red alga *P. cruentum*, the PUFA eicosapentaenoic acid (EPA, 20:5ω3) apparently fulfills a role similar to that of 18:3ω3 in cyanobacteria and *Arabidopsis* (3). Since the EPA content increases at low temperatures we have hypothesized that by selection and screening of chill-sensitive mutants of this alga it would be possible to obtain EPA-deficient mutant which may aid in the elucidation of EPA biosynthesis. Based on the data we have obtained we suggest that triglycerides participate in the eukaryotic pathway of EPA biosynthesis in *P. cruentum*. The HZ3 mutant of *P. cruentum* is the first mutant of any, higher or lower, plant which appear to be deficient in its ability to hydrolyze TAGs.

RESULTS

Lipid and fatty acid composition

The lipid composition of the HZ3 mutant demonstrated a three fold increase in the proportion of TAGs and a corresponding decrease in that of PC in comparison to the wild type (Table I). Moreover, the proportion of EPA of the mutant decreased from 41.1% (of total fatty acids) in the wild type to 26.5%. These alterations were most noticeable in MGDG where the proportion of EPA was reduced from 56 to 45.5%. The molecular species composition of MGDG demonstrated a decrease in the proportion of the major eukaryotic species 20:5/20:5 which was accompanied by increases in the prokaryotic species 18:2/16:0, 20:4/16:0 and 20:5/16:0 (Table I). DGDG which is entirely prokaryotic, was not significantly affected. We have interpreted this as an indication that the mutation affected the eukaryotic pathway.

TABLE 1. Molecular species composition of galactolipids of wild type and HZ3 mutant of *Porphyridium cruentum*.
Galactolipids were isolated by 2-D TLC. Molecular species of galactolipids were separated by reverse phase chromatography and are arranged in the order they were eluted from the chromatographic column.

Culture	Lipid	Molecular species composition (% of total)					
		20:5/20:5	20:4/20:5	20:4/20:4	20:5/16:0	20:4/16:0	18:2/16:0
WT	MGDG	31.4	4.2	1.0	50.0	3.2	10.3
HZ3	MGDG	19.1	5.0	1.0	52.4	9.5	13.1
WT	DGDG	-	-	-	88.2	3.1	8.6
HZ3	DGDG	-	-	-	86.4	6.4	7.3

Incorporation of [1-^{14}C]linoleic acid

Following the incorporation of [1-^{14}C]18:2, most of the label was introduced into PC and TAGs (Fig. 1, top) in both the wild type and the HZ3 mutant. However, in the wild type, these lipids turned over their counts in favor of chloroplastic lipids (Fig. 1, bottom), while in the mutant only PC was significantly turned over. The percent of label incorporated into TAGs of the mutant was stable and decreased slightly only after 22 hrs (Fig. 1).

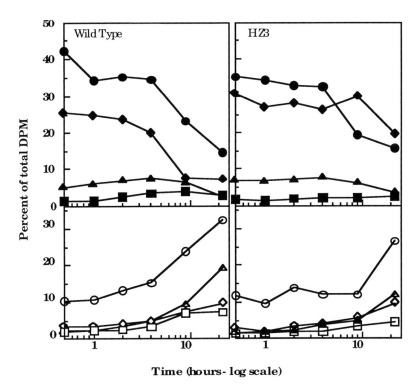

Figure 1. Redistribution of radioactivity in lipids of wild type and mutant of *P. cruentum* after labeling with [1-^{14}C]linoleic acid. Lipids were separated by 2-D TLC. ●, PC; ◆, TAG; ■, PE; ▲, PI; ○, MGDG; △, DGDG; ◇, SQDG; □, PG.

The distribution of label in total fatty acids was only slightly modified (data not shown). However, in TAGs, 18:2 retained most of its label even after 22 hrs while that of the wild type lost most of its counts (Fig. 2). Similarly, the TAGs of the mutant accumulated more labeled AA than those of the wild type.

The molecular species analysis of MGDG of the mutant revealed a delay and eventually a decrease, in comparison to the wild type, in the accumulation of all the eukaryotic species (Fig. 3). The share of the labeled prokaryotic species 20:4ω6/16:0 and 20:5ω3/16:0 was similar to that of the wild type while that of 18:2/16:0 increased.

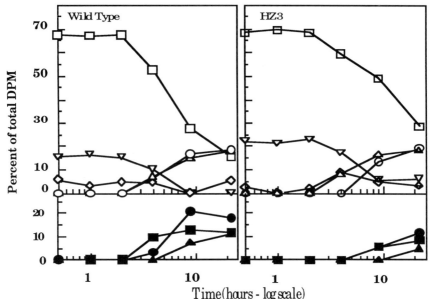

Figure 2. Redistribution of radioactivity in molecular species of MGDG of wild type and mutant after labeling with [1-^{14}C]linoleic acid. □, 18:2/16:0; ○, 20:4/16:0; ▽, 20:4/18:2; △, 20:5/16:0; ◇, 20:5/18:2; ■, 20:4/20:4; ●, 20:4/20:5; ▲ 20:5/20:5

DISCUSSION

Using chill-sensitivity as a selection tool we isolated a mutant of *P. cruentum* deficient in EPA production. The lipid and fatty acid composition of the mutant, which showed a reduced level of eukaryotic molecular species of MGDG support our hypothesis that the mutant is deficient in the eukaryotic pathway.

In oil plants, TAGs share a common DAG pool with phospholipids, primarily with PC. However, the conversion of DAGs to TAGs is generally considered to be unidirectional (4). Our radiolabeling studies clearly show that in addition to PC, there is a significant contribution of TAGs to the synthesis of chloroplastic lipids of *P. cruentum*. The substantial turnover of PC into chloroplastic lipids in the mutant indicate that the eukaryotic pathway is functional. However, TAGs accumulated a higher percentage of the initial label and were severely limited in their ability to turn over the label. We thus attribute the lesion to a deficiency in the ability of the mutant to hydrolyze TAGs to DAGs that can be utilized in the eukaryotic pathway.

At present, we have no data to support any hypothesis concerning the mechanism by which the reutilization of TAGs may take place. Nonetheless, two possibilities could be conceived 1) lipase activity and 2) a reversible DAG acyl transferase (DAGAT).

REFERENCES

1. **Wada H, Murata N** (1989) *Synechocystis* PCC 6803 mutants defective in desaturation of fatty acids. Plant Cell Physiol 30: 971-978
2. **Wada H, Gombos Z, Sakamoto T, Murata N** (1992) Genetic manipulation of the extent of desaturation of fatty acids in membrane lipids in the cyanobacterium *Synechocystis* PCC6803. Plant Cell Physiol 33: 535-540
3. **Cohen Z, Vonshak A, Richmond A** (1988) Effect of environmental conditions of fatty acid composition of the red alga *Porphyridium cruentum*: correlation to growth rate. J Phycol **24**: 328-332
4. **Browse J, Somerville CR** (1991) Glycerolipid synthesis: Biochemistry and regulation. Ann Rev Plant Mol Biol 42: 467-506

ELUCIDATION OF THE BIOSYNTHESIS OF EICOSAPENTAENOIC ACID (EPA) IN THE MICROALGA *PORPHYRIDIUM CRUENTUM*

I. Khozin, D. Adlerstein, C. Bigogno and Z. Cohen
The Laboratory for Microalgal Biotechnology, Jacob Blaustein Institute for Desert Research, Ben-Gurion University of the Negev, Sde-Boker Campus 84990, Israel

INTRODUCTION

The biosynthetic pathways in algae that produce PUFAs of no more than 18 carbon atoms are presumed to be similar to those of higher plants (1). However, the biosynthesis of C_{20} PUFAs in various algae appears to be different. The few detailed studies suggest that the biosynthesis of eicosapentaenoic acid (EPA, 20:5ω3) from C_{18} precursors is entirely cytoplasmatic (2 - 4). Feeding external fatty acids to *P. cruentum* indicated the existence of two possible pathways as outlined in sequences I and II, respectively (unpublished data).

18:2ω6———> 18:3ω6———> 20:3ω6———> 20:4ω6———>20:5ω3   I
18:2ω6———> 18:3ω3———> 18:4ω3———> 20:4ω3———>20:5ω3   II

In this study, we have attempted to elucidate the pathways of EPA biosynthesis at the fatty acid and lipid level in *P. cruentum* by pulse labeling with various radioactive fatty acid precursors.

RESULTS

After a 30 minutes pulse with [1-^{14}C]linoleic, [1-^{14}C]α-linolenic or [1-^{14}C] arachidonic acid, PC and TAG were the highest labeled lipids (Fig. 1), however, their counts decreased rapidly in favor of MGDG, DGDG and SQDG. Other phospholipids remained minor throughout the experiments. The changes in the labeling pattern suggest a flux of fatty acids from PC to chloroplastic lipids, a characteristics of a eukaryotic pathway.

The 18:2 labeling kinetics suggested a turnover of 18:2 to 20:4ω6 and possibly to 20:5ω3 (sequence I). In PC, 18:2 was metabolized mainly via an ω6 pathway to 20:4ω6, however 20:5 was only scarcely and lately labeled (Fig. 2A). The pattern of label distribution among PC molecular species (Fig. 2C), revealed the existence of a cytoplasmic ω6 pathway according to sequence III :

20:4ω6/18:2*———>20:4ω6/18:3ω6*———>20:4ω6/20:3ω6*———>20:4ω6*/20:4ω6*   III
(16:0)   (16:0)   (16:0)   (16:0)

In fatty acids of MGDG, initially 18:2 and later only 20:4ω6 and 20:5, were significantly labeled (Fig. 2B). There was no decrease in 18:2/16:0 MGDG compatible

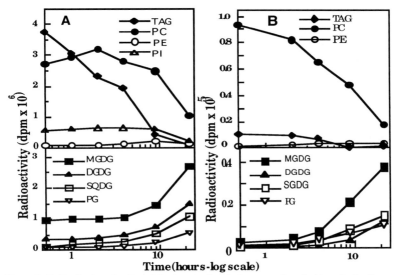

Figure 1. Redistribution of radioactivity in cytoplasmic (upper part) and chloroplastic (lower part) lipids of *P. cruentum* after labeling with [1-^{14}C]linoleic acid (A) and [1-^{14}C] arachidonic acid (B).

with the increase of label in 20:4ω6/16:0 and 20:5/16:0 (data not shown). The first eukaryotic molecular species to appear was 20:4ω6/20:4ω6 followed by its putative ω3 desaturation products 20:4ω6/20:5ω3 and 20:5ω3/20:5ω3 (data not shown). We didn't find any other molecular species which could have been considered as a precursor for either the prokaryotic or the eukaryotic C20-containing species. Labeling kinetics of the molecular species of MGDG and PC suggested that eukaryotic 20:4ω6/20:4ω6 DAG moieties as well as 20:4ω6 for prokaryotic MGDG molecular species are imported from the cytoplasm and only the ω3 desaturation takes place in the chloroplast according to sequences IV and V:

20:4ω6/16:0 ——> 20:5ω3/16:0 IV
20:4ω6/20:4ω6 ——> 20:4ω6/20:5ω3 ——> 20:5ω3/20:5ω3 V

Following a label with [^{14}C]AA (20:4ω6), EPA accumulation in MGDG was evident already after 2 hrs, whereas AA remained the only labeled fatty acid in PC even after 30 hrs (data not shown). It appears, therefore, that while AA is the major product of fatty acid biosynthesis in PC, its further desaturation to EPA in the cytoplasm, is either insignificant or nonexistent. The initial molecular species composition of MGDG was dominated by 20:4ω6*/16:0, 20:4ω6*/20:4ω6* and 20:4ω6*/20:5*, which were gradually desaturated to 20:5ω6/16:0, 20:4ω6*/20:5* and 20:5*/20:5*.

The fatty acid analysis of total lipids and PC after labeling with [^{14}C]18:3ω3 (data not shown) revealed all the ω3 fatty acids depicted in sequence VI:

18:3ω3 ——> 18:4ω3 ——> 20:4ω3 ——> 20:5ω3 VI

suggesting that the enzymatic sequence necessary to convert 18:3ω3 to EPA, in PC, is the same one required for the production of AA from 18:2, i.e. a Δ6 desaturase, an elongation system and a Δ5 desaturase. In MGDG, after a lag of 2 hours, EPA rapidly accumulated and eventually took over. Once again, we did not observe any possible intermediate in either the fatty acid or the molecular species analysis of MGDG.

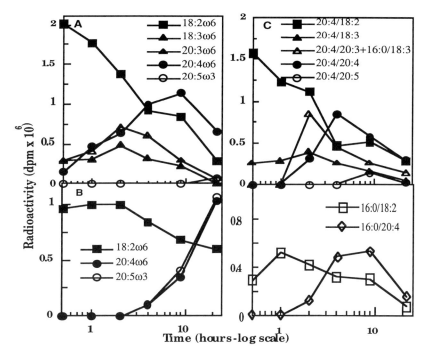

Figure 2. Redistribution of radioactivity in fatty acids of PC (A), MGDG (B) and in the molecular species of PC (C) after labeling with [1-^{14}C]linoleic acid.

CONCLUSIONS

Our data show that the end products of EPA biosynthesis are galactolipids of eukaryotic and prokaryotic structure. However, we suggest that both types of molecular species are formed in at least 2 pathways which involve cytoplasmic and chloroplastic lipids. In the major ω6 pathway, 18:2 bound PC is sequentially converted to 20:4ω6-PC by a sequence that include a Δ6 desaturase, an elongation system and a Δ5 desaturase. In the minor ω3 pathway, 18:2 bound PC is presumably desaturated to 18:3ω3 which is sequentially converted by the enzymatic sequence of the ω6 pathway to 20:5ω3-PC. The products of both pathways are exported, as their diacylglycerol moieties to the chloroplast to be galactosylated into the respective MGDG molecular species. The prokaryotic species of MGDG, 20:4ω6/16:0 apparently import 20:4ω6 for its *sn*-1 position from extraplastidial lipids. The 20:4ω6 in both eukaryotic and prokaryotic species of MGDG can be further desaturated to EPA by a chloroplastic Δ17 desaturase.

REFERENCES

1. Norman HA, Smith LA, Lynch DV, Thompson GA (1985) Effects of low-temperature stress on the metabolism of phosphatidylglycerol molecular species in *Dunaliella salina*. Arch Biochem Biophys **242**: 157-167
2. Arao T, Yamada M (1994) Biosynthesis of polyunsaturated fatty acids in the marine diatom, *Phaeodactylum tricornutum*. Phytochemistry **35**: 1177-1181
3. Henderson JR, Mackinlay EE (1991) Polyunsaturated fatty acid metabolism in the marine dinoflagelate *Crypthecodinium cohnii*. Phytochemistry **30**: 1781-1787
4. Schneider JC, Roessler P (1994) Radiolabeling studies of lipids and fatty acids in *Nannochloropsis* (Eustigmatophyceae), an oleaginous marine alga. J Phycol **30**: 594-598

IMMOBILIZATION OF HYDROPEROXIDE LYASE FROM CHLORELLA FOR THE PRODUCTION OF C13 OXO-CARBOXYLIC ACID

ALBERTO NUÑEZ, THOMAS A. FOGLIA, AND GEORGE J. PIAZZA
Eastern Regional Research Center, ARS, USDA, 600 East Mermaid Lane, Wyndmoor, PA 19038, USA

1. Introduction

The hydroperoxide, 13-(S)-hydroperoxy-9(Z),11(E)-octadecadienoic acid (HPOD), obtained from the lipoxygenase (LOX, EC 1.13.11.12) catalyzed reaction of oxygen with linoleic acid (LA), can be cleaved enzymatically to an oxo-carboxylic acid by hydroperoxide lyase (HPLS). Depending upon the source of HPLS, the enzymatic cleavage of (HPOD) produces either a C12 or C13 oxo-carboxylic acid, and a C6 or C5 fragment respectively. Membrane-bound HPLS from higher plants produces the C12 oxo-carboxylic acid (1), whereas the water soluble HPLS from the unicellular algae *Chlorella* and *Oscillatoria* sp. gives the C13 acid (2,3). The C13 oxo-carboxylic acid can be chemically converted to a C13 dicarboxylic acid. This dicarboxylic acid can be used to produce a polyamide similar to nylon 13,13, a polymer that has excellent dimensional stability, low affinity for moisture, high dielectric constant and good melting properties (4).

To facilitate the reuse of HPLS, ways of immobilizing this enzyme are needed. While there are a variety of published methods for the immobilization of LOX (5-7), there are no reports concerning the immobilization of HPLS.

2. Result and Discussion

Because HPLS from algae is water soluble rather than membrane-bound as in higher plants, algae were considered attractive sources for obtaining HPLS. Acetone powder extracts from *Chlorella pyrenoidosa* and *fusca* that contained HPLS were partially purified by chromatography on DEAE Sepharose CL-6B (8). Five commercially available gels were evaluated for their ability to immobilize the purified HPLS preparations (Table 1). Enzyme assays were performed by following the decrease in absorption at 234 nm (9). This method is not specific to HPLS, because it does not differentiate between HPLS and hydroperoxide dehydrase (HD), which will also cause a decrease in absorbance at 234 nm. A specific spectrophotometric assay for HPLS based upon the ability of yeast alcohol dehydrogenase (ADH) to reduce aldehydes to alcohols in the presence of NADH has been reported by Vick (9). Initial rates, measured at 234 nm, were comparable with those determined with ADH and NADH at 340 nm, indicating that HD was not contaminating our HPLS preparations.

The enzyme activity is represented by the percent yield of product in phosphate buffer at pH 7.0 according to Equation (1), where A_t is the absorption after 5 min, and A_0 is the absorption of a sample identically treated with gel that contained no immobilized HPLS.

$$Yield\ \% = \frac{|A_0 - A_t|100}{A_0} \qquad (1)$$

It was found that Reacti-Gel (6X), Affi-Gel 10, 15, 102 and 501 could bind 60-90% of the available protein. However, HPLS activity was only detected when the enzyme was immobilized on Affi-Gels 10, 15 and 501 (Table 1). Moreover, the activity was lost after the Affi-Gel 15 was used one time. Better retention of HPLS activity was obtained by immobilization on Affi-Gel 10 and 501. HPLS activity was retained at a higher level with both of these gels, even after two reaction cycles, with HPLS from *C. fusca*.

The stability of immobilized HPLS (on Affi-Gel 10 and 501) during storage at 4 °C for several months was determined. Fig. 1 shows a plot of the percent yield of product after HPLS was stored up to 4 months. The data show that there was no detected loss of activity over this period.

Table 1 Comparison of the capacity of commercial cross linked agarose gels to covalently couple with HPLS from *Chlorella*

Gel	Chlorella Sp	% Protein immobilized	1st Cycle (Yield %)	2nd Cycle
Reacti-Gel	Fusca	56.7	0	---
Affi-Gel 10	Pyrenoidosa	83.2	55.3	5.7
	Fusca	79.4	56.7	51.1
Affi-Gel 15	Pyrenoidosa	82.2	21.7	4.9
	Fusca	59.6	32.9	10.6
Affi-Gel 102	Pyrenoidosa	82.7	0	---
	Fusca	61.2	0	---
Affi-Gel 501	Pyrenoidosa	87.2	67.1	21.5
	Fusca	82.1	59.0	65.0
	Fusca (immob. x 2)	86.2	65.6	66.0

Also, product yields with repeated use of the immobilized preparations were determined (Figure 2). HPLS immobilized on Affi-Gel 501 could be used with little loss of activity up to five cycles in phosphate buffer at pH 7.0. After five recycles, this preparation began to lose activity. After the seventh recycle the gel retained only 11% of its original activity. The Affi-Gel 10 preparation had a lower capacity for recycling, losing 85% of its original activity after only five recycles.

Among the gel tested Affi-Gel 501 appears to be the most promising. In addition to its better capacity for recycling, the reaction of sulfhydryl groups with the mercaptide-forming ligand of Affi-Gel 501 is reversible, and this gel can be regenerated by washing the protein off with mercuric acetate. The regenerated gel gave an immobilized preparation that had an activity equivalent to that obtained with fresh Affi-Gel 501 as shown in the last entry of Table 1.

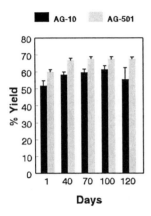

Fig. 1: Stability of HPLS from *C. fusca* storage at 4°C.

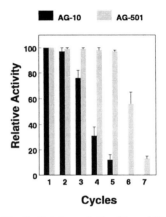

Fig. 2: Relative activity of immobilized HPLS from *C. fusca* with reuse.

Enzymatic activity for the immobilized preparation on Affi-Gel 501 was determined from pH 5 up to pH 9, with maximal activity at pH 6.5. These results are consistent with an earlier study of HPLS from the alga *Oscillatoria*, which showed maximal activity at pH 6.4 (3).

References:

1. Gardner, H. W. (1991) Recent investigations into the lipoxygenase pathway of plants, *Biochim. Biophys. Acta* **1084**, 221-239.
2. Vick, B. A. And Zimmerman, D. C. (1989) Metabolism of fatty acid hydroperoxide by *Chlorella pyrenoidosa*, *Plant Physiol.* **90**, 125-132.
3. Andrianarison, R. H. Beneytout, J. L. and Tixier, M. (1989) An enzymatic conversion of lipoxygenase products by a hydroperoxide lyase in blue-green algae (*Oscillatoria* sp), *Plant Physiol.* **91**, 1280-1287.
4. Van Dyme D. L. and Blase, M. G. (1990) Process design, economic feasibility, and market potential for nylon 1313 produced from erucic acid, *Biotechnol. Prog.* **6**, 273-276.
5. Parra-Diaz, D., Brower, D. P., Medina, M. B. and Piazza, G. J. (1993) A method for immobilization of lipoxygenase, *Biotechnol. Appl. Biochem.* **18**, 359-362.
6. Battu. S., Cook-Moreau, J.,and Beneytout, J. L., (1994) Stabilization of potato tuber lipoxygenase on talc, *Biochim. Biphys. Acta* **1211**, 270-276.
7. Battu, S., Rabinovitch-Chable, H. And Beneytout, J. L. (1994) Effectiveness of talc as adsorbent for purification and immobilization of plant lipoxygenases, *J. Agric. Food Chem. Vol.* **42**, 2115-2112.
8. Nuñez, A, Foglia, T. A. and Piazza, G. J. (1995) Improved method for extraction of hydroperoxide lyase from *Chlorella*, *Biotechnol. Techniques* **9**, 613-616.
9. Vick, B. A. (1991) A spectrophotometric assay for hydroperoxide lyase, *Lipids* **76**, 315-320.

ALLENE OXIDE CYCLASE FROM CORN: PARTIAL PURIFICATION AND CHARACTERIZATION

JÖRG ZIEGLER[1]; MATS HAMBERG[2] AND OTTO MIERSCH[1]
[1]Institute of Plant Biochemistry, Dept. of Hormone Research,
Weinberg 3, D-06120 Halle, Germany
[2]Karolinska Institute, Dept. of Physical Chemistry 2,
S-17177 Stockholm, Sweden

1. Introduction

In plants, the oxylipin pathway gives rise to several oxygenated fatty acid derivatives such as hydroxy- and keto fatty acids as well as volatile aldehydes and cyclic compounds, which are, in part, physiologically active [1]. Among these, jasmonic acid is discussed as signalling molecule during several stress responses, wounding, senescense and plant pathogen interactions [2].
The initial step for the biosynthesis of jasmonic acid and of the other oxygenated fatty acids is the lipoxygenase catalyzed peroxygenation of linoleic or linolenic acid to give 9- or 13-hydroperoxy fatty acids. From these compounds the oxylipin pathway is diverged into 3 major routes:
1.: the formation of ω-oxo fatty acids and volatile aldehydes by hydroperoxide lyase [3].
2.: the generation of polyhydroxylated fatty acids and epoxyalcohols by a peroxygenase which uses the hydroperoxy fatty acids as cosubstrate to epoxidize another fatty acid molecule [4,5].
3.: the dehydration of the hydroperoxy fatty acids by allene oxide synthase to give unstable allene oxides which are spontaneously hydrolyzed to α- and γ- ketols [6].
The last route is important for the formation of jasmonic acid. The action of allene oxide synthase on 13-hydroperoxy linolenic acid initially results in the generation of 12,13-epoxy-octadecatrienoic acid. This unstable epoxide is either chemically hydrolyzed to α- and γ-ketols and racemic 12-oxo-phytodienoic acid or, in the presence of allene oxide cyclase (AOC), is further converted to enantiomeric pure 12-oxo-PDA [7]. The ring double bond of PDA is then reduced in a NADPH dependent reaction by 12-oxo-PDA reductase and after shortening of the side chain containing the carboxy group by 3 rounds of ß-oxidation, the biosynthesis of jasmonic acid is completed.
To understand the flow of metabolites through this pathway it is necessary to investigate the regulation of the participating enzymes. In this report we focus our interest on the partial purification and characterization of allene oxide cyclase.

2. Results and Discussion

Allene oxide cyclase was purified over 2000-fold with a recovery of 3.1% by different chromatographic techniques (Table 1). Especially hydrophobic interaction chromatography proved to be very efficient during purification. This was due to the strong binding of AOC on these columns and might be an indication for a rather hydrophobic surface of this soluble protein. The purest fraction from the Mono Q column showed one major and one minor protein band in the range of 20 kD on SDS-PAGE (Fig.1). These two bands are the only ones matching the activity profiles on Mono Q and other matrices used during purification, suggesting that one or both bands correspond to AOC. Since size exclusion chromatography gave a native molecular size of 47 kD for AOC, the enzyme probably possesses a dimeric structure. Figure 1 also shows the almost complete removal of impurities by chromatography on phenyl-superose. This stresses the importance of hydrophobic interaction chromatography for purification of this enzyme.

TABLE 1. Purification of allene oxide cyclase

	spec.activity nmol PDA/mg	total activity nmol PDA	purification fold	recovery %
crude extract	11.7	16098	1	100
octyl-sepharose	666	8200	56	50
hydroxyapatite	2032	2133	173	13
phenyl-superose	11684	1110	998	6.8
Mono Q	24900	498	2128	3.1

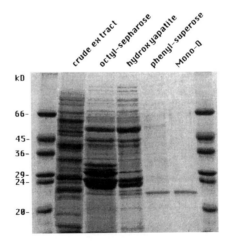

Figure 1: Protein pattern of the single purification steps

Figure 2 shows the influence of metabolites of the oxylipin pathway on allene oxide cyclase activity. 12-oxo-PDA up to 500 µM has no impact on enzyme activity, whereas jasmonic acid seems to slightly stimulate the enzyme at a concentration above 40 µM. This argues against a feedback regulation of AOC by its product. 12,13-epoxy octadecenoic acid strongly inhibits AOC activity with half maximum inhibition at 20 µM. The methyl ester of this compound as well as the positional isomer 9,10-epoxy octadecenoic acid both have no influence on cyclase activity up to 100 µM. Because of the high similarity to the natural substrate 12,13-epoxy octadecatrienoic acid the inhibition by 12,13-epoxy octadecenoic acid seems to be competitive. Since the methyl ester does not affect the activity, a free carboxy group is essential for binding of the inhibitor and perhaps also for binding of the substrate. The inhibitor is one of the products of the peroxygenase pathway and one might speculate that the flow of metabolites through the oxylipin pathway could also be regulated by the interaction of metabolites from one branch of the pathway with enzymes of another branch.

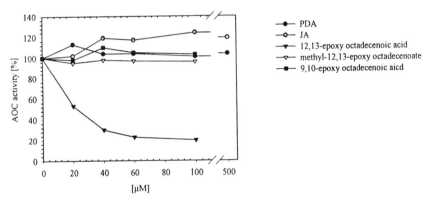

Figure 2: Influence of metabolites of the oxylipin pathway on AOC activity

3. References:

1. Farmer, E.E.: Fatty acid signalling in plants and their associated microorganisms, *Plant Mol. Biol.* **26** (1994), 1423-1437
2. Sembdner, G. and Parthier, B.: The biochemistry and the physiological and molecular actions of jasmonates, *Annu. Rev. Plant Physiol. Plant Mol. Biol.* **44** (1993), 569-589
3. Shibata, Y., Matsui, K., Kajiwara,T., Hatanaka, A.: Purification and properties of fatty acid hydroperoxide lyase from green bell pepper fruits, *Plant Cell Physiol.* **36**(1) (1995), 147-156
4. Blée, E. and Schuber, F.: Biosynthesis of cutin monomers: involvement of a lipoxygenase/ peroxygenase pathway, *The Plant Journal* **4**(1) (1993), 113-123
5. Hamberg, M. and Hamberg, G.: Hydroperoxide-dependent epoxidation of unsaturated fatty acids in the broad bean (Vicia faba L.), *Arch. Biochem. Biophys.* **283**(2) (1990), 409-416
6. Hamberg, M. and Gardner, H.W.: Oxylipin pathway to jasmonates: biochemistry and biological significance, *Biochim. Biophys. Acta* **1165** (1992), 1-18
7. Hamberg, M. and Fahlstadius, P.: Allene oxide cyclase: a new enzyme in plant lipid metabolism, *Arch. Biochem. Biophys.* **276**(2) (1990), 518-526

Section 2:

Glycerolipid Biosynthesis

HOW IS SULPHOLIPID METABOLISED?

C.E. PUGH, A.B. ROY, G.F. WHITE AND J.L. HARWOOD
School of Molecular and Medical Biosciences,
University of Wales College of Cardiff, Cardiff CF1 3US, U.K.

1. Introduction

All organisms carrying out oxygenic photosynthesis contain the plant sulpholipid, sulphoquinovosyldiacylglycerol, as a major component of their thylakoid membranes. The lipid was first discovered by Benson and co-workers, and, in some marine algae, may represent up to half of the total lipids. Thus, the plant sulpholipid is a very significant membrane component and makes a major contribution to the sulphur cycle [1,2].

In spite of its importance, research into sulpholipid metabolism has been sporadic and neither the routes for synthesis nor for degradation have been elucidated. Most experiments on synthesis have centred on the so-called sulphoglycolytic pathway for which some theoretical problems exist [3,4]. Another possible route for the formation of sulphonic acids involves the addition of sulphite to double bonds. A facile chemical synthesis of sulphoquinovose uses this method and there is some experimental evidence for the involvement of sulphite in sulpholipid synthesis [see 5]. Accordingly, we began work on sulpholipid synthesis with this possible reaction strongly in our minds.

2. The proposed pathway for synthesis

In our scheme (see Figure 1) we envisage the alternatives of UDP-glucose or UDP-galactose acting as precursors. These substrates are interconverted via a 4-keto intermediate by UDP-galactose 4-epimerase which has sequence similarities to a protein involved in sulphoquinovosyldiacylglycerol formation in *Rhodobacter sphaeroides* [6]. By elimination of water, UDP-6-deoxy-4-ketoglucose-5-ene would be formed. Alternatively, this latter intermediate could be produced by UDP-glucose oxidoreductase initially abstracting a hydride equivalent from C-4 followed by stabilisation through the elimination of water. Such a reaction is established for the formation of 6-deoxysugars like rhamnose. Addition of sulphite to the UDP-6-deoxy-4-ketoglucose-5-ene would be expected to occur easily [5] and would lead, after reduction, to production of UDP-sulphoquinovose which has been demonstrated clearly to be used in the final stage of sulphoquinovosyldiacylglycerol formation in plants [7]. Aspects of this proposed scheme and the merits of alternatives are discussed by Pugh *et al.* [5].

3. Experimental evidence

Using chloroplast preparations from pea leaves we have established an experimental system capable of good rates of sulphoquinovosyldiacylglycerol formation. Labelling from $^{35}SO_4^{2-}$ is stimulated by UTP but not significantly by other nucleotides and we have demonstrated incorporation of radioactivity from both UDP-[^{14}C]glucose and UDP-[^{14}C]galactose. Incorporation of radioactivity from the latter is, however, complicated by substantial competition from galactolipid synthesis in the chloroplast preparations. Addition of methyl α-glucoseenide, as a potential acceptor of sulphite, significantly stimulated incorporation of $^{35}SO_4^{2-}$ into sulphoquinovosyldiacylglycerol. Moreover, addition of an enzyme preparation capable of forming UDP-6-deoxy-4-ketoglucose-5-ene (as an intermediate in rhamnose synthesis) also increased the labelling of sulphoquinovosyl-diacylglycerol from UDP-[^{14}C]glucose. Together with the previously published work on sulphoquinovosyldiacylglycerol formation or on analogous reactions, these results are fully in keeping with our proposed pathway [5].

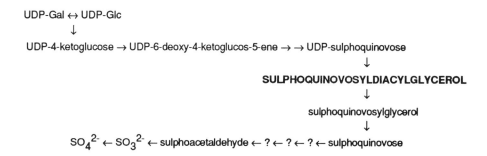

Figure 1. Proposed pathways for the synthesis and degradation of sulpholipid.

4. Catabolism of sulphoquinovosyldiacylglycerol and sulphoquinovose

During plant senescence and/or leaf fall, sulphoquinovosyldiacylglycerol has to be broken down in large amounts to yield, eventually, inorganic sulphate for re-uptake by growing plants. This pathway is a significant element of the sulphur cycle in

Nature. However, recent interest in this phenomenon has been stimulated by the growing problem of sulphur availability in soils. Plants contain active acyl hydrolases which rapidly deacylate sulphoquinovosyldiacylglycerol. One isoform in *Phaseolus multiflorus* was found to have a good specificity towards sulphoquinovosyldiacylglycerol [3]. In addition, experiments with [^{35}S]sulphoquinovosyldiacylglycerol, as well as the identification of various sulphonic acids in plants or algae, suggest that cleavage of glycerol from sulphoquinovose also takes place at good rates. So far as we can tell, plants are not capable of splitting the carbon-sulphur bond at significant rates. However, we have isolated a soil bacterium (*Pseudomonas* species, tentatively identified as *Ps. putida* Type A) that is capable of good rates of growth on sulphoquinovose which it can use as a carbon source, thereby initiating biodegradation of sulphoquinovose. The final catabolic step releasing inorganic sulphur is as yet unknown but it may well be analogous to the sulphoacetaldehyde sulpholyase reaction which completes the bacterial degradation of the structurally related C_2 sulphonates, taurine and isethionate, to acetate and sulphite [8].

The intermediate reactions between sulphoquinovose and sulphoacetaldehyde and the relative contributions of potential intermediates like sulphopyruvate, sulpholactate or sulphoacetate, are unknown at present. The sulphite released would be oxidised either by sulphite oxidases in microorganisms or spontaneously in aerated soils to yield inorganic sulphate, and so provide sulphur for uptake by plants and assimilatory reduction.

5. Acknowledgement

We are grateful to Zeneca Agrochemicals and B.B.S.R.C. for financial support.

6. References

1. Harwood, J.L., and Nicholls, R.G.: *Biochem. Soc. Trans.* **7** (1979), 440-447.
2. Heinz, E. (1993), in L.J. De Kok (ed), *Sulphur Nutrition and Assimilation in Higher Plants* Academic Press, New York, pp. 163-178,
3. Harwood, J.L. (1980) in P.K Stumpf and E.E Conn (eds), *The Biochemistry of Plants* **4**, Academic Press, New York, pp. 301-320.
4. Mudd, J.B., and Kleppinger-Sparace, K.F. (1987) in P.K Stumpf and E.E. Conn (eds), *The Biochemistry of Plants* **9**, Academic Press, New York, pp 275-289.
5. Pugh, C.E., Roy, A.B., Hawkes, T., and Harwood, J.L.: *Biochem. J.* **309** (1995), 513-519.
6. Benning, C., and Somerville, C.R.: *J. Bacteriol.*. **174** (1992), 6479-6487.
7. Seifert, U., and Heinz, E.: *Bot. Acta* **105** (1992), 197-205.
8. Kondo, H., and Ishimoto, M.: *J. Biochem., Tokyo* **78** (1975), 317-325.

BIOCHEMICAL CHARACTERIZATION OF COTTONSEED MICROSOMAL N-ACYLPHOSPHATIDYLETHANOLAMINE SYNTHASE

KENT D. CHAPMAN and ROSEMARY S. McANDREW
Department of Biological Sciences
University of North Texas
Denton, TX 76203

1. Introduction

N-Acylphosphatidylethanolamine (NAPE) is an unusual derivative of the common membrane phospholipid, phosphatidylethanolamine (PE), with a third fatty acyl group attached through an amide linkage to the amino group of ethanolamine. NAPE has been reported to occur in various amounts in lipid extracts from a variety of seeds (Schmid et al., 1990). In reports where isolated NAPE was subjected to rigorous chemical analyses (analyses of degradative products and IR spectroscopy), it was estimated to account for approximately 4 mol% of lipid P in wheat (Bomstein 1969), 5% in pea, 3.5% in soybean, and 13.3% in oat (Dawson et al., 1969). Recent data with different tissues and organs of cotton seedlings and cotyledons at several different developmental stages indicated the NAPE was present in all parts of cotton plants at concentrations between 1.9 and 3.2 mol% of the total phospholipids (Chapman and Sprinkle, 1996). Cottonseed NAPE was characterized structurally by FAB-MS/MS analyses (Chapman and Moore, 1993a; Sandoval et al., 1995), and the major molecular species were identified as 16:0/18:2 PE (N-palmitoyl), 16:0/18:2 PE (N-linoleoyl), and 18:2/18:2 PE (N-palmitoyl).

In animals and microorganisms, NAPE appeared to be synthesized by a Ca^{2+}-dependent transacylase activity whereby a fatty acid is transferred from the sn-1or -2 O-acyl position of a phospholipid to the ethanolamine head group of PE (Schmid et al., 1990). Evidence indicated that transacylation could occur intermolecularly (with PE, PC or cardiolipin able to serve as the acyl donor), or intramolecularly. Most notably, free fatty acids or acyl moieties from fatty acylCoA were not incorporated into NAPE. To date, this transacylase activity has not been solubilized from any membrane source in an active form.

By contrast, it appeared that NAPE was synthesized in cotyledons of cotton seedlings by a direct acylation of PE with free fatty acids (Chapman and Moore, 1993b). Reconstitution of NAPE biosynthesis in cottonseed microsomes showed that free palmitic acid was incorporated into the N-position of NAPE two or three orders of magnitude more rapidly than from palmitoylCoA or from dipalmitoylPC, respectively (Chapman and Moore, 1993b). Time courses in vitro showed that the synthesis of NAPE in cottonseed microsomes from fatty acylCoA or PC was preceded by the accumulation of free fatty acids (Chapman and Moore, 1993b; Chapman et al., 1995a), suggesting that free fatty acids released into membranes could be incorporated into NAPE. This cottonseed microsomal NAPE synthase activity was solubilized from cottonseed membranes in dodecylmaltoside (DDM; Chapman and Moore, 1993b) and recently purified to homogeneity by immobilized artificial membrane (IAM) chromatography (Cai, et al., 1995). Here we report some kinetic properties of the purified NAPE synthase enzyme. In particular, we have examined the activity of this enzyme towards palmitic and linoleic acids, the major fatty acids present at the N-position of cottonseed NAPE as well as the major fatty acids present in cottonseed membranes. Preliminary evidence with site-specific inhibitors indicates that serine, cysteine, and/or histidine residues may be involved in catalysis. These data are consistent with the notion that NAPE synthase catalyzes the direct acylation of PE with free fatty acids via a CoA-independent mechanism, possibly by serving itself as an acyl intermediate. Based on its catalytic properties, it seems reasonable to speculate that NAPE synthase may function as a free fatty acid scavenger in membranes of plant cells.

2. Methods

2.1. ENZYME PURIFICATION AND ANALYSES

Cotyledons of germinated cotton (*Gossypium hirsutum* L.) seeds (18-h) were homogenized by a chopping in a medium consisting of 400 mM sucrose, 100 mM K-phosphate, pH 7.2, 10 mM KCl, 1 mM EDTA, 1 mM $MgCl_2$, 1 mM benzamidine HCl, 0.5 mM EGTA, 0.001 mM leupeptin, and 0.001 mM pepstatin A. Microsomal membranes were isolated by differential centrifugation and solubilized in 0.2 mM DDM as previously described (McAndrew et al., 1995). NAPE synthase was purified to homogeneity by a two-step IAM chromatography procedure employing two different PE surfaces. DDM-solubilized microsomal proteins were concentrated by ultrafiltration to approximately 0.4 mg/ml and injected (approx. 2 mg protein total) onto a column containing an immobilized analogue of PE lacking the glycerol backbone (designated ^{8G}IAM.PE; Ong et al., 1995). NAPE synthase-enriched fractions collected from the eluent were pooled, concentrated, and injected (approximately 0.1 mg total protein) onto a second column containing another immobilized analogue of PE with the glycerol backbone ether-linked to the acyl moiety (designated ^{ether}IAM.PE; Cai et al., 1995). NAPE synthase was bound to the second column and was specifically eluted with mobile phase containing 0.2 mM dimyristoylPE, 0.08 mM dioleoylPE and 1 mM DDM. Fractions were analyzed for NAPE synthase activity as previously described (McAndrew, et al., 1995). Protein content was estimated using bovine serum albumin as a standard (fluorometrically with OPA reagent, Pierce). Enzyme purity was evaluated by examining the protein composition of column fractions in silver-stained, Tricine-SDS gels. Purified NAPE synthase fractions were pooled for kinetics and inactivation experiments.

3. Results and Discussion

Purified cottonseed NAPE synthase enzyme exhibited non-Michaelis-Menten biphasic kinetics with respect to the free fatty acid substrates, palmitic and linoleic acids. Kinetic parameters for the two saturable sites were calculated from various transformations (*e.g.*, double-reciprocal and Hill plots; Cornish-Bowden, 1995) of initial velocity/ substrate concentration data and are summarized in TABLE 1. Preliminary experiments with several group-specific modifiers indicated that NAPE synthase was progressively inactivated by increasing concentrations of 5,5'-dithiobis(2-nitrobenzoic acid) (DTNB), diisopropyl fluorophosphate (DFP), phenylmethylsulfonylfluoride (PMSF), diethylpyrocarbonate (DEPC) (TABLE 2). These results suggest that NAPE synthase may form a thioester- or ester-intermediate through a cysteine or serine residue, respectively, and a histidine residue may participate in catalysis as well.

Recently, two groups independently reported a membrane-bound enzyme activity in bovine brain that catalyzes the formation of an amide bond between the carboxyl group of arachidonic acid and the amino group of ethanolamine in an ATP- and CoA-independent manner (Devane and Axelrod, 1994; Kruszka and Gross, 1994). This activity was reported to be sensitive to inhibition by DTNB (Kruszka and Gross, 1994) and PMSF (Devane and Axelrod, 1994). It is possible that the cottonseed NAPE synthase enzyme and the bovine microsomal enzyme share a common mechanism. Future work with the purified enzyme should help to clarify the precise mechanism of this unusual enzyme activity.

TABLE 1. Summary of Kinetic Parameters of Purified Cottonseed NAPE Synthase.

	$S_{0.5}$ μM	V_{max} μmol/min/mg protein	k_{cat} mol/sec
"High-affinity site"			
Linoleic acid	12	1.83	2.0
Palmitic acid	17.5	2.08	2.2
"Low-affinity site"			
Linoleic acid	65	5.56	6
Palmitic acid	55	5.56	6

TABLE 2. Inhibitor Sensitivities of Cottonseed NAPE Synthase Activity. Values are reported as % remaining enzyme activity assayed after 20 min incubation with the inhibitors at 1 and 10 mM final concentration.

Inhibitor/concentration		Target amino acid	% remaining activity
NO Inhibitor		NA	100% (5.4 μmol/min/mg)
DTNB	1 mM	Cys	74%
	10 mM	Cys	43%
DFP	1 mM	Ser	68%
	10 mM	Ser	18%
PMSF	1 mM	Ser	71%
	10 mM	Ser	17%
DEPC	1 mM	His	82%
	10 mM	His	14%

4. References

Bomstein R. A. (1965) A new class of phosphatides isolated from soft wheat flour *Biochem Biophys Res Comm 21*, 49-54.
Cai SJ, McAndrew RS, Leonard BP, Chapman KD, Pidgeon C. (1995). Rapid purification of cottonseed membrane-bound *N*-Acylphosphatidylethanolamine synthase by immobilized artificial membrane chromatography. *J Chromatography 696:* 49-62.
Chapman KD, Lin I, Decease AD. (1995). Metabolism of cottonseed microsomal *N*-Acylphosphatidylethanolamine. *Arch Biochem Biophys 318:* 401-407.
Chapman, KD and Moore, T.S., Jr. (1993a) *N*-Acylphosphatidylethanolamine synthesis in plants: occurrence, molecular composition, and phospholipid origin *Arch. Biochem. Biophys. 301*, 21-33.
Chapman, KD and Moore, T.S., Jr. (1993b) A newly-discovered acyltransferase that synthesizes *N*-Acylphosphatidylethanolamine in cottonseed microsomes. *Plant Physiol, 102(3),* 761-769.
Chapman, KD and Moore, T.S., Jr. (1994) Isozymes of cottonseed microsomal N-Acylphosphatidylethanolamine synthase: detergent solubilization and electrophoretic separation of active enzymes with different properties. *Biochim Biophys Acta 1211:*29-36.
Chapman KD and Sprinkle WB. 1996. Developmental, tissue-specific, and environmental factors regulate the biosynthesis of *N*-Acylphosphatidylethanolamine in cotton (*Gossypium hirsutum* L.) *Journal of Plant Physiology, in press.*
Cornish-Bowden, A. (1995) *Fundamentals of Enzyme Kinetics*, Portland Press, London.
Dawson, R.M.C., Clarke, N., and Quarles, R.H. (1969) *N*-Acylphosphatidylethanolamine, a phospholipid that is rapidly metabolized during the early germination of pea seeds. *Biochem. J. 114*, 265-270.
Devane, W.A. and Axelrod, J. (1994) Enzymatic synthesis of anandamide, an endogenous ligand for the cannabinoid receptor, by brain membranes. *Proc. Natl. Acad. Sci. 91:* 6698-6701.
Frentzen M. (1993) Acyltransferases and triacylglycerols. *In* T.S. Moore, Jr., ed. Lipid Metabolism in Plants. CRC Press, Boca Raton, pp 195-231.
Kruszka, K.K. and Gross, R.W. (1994) The ATP- and CoA- independent synthesis of arachidonylethanolamine. *J. Biol. Chem. 269(20):* 14345-14348.
McAndrew RS, Leonard BP, Chapman KD. 1995. Photoaffinity labeling of cottonseed microsomal *N*-Acylphosphatidylethanolamine synthase protein with a substrate analogue, 12-[(4-azidosalicyl)amino]dodecanoic acid. *Biochim Biophys Acta 1256:* 310-318.
Ong, S., Liu, H., Qui, X., Bhat, G, and Pidgeon, C. (1995) Membrane partition coefficients chromatographically measured using immobilized artificial membrane surfaces. *Anal Chem 67:* 755-762.
Sandoval JA, Huang Z-H, Garrett DC, Gage DA, Chapman KD. 1995 *N*-Acylphosphatidylethanolamine in dry and imbibing cottonseeds: amounts, molecular species and enzymatic synthesis. *Plant Physiol 109:* 269-275.
Schmid, HHO, Schmid, PC, and Natarajan, V (1990) *N*-acylated glycerophospholipids and their derivatives. *Prog. Lipid Res. 29*,1-43.

PHOSPHATIDYLCHOLINE BIOSYNTHESIS IN SOYBEANS: THE CLONING AND CHARACTERIZATION OF GENES ENCODING ENZYMES OF THE NUCLEOTIDE PATHWAY

DAVE E. MONKS, JOHN H. GOODE, POLLY K. DINSMORE and RALPH E. DEWEY. *Department of Crop Science, North Carolina State University, Raleigh, NC 27695.*

Introduction

As the predominant phospholipid constituent of most eukaryotic cellular membranes, phosphatidylcholine (PC) plays a vital role in cellular structure and physiology. PC is also the substrate for C_{18} fatty acid polyunsaturation, and is therefore additionally involved in the processes that determine the polyunsaturation content of both membrane lipids and seed storage oils. Two distinct pathways for PC biosynthesis have been thoroughly characterized in both animals and yeast: (a) the methylation pathway, whereby PC is produced by the sequential methylation of phosphatidylethanolamine (PE); and (b) the nucleotide pathway, a three-step enzymatic process where free choline is incorporated into PC via the enzymes choline kinase (CKase), cholinephosphate cytidylyltransferase (CCTase) and cholinephosphotransferase (CPTase; Figure 1). In higher plants, however, recent studies of PC biosynthesis have provided compelling evidence suggesting that the majority of PC is produced through a single biosynthetic pathway possessing elements of both the methylation and nucleotide pathways (1). The isolation and characterization of the genes encoding enzymes of this consolidated pathway will accelerate the study of PC biosynthesis in higher plants and help elucidate the processes by which plants regulate the production of this important phospholipid at both the molecular and cellular levels.

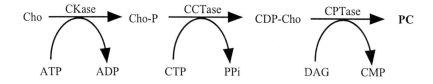

Figure 1: The nucleotide pathway of PC biosynthesis. Cho, choline; P, phosphate; CKase, choline kinase; CCTase, cholinephosphate cytidylyltransferase; CPTase, cholinephosphotransferase.

Choline Kinase

An expressed sequence tag (EST) from Arabidopsis that displayed sequence homology to mammalian and yeast CKases was used to isolate CKase-like cDNAs from soybean (2). Two distinct cDNAs, designated GmCK1 and GmCK2, possessed full-length open reading frames; the third unique cDNA (GmCK3) was not full-length. The predicted amino acid sequences of GmCK1 and GmCK2 share 81% identity; the predicted polypeptide of the incomplete GmCK3 sequence shares only 63% identity with both GmCK1 and GmCK2. To facilitate the characterization of the CKase reaction, we expressed GmCK1 in yeast and GmCK2 in yeast and *E. coli*. Both isoforms showed very high levels of CKase activity when expressed in these heterologous systems. Unlike the CKase enzymes of yeast and animals, the two soybean CKase isoforms demonstrated negligible ethanolamine kinase activity. Other potential substrates for the CKase enzyme include monomethylethanolamine (MME) and dimethylethanolamine (DME). A combination of competitive inhibition studies and direct enzyme assays revealed that neither isoform can effectively utilize MME as a substrate, and DME can act as a substrate only for the GmCK2 encoded enzyme. Our results indicate that CKase is encoded by a small multigene family comprised of two or more distinct isoforms. Once a full-length cDNA for GmCK3 has been obtained, we will determine whether this cDNA represents yet another CKase isoform or, alternatively, encodes a closely related function such as ethanolamine kinase.

Cholinephosphate Cytidylyltransferase

The conversion of cholinephosphate to CDP-choline is catalyzed by the second enzyme of the nucleotide pathway, CCTase. In both plants and animals this step has been demonstrated as being rate limiting in PC biosynthesis. Using an Arabidopsis EST showing significant homology to yeast and mammalian CCTase genes as a heterologous hybridization probe, we have isolated two distinct CCTase-like cDNAs from soybean (D.E. Monks, unpublished results). Once we have demonstrated that these cDNAs encode functional CCTases, they will provide a valuable tool in addressing questions regarding the regulation of this pivotal enzyme and the mechanisms by which plants control the flow of metabolites into PC.

Aminoalcoholphosphotransferase

Aminoalcoholphosphotransferases (AAPTases) utilize diacylglycerols and cytidine diphosphate (CDP)-aminoalcohols as substrates in the synthesis of both PC and PE. We isolated a soybean cDNA, designated AAPT1, encoding an AAPTase that demonstrates high levels CDP-choline:*sn*-1,2-diacylglycerol cholinephosphotransferase activity by the complementation of a yeast strain deficient in this function (3). The deduced amino acid sequence of the soybean cDNA showed nearly equal similarity to each of the two characterized AAPTase sequences from yeast, cholinephosphotransferase and ethanolaminephosphotransferase. Moreover, in vivo labelling studies of yeast cells expressing the soybean AAPT1 gene, and competitive inhibition assays demonstrated that the AAPT1 gene product efficiently functions as both a choline and ethanolaminephosphotransferase. Using the AAPT1 cDNA as a heterologous hybridization probe, we have subsequently isolated and characterized two AAPTases from *Arabidopsis thaliana*, designated AtAAPT1 and AtAAPT2. Like the soybean AAPT1 gene product, both of the Arabidopsis enzymes can utilize CDP-choline and CDP-ethanolamine as substrates (J.H. Goode, unpublished results).

Table 1: Predicted amino acid sequence identities/similarities between enzymes of the nucleotide pathway of soybean (or Arabidopsis) and yeast.

	Soybean Gene	Yeast Gene	% identity	% similarity
Choline Kinases	GmCK1	CK	31	59
	GmCK2	CK	32	58
Cholinephosphate Cytidylyltransferases	GmCCT1	CCT	43	64
	GmCCT2	CCT	52	70
Aminoalcohol-phosphotransferases	AAPT1	CPT	32	57
		EPT	33	59
	Arabidopsis Gene			
	AtAAPT1	CPT	33	58
		EPT	33	56
	AtAAPT2	CPT	33	57
		EPT	32	56

Methyltransferases

In nature, where free choline is limiting, the plant must rely on the methylation of the ethanolamine head group to generate PC. Another critical component of PC biosynthesis, therefore, is the methyltransferase(s) responsible for this sequential methylation of ethanolamine. The availability of the ESTs has facilitated our current efforts to isolate and characterize a higher plant representative of this important class of enzymes (P.K. Dinsmore, unpublished results).

References
1. Kinney A.J. (1993). Phospholipid head groups. *In* T.S. Moore, ed, Lipid Metabolism in Plants. CRC Press, Boca Raton, FL, pp.259-284.
2. Monks D.E., Goode J.H., Dewey R.E. (1996). Characterization of soybean choline kinase cDNAs and their expression in yeast and *Escherichia coli*. Plant Physiology 110: 1197-1205.
3. Dewey R.E., Wilson R.F., Novitsky W.P., Goode J.H. (1994). The *AAPT1* gene of soybean complements a cholinephosphotransferase-deficient mutant of yeast. Plant Cell 6: 1495-1507.

PHOSPHOLIPID METABOLISM BY CASTOR MICROSOMES

J.T. LIN, C.L. WOODRUFF, O.J. LAGOUCHE and T.A. McKEON
Western Regional Research Center, Agricultural Research Service, U. S. Department of Agriculture, Albany, California 94710, USA

1. Abstract

Intact metabolites of 2-[^{14}C]oleoyl-PC have been separated and identified by HPLC after incubation with castor microsomes. The metabolic pathway of phospholipids was shown to be 2-oleoyl-PC → 2-ricinoleoyl-PC → ricinoleate → triglycerides (e. g. triricinolein). The phospholipase A_2 strongly favored release of ricinoleate over other fatty acids from PC.

2. Introduction

Ricinoleate has many industrial uses, but the only commercial source is castor bean, which is hazardous. It is thus desirable to produce ricinoleate in a transgenic plant. The cDNA for oleoyl-12-hydroxylase, the key enzyme in the biosynthesis of ricinoleate, has been cloned recently (1). In order to reach the 90% ricinoleate level provided by castor bean, it will require understanding of ricinoleate production, the enzymes that move it into triglyceride (TG) and the enzymes that keep oleate available as hydroxylase substrate. We have recently developed an enzyme assay method to characterize oleoyl-12-hydroxylase using the putative substrate, 1-acyl-2-oleoyl-*sn*-glycero-3-phosphocholine (2). We report here the *in vitro* metabolism of 2-oleoyl-PC in the microsomes isolated from immature castor bean.

3. Experimental Procedures

3.1. MICROSOMAL INCUBATION

The incubations with the microsomal fraction from the endosperm of immature castor bean (92 µg of protein) were carried out with 1-palmitoyl-2-[1-^{14}C]oleoyl-*sn*-glycero-3-phosphocholine (0.125 µCi, 2.16 nmol, Amersham) as described (2). The radioactive substrate in 20 µL of ethanol was added last into a screw-capped tube followed by immediate mixing. The mixture was incubated in a shaking water bath for 30 min at 22.5°C and stopped by addition of 3.75 mL of chloroform/methanol (1:2, vol/vol).

3.2. HIGH-PERFORMANCE LIQUID CHROMATOGRAPHY (HPLC)

The HPLC system used absorbance at 205nm to detect eluted lipids and radioactive-flow detection for quantitation (2,4). Lipid classes were separated by a silica column (25 cm x 0.46 cm, 5 µm, Spherisorb S5W, Phase Separations) according to Singleton and Stikeleather (3) with a linear gradient starting with isopropanol/hexane (4:3, vol/vol) to isopropanol/ hexane/

water (4:3:0.75, vol/vol/vol) in 20 min then isocratically for 15 min at the flow rate of 1 mL/min. The HPLC of the methyl esters of fatty acids and of free fatty acids was performed as previously described (4). The HPLC of TG molecular species was performed using a C_{18} column and a linear gradient starting with 100% methanol to 100% isopropanol in 40 min at a flow rate of 1 mL/min. The HPLC of PC molecular species used a C_8 column (25 cm x 0.46 cm, 5 µm, Ultrasphere C8, Beckman) with a linear gradient of 90-100% methanol (containing 0.1% of NH_4OH) in 40 min at a flow rate of 1 mL/min.

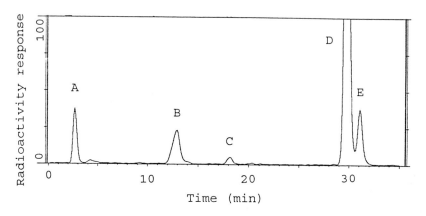

Fig. 1. Radiochromatogram of the separation of lipid classes after the microsomal incubation of 2-[14C]oleoyl-PC.

4. Results and Discussion

It is generally accepted that oleoyl-12-hydroxylase catalyzes the reaction from 2-oleoyl-PC to 2-ricinoleoyl-PC (5). We have recently characterized oleoyl-12-hydroxylase in micro-somes from the endosperm of immature castor bean using the putative substrate, 2-[^{14}C]-oleoyl-PC (2). Our results support the hypothesis that the actual substrate of oleoyl-12-hydroxylase is 2-oleoyl-PC. In our previous study, the enzyme activity was measured by the radioactivity of ricinoleate after methanolysis of the total lipids from the incubation products. In this report, for the purpose of establishing the biosynthetic pathway, radio-active intact lipid metabolites of 2-[^{14}C]oleoyl-PC were separated by HPLC and their radioactivity quantified.

Figure 1 is the radiochromatogram of the separation of lipid classes after the microsomal incubation of 2-[^{14}C]oleoyl-PC. Using various lipid standards and co-chromatography, the radioactive peaks (A-D) have been identified as TG and fatty acids (A) at retention time 2.7 min, PA (B) at 13.0 min, PE (C) at 18.2 min and PC (D) at 29.8 min. Peak E at 31.1 min was identified as 2-[^{14}C]Ricinoleoyl-PC by the methanolysis of the peak and then the HPLC of the methyl esters of fatty acids. The percentages of the radioactivity of each peak were about 6% (A), 7% (B), 1% (C), 78% (D), and 8% (E).

Our previous report (2) showed that the only metabolites (fatty acids) of oleate at *sn*-2 of 2-[^{14}C]oleoyl-PC were ricinoleate and linoleate. Each radioactive lipid class shown in Fig. 1 (e.g. TG and free fatty acid, PA, PE, PC) was transmethylated and the esters separated by HPLC on a 5 cm, C_{18} column. All lipid classes contained radioactive ricinoleate, linoleate and oleate. The percent radioactivity of fatty acids in each lipid class is shown in Table 1.

TABLE 1
14C-FATTY ACID COMPOSITION OF LIPIDS

Lipid	Ricinoleate	Linoleate	Oleate
TG & FFA	40%	4%	56%
PA	6%	4%	90%
PE	52%	4%	44%
PC	6%	3%	91%

In this microsomal incubation, the conversion of PC to free fatty acids by phospholipase A_2 strongly favored ricinoleate among the three fatty acids on the *sn*-2 position of PC. The conversion of PC to PE also showed preference for ricinoleate. There was little preference for ricinoleate in the conversion of PC to PA by phospholipase D.

Radioactive TG's and free fatty acids in Fig. 1 (peak A, 2-5 min) were mixed with castor oil and separated by HPLC. The radioactivity of TG's was about 88% that of peak A (2-5 min). The radioactivity was associated with the three main TG's in castor oil and their retention times were 7.9 min (triricinolein), 13.9 min and 15.6 min. Their radioactivities were about 39%, 32% and 29% respectively. These radioactive TG's and free fatty acids (peak A, 2-5 min) were also co-chromatographed as free fatty acids with ricinoleate, lin-oleate and oleate on HPLC. Radioactivity in free fatty acids was about 18% of that of peak A and it was associated with ricinoleate (63%), linoleate (7%) and oleate (30%). The phospholipase A_2 strongly favored the hydrolysis of 2-ricinoleoyl-PC as shown by others (5).

Microsomal incubations with 2-[14C]oleoyl-PC for up to two hours showed continuous increase in labelling of TG, PE and PA, while that in ricinoleoyl-PC was maximum at about 30 min. Radioactive ricinoleoyl-PC purified by C_8 HPLC was also incubated with microsomes. No radioactive PC, PE and PA were detected in the products and the main metabolite was triricinolein, indicating that 2-ricinoleoyl-PC was metabolized rapidly.

Our results are consistent with the proposed (6) metabolic pathway of phospholipids in castor microsomes:

2-oleoyl-PC → 2-ricinoleoyl-PC → ricinoleate → TG (mostly triricinolein).

5. References

1. Van de Loo, F.J., Broun, P., Turner, S. and Somerville, C. (1995) An oleate 12-Hydroxylase from *Ricinus communis* L. is a fatty acid acyl desaturase homolog, *Natl. Acad. Sci. USA* **92**, 6743-6747.
2. Lin, J.T., McKeon, T.A., Goodrich-Tanrikulu, M. and Stafford, A.E. (1996) Characterization of oleoyl-12-hydroxylase in castor microsomes using the putative substrate, 1-acyl-2-oleoyl-*sn*-glycero-3-phosphocholine, *Lipids* **31**, 571-577.
3. Singleton, J.A. and Stikeleather, L.F. (1995) High-performance liquid chromatography analysis of peanut phospholipids. II. Effect of postharvest stress on phospholipid composition, *J. Am. Oil Chem. Soc.* **72**, 485-488.
4. Lin, J.T., McKeon, T.A. and Stafford, A.E. (1995) Gradient reversed-phase high-performance liquid chromatography of saturated, unsaturated and oxygenated free fatty acids and their methyl esters, *J. Chromatogr.* **699**, 85-91.
5. Stahl, U., Banas, A. and Stymne, S. (1995) Plant microsomal phospholipid acyl hydrolases have selectivities for uncommon fatty acids, *Plant Physiol.* **107**, 953-962.
6. Bafor, M., Smith, M.A., Jonsson, L., Stobart, K. and Stymne, S. (1991) Ricinoleic acid biosynthesis and triacylglycerol assembly in microsomal preparations from developing castor bean (*Ricinus communis*) endosperm, *Biochem. J.* **280**, 507-514.

BIOSYNTHESIS OF DIACYLGLYCERYL-N,N,N,-TRIMETHYLHOMOSERINE (DGTS) IN *RHODOBACTER SPHAEROIDES* AND EVIDENCE FOR LIPID-LINKED N-METHYLATION

Markus Hofmann and Waldemar Eichenberger
Department of Biochemistry, University of Berne, Switzerland

Introduction

Diacylglyceryl-N,N,N-trimethylhomoserine (homoserine lipid, DGTS, see Fig. 2) is a common lipid constituent of ferns, mosses and most green algae (1). Attempts to elucidate the biosynthetic pathway of DGTS with several eukaryotic organisms revealed methionine to be the precursor of both the C_4 backbone and the N-methyl groups of the polar part (2). Only recently, it was reported that the photosynthetic bacterium *Rhodobacter sphaeroides* produces DGTS under phosphate-limiting growth conditions (3). This offered the possibility to further investigate the biosynthetic route of DGTS formation.

Experimental

Rhodobacter sphaeroides wt 2.4.1, obtained from Dr. C. Benning from the Institut für Genbiologische Forschung Berlin, was cultivated in Sistrom's medium containing 0,1 mM phosphate (3). Cells were incubated with [1-^{14}C]methionine (Amersham) or [methyl-^{14}C]-methionine (Du Pont) for 20 min (pulse) and then allowed to grow in fresh medium for 120 min (chase). When required cells were preincubated for 1 hour with the inhibitor 3'-deazaadenosine (160 µM) (4). The identification of compound 3 and its transformation to DGTS by *Ochromonas danica* was achieved according to (5)and (6).

Results

When cells of *R. sphaeroides* grown under phosphate-limiting conditions were incubated with [1-^{14}C]methionine in a pulse-chase experiment, the label was very selectively incorporated into DGTS and three unknown compounds 1-3. In compound 1, the label was highest at the end of the pulse and then steadily decreased during the chase period. In compounds 2 and 3 which could not be properly separated by TLC, the label first increased and then decreased suggesting a turnover process. In DGTS, the radioactivity continuously accumulated with time (Fig. 1a).
After deacylation almost 100% of the label was recovered from the polar part of DGTS and compounds 1-3. These results confirm that methionine is used as a precursor for the formation of DGTS by the bacterium, too, and suggest compounds 1-3 to be intermediates of this process.
In a pulse-chase with L-[*methyl*-^{14}C]methionine which acts as methyl donor in DGTS biosynthesis in algae (2), the radioactivity increased in DGTS with time. In compounds 2 and 3, the label

decreased suggesting that these are partially *N*-methylated precursors of DGTS (Fig. 1b). Since compound 1 was labelled with L-[1-^{14}C]- but not with L-[*methyl*-^{14}C]methionine, it is supposed to be the non-methylated substrate for a *N*-methylation sequence leading to DGTS. This view is also favoured by the fact that the incorporation of L-[1-^{14}C]methionine into DGTS and compounds 2 and 3 was reduced in the presence of the methyltransferase inhibitor 3'-deazaadenosine (4), whereas an accumulation of label in compound 1 was observed (Fig. 1c).

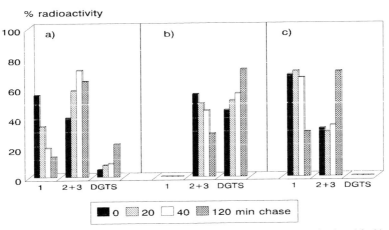

Fig.1: Pulse-chase experiments. Cells grown under phosphate-limiting conditions were incubated for 20 min with the corresponding substrates followed by a 120 min chase. a) L-[1-^{14}C]methionine; b) L-[*methyl*-^{14}C]methionine; c) L-[1-^{14}C]methionine in the presence of 160 µM 3'-deazaadenosine.

These results strongly suggested compound 3 to be the final partially methylated intermediate which is converted into DGTS. It could be identified as diacylglyceryl-*N,N*-dimethylhomoserine by co-chromatography in two different solvent systems with the corresponding reference which was obtained semisynthetically from DGTS (5). A direct transformation of compound 3 into DGTS was achieved by the DGTS-forming alga *Ochromonas danica* (Chrysophyceae) which is capable to incorporate and process intact lipids (6).

Discussion

The presence of the *N,N*-dimethyl lipid in *R. sphaeroides* and its transformation to DGTS by *O. danica* strongly suggest that the *N*-methylation is the last step in the biosynthesis of DGTS. Evidently, this process includes two preceeding *N*-methylation steps with a non-methylated amino lipid and a *N*-monomethyl lipid as intermediary products (Fig. 2). This is in accordance with the presence of the other labelled compounds 1 and 2. Compound 1 is labelled with L-[1-^{14}C]-, but not with L-[*methyl*-^{14}C]methionine, as expected for a non-methylated intermediate. Compound 2, in contrast, is labelled with both L-[1-^{14}C]- and L-[*methyl*-^{14}C]methionine, as expected for a *N*-monomethyl lipid. The labelling kinetics of the two compounds, as compared to DGTS, are in accordance with their intermediary role. Based on these results, a biosynthetic pathway for DGTS can be proposed which starts with a transfer of the C_4-α-amino acid moiety from methionine (probably S-adenosylmethionine, SAM), to a free or activated diacylglycerol (DAG), giving a non-methylated aminolipid (Fig.2). This in turn, is used as a substrate for a three-step N-methylation reaction leading through N-methyl- and N,N-dimethylaminolipid to DGTS. Consequently, the N-methylation occurs as the last step on the preformed lipid molecule and is therefore a lipid-linked process.

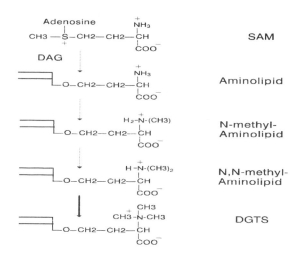

Fig.2: Tentative pathway of DGTS biosynthesis.
SAM = S-adenosylmethionine, DAG = diacylglycerol, DGTS = diacylglyceryl-N,N,N-trimethylhomoserine.

Acknowledgements

This work has been supported by the Swiss National Science Foundation (Grant 31-36078.92). We are indebted to Dr. C. Benning for the gift of the bacterial strains.

References

1. Eichenberger, W. 1993. Betaine lipids in lower plants. Distribution of DGTS, DGTA and phospholipids, and the intracellular localization and site of biosynthesis of DGTS. Plant Physiol. Biochem. 31(2):213-221.
2. Sato, N. 1992. Betaine lipids. Bot. Mag. Tokyo. 105:185-197.
3. Benning, C., Z.H. Huang, D.A. Gage, and C.R. Somerville. 1995. Accumulation of a novel glycolipid and a betaine lipid in cells of *Rhodobacter sphaeroides* grown under phosphate limitation. Arch. Biochem. Biophys. 327:103-111.
4. Haydn Pritchard, P., P.K. Chiang, G.L. Cantoni, and D.E. Vance. 1982. Inhibition of phosphatidylethanolamine N-methylation by 3'-deazaadenosine stimulates the synthesis of phosphatidylcholine via the CDP-choline pathway. J. Biol. Chem. 257(11):6362-6367.
5. Eichenberger, W., and A. Boschetti. 1978. Occurrence of 1(3),2-diacylglyceryl-(3)-O-4'-(N,N,N-)trimethyl)-homoserine in *Chlamydomonas reinhardtii*. FEBS Lett. 88:201-204.
6. Vogel, G., and W. Eichenberger. 1992. Betaine lipids in lower plants. Biosynthesis of DGTS and DGTA in *O. danica* (Chrysophyceae) and the possible role of DGTS in lipid metabolism. Plant Cell Physiol. 33:427-436.

DE NOVO BIOSYNTHESIS OF EUKARYOTIC GLYCEROLIPIDS IN CHLOROPLASTS OF TRANSGENIC TOBACCO PLANTS

D. WEIER AND M. FRENTZEN
Universität Hamburg, Institut für Allgemeine Botanik,
Ohnhorststr. 18, D-22609 Hamburg, Germany

1. Introduction

The *sn*-1-acylglycerol-3-phosphate acyltransferase (EC 2.3.1.51, LPA-AT) catalyses the second step of *de novo* glycerolipid synthesis, namely the acylation of *sn*-1-acylglycerol-3-phosphate to *sn*-1,2-diacylglycerol-3-phosphate. In plant cells, this enzyme is located in different subcellular compartments. The plastidial enzyme is characterized by its ability to utilize ACP thioesters and its specificity and selectivity for palmitoyl-ACP. Due to the properties of this enzyme, glycerolipids with prokaryotic fatty acid patterns are formed in plastids. The respective LPA-AT of *Escherichia coli*, encoded by the *pls*C gene (Coleman, 1992), can also use ACP thioesters as acyl donors, but in contrast to the plastidial enzyme the bacterial acyltransferase prefers oleoyl-ACP. With regard to these properties one aim of our work is the functional expression of the *E. coli* LPA-AT in chloroplasts of transgenic plants in order to achieve *de novo* biosynthesis of eukaryotic glycerolipids within plastids and thereby decreasing the proportion of nonfluid lipid species in the plastidial membrane.

2. Results and Discussion

For plant transformation the *E. coli pls*C gene was cloned under the control of the strong promoter of the *rbc*S1 gene from potato in frame to the plastidial targeting sequence of the same gene (Fritz et al., 1993). This chimeric construct was inserted into the binary plant vector pCV720 (Koncs et al., 1987) which provides the hygromycin-B resistance gene as the selectable marker. The ß-glucuronidase (GUS) gene, with an integral plant intron (Vancanneyt et al., 1990) was included as an additional screenable marker. The resulting construct pDW5.4 was introduced into *Agrobacterium tumefaciens* GV3101::pMP90RK and used to transform *Nicotiana tabacum* W38 by the leaf disk method. Regenerated hygromycin-B resistant shoots were screened for GUS expression. A subsample of the GUS-positive plants was also analyzed by RT-PCR to confirm the transcription of the introduced *E. coli* gene.

GUS-positive as well as RT-PCR-positive plants were used for lipid analysis, especially the analysis of the fatty acid composition of phosphatidylglycerol (PG). To this end, PG was separated by thin layer chromatography and subjected to enzymatic hydrolysis either with *Rhizopus arrhizus* lipase or phospholipase A_2 followed by gaschromatographic analysis of fatty acid methyl esters.

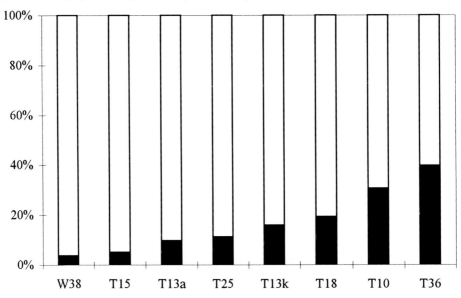

Fig. 1. Fatty acid composition at the *sn*-2 position of phosphatidylglycerol from wildtype W38 and transgenic tobacco plants (dark bars: mol% C_{18} fatty acids; light bars: mol% C_{16} fatty acids).

The fatty acid composition at the *sn*-2 position of PG from the transgenic tobacco plants clearly displayed patterns different to that of the wild type (Fig. 1). The proportion of C_{18} fatty acids at the *sn*-2 position of PG increased from about 5 mol% in the wild type up to about 40 mol% in the transgenic plants (Fig. 1). Four out of 30 analysed transgenic plants were found to have a C_{18} fatty acid proportion of 30 to 40 mol%. Thus, the fatty acid profiles of PG from the transgenic plants exhibited a more or less distinct shift towards a eukaryotic fatty acid pattern. Fatty acid analysis further revealed that not only the fatty acid composition at the *sn*-2 but also at the *sn*-1 position of PG was altered. In transgenic plants T10 and T36 (Fig. 1) the appreciable decrease in the proportion of C_{16} fatty acids at the *sn*-2 position of PG was partially counterbalanced by an increase in the proportion of saturated acyl groups at the *sn*-1 position. In that way the formation of PG species carrying unsaturated C_{18} fatty acids in both positions was largely prevented. In contrast to the distinct differences observed in the fatty acid composition of PG from the transgenic tobacco plant T36 compared to the wild type, hardly any alterations in the membrane lipid compositions were noted apart from a slight decrease in the content of monogalactosyldiacylglycerol and a slight increase in the content of both phosphatidylcholine and -ethanolamine.

In summary the presented data clearly show that the bacterial LPA-AT is functionally expressed in plastids and effecitvely competes with the plastidial enzyme. Hence, transgenic plants have been developed which are capable of *de novo* biosynthesis of eukaryotic glycerolipids within plastids. In spite of this ability the formation of PG species carrying unsaturated fatty acids at both positions was largely prevented. Whether this is due to the substrate species selectivities of the enzymes involved in PG synthesis, is currently under investigation.

3. Acknowledgments

We are grateful to J. Coleman, F.P. Wolter, C. Koncs and K. Düring for providing us with the *pls*C gene, the *rbc*S1 gene, the vector pCV720 and the GUS-intron construct, respectively. This work was financially supported by the Bundesministerium für Bildung, Wissenschaft, Forschung und Technologie (Förderkennzeichen: 0316601).

4. References

Coleman, J. (1992) Characterization of the *Escherichia coli* gene for 1-acyl-*sn*-glycerol-3-phosphate acyltransferase (*pls*C), *Mol. Gen. Genet.* **232**, 295-303.

Fritz, C.C., Wolter, F.P., Schenkenmeyer, V., Herget, T., Schreier, P.H. (1993) The gene family encoding the ribulose-(1,5)-bisphosphate carboxylase/oxygenase (Rubisco) small subunit of potato, *Gene* **137**, 271-274.

Koncz, C., Olsson, O., Langridge, W.H.R., Schell, J., Szalay, A.A. (1987) Expression ans assembly of functional bacterial luciferase in plants, *Proc. Natl. Acad. Sci. USA* **84**, 131-135.

Vancanneyt, G., Schmidt, R., O'Connor-Sanchez, A., Willmitzer, L., Rocha-Rosa, M. (1990) Construction of an intron-containing marker gene: Splicing of the intron in transgenic plants and its use in monitoring early events in *Agrobacterium*-mediated plant transformation, *Mol. Gen. Genet.* **220**, 245-259.

FUNGAL OIL BIOSYNTHESIS

FRANCES JACKSON, LOUISE MICHAELSON, THOMAS FRASER, GARETH GRIFFITHS AND KEITH STOBART.
University of Bristol
School of Biological Sciences, Bristol BS8 1UG, UK.

Introduction

γ-linolenic acid (GLA; 18:3 Δ6,9,12) is produced by only a few higher plant species and is also found in the oil of some fungi. Studies on the fungus *Mucor circinelloides* indicated that the major site for Δ6 desaturation was position sn-2 of PI [1]. It also has been proposed that in the fungus *Mucor* INMI Δ6 desaturation can occur on PC, PE and acyl-CoA [2]. We have studied fatty acid synthesis and oil formation in *Mucor circinelloides*, a fungus which accumulates oil rich in polyunsaturated fatty acids.

Materials and Methods

Mucor circinelloides was grown in liquid culture in a supplemented glucose medium at 25°C with shaking at 200 rpm and cells were harvested after 48 hours. Microsomes were prepared using a method devised for yeast [3]. The resultant pellet was resuspended in 0.1 M phosphate buffer, pH 7.2. The activity of the acyltransferase LPCAT, was measured in a continuous assay using Ellmens reagent, DTNB, following the change in absorbance at 409 nm at 25°C

Results

FATTY ACID DESATURATION

Δ9 desaturation was assayed by following the abstraction of ^{3}H from [9,10 ^{3}H] stearoyl-CoA to water [4]. Microsomes catalysed the desaturation of [9,10 ^{3}H] stearoyl-CoA on addition of reductant (Figure 1). The desaturase enzyme was cyanide sensitive but CO insensitive. The addition of soluble cytochrome b_5 stimulated the reaction nearly two-fold. There was no soluble Δ9 desaturase activity using either CoA or ACP as substrate.

Microsomes catalysed the transfer of oleate from [^{14}C]oleoyl-CoA to the sn-2 position of PC. Desaturation of oleate to linoleate was observed on addition of reductant with NADH preferred to NADPH. There was little production of GLA. To show conclusively that the Δ12 desaturase was acting on oleate esterified to PC, microsomes were incubated with [^{14}C]oleoyl-CoA, repelleted and washed to remove any remaining acyl-CoA. The labelled microsomes were incubated with NADH and desaturation of in situ labelled oleoyl-PC to produce linoleoyl-PC was observed (Figure 2).

No significant Δ6 desaturation was observed in the above experiments. However, *in vivo* studies using [^{14}C] acetate indicated that the Δ6 desaturase also utilised the complex lipid PC as a substrate. In order to try and observe *in vitro* desaturation an exogenous lipid substrate was used. Microsomes catalysed the desaturation of linoleate on exogenously supplied 1-palmitoyl, 2[^{14}C]-linoleoyl-PC (Figure 3).

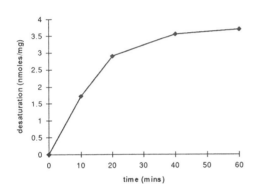

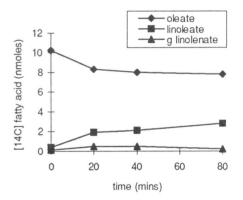

Figure 1. Desaturation of [9,10 ^{3}H] stearoyl-CoA

Figure 2. Desaturation of in-situ labelled [^{14}C]oleoylPC.

Microsomes (0.2mg) were incubated with [9,10 ^{3}H] stearoyl-CoA and NADH (200 nmoles) and the reaction was terminated at regular intervals.

Microsomes were incubated with [^{14}C] 18:1 CoA for 10 mins. The reaction mixture was diluted and membranes pelleted. After washing, the membranes were incubated with NADH and analysed at intervals.

LPCAT ACTIVITY

The activity of LPCAT, the enzyme which catalyses the transfer of acyl groups from CoA to PC, was assayed as described in the methods. The activity was much higher both when oleoyl and linoleoyl-CoA were supplied than saturated acyl-CoA species (Table 1).

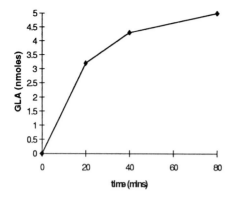

Figure 3. Desaturation of exogenously supplied 1-palmitoyl, 2[^{14}C]-linoleoylPC.

Table 1. Activity of LPCAT using different acyl-CoA donors.

Microsomes were incubated with 1-palmitoyl, 2-[^{14}C]-linoleoyl-PC, 0.1% Triton X-100 and NADH and the distribution of the radioactivity in the fatty acids of PC at intervals was determined.

The activity of LPCAT was measured using palmitoyl-lysoPC as the acceptor and acyl donors as described in the table.

Conclusions

Mucor circinelloides microsomal membranes catalysed Δ9, Δ12 and Δ6 desaturation. The Δ9 desaturase enzyme utilised stearoyl-CoA, preferred NADH to NADPH and was inhibited by cyanide. Oleate was transferred to the sn-2 position of PC where it was desaturated to linoleate by the Δ12 desaturase enzyme. *In vivo* and *in vitro* studies indicated that the Δ6 desaturase also utilised the complex lipid PC as substrate. LPCAT was more active with unsaturated than saturated acyl-CoA species supporting the results for substrates of the desaturase enzymes.

Acknowledgements

M.circinelloides was a kind gift from Professor Herbert, Dundee University. F. Jackson and L. Michaelson acknowledge the receipt of BBSRC studentships.

References
1. Kendrick, A. & Ratledge, C. (1992a) Eur. J. Biochem. 209, 667 - 673.
2. Funtikova, N. S. & Zinchenko, G. A. (1991) Mikrobiologiya 60, 837-841.
3. Murcott, T., McNally, T., Allen,S., Fothergill-Gilmore, L.& Muirhead, H.(1991) Eur. J. Biochem. 198, 513-519.
4. Shanklin, J. & Somerville, C. (1991) Proc. Natl. Acad. Sci. USA **88**, 2510-2514.

TRITON X-100 SOLUBILIZATION OF DIACYLGLYCEROL ACYLTRANSFERASE FROM THE LIPID BODY FRACTION IN AN OLEAGINOUS FUNGUS

YASUSHI KAMISAKA
National Institute of Bioscience and Human-Technology, Tsukuba-shi, Ibaraki-ken 305, Japan

1. Introduction

Diacylglycerol acyltransferase (DGAT) activity, which contributes to triacylglycerol (TG) accumulation and the fatty acid composition of TG, was much higher in the lipid body fraction than in the membrane fraction in an oleaginous fungus, *Mortierella ramanniana* var. *angulispora* [1]. The subcellular distribution of DGAT may reflect lipid body formation and regulation of TG accumulation. Since this fungus efficiently produces TG containing γ-linolenic acid [2], elucidation of TG accumulation mechanism is also of great interest to improve the quality of TG production by this fungus. In this report, DGAT activity was solubilized with Triton X-100 from the lipid body fraction in this fungus and lipid requirement was investigated for the solubilized DGAT activity. The solubilized DGAT activity was increased by anionic phospholipids such as phosphatidic acid (PA) and phosphatidylserine (PS).

2. Materials and Methods

Fungal cells were homogenized with a Braun Homogenizer and the lipid body fraction was obtained as a floating fraction after sucrose density gradient centrifugation as described [1]. The lipid body fraction was solubilized by incubating with 10 mM phosphate buffer (pH 7.0), 0.15 M KCl, 1 mM EDTA, 1 mM phenylmethylsulfonyl fluoride, 1 mM dithiothreitol, 0.1% (for electron microscopy) 0.5% (for lipid requirement experiment) Triton X-100 for 1 h at 4°C. After the incubation, the sucrose concentration of the mixture was adjusted to 0.4 M and upon the mixture 10 mM phosphate buffer (pH 7.0) containing 0.15 M KCl and 0.3 M sucrose was overlayered. The discontinuous sucrose gradient was centrifuged at 58,000 x g for 3 h and the remaining soluble fraction served as the Triton X-100 extract.

The assay mixture for DGAT activity contained 10 mM phosphate buffer (pH 7.0), 150 mM KCl, 3.4 μM (0.2 μCi/mL) [1-^{14}C]oleoyl-CoA, 1.0 mM 1,2-diolein, various concentrations of Triton X-100, exogenous lipids and the detergent extract (0.30 μg protein) in a final volume of 100 μL. The reaction was carried out at 30°C for 5 min, and DGAT activity was calculated from radioactivity in TG as described [1].

For electron microscopy, the preparations were fixed in 1% glutaraldehyde and 1% OsO_4, and post-fixed in 1% OsO_4. The fixed preparations were dehydrated through a graded ethanol series and were embedded in Epon 812 resin. The sections were sequentially stained with 4% uranyl acetate and 0.4% lead citrate and viewed under the electron microscope.

3. Results and Discussion

Lipid bodies in *Mortierella ramanniana* var. *angulispora* could be observed as particles which had a diameter of about 0.5-3 μm by electron microscopy. Electron microscopy confirmed that the lipid body fraction was mainly consisted of these lipid bodies, which had been already shown by several biochemical studies [1]. Fractions obtained after solubilization of the lipid body fraction with 0.1% Triton X-100 was also observed by electron microscopy. Triton X-100 was chosen for the solubilization of DGAT activity since preliminary studies showed that it was the most efficient among several detergents. The floating fraction after the ultracentrifuge of the solubilization mixture with Triton X-100 contained lipid body-like structures, which were likely to be larger than those found in the lipid body fraction. On the other hand, in the pellet after the ultracentrifuge of the solubilization mixture, wrinkled membranous structures, which may be similar to images obtained form delipidated plant lipid bodies by organic solvents [3], could be observed. These observations suggest that lipid bodies maintain its integrity in spite of its enlargement and that some parts of lipid body membranes were sedimented into the pellet after the solubilization. Thus, DGAT in the lipid body fraction is likely to be located in lipid body membranes, but not embedded in lipid bodies. Finally, solubilization of DGAT activity in the Triton X-100 extract was confirmed by filtration through a 0.22 μm membrane filter and by high speed centrifugation at 100,000 x g. Both treatments did not significantly affect DGAT activity in the Triton X-100 extract.

Although Triton X-100 was found to be the most efficient for solubilization of DGAT activity from the lipid body fraction, the recovery was quite low. Anionic phospholipids such as PA and PS were found to increase DGAT activity in the Triton X-100 extract. The activation by these phospholipids was varied by Triton X-100 concentration in the assay mixture and optimal activation was obtained when 0.2% Triton X-100 was added (Figure 1.). Lower concentrations of Triton X-100 were not be enough to form Triton X-100 mixed micelles, which might provide good environment for the activation of DGAT by these phospholipids. The decrease in the activation at higher concentrations of Triton X-100 may be due to surface dilution of these phospholipids in Triton X-100 mixed micelles [4]. The activation was not inhibited by addition of EDTA, suggesting that Ca^{2+} or other metal ions are not involved in the activation by these anionic phospholipids (Figure 2.). Cardiolipin also increased DGAT activity, but lysophosphatidic acid, glycerol-3-phosphate and phosphatidylinositol were less effective for the activation, indicating that negative charge is not enough to account for the activation (Figure 2.). Major phospholipids such as phosphatidylcholine and phosphatidylethanolamine slightly increased DGAT activity whereas neutral lipids had no stimulatory effects.

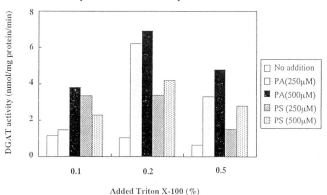

Figure 1. Effect of PA and PS on DGAT activity in the 0.5% Triton X-100 extract.

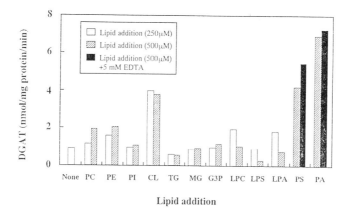

Figure 2. Effect of various lipids on DGAT activity in the 0.5% Triton X-100 extract. CL, cardiolipin; G3P, glycerol-3-phosphate; LPC, lysophosphatidylcholine; LPS, lysophosphatidylserine; LPA, lysophosphatidic acid.

Under the assay condition, PA specifically increased incorporation of [1-^{14}C]oleoyl-CoA into TG or DGAT activity, but not into other lipids. The result suggests that the activation of DGAT by PA is not due to activation or inhibition of other enzyme activities by PA, but due to a direct effect on DGAT or the interaction between DGAT and its substrates. Based on PA dependency of DGAT activity, the PA concentration causing half-maximal activity can be calculated to be 1.0 mol % in Triton X-100 mixed micelles. The value suggests that DGAT is activated by small numbers of PA molecules [5].

Although microsomal phospholipids were reported to increase DGAT activity in rat liver [6], the activation of DGAT by anionic phospholipids has not been reported yet. This adds a new instance where phospholipids and diacylglycerol (DG) interact with enzymes. It would be of great interest if the *in vitro* activation of DGAT by these anionic phospholipids is physiologically involved in regulation of DGAT and TG accumulation. In fact, DGAT is located at a branching point of *de novo* lipid biosynthesis, converting DG into TG, whereas DG choline- and ethanolaminephosphotransferases, enzymes converting DG into phospholipids are activated by phosphatidylcholine rather than anionic phospholipids [7]. Although further studies are required, the present results raise the possibility that the channeling of DG into TG or phospholipids is regulated by phospholipids surrounding DG-utilizing enzymes.

4. References

1. Kamisaka, Y. and Nakahara, T. (1994) Characterization of the diacylglycerol acyltransferase in the lipid body fraction from an oleaginous fungus, *J. Biochem.* **116**, 1295-1301.
2. Nakahara, T., Yokochi, T., Kamisaka, Y. and Suzuki O. (1992) Gamma-linolenic acid from genus *Mortierella*, in D.J. Kyle, and C. Ratledge (eds.), *Industrial applications of single cell oils*, American Oil Chemists' Society, Champaign, pp. 61-97.
3. Murphy, D.J. and Cummins, I. (1989) Seed oil-bodies: Isolation, composition and role of oil-body apolipoproteins, *Phytochemistry* **28**, 2063-2069.
4. Carman, G.M., Deems, R.A. and Dennis, E.A. (1995) Lipid signaling enzymes and surface dilution kinetics, *J. Biol. Chem.* **270**, 18711-18714.
5. Hjelmstad, R.H. and Bell, R.M. (1991) Molecular insights into enzymes of membrane bilayer assembly, *Biochemistry* **30**, 1731-1739.
6. Polokoff, M.A. and Bell, R.M. (1980) Solubilization, partial purification and characterization of rat liver microsomal diacylglycerol acyltransferase, *Biochim. Biophys. Acta* **618**, 129-142.
7. Hjelmstad, R.H. and Bell, R.M. (1991) *sn*-1,2-Diacylglycerol choline- and ethanolaminephosphotransferases in *Saccharomyces cerevisiae*: Mixed micellar analysis of the *CPT1* and *EPT1* gene products, *J. Biol. Chem.* **266**, 4357-4365.

BINDING OF LIPIDS ON LIPID TRANSFER PROTEINS

F. GUERBETTE, A. JOLLIOT, J.- C. KADER AND M. GROSBOIS
Université Pierre et Marie Curie (Paris 6), Laboratoire de Physiologie Cellulaire et Moléculaire, CNRS - URA 2135, Case 154, 4 place Jussieu, F -75252 Paris Cedex 05, France.

1. Introduction

LTPs transfer phospholipids between membranes "*in vitro*" (1). Plant LTPs are still in search of an "*in vivo*" role. These proteins share in common structural and functional properties (1). NMR and X-ray crystallography have been carried out on plant LTPs. These proteins comprise a single compact domain with four alpha-helices and a long C-terminal region. They have a tunnel-like hydrophobic cavity able to accomodate fatty acids (2-3). A complex between maize LTP and palmitic acid has been indeed recently crystallized (2). This recent model of ns-LTPs gives a basis for understanding the function of LTPs in relation to the transfer and binding of molecules containing acyl chains (PLP, fatty acids, acyl-CoA).

Various cDNAs coding for LTPs contained a signal peptide suggesting a possible secretion of these proteins (4-5). Though it was of interest to study the interaction of LTPs with other ligands than PLP used to define this class of proteins (6), especially fatty acid alcohols monomers of cutin. The first step in this experimentation was to find acyl chain containing ligands of this isoform of maize LTP forming complexes stable enough to be studied. Lyso-PC and acyl-CoA derivatives were compared, to determine the efficiency of these two kinds of lipophilic compounds as ligands.

2. Material and methods

Maize seeds (*Zea mays* L., cv. Mona) were a gift from Pioneer France-Maïs. Steps of purification were as define previously (7). Purified LTP used was the isoform fully characterized (8).

Lyso PC binding assays : the binding of lyso-PC was determined after separation of LTP / lyso-PC complex from unbound LTP and ligand by gel filtration (Sephadex G 100). LTP elution was followed by ELISA according to (9).

Acyl-CoA binding assays : the binding of (^{14}C) oleoyl- CoA to LTP was estimated after separation of LTP ligand / complex by FPLC on a Mono S column according to protocol described in (10).

PC transfer activity of LTP complexes : LTP complexes were formed either with lyso-PC or oleoyl-CoA under the conditions described above and PC tranfer activity was tested from liposomes to mitochondria as in (11).

Abbreviations : ELISA, enzyme link immunosorbent assay; ns-LTP, non specific Lipid Transfer Protein; PC, phosphatidylcholine; PLP, phospholipid.

3. Results

After incubation of LTP with lyso-PC and separation on Sephadex G100, ^{3}H radioactivity corresponding to lyso-PC was observed in LTP containing fractions detected by ELISA (Fig:1) After molecular sieving LTP elution peak divided into two parts, LTP / lyso-PC complex being more retained than free LTP (58 % of initial radioactivity of lyso-PC was found associated to LTP).

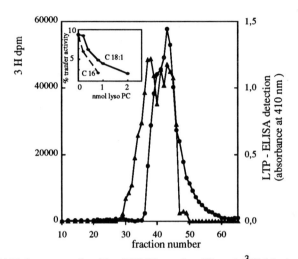

Fig 1 : Sephadex G 100 chromatography of lysoPC-LTP complex; 73 nmol (^{3}H) 1 hydroxy-2 linoleoyl sn glycerol (880 x 10^3 dpm) incubated 30 min at 30°C with 100 nmol pure LTP.
(O) ^{3}H dpm corresponded to (^{3}H)lysoPC detected in the different fractions (4 ml).
(Δ) Specific ELISA detection of LTP by affinity purified specific IgGs directed against this maize LTP.
Insert : Effect of lysoPC binding on LTP transfer activity. The complex was formed previously to transfer assay either with 1 hydroxy-2 oleoyl-sn-glycerol (C18:1) or 1 hydroxy-2 palmitoyl-sn-glycerol (C16). Transfer activity tested with 50 nmol LTP.

The pattern of elution caracterized by two peaks was also observed with LTP/PC complexes (11). According to recent models of LTP structure (3), loading of an acyl moeity produces a conformational modification of LTP that could account for the lagging of the complex. Because the tunnel like hydrophobic cavity binds only to the acyl chain, the residual part of lyso-PC molecule extends outside the complex. Elution shift could originate from differences in the interactions between Sephadex and free or complexed LTP.

PC transfer activity was impaired by a pretreatment of LTP (50 nmol) by lyso-PC : 83 % of inhibition of transfer activity was observed with 2 nmol lyso-C18:1 and 1 nmol lyso-C16 respectively (Fig:1-Insert). This suggests that the complex is more stable with lyso derivatives than with PC. As a matter of fact, only a weak binding to PC has been yet reported (11).

LTP/oleoyl-CoA complex separated by FPLC on Mono S (Fig:2) showed a binding activity corresponding to 33 % of the initial radioactivity of oleoyl-CoA in the assay. Binding of oleoyl-CoA on LTP before PC tranfer assay led to only 25 % inhibition of transfer (100 nmol LTP, 50 nmol oleoyl-CoA) (Fig:2- Insert). In contrast, a rapeseed LTP was unable to transfer PC after oleoyl CoA loading (10).

There are growing evidence that different isoforms of ns-LTPs are associated to different functions in relation to binding and transfer of acyl-chain components. This preliminary work shows that a ns-LTP isoform, purified from maize seeds, exhibited a greater binding affinity to lyso-PC and oleoyl-CoA than to PLP if we compare with previous results obtained with PC (11). Phospholipid/LTP complexes are classically supposed to be involved in transfer activity between membranes, supposing a loosy strength of association facilitating loading as well as delivery of the ligand. Lyso-PC and acyl-CoA - intermediates in biosynthesis - appeared to be more efficient acyl-group containing ligands of maize seed LTP. From this , lyso-PC and acyl-CoA/LTP complexes containing different kinds of acyl in relation to cutin synthesis will be used to study the formation of LTP / ligand complexes and the possible role of LTP in this biosynthesis.

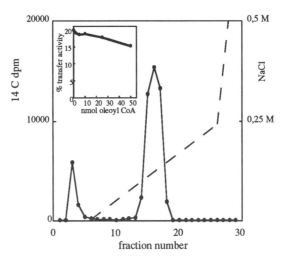

Fig 2 : FPLC chromatography on mono S column of oleoyl-CoA-LTP complex; 1,8 nmol (^{14}C)oleoyl-CoA (278 x 10^3 dpm incubated in melting ice with 39 nmol pure LTP during 30 min under agitation. Elution by NaCl gradient (0 to 250 mM). Fractions 1 ml. Insert : Effect of oleoyl-CoA binding on LTP transfer activity. Transfer activity tested with 100 nmol LTP.

References

1. Kader J.C. (1996) *Annu. Rev. Plant Physiol. Plant Mol. Biol.*, in the press.
2. Shin D.H., Lee I.Y., Hwang K.Y., Kim K.K. and Such S.W. (1995) *Structure*, **3**, 189-199.
3. Gomar J., Petit M.C., Sodano P., Sy D., Marion D., Kader J.C., Vovelle F. and Ptak M. (1996)*Protein Science* , **5**, 565-577.
4. Sterk Pi.,Booij H., Schellekens G.A., Van Kammen A. and De Vries,S. (1991) *The Plant Cell*, **3**, 907-921.
5. Pyee J., Yu H. and Kolattukudy P.E. (1994) *Arch. Biochem. Biophys.*, **331**, 460-468.
6. Kader J.C. (1990) in H.J. Hilderson (ed.) Subcellular Biochemistry Plenum Publishing Corporation, 69-111.
7. Douady D., Guerbette F., Grosbois M. and Kader J.C. (1985) *Physiol. Vég.*, **4**, 373-380.
8. Tchang F., This P., Stiefel V., Arondel V., Morch M.D., Pages M., Puigdomenech P., Grellet F., Delseny M., Bouillon P., Huet J.C., Guerbette F., Beauvais-Cante F., Duranton H., Pernollet J.C. and Kader J.C. (1988) *J. Biol. Chem.*, **32**, 16849-16855.
9. Grosbois M., Guerbette F., Douady D. and Kader J.C. (1987) *Biochim. Biophys. Acta* , **917**, 162-168.
10. Østergaard J., Vergnolle C., Schoentgen F. and Kader J.C (1993) *Biochim. Biophys. Acta* , **1170**, 109-117.
11. Grosbois M., Guerbette F., Jolliot A., Quintin F. and Kader J.C. (1993) *Biochim. Biophys. Acta* ,**1170**, 197-203.

VARIATIONS IN UDGT ACTIVITY IMPOSED BY GROWTH CONDITIONS REFLECT INTERNAL AVAILABILITY OF SUBSTRATES

A. LIVNE and A. SUKENIK
National Institute of Oceanography, Israel Oceanographic and Limnological Research, P.O. Box 8030, Haifa 31080, Israel.

1. Introduction

Lipid content and composition in unicellular algae are regulated to a large extent by environmental conditions (Roessler 1990). Temperature, nutrient level, light intensity, salinity and other environmental factors have been examined in regard to lipid metabolism and fatty acid. Under subsaturating light conditions, unicellular algae such as *Nannochloropsis* preferentially synthesize galactolipids which are enriched with eicosapentaenoic acid (C20:5ω3) (Sukenik et al. 1989).

In this study we evaluate the effect of growth irradiance level on the *in vitro* activity of UDGT, an enzyme which catalyzes the final step in the formation of the major glycerolipids in this alga, MGDG. This study represents an initial step in our effort to elucidate the mechanism which regulates the synthesis of structural glycerolipids such as MGDG in unicellular algae.

2. Materials and Methods

Alga and growth conditions - *Nannochloropsis* sp. is currently kept in our collection as strain NANNOP-IOLR. Cultures were grown in 2.6 l turbidostats fed with filtered natural sea water enriched with nutrients as previously described (Sukenik et al. 1989). Cells were grown at 35 and 300 μmol quant $m^{-2} s^{-1}$. Steady state cultures were exposed to a one-step shift in irradiance level (Sukenik et al. 1990). During steady state growth prior to the light transition, and at selected times thereafter, the cultures were sampled for measuring the *in vitro* activity of UDGT.

Preparation of membranes - Cells were harvested and broken by grinding with a Teflon pestle and an equal volume of glass beads (<105 mm) in Tris Acetate buffer (50 mM pH 7.2) containing 0.3 M sucrose, 5 mM $MgCl_2$, 0.1 M Na Acetate, 1 mM EDTA, 2 mM DTT, 1 mM PMSF and 1 mg ml^{-1} BSA (Teucher and Heinz 1991). Membranes were dissolved in 6 mM CHAPS (Marechal et aL 1991) followed by mixing for 30 min at 4 °C.

Enzymatic assay - The *in vitro* activity of UDP-galactose diacylglycerol galactosyl-transferase (UDGT) was assayed by following the formation of monogalactosyl diacylglycerol as described by Teucher and Heinz (1991). The reaction was initiated by the successive addition of UDP-[1-^{14}C]galactose and crude membrane preparation, and incubated at 20°C for 20 min.

3. Results and Discussion

***In vitro* activity of UDGT** - The optimal conditions for the *in vitro* activity of UDGT in crude membranes were pH 7.2 and 20°C. The galactosylation reaction was linear with time for the first 20 min. The UDGT activity was saturated with 1 mM UDP galactose. Other cofactors such as magnesium and DTT were required. Addition of 0.1 M Na Acetate increased the measured *in vitro* activity and preserved the activity for several days. No specificity for a certain DAG molecular species (i.e. 16:/16:0, 18:0/18:0 or 18:1/18:1) was observed, and the synthesis rate of MGDG was apparently independent of the concentration of external DAG in the assay.

Effect of growth conditions on the *in vitro* activities - The *in vitro* activity of UDGT was measured in crude membranes prepared from cells cultivated under different environmental conditions of irradiance levels and temperatures. In low light grown cells, the specific activity of UDGT ranged between 2.4 and 3.9 pmol MGDG mg protein^{-1} min^{-1} with a relatively little effect of temperature. In high light grown cultures, the specific activity of UDGT was significantly reduced to about 0.8 pmol MGDG mg protein^{-1} min^{-1}, whereas the growth temperature had no effect on the *in vitro* specific activity (data not shown).

Variations in UDGT activity in response to transition in irradiance levels - Following a one-step transition in growth irradiance level from HL to LL, the *in vitro* specific activity of UDGT only slightly increased during the first 24 hours (Fig. 1). This lag period was followed by a rapid increase in the enzyme activity, and within the next 48 hours a new, high steady state value was established. No such lag in the response of UDGT activity was observed when cultures were transferred from LL to HL. In that case a gradual decrease in the *in vitro* specific activity of UDGT was recorded until a new low steady state level was obtained after 72 hours (Fig. 1).

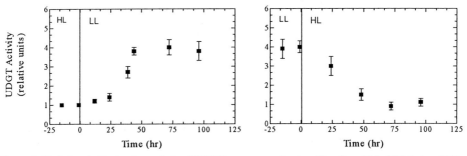

Figure 1: Variations in the *in vitro* activity of UDGT upon a one-step transition from high (HL) to low (LL) irradiance levels (left panel) and from low to high irradiance level. UDGT specific activity of steady state LL cells was 3.2 pmol MGDG mg protein^{-1} min^{-1} and for steady state HL cells 0.8 pmol MGDG mg protein^{-1} min^{-1}.

UDGT *in vitro* activity can be enhanced by the addition of inactivated cellular extract - The reduction in the *in vitro* specific activity of UDGT in HL grown cells can be attributed to a decrease in the cellular content of this enzyme. However, due to a lack of response to externally added DAG, the availability of specific DAG substrates that are presumably extracted from the cells during membrane preparation may affect the measured *in vitro* activity. To evaluate this point, crude membranes (ca. 2 µg protein/µl) were treated at 56 °C for 10 min. (in order to abolish the activity of UDGT) and were added to enzymatic assays performed with crude extract. The inactivated crude membranes from LL grown cells slightly increased the *in vitro* activity of UDGT of HL cells but not in crude membrane preparations from LL cells. Furthermore, the enhanced UDGT activity in HL cells was observed only when thermally inactivated membranes from LL cells were used. The enhanced UDGT activity was further amplified when thermally inactivated LL membrane preparations were added to crude extracts from LL cells that were acclimated to HL conditions for 24 or 48 hrs (Fig. 2). These results clearly demonstrate that cultures in a transition state of acclimation to HL irradiance lack certain components that are required for optimal activity of UDGT. These components are stable in high temperature and gradually increase the *in vitro* activity of UDGT until saturation is exceeded (Fig. 3). The results indicate that the low activity of UDGT in HL grown cells can be attributed to low availability of an unidentified substrate that is apparently available in crude extracts from LL cells. A decrease in the cellular abundance of UDGT only slightly contributes to the reduction in the UDGT *in vitro* activity in HL cells.

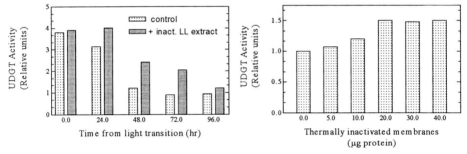

Figure 2: The effect of addition of thermally inactivated extract from LL grown cells on the *in vitro* activity of UDGT in *Nannochloropsis* cultures exposed to a one-step transition from low to high irradiance level.

Figure 3: Enhancement of UDGT *in vitro* activity by the addition of different quantities of thermally inactivated membrane preparation from LL grown cells. The assay was performed with LL cells acclimated for HL conditions for 24 hrs.

4. References

Marechal, E., Block, M.A., Joyard, J and Douce, R. (1991) Purification of UDP-galactose: 1,2-diacylglycerol galactosyltransferase from spinach chloroplast enevelope membranes. *C.R.Acad. Sci.Paris, t.* **313 Serie III,** 521-528.

Roessler, P.G. (1990) Environment control of glycerolipid metabolism in microalgae: commercial implications and future research directions. *J. Phycol.* **26,** 393-399.

Sukenik, A., Carmeli, Y. and Berner, T. (1989) Regulation of fatty acid composition by irradiance level in the eustigmatophyte *Nannochloropsis* sp. *J. Phycol.* **25,** 686-692.

Sukenik, A. Bennett, J., Mortain-Bertand, A. and Fakowski, P.G. (1990) Adaptation of the photosynthetic apparatus to irradiance in *Dunaliella tertiolecta* - A kinetic study. *Plant. Physiol.* **92,** 891-898.

Teucher, T. and Heinz, E. (1991) Purification of UDP-galactose: diacylglycerol galactosyl transferase from chloroplast envelopes of spinach (*Spinacia oleracea* L.). *Planta* **184,** 319-326.

THE ROLE AND PARTIAL PURIFICATION OF CTP: ETHANOLAMINEPHOSPHATE CYTIDYLYLTRANSFERASE IN POSTGERMINATION CASTOR BEAN ENDOSPERM

FUQIANG TANG AND THOMAS S. MOORE, JR.
Department of Plant Biology, Louisiana State University
Baton Rouge, LA 70803-1705, USA

The nucleotide pathway is the major pathway of phosphatidylethanolamine (PE) headgroup biosynthesis by plants (Sparace, *et al.*, 1981). It is catalyzed in sequence by ethanolamine kinase (EK), ethanolaminephosphate cytidylyltransferase (ECT) and ethanolamine phosphotransferase (EPT). EK has been characterized and purified from spinach leaves (Macher and Mudd, 1976) and soybean seeds (Williams and Harwood, 1994), EPT has been characterized in several plants (10), but little information is available about ECT. The only plant tissue in which ECT has been characterized is castor bean endosperm, where approximately 80% of the activity is in the outer mitochondrial membrane and the remainder in the endoplasmic reticulum (ER; Wang and Moore, 1991). ECT is soluble in mammals and yeast (Sundler, 1975).

The ER is the primary site for PE biosynthesis in this tissue (Sparace, *et al.*, 1981), and so the ER-bound ECT is a component of this synthesis. Mitochondrial ECT is too active to simply be supplying substrate to the ethanolamine phosphotransferase of that organelle. These points have been taken to indicate a function for the mitochondrial enzyme different from that of the ER, but that role remains undefined. For this reason we have examined the relative activities of the PE nucleotide synthesis pathway enzymes and partially purified the mitochondrial ECT.

1. Methods

1.1 DEVELOPMENT

Endosperm halves were collected, homogenized, fractionated and enzymes assayed at selected times following germination of castor beans in moist vermiculite in the dark at 30°C. For the ethanolamine kinase (EK) assay the reaction mixture contained in a final volume of 50 µl: 100 mM Tris-HCl (pH8.3), 5 mM ATP, 10 mM $MgCl_2$, 563 µM [2-^{14}C]ethanolamine. The reaction was carried out at 37°C for 1 h, terminated with 80 µl of 60% trichloroacetic acid, and phosphoethanolamine separated from ethanolamine by ascending paper chromatography. Ethanolamine-phosphate cytidylyltransferase (ECT) was prepared from germinated castor bean endosperm and assayed as described by

Wang and Moore (1991). Ethanolamine-phosphotransferase (EPT was measured as previously described (Sparace, *et al.*, 1981).

1.2 PURIFICATION

Mitochondrial membranes were prepared by diluting thawed mitochondria to a final protein concentration of 2 mg/mL with 20 mM Tris-HCl buffer containing 2 mM DTT and 1 mM EDTA. This suspension was homogenized with a Potter homogenizer for 10 min. The homogenized mitochondria were centrifuged at 100,000 X g for 1h, the pellet washed, and then suspended in fresh buffer and re-centrifuged. The washed mitochondrial membrane pellet was resuspended in 100 mM potassium phosphate buffer (pH7.2) containing 0.5 M KCl, 1 mM EDTA and 1 mM DTT. A suspension containing protein at a concentration of about 30mg/ml was incubated with octyl-β-D-glucopyranoside (OG), CHAPSO or cholate. Solubilization was carried out for 1 h at 4°C with constant stirring followed by centrifugation at 105000 X g for 1 h.

Initial gel filtration was on a Sepharose CL-4B column equilibrated with 100 mM potassium phosphate buffer (pH7.2), 0.5 M KCl, 1 mM EDTA, 1 mM DTT and containing CHAPSO at either 0.5 or 1.0% (w/v). Sephacryl S-200 columns were equilibrated and eluted with the same buffer solution containing 1.5% CHAPSO.

For sucrose density gradients, solubilized membrane protein was layered onto a 10 mL, 5-30% (w/v) gradient containing 100 mM potassium phosphate buffer (pH 7.2), 0.5 M KCl, 1 mM EDTA and 1 mM DTT. Sedimentation was at 500,000 x g in a T-875 rotor for 17 h at 4° C in a Sorvall ultracentrifuge.

PEG fractionation was accomplished by bringing solubilized protein to 10% PEG 8000 by addition of a 50% (w/v) stock solution with continuous stirring. The PEG was prepared in the same buffer as used for the ECT solubilization, which contained 1.5% CHAPSO. The precipitated proteins were sedimented by centrifugation (12000g; 20 min). The supernatant was removed, and brought to 19% PEG, followed by centrifugation. The 10-19% PEG 8000 cut was suspended in solubilization buffer with 1.5% CHAPSO.

2. Results

Within the four day postgermination period, EK activity remained relatively constant and then slowly declined. Total ECT activity increased until the fourth day, but ECT activities in the ER and mitochondria behaved differently, with the ER activity reaching its peak at day 3, followed by a sharp decline; the mitochondrial activity increased until day 4, after which a slow decrease began. EPT increased during the first two days postgermination, after which it declined. The lowest activity of EK during postgermination was over 5-fold more than the maximum activity of ECT at any time. The total activity of EPT at day 2 was at least 3-times that of the ECT of the ER. These results, along with the kinetic characterics of the three enzyme, indicate that ECT is the primary candidate for regulation of PE synthesis by the nucleotide pathway.

Mitochondrial ECT was solubilized with 1.5% CHAPSO and purified approximately 118-fold by PEG precipitation, chromatography on Sephacryl S-200, and then sucrose gradient centrifugation. This unusual procedure resulted from a need for the continuous presence of both salt and detergent to maintain activity. SDS-PAGE resulted in a prominent protein band at 34.7 kDa, which paralleled in concentration the ECT activity in fractions from either sucrose gradient centrifugation or gel filtration on the Sephacryl S-200 column. Addition of pure phospholipids to ECT preparations prior to assaying the activity resulted in stimulation of the activity at all concentrations tested, with phosphatidyldimethylethanolamine being most effective, followed by phosphatidylmethylethanolamine, phosphatidylcholine, and PE in that order.

3. Conclusions

ECT of the ER probably regulates PE headgroup biosynthesis by the nucleotide pathway. This is supported by its activity throughout development being consistent lower than the other key enzymes, and the Km being higher than those for either EK or EPT. This suggests abundant production of ethanolamine-P and ready use of any CDPethanolamine produced.

A major difficulty encountered in these studies was a need for the continuous presence of detergent and high salt concentrations in order to prevent aggregation of the enzyme. This requirement reduced the available procedures for purification, thus leading to an unusual protocol for purifying ECT which resulted in a 118-fold purification from mitochondrial membranes. The primary component of the purified protein was a 34.7 kDa fraction, as determined by SDS PAGE. Purified mammalian ECT exhibits a single, 49.6 kDa protein band on SDS-PAGE, but on gel filtration was found to be 100-120kDa (Sundler, 1975; Tijburg, *et al.* 1992.

Acknowledgement. This research was supported by grant 93-37304-9368 from NRICGP/USDA to TSM.

4. References

Macher B.A. and Mudd J.B. (1976) Partial purification and properties of ethanolamine kinase from spinach leaf, *Arch. Biochem. Biophys.* **177**: 24-30.

Sparace S.A., Wagner L.K. and Moore T.S. (1981) Phosphatidylethanolamine synthesis in castor bean endosperm, *Plant Physiol.* **67**:922-925.

Sundler, R. (1975) Ethanolamine cytidylyltransferase. Purification/characterization of the enzyme from rat liver, *J. Biol. Chem.* **250**:8585-8590.

Tijburg L.B.M., Houweling M., Geelen M.J.H., and van Golde L.M.G. (1993) Ethanolaminephosphate cytidylyltransferase, *Meth. Enzymol* **209**: 258-263.

Wang X. and Moore T.S. (1991) Phosphatidylethanolamine synthesis by castor bean endosperm. Intracellular distribution and characteristics of CTP:ethanolamine-phosphate cytidylyltransferase, *J. Biol. Chem.* **266**:19981-19987.

Williams M. and Harwood J.L. (1994) Alternative pathways for phosphatidylcholine synthesis in olive (*Olea europaea* L.) callus cultures, *Biochem. J.* **304**: 463-468.

IDENTIFICATION OF A REGION IN STROMAL GLYCEROL 3-PHOSPHATE ACYLTRANSFERASE WHICH CONTROLS ACYL CHAIN SPECIFICITY

S.R. FERRI, O. ISHIZAKI-NISHIZAWA, M. AZUMA and T. TOGURI
Central Laboratories for Key Technology, Kirin Brewery Co., Ltd., Fukuura, Kanazawa, Yokohama 236, Japan

The less fluid phosphatidylglycerols, containing only saturated and trans-monounsaturated acyl chains, had been reported to be in higher proportions in chilling sensitive plants [1, 2, 3]. The stromal glycerol-3-phosphate acyltransferase (GPAT, EC 2.3.1.15), which transfers the acyl chain from palmitoyl- or oleoyl-acyl carrier protein (16:0- or 18:1-ACP) to the *sn*-1 position of glycerol 3-phosphate (G3P)(reviewed in [4]), was thought to control this lipid distribution and was therefore perceived as playing a major role in determining a plant's ability to tolerate chilling temperatures.

Although the mature GPATs from spinach [5] and squash [6] are highly conserved (66% identical), important differences in their activities have been reported (summarized in [4]). The GPATs from spinach and pea, both chilling-tolerant plants, were reported to prefer 18:1 esters of CoA and ACP over 16:0 esters, but only when both substrates were present simultaneously in the reaction. On the other hand, the two acyl chains could generally not be distinguished by the enzyme from squash, a chilling-sensitive plant. The pea and spinach GPATs were thought to have acyl chain selectivity, but not specificity, and the squash enzyme was thought to have neither [4].

In the present study, the spinach and squash GPAT activities were found to possess very different preferences for the two acyl chains when they were offered individually. By creating chimeras with the two genes, we set out to determine which structural features are responsible for their different acyl chain specificities.

Materials and Methods

Appropriate restriction endonuclease sites were added by PCR to the amino termini of the mature GPATs from spinach (SP) and squash (SQ). The modified genes were inserted into pET-3 expression vectors (Novagen) with the resulting proteins having 2 (SP) or 13 (SP-L and SQ-L) amino acids from the φ10 fragment of T7 fused to the amino terminus. A series of chimeras were created by using conserved *Hin*dIII and *Kpn*I restriction endonuclease sites which divide the spinach and squash GPAT genes in three segments of roughly identical length, corresponding to 126, 119 and 124 amino acids (from amino to carboxyl terminus). The second letters of "SPINACH" and "SQUASH" were used to describe the origins of the thirds in labelling the resulting chimeras; for example, PPQ represents a chimera with the amino two thirds originating from the spinach GPAT and the carboxyl third originating from the squash enzyme. All chimeras, as well as the wild type spinach (PPP) and squash (QQQ) GPAT genes, were inserted into the pET-17b expression vector (Novagen) so as to yield proteins of the same length as the natural mature GPATs. In the chimera P2P, residues 126-197 of the middle third originate from spinach and residues 198-245 from squash.

Expression of the constructs in *E. coli* (BL21(DE3) and BL21(DE3)pLysS; Novagen) was performed according to the manufacturer's recommendation, except that the temperature was shifted to 30°C for 3h after induction. Cell pellets were lysed, centrifuged at 100 000 g for 1 h at 4°C and kept on ice until assayed for GPAT activity.

The standard GPAT assay, which consists of measuring the rate of transfer of the acyl chain from acyl-CoA to [U-^{14}C]G3P was performed as previously described [7], with minor modifications. Cell-free extract (~1µg) was incubated at 24°C in 80 µl of 0.25 M Hepes (pH 7.4), 6 mg/ml bovine serum albumin, 0.3 mM [U-^{14}C]G3P (0.9 Ci/mol), and 0.4 mM 16:0- or 18:1-CoA. Acylated [U-^{14}C]G3P was extracted with 2.3 ml chloroform-methanol (1:1) and 1 ml 1 M KCl, 0.2 M H_3PO_4 and counted.

Results and Discussion

Extract containing SP was found to have the highest activity, using 18:1-CoA (300 pmol/min/µg) 1.5 times faster than 16:0-CoA (Fig. 1). By extending the leader sequence from 2 to 13 amino acids, its specificity was only slightly affected, transferring 18:1 (220 pmol/min/µg) 1.3 times faster than 16:0. In contrast, the squash GPAT (SQ-L) used 16:0-CoA (200 pmol/min/µg) over 3 times faster than 18:1-CoA. The wild-type GPATs PPP and QQQ had almost identical relative activities as their longer counterparts, using their preferred substrate at rates of 310 and 460 pmol/min/µg, respectively. The difference between SQ-L and QQQ absolute activities is mainly due to an unusually low expression level for SQ-L, as determined by SDS-PAGE (data not shown). Because PPP and QQQ are of equal length to the natural mature enzymes, our results indicate that small additions to the amino terminal have little effect on absolute activity and specificity. A similar addition of 23 amino acids to pea GPAT had previously been reported not to abolish enzyme activity or fatty acid selectivity [8].

Substitution of the amino third of GPAT appears to have abolished its specificity, with both PQQ and QPP only slightly preferring 16:0-CoA. Substitution of the central third of PPP to that of squash changed its specificity to almost that of the wild-type squash GPAT. Substitution of the central portion of QQQ also reversed its specificity, albeit to a level much greater than that of the wild-type spinach enzyme. Surprisingly, substitution of the carboxyl third of PPP to that of squash also caused a drastic increase in its preference for the 18:1 acyl chain. However, the QQQ to QQP substitution had no detectable affect on its specificity, indicating the carboxyl third is not important in determining fatty acid specificity. The enhancement in 18:1 specificity resulting from the QPP to QPQ and PPP to PPQ substitutions seems to be caused by a conformational change which by chance drastically reduces the enzyme's ability to use 16:0-CoA, without affecting its ability to use 18:1-CoA.

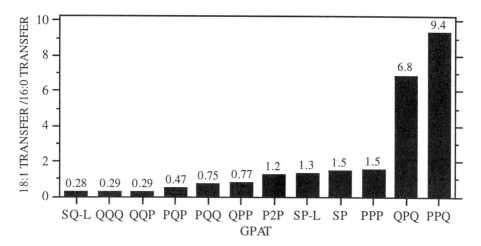

Figure 1. Relative activities of wild-type and chimeric GPATs with 18:1-CoA or 16:0-CoA. Results are expressed as fractions.

The central third seems most likely to be responsible for the fatty acyl specificities in the native enzymes. Indeed, the XPX GPATs all have greater preferences for 18:1 than any XQX enzyme (Fig.1). By dividing the central third approximately halfway, the chimera P2P was created whose specificity was very similar to that of the wild-type spinach enzymes. Although, most substitutions had some affect, the amino half of the central third, between residues 126 and 197 of the 370 amino-acid protein, appears to contain the structural feature(s) which dictates GPAT's fatty acid specificity.

The correlation between the plant's chilling tolerance and its GPAT's acyl chain specificities, as demonstrated by our results, is consistent with the enzyme's important role in chilling tolerance. Increasing our understanding of the relationships between structure and function will hopefully be beneficial in our quest to improve plant chilling tolerance through genetic engineering.

References

1. Murata, N. Sato, N., Takahashi, N., and Hamazaki, Y (1982) *Plant Cell Physiol.* **23**; 1071-1079.
2. Murata, N. (1983) *Plant Cell Physiol.* **24**; 81-86.
3. Roughan, P.G. (1985) *Plant Physiol.* **77**, 740-746.
4. Frentzen, M. (1993) Acyltransferases and triacylglycerols, in T.S. Moore, Jr. (ed.), *Lipid Metabolism in Plants*, CRC Press, Boca Raton, pp. 195-231.
5. Ishizaki-Nishizawa, O., Azuma, M., Ohtani, T., Murata, N. and Toguri, T. (1995)(EMBL X77370), *Plant Physiol.* **108**, 1342.
6. Ishizaki, O., Nishida, I., Agata, K., Eguchi, G., and Murata, N. (1988) *FEBS Lett.* **238**, 424-430.
7. Bertrams, M. and Heinz, E. (1981) *Plant Physiol.* **68**, 653-657.
8. Frentzen, M., Peterek, G. and Wolter, F.P. (1994) *Plant Sci.* **96**, 45-53.

THE PURIFICATION AND CHARACTERISATION OF PHOSPHATIDATE PHOSPHATASE FROM AVOCADO

MATT PEARCE AND ANTONI R. SLABAS
University of Durham
Department of Biological Sciences, South Road, Durham, DH1 3LE,
U.K.

1. Introduction

Phosphatidate phosphatase (EC 3.1.3.4) catalyses the dephosphorylation of phosphatidic acid to diacylglycerol prior to the final acylation step of the Kennedy Pathway. Unlike the acyltransferases, phosphatidate phosphatase (PAP) has received very little attention in plant lipid research. PAP appears to be the rate limiting factor in triacylglycerol biosynthesis, and has been shown to be controlled by translocation to and from the endoplasmic reticulum [1]. As well as in triacylglycerol biosynthesis, both phosphatidic acid (PA) and diacylglycerol (DAG) are intermediates for the major phospholipids. Therefore PAP lies at a theoretical branchpoint between triacylglycerol and phospholipid biosyntheses, the production of these two lipid species may be governed by the extent of translocation of PAP within the cell. It remains to be seen whether PAP plays a role in signal transduction as has been suggested in animal systems, where it plays a pivotal role by regulating the amounts of both lipid second messengers, PA and DAG. It has been found that there are several isoforms of PAP in both yeast and animals, these appear to serve the different functions.
PAP has been purified from yeast [2], and animal tissue [3,4,5]. We have purified microsomal PAP to homogeneity from avocado (*Persea Americana*) and report here enzymological properties for solubilised PAP.

2. Results

2.1. ASSAYING FOR PAP ACTIVITY

PAP activity was routinely measured by a procedure based on the assay of Lin and Carman(1989) [2], by following the formation of water soluble [^{32}P]Pi from chloroform soluble [γ-^{32}P]PA.

2.2. IDENTIFICATION OF PAP

During a chromatographic purification, PAP was identified on a 10% SDS-PAGE gel by loading the pooled active fractions alongside the preceding non-active fractions (Figure 1). The band at 49KDa was therefore positively identified as PAP.

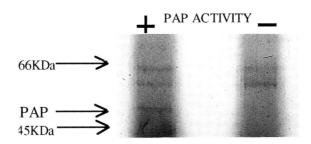

Figure 1. The identification of PAP with SDS-PAGE

2.3. THE PURIFICATION OF PAP

PAP was purified to homogeneity from washed avocado microsomes, using a 5-step procedure. This entailed CHAPS solubilisation; ion exchange, affinity and hydrophobic interaction chromatography. The resulting protein band was approximately 49KDa on silver-stained SDS-PAGE, as shown in Figure 2 below:

Figure 2. Purified PAP at approximately 49KDa

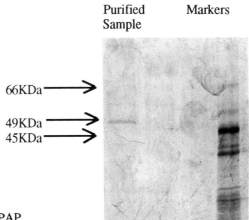

2.4. THE PROPERTIES OF PAP

Little is known about plant PAP. It has recently been characterised in microspore-derived cell cultures [6] and in spinach chloroplasts [7]. The biological properties of CHAPS-solubilised PAP were investigated and some of these are presented in Table 1.

TABLE 1. The properties of partially purified avocado PAP

pH Optimum of 6.0
Triton X-100 optimum at PA:Triton X-100 1:4 (20mol%)
Activity enhanced with 1mM Mg^{2+}
Dithiothreitol (disulphide reducing agent) independent
6-10% of phosphatase activity in microsomes
Pure PAP is stable at $4^{\circ}C$ for at least 1 month
PAP is stable to at least 3 rounds of freeze-thaw

4. Discussion

We have positively identified PAP and the molecular weight was further confirmed with native PAGE. The reproducible purification will enable more detailed characterisation and we are currently trying to obtain amino-acid sequence data. Our characterisation findings appear to corroborate those found in *Brassica Napus* microspore-derived cultures[6].

5. Acknowledgements

M.P. would like to thank the Gatsby Charitable Foundation for the research studentship.

6. References

1. Ichihara, K., Murota, N. and Fujii, S. (1990) Intracellular translocation of phosphatidate phosphatase in maturing safflower seeds: a possible mechanism of feedforward control of triacylglycerol synthesis by fatty acids, *Biochim. Biophys. Acta* **1043**, 227-234
2. Lin, Y-P. and Carman, G.M. (1989) Purification and characterisation of phosphatidate phosphatase from *Saccharomyces cerevisiae*, *Journal of Biological Chemistry* **264**, 8641-8645
3. Kanoh, H., Imai, S., Yamada, K. and Sakane, F. (1992) Purification and properties of phosphatidic acid phosphohydrolase from porcine thymus membranes, *Journal of Biological Chemistry* **267**, 25309-25314
4. Fleming, I.N. and Yeaman S.J. (1995) Purification and characterisation of N-ethylmaleimide-insensitive phosphatidic acid phosphohydrolase (PAP2) from rat liver, *Biochemical Journal* **308**, 983-989
5. Siess, E.A. and Hofstetter M.M. (1996) Identification and phosphatidate phosphohydrolase purified from rat liver membranes on SDS-polyacrylamide gel electrophoresis, *FEBS Letters* **381**, 169-173
6. Kocsis, M.G., Weselake, R.J., Eng, J.A., Furukawastoffer, T.L. and Pomeroy, M.K. (1996) Phosphatidate phosphatase from developing seeds and microspore derived cultures of *Brassica Napus*, *Phytochemistry* **41**, 353-363
7. Malherbe, A., Block, M.A., Douce, R. and Joyard, J. (1995) Solubilisation and biochemical properties of phosphatidate phosphatase from spinach chloroplast envelope membranes, *Plant Physiol. Biochem.* **33**, 149-161

ANALYSIS OF PLANT CEREBROSIDES BY C_{18} AND C_6 HPLC

B.D. Whitaker

Hort. Crops Quality Lab, Agricultural Research Service, USDA Beltsville Agricultural Res. Center, Beltsville, MD 20705, USA

HPLC methods using a C_{18} (Cahoon and Lynch 1991) or a C_6 (Whitaker 1996) column have been devised for analysis of plant CBs. The methods were compared with respect to detection limits and separation based on fatty-acid (FA) chain length and sphingoid (SP) hydroxylation, unsaturation, and *cis* versus *trans* isomerization. Results with CBs from apple and melon fruits were as follows: 1) The C_6 method, which uses a gradient of MeCN in water with UV monitoring at 205nm, was about 3-fold more sensitive than the C_{18} method (isocratic MeCN:MeOH, with UV detection at 210nm); 2) C_6 was comparable to C_{18} for separation according to FA chain length and SP hydroxylation and unsaturation; 3) C_{18} gave near baseline resolution of CBs with 8-*cis* and 8-*trans* SP isomers, whereas C_6 gave no or partial separation; 4) Total run times were 60 min on C_{18} versus 40 min on C_6 and CB retention times differed substantially. Overall, the C_6 method is suitable for routine quantification and characterization of plant CBs, particularly when the amount of sample is limiting. However, for separation and identication of CB molecular species, the C_{18} method is clearly superior.

1. Introduction

Cerebrosides (CBs) are glycosphingolipids composed of a hexose, a long-chain aminoalcohol (sphingoid; SP), and an amide-linked long-chain fatty acid (FA). Plant CBs received little note prior to the finding that they can comprise a major fraction of plasma membrane and tonoplast lipids (Lynch 1993). Reports have implicated CBs to play a role in chilling injury (Yoshida *et al.* 1988) and in acclimation to cold and water-deficit stress (Lynch and Steponkus 1987, Norberg and Liljenberg 1991).

CBs from all plant sources examined thus far have included glucose as the sole hexose in ß-1 linkage with the 1-hydroxyl of the SP. Plant CBs also include mostly 2-

hydroxy (2h) FA, saturated or monoenoic, ranging from 14 to 26 carbons in length. The major SPs are typically the dihydroxy bases 8-sphingenine and 4,8-sphingadienine (d18:1 and d18:2), and the trihydroxy bases 4-hydroxysphinganine and 4-hydroxy-8-sphingenine (t18:0 and t18:1). The 4,5 double bond in SPs is generally *trans*, whereas both 8-*cis* and 8-*trans* isomers are common (Lynch 1993).

2. Materials and Methods

Total lipid extracts were separated on a silicic acid column. CBs eluted with the second glycolipid fraction (GL-2) in $Me_2CO:MeOH$, 20:1. GL-2 was separated by TLC on silica gel 60 in $CHCl_3:Me_2CO:MeOH:HOAc:H_2O$, 10:4:2:2:1, and the CB band was scraped and eluted. CB fractions were dissolved in MeOH for HPLC analysis.

A Waters 600MS HPLC, 490MS detector, and Maxima 820 Workstation were used for CB analyses. The HPLC columns used were a C_6 reverse-phase on 5μ Spherisorb, 4.6mm i.d. x 150mm, and a C_{18} reverse-phase on 5μ Spherisorb, 4.6mm i.d. x 250mm. C_6-HPLC separation of CBs was effected by a linear gradient of $MeCN-H_2O$ with UV detection at 205nm. The gradient started at a 63:37 ratio, increased to 80:20 over 35 min, then returned to 63:37 from 35 to 40 min, with a constant flow of 0.9 ml min^{-1}. C_{18}-HPLC separation of CBs was isocratic using MeCN:MeOH, 6:4, at a flow of 1.0 ml min^{-1}. UV detection was at 210nm.

3. Results and Discussion

C_{18} HPLC resolved the major CBs from apple into pairs ("a" and "b" peaks in *Fig.1a*), whereas the C_6 method did not resolve CB1 and the "b" component of CBs 5, 6, and 7 appeared as a shoulder trailing the "a" peak (*Fig.1b*). This same trend was noted for CBs from melon (not shown). IR data indicated that "a" and "b" peaks are 8-*cis* and 8-*trans* SP isomers of the same CB. After HCl methanolysis, GC and GC-MS identified CB1 as d18:2/2h16:0, and CBs 5, 6, and 7 as species with t18:1 amide-linked to 2h22:0, 2h23:0, and 2h24:0, respectively. Thus, the broad separation of CB1 from CB5 is based on SP hydroxylation and unsaturation as well as FA chain length, whereas FA chain length alone is the basis of separation of CBs 5, 6, and 7. Similar pairings of SP and 2hFA moieties were noted in melon, and previously in bell pepper and tomato (Whitaker 1996), CBs. All major CBs eluted in 40 min with C_6 and in 60 min with C_{18} HPLC. Sitosteryl glucoside coeluted with CB1b from the C_{18} column but preceded CB1 by 3-4 min on the C_6 column. Thus, steryl glycosides and CBs could be analyzed at the same time using C_6 HPLC. Because the absorbance maximum of CBs is at 204nm, the C_6 method gave 3-fold greater peak area per unit mass, with a detection limit of about 1 nmol. Perhaps the detection limit of the C_{18} method could be improved by increasing the % MeCN in the mobile phase, allowing detection at 205nm. Also, a 250mm C_6 column might give better resolution of *cis* and *trans* CB isomers.

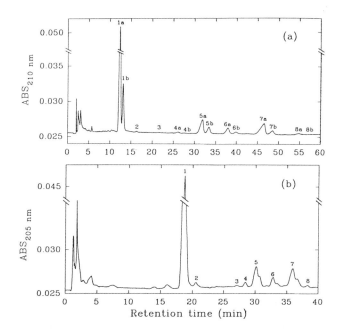

Figure 1. C_{18} (a) and C_6 (b) HPLC chromatograms of apple cortical tissue CBs. Total amounts of CBs injected were 54 nmol for C_{18} and 27 nmol for C_6 HPLC.

4. References

Cahoon, E.B. and Lynch, D.V. (1991) Analysis of glucocerebrosides of rye (*Secale cereale* L. cv Puma) leaf and plasma membrane. Plant Physiol. **95**, 58-68.

Lynch, D.V. (1993) Sphingolipids, in T.S. Moore, Jr. (ed.), *Lipid Metabolism in Plants*, CRC Press, Boca Raton, Florida, pp. 285-308.

Lynch, D.V. and Steponkus, P.L. (1987) Plasma membrane lipid alterations associated with cold acclimation of winter rye seedlings (*Secale cereale* L. cv Puma). Plant Physiol. **83**, 761-767.

Norberg, P. and Liljenberg, C. (1991) Lipids of plasma membranes prepared from oat root cells: effects of induced water-deficit tolerance. Plant Physiol. **96**, 1136-1141.

Yoshida, S., Washio, K., Kenrick, J., and Orr, G. (1988) Thermotropic properties of lipids extracted from plasma membrane and tonoplast isolated from chilling-sensitive mung bean (*Vigna radiata* [L.] Wilczek). Plant Cell Physiol. **29**, 1411-1416.

Whitaker, B.D. (1996) Cerebrosides in mature-green and red-ripe bell pepper and tomato fruits. Phytochemistry **42**, 627-632.

Section 3:

Membranes

THE ROLE OF MEMBRANE LIPIDS IN THE ARRANGEMENT OF COMPLEXES IN PHOTOSYNTHETIC MEMBRANES

PETER J. QUINN
Division of Life Sciences, King's College London, Campden Hill, London W8 7AH, United Kingdom.

1. Introduction

The fluid-mosaic model of Singer and Nicolson (1972) states that the lipids arranged in a bilayer represent the fluid matrix in which the membrane proteins are distributed in a random manner within the bilayer plane. The absence of long-range order is said to result from the fact that there are no long-range interactions intrinsic to the lipid matrix that can influence the distribution of protein molecules. In situations where long-range order does exist it is believed to arise from either a summation of short-range specific interactions amongst the integral membrane proteins to induce the formation of unusually large two-dimensional arrays or that an agent extrinsic to the membrane interacts with integral membrane proteins to produce clustering of these proteins in a limited area of the membrane surface. Specialised membrane junctions and synaptic membranes are cases in point.

With respect to the distribution of intrinsic membrane proteins, the photosynthetic membrane of higher plant chloroplasts provides an excellent example with which to examine the model particularly in regard to the role of membrane lipids in the arrangement of complexes in the membrane. The thylakoid membrane is segregated laterally into grana lamellae, in which the membrane surfaces are closely associated in a stacked arrangement, and stroma lamellae where the membrane is unstacked. The five main oligomeric complexes are largely segregated within the lamellae. Most of the photosystem-II, for example, is located in the grana membranes while photosystem-I and ATP-synthase are predominantly situated in the stroma lamellae. Cytochrome b_6/f, however, is found in both stacked and unstacked membranes. When examined by freeze-fracture electron microscopy, the complexes, believed to be the membrane-associated particles in the respective exoplasmic (EF) and protoplasmic (PF) fracture faces, are randomly distributed in the plane of the membranes thermally quenched from about 20°C.

A consistent feature of the lipid composition of higher plant thylakoid membranes is the high proportion of polyenoic molecular species. It is well known that the unsaturated bonds of these lipids are largely responsible for the fluid character of the membrane bilayer matrix (Quinn et al., 1989). Previous freeze-fracture studies of fatty acid defective mutants of *Arabidopsis thaliana*, JB67 and LK3, which have markedly reduced polyenoic molecular species of lipids showed that the complexes of the thylakoid membrane were aligned in characteristic arrays in the grana stacks (Tsvetkova et al., 1994). The alignment of particles was associated with an apparent decrease in membrane associated particle size distribution and a marked increase in packing density of particles exposed in both the exoplasmic and ectoplasmic fracture faces of the photosynthetic membrane. To check the role of lipid fluidity in membrane-associated particle alignment, the effect of temperature on particle distribution was investigated (Tsvetkova et al., 1995). Chloroplasts were isolated from wild-type *Arabidopsis thaliana* and JB67 and LK3 mutant strains and the fracture planes in the different surfaces of grana and stroma lamellae were examined in samples thermally quenched from temperatures ranging from 4°C to 50°C.

2. Temperature Effects on Membrane Ultrastructure

The distribution of membrane-associated particles in chloroplast thylakoids of wild-type *Arabidopsis thaliana* thermally quenched from 20°C is random on all exposed fracture faces (Fig. 1A). This feature is seen in chloroplasts suspended in Hepes, Mops and phosphate buffers (but not Tricine) at pH values ranging from 5 to 8. By contrast, the membrane-associated particles of appressed membranes of the grana stacks in chloroplasts isolated from JB67 and LK3 mutant strains of *Arabidopsis* had large areas of particles aligned in characteristic arrays. The repeat distances between, and along the rows of particles were approximately 18nm and 23nm, respectively. This is shown for the LK3 mutant in Fig. 1B; the same appearance of the fracture faces was observed in the JB67 mutant thermally quenched from 20°C. The formation of particle arrays was seen in all buffers and was also unaffected by pH between 5 and 8. Arrays were not found in in chloroplasts suspended in destacking media or in stroma lamellae.

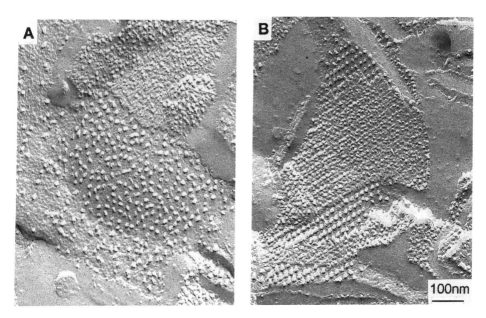

Figure 1. Electron micrographs of freeze-fracture replicas prepared from chloroplasts suspended in 20mM Hepes buffer (pH7.6) isolated from (A) wild-type *Arabidopsis thaliana* and (B) LK3 mutant strain thermally equilibrated and quenched from 20°C

When chloroplasts from wild-type *Arabidopsis* were equilibrated at 4°C and thermally quenched from this temperature areas of both the EFs and PFs fracture faces showed particle arrays similar in dimensions as those observed in the mutant strains thermally quenched from 20°C. This suggests that fluidity of the membrane may be required to distribute the oligomeric complexes of photosystem-II randomly in the plane of the membrane. The role of unsaturated bonds of the lipids forming the bilayer matrix in maintaining fluidity was confirmed by heating chloroplasts isolated from fatty acid defective mutant strains. Fig. 2C shows a region of the stacked membranes from the LK3 mutant thermally quenched from 50°C It can be seen that the arrays of particles have disappeared and the distribution of particles becomes random in the plane of the bilayer.

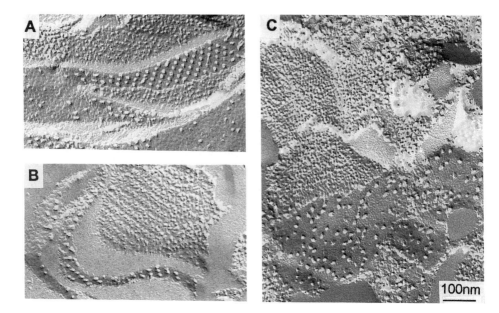

Figure 2. Electron micrographs of freeze-fracture replicas prepared from chloroplasts suspended in 20mM Hepes buffer (pH7.6) isolated from (A,B) wild-type *Arabidopsis thaliana* thermally equilibrated and quenched from 4°C showing areas of EF_s and PF_u respectively and (C) LK3 mutant strain thermally equilibrated and quenched from 50°C.

3. Conclusions

The random distribution of oligomeric complexes in the stacked region of chloroplasts is maintained by thermal motion in the membrane lipid matrix. When fluidity is reduced by lowering the temperature or reducing the level of unsaturation of the membrane lipids the complexes become aligned and pack more densely in the membrane. Areas of the PF_s and EF_s where particle alignment occurs tend to be complimentary suggesting that the photosystem-II core complexes and their associated light-harvesting chlorophyll-protein complexes interact across appressed membranes thereby coordinating the domains of particle alignment.

Acknowledgements: The work was funded by a grant from the Agricultural and Food Research Council. Collaborators in the studies are listed in the references.

References:

Singer, S.J. and Nicolson, G.L. (1972) The fluid mosaic model of the structure of cell membranes. *Science,* **175,** 720-731.

Tsvetkova, N.M., Brain, A.P.R. and Quinn, P.J. (1994) Structural characteristics of thylakoid membranes of *Arabidopsis* mutants deficient in lipid fatty acid desaturation. *Biochim. Biophys. Acta,* **1192,** 263-271.

Tsvetkova, N.M., Brain, A.P.R., Apostolova, E.G., Williams, W.P. and Quinn, P.J. (1995) Factors influencing PS II particle array formation in *Arabidopsis* chloroplasts and the relationship of such arrays to the thermostability of PS II. *Biochim. Biophys. Acta,* **1228,** 201-210.

Quinn, P.J., Joo, F. and Vigh, L. (1989) The role of unsaturated lipids in membrane structure and stability. *Prog. Biophys. Mol. Biol.,* **53,** 71-103.

LIPIDS IN CELL SIGNALLING: A REVIEW

P. MAZLIAK
Laboratoire de Physiologie cellulaire et moléculaire des plantes,
URA CNRS 2135, Université P. et M. Curie Tour 53, 4 Place Jussieu
75252 Paris cedex 05 - France

Many lipid-derived products are now implicated as mediators or second messengers in signal transduction. The idea that membrane lipids are indeed true "signal reservoirs" is presently more and more widely accepted.

1. Inositol trisphosphate as an intracellular second messenger.

Twelve years ago, in a famous review article, Berridge and Irvine [1] collected many data to prove that, in animal cells, phosphatidylinositol-4,5-bisphosphate (PIP_2) was hydrolyzed to diacylglycerol (DAG) and inositol trisphosphate (IP_3) as a part of a signal transduction mechanism controlling a variety of cellular processes. The IP_3 released into the cytoplasm mobilized calcium from internal stores [2], whereas DAG activated protein kinase C. In animal tissues many agonists (acetylcholine, histamine, neuropeptides...) bound to membrane receptors (R_1) which used a GTP-binding protein (G) to activate a phosphodiesterase (inositol-specific, phospholipase C). Most of the enzymes identified in this inositol lipid sequence have been defined by molecular cloning in animal tissues and some in plants as well [3, 4]. However no enzyme equivalent to a protein kinase C has been so far identified in plants [5] ; although the functions of G proteins in plants are yet to be determined, molecular cloning has clearly demonstrated that plants, as animals, have both heterotrimeric and small G proteins [6].

There has been numerous efforts these last years to demonstrate the functioning of a signal system involving polyphosphoinositides in plant cells but, unfortunately, they were not entirely conclusive. In 1985 Boss and Massel were the first researchers to state in a paper : polyphosphoinositides are present in plant tissue culture cells [7]. Since that time nearly all the intermediates of the cycle have been evidenced in many different plant tissues [5]. However, it must be recognized that PIP_2, a key intermediate in the cycle, is very often apparently absent from plant tissues or present in so tiny amounts that it is difficult to follow the quantitative variations of this compound. Instead, lyso compounds (lyso PI and lyso PIP) are often present in plant tissues and have been proposed to play some role, directly, as signalling metabolites [8].

In a recent review Coté and Crain [5] have pointed to three phenomena which they thought particularly convenient to study the physiological role of phosphoinositides in plants. These were : 1) the turgor changes in stomatal guard cells, particularly after a treatment with abscisic acid which provokes the closure of stomata, 2) the osmotic stress in *Dunaliella salina* or carrot cell cultures and 3) the deflagellation of *Chlamydomonas reinhardtii* after exposure of the alga to acid pH, elevated temperature or ethanol. Nevertheless the rates of apparition of inositolphosphates after stimulation have been found very different in various plant tissues (from 30 sec to 30 min).

Boss *et al* [9] have proposed a role for inositol phospholipids in the regulation of cytoskeletal structure and Pia Harryson [10] has underlined that plant cells always contain high levels of inositol phosphates which are metabolized into cell-wall polysaccharides ; these compounds may interfere with signal transduction mechanisms.

2. Phosphatidylcholine hydrolysis

In animal cells accelerated turnover of phosphatidylcholine (PC) was first described as a wide spread cellular response to phorbol diester treatment and related agonists [11].

The agonists that hydrolyze PC also promote PI-bisphosphate hydrolysis in their target cells. Thus a typical response pattern is a biphasic increase in DAG, with an initial, rapid, transient peak followed by a more slowly developing, but prolonged accumulation. The first peak is due to PIP_2 hydrolysis ; the second phase is due to PC hydrolysis.

However, the major phorbol diester-activated enzymatic event is the phospholipase D (PLD)-catalyzed cleavage of PC, giving phosphatidic acid (PA) and bis-phosphatidic acid (bis PA), thus provoking cell proliferation. Direct evidence for phospholipase D signalling in plants was recently provided [12].

3. Fatty acid signalling in plants

Phospholipases A_1 and A_2 release free fatty acids which are the precursors of a myriad of oxygenated compounds derived from arachidonic acid in animal cells and from linoleic or linolenic acids in plant cells (jasmonic acid or traumatic acid for examples) [13].

4. Sphingolipid signalling

Ceramide and sphingosine-1-phosphate are formed as a result of the hydrolysis of membrane sphingolipids. Recent work has often emphasized the antiproliferative and differentiation-inducing effects of ceramides and sphingolipid metabolites have been also implicated in driving cell death by apoptosis. No striking result on the regulatory role of sphingolipid metabolites in plant cells has been published so far.

5. Conclusion : Interconnections of lipid second messengers.

The study of cell signalling *via* lipid molecules has been an exciting area of research in recent years. Two main lines of investigation are appearing presently, in animal tissues : 1°) there seems to be real interconnections between all cellular systems producing lipid second messengers. 2°) Another trend of thought is that phosphoinositide cycling and phosphatidylcholine degradation generate lipid second messengers that activate cell growth, while sphingolipid catabolic products would be linked to programmed cell death (apoptosis).

It is much too early to decide whether this beautiful scheme can be applied to the life of plant cells.

References

1. Berridge, M.J. and Irvine R.F. Inositol trisphosphate, a novel second messenger in cellular signal transduction, *Nature* **312** (1984), 315-321.
2. Berridge, M.J. Inositol triphosphate and calcium signalling, *Nature* **361** (1993), 315-325.
3. Drøbak, B.K. The plant phosphoinositide system, *Biochem. J.* **288** (1992), 697-712.
4. Yamamoto, Y.T., Conkling, M.A., Sussex, I.M. and Irish, V.F. An Arabidopsis cDNA related to animal phosphoinositide-specific phospholipase C genes, *Plant Physiology* **107** (1995), 1029-1030.
5. Cote, G.G. and Crain, R.C. Biochemistry of phosphoinositides, *Ann. Rev. Plant Physiol. Plant Mol. Biol*, **44** (1993), 333-356.
6. Ma, H. GTP-binding proteins in plants ; new members of an old family, *Plant molecular biology* **26** (1994), 1611-1636.
7. Boss, W.F. and Massel, M.O. Polyphosphoinositides are present in plant tissue culture cells, *Biochem. Biophys. Res. Commun* **132** (1985), 1018-1023.
8. Wheeler, J.J. and Boss, W.F. Inositol lysophospholipids, in D.J. Morre, W.F. Boss and F.A. Loewus (eds.), *Inositol metabolism in plants*, Wiley-Liss, New York, 1990, pp. 163-172.
9. Boss, W.F., Yang, W., Tan, Z. and Cho, M. Regulation of phosphatidyl-inositol-4-kinase by protein phosphorylation : a plasma membrane-cytoskeletal connection, in P. Mazliak and J.C. Kader (eds.), *Plant Lipid Metabolism*, Kluwer Academic Publishers, Dordrecht, 1995, pp. 219-223.
10. Harryson, P.: *Lipid transfer and signal transduction in plants : a study of inositol phospho-lipids*. Ph. D. Thesis, University of Göteborg, Sweden, 1995.
11. Exton, J.H. Phosphatidylbreakdown and signal transduction, *Bioch. Biophys. Acta* **1212** (1994), 26-42.
12. Munnik, T., Arisz, S.A., De Vrije, T. and Musgrave, A. G protein activation stimulates phospholipase D signaling in plants, *Plant Cell* **7** (1995), 2197-2210.
13. Farmer, E.E. Fatty acid signalling in plants and their associated microorganisms, *Plant molecular Biology* **26** (1994), 1423-1437.
14. Michell, R.H. Wakelam, M.J.O. Sphingolipid signalling, *Current Biology* **4** (1994), 370-373.

RELEASE OF LIPID CATABOLITES FROM MEMBRANES BY BLEBBING OF LIPID-PROTEIN PARTICLES

J.E. THOMPSON, C.D. FROESE, Y. HONG, K.A. HUDAK AND M. D. SMITH
Department of Biology, University of Waterloo, Waterloo, Ontario, Canada N2L 3G1

1. Introduction

Pulse-chase radiolabelling experiments have provided clear evidence for turnover of membrane lipids. Membrane lipid catabolites, however, have a propensity to destabilize bilayer structure, and this confounds an otherwise balanced process. Free fatty acids, for example, act as detergents and also phase-separate within bilayers [1]. It would seem likely, therefore, that lipid catabolites are removed from membranes such that they do not accumulate and compromise the structural integrity of the bilayer. Herein, we describe evidence indicating that lipid catabolites are voided from membranes by blebbing of lipid-protein particles that appear to be generically related to oil bodies, and propose that blebbing is an integral feature of membrane turnover.

2. Lipid-Protein Particles

Osmiophilic lipid-protein particles enriched in membrane lipid catabolites have been isolated from the cytosol of carnation petals and of cotyledons (*Phaseolus vulgaris*) by ultrafiltration or flotation centrifugation [2,3]. The particles are spherical in nature and range from 150 to 300 nm in diameter [Fig. 1]. Osmiophilic particles of similar size and shape are also detectable in the cell cytoplasm [3]. The particles contain phospholipid, but are enriched (10- to 100- fold) by comparison with membranes in free fatty acids, triacylglycerols, and steryl and wax esters [Fig. 2].

Several observations support the contention that cytosolic lipid-protein particles originate from membranes. In particular, they contain phospholipid with the same fatty acids, albeit in different proportions, found in phospholipid of corresponding microsomal membranes [3]. In addition, lipid-protein particles with essentially similar properties can be generated *in vitro* from isolated microsomal membranes under conditions in which phospholipid catabolism has been activated by the addition of Ca^{2+} [2,3]. The finding that the particles contain high concentrations of free fatty acids and steryl and wax esters suggests that they are formed at specific sites in membranes. Indeed, it has been proposed that catabolites and metabolites of membrane lipids phase-separate within the plane of the bilayer, forming sites from which lipid-protein particles are blebbed [2].

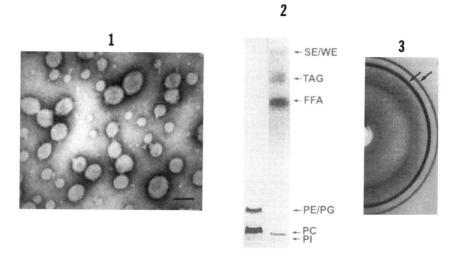

Figure 1. Electron micrograph of cytosolic lipid-protein particles isolated from carnation petals and stained with uranyl acetate. Bar = 100 nm.

Figure 2. Thin layer chromatogram of total lipids extracted from microsomal membranes (lane 1) and cytosolic lipid-protein particles (lane 2) isolated from cotyledon tissue of *Phaseolus vulgaris*. PC, phosphatidylcholine; PE, phosphatidylethanolamine; PG, phosphatidylglycerol; PI, phsophatylinositol; FFA, free fatty acids; TAG, triacylglycerol; SE/WE, steryl and wax esters.

Figure 3. Wide angle X-ray diffraction pattern recorded at 25°C from a hydrated mixture of free fatty acids and steryl and wax esters isolated from cytosolic lipid-protein particles of *Phaseolus vulgaris* cotyledon tissue. Arrows denote X-ray reflections derived from gel phase packing.

Cytosolic lipid-protein particles appear to have properties in common with oil bodies of seeds, which are formed by blebbing from the endoplasmic reticulum and are enriched in triacylglycerol [4]. The particles resemble oil bodies in that they can be isolated by flotation, appear to be circumscribed by a monolayer of phospholipid and contain triacylglycerol, but are distinguishable from oil bodies in that they contain less triacylglycerol. For example, phospholipid and diacylglycerol constitute ~60% of the total fatty acids in cytosolic particles isolated from cotyledon tissue of freshly hydrated *Phaseolus vulgaris* seeds and only 16% of the total fatty acids in oil bodies from the same tissue, whereas triacylglycerol constitutes 82% of the total fatty acids in the oil bodies and only 33% of the total in the cytosolic particles [unpublished data]. As well, cytosolic lipid-protein particles contain a protease-protected 17 kDa protein that resembles oil body oleosin (unpublished data). Thus cytosolic lipid-protein particles may be generically related to oil bodies but, rather than storing triacylglycerol, appear to serve as a vehicle for removal of lipid catabolites and metabolites from membranes that would otherwise destsabilize the bilayer.

Lipid-protein particles analagous to those present in the cell cytosol have also been isolated from chloroplasts by ultrafiltration or flotation centrifugation of thylakoid-free stroma. These particles range from 150 to 300 nm in diameter, are uniformly osmiophilic and can be generated *in vitro* by illumination of isolated thylakoids [5]. Their lipid composition is analagous to that of cytosolic lipid-protein particles in that they are enriched in free fatty acids and in steryl and wax esters. However, inasmuch as they originate from thylakoids they contain galactolipids rather than phospholipids, and are also enriched in catabolites of photosynthetic thylakoid proteins [5]. Thus blebbing of lipid-protein particles from thylakoids appears to be a means of removing destabilizing lipid and protein catabolites.

3. Impairment of Blebbing with Advancing Senescence

X-ray diffraction studies have provided evidence for lipid phase separations leading to coexisting liquid-crystalline and gel phase domains in membranes of senescing tissue [2]. These separated phases render membranes leaky because of packing imperfections at the phase boundaries, and contribute to loss of membrane function during senescence. Of particular interest is the finding that these phase separations appear to be attributable to impairment of blebbing from senescing membranes and a consequent accumulation of lipid catabolites and metabolites in the membrane bilayers. Indeed, blebbing of lipid-protein particles in senescing tissue is only ~20% of that in corresponding young tissue [unpublished data]. As well, the free fatty acids and steryl and wax esters enriched in lipid-protein particles are capable of forming gel phase domains [Fig. 3], which are identical to the gel phase domains of lipid present in senescing membranes [2]. It would appear, therefore, that gel phase-forming products of lipid catabolism and metabolism are normally voided from membranes by blebbing of lipid-protein particles. With advancing senescence, however, blebbing becomes impaired and these products accumulate and phase-separate within the bilayer, thus contributing to membrane dysfunction.

4. References

1. Welti, R., and Glaser, M.: Lipid domains in model and biological membranes, *Chem. Phys. Lipids* 73 (1994), 121-137.
2. Yao, K., Paliyath, G., and Thompson, J.E.: Nonsedimentable microvesicles from senescing bean contain gel phase-forming phospholipid deegradation products, *Pl. Physiol.* 97 (1991), 502-508.
3. Hudak, K.A., and Thompson, J.E.: Flotation of lipid-protein particles containing triacylglycerol and phospholipid from the cytosol of carnation petals, *Physiol. Plant.* (in press).
4. Huang, A.H.C.: Oil bodies and oleosins in seeds, *Ann. Rev. Pl. Physiol. Pl. Mol. Biol.* 43 (1992), 177-200.
5. Ghosh, S., Hudak, K.A., Dumbroff, E.B., and Thompson, J.E.: Release of photosynthetic protein catabolites by blebbing from thylakoids, *Pl. Physiol.* 106 (1994) 1547-1553.

CHLOROPLAST MEMBRANE LIPIDS AS POSSIBLE PRIMARY TARGETS FOR NITRIC OXIDE – MEDIATED INDUCTION OF CHLOROPLAST FLUORESCENCE IN *PISUM SATIVUM* (*ARGENTUM* MUTANT)

Y.Y. LESHEM, E. HARAMATY, Z. MALIK AND Y. SOFER
Department of Life Sciences, Bar-Ilan University
Ramat Gan 52900, Israel

1. Introduction

Recent research suggests that endogenous NO, a free radical gas, has growth regulatory action in plants as well as in mammals [3]. While low concentrations inhibit ethylene evolution, little is known of the mode of action of its inhibitory effects at high concentrations as would be obtained under prolonged exposure to stress. The latter aspect was addressed by use of advanced fluorescence techniques, surface tension tensiometry and LOX assay, thereby monitoring possible NO effects on chloroplast membrane lipids. A further aim was to ascertain whether NO or its protonated peroxynitrite adduct, HOONO, formed by the action of superoxide (O_2^-) on NO [7], is the biologically active species.

2. Methods

Pisum sativum (*argentum* mutant) leaves, from which lower epidermis leaf strips can be easily detached, were used throughout. Method of treatment of these or of plantlets was as detailed elsewhere [4]. In experiments with epidermal strips, NO was generated chemically [3]. In the others, NO was obtained by employment of S-nitroso-N-acetylpenicilamine (SNAP) which in aqueous solution releases NO [2].

Initial fluorescence scans on epidermal strip guard cell chloroplasts in a single stomate were made using the SpectraCube 1000 system [5] after which NO treated and untreated strips were photographed (Fuji ASA/100 - color film) by an Olympus-Vanox AH-3 Fluorescent Microscope - excitation wavelength – 410-490 nm, emission – 669 nm, exposure time – 18 sec. Developed color slides were subsequently scanned with a computer-linked adapted Li-Cor Spectroradiometer Model 1800, which enabled quantitative composite fluorescence readings at 669 nm.

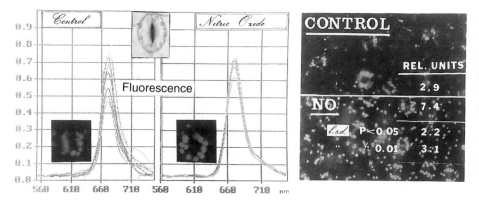

Figure 1 - left — Chlorophyll fluorescence of chloroplasts of a single stomate of NO treated and non-treated pea foliage, as measured by the SpectraCube 1000 system. Each curve is the red fluorescence spectrum of a chloroplast; *right* — Large scale sample fluorescence at 669 nm of pea foliage epidermal strips as assessed by a Li-Cor Spectroradiometer. 9 replicate means each of ± 135 chloroplasts.

Wilhelmy tensiometry was performed on MGDG and PLPC monolayers as detailed elsewhere [1]. Monolayer molarity was well below the CMC which ordinarily would preclude MGDG monolayer orientation. Treatments were as seen in Figure 2. To prevent $O_2^- \rightarrow O_2$ conversion, the former was applied in a crown-ether caged form of KO_2^- [1]. Lipoxygenase (LOX) was assayed on 14-day-old plants which were sprayed twice weekly with 10^{-3} or 10^{-5} M SNAP in 0.01 M phosphate buffer pH 6.5, containing 10^{-5} M $CaCl_2$ and 0.01% Tween 20. Controls were sprayed with buffer alone. LOX assay was that of Pinsky et al. [6].

3. Results and Discussion

NO induces a highly significant increase of chloroplast fluorescence [Figure 1]. This implies an impairment of photosynthetic efficiency at the PS II site as may occur during stress [9]. That this may be the result of photosynthetic membrane rigidity increase is borne out since NO markedly increased surface tension of MGDG monolayer, while only marginally that of PLPC [Figure 2].
This effect is likewise evoked by O_2^-; however the combination of NO and O_2^- is not additive. This

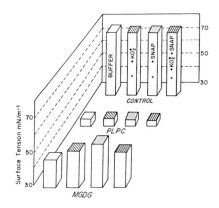

Figure 2. Effect of NO, O_2^- (and their combination to form HOONO) on surface tension of MGDG and PLPC monolayers. Four replicate means. See text for details.

TABLE 1. Lipoxygenase activity in pea foliage treated with SNAP

SNAP concentration	LOX activity $\mu l\ O_2$/min/mg protein
Untreated control	10.20
10^{-4} M	8.11
10^{-3} M	4.58
l.s.d. $p > 0.05$	1.90

may be interpreted that NO and not the peroxynitrite is the active factor. That nature of NO effect on MGDG and not on PLPC remains to be elucidated.

Overall results suggest that NO acts by inducing biophysical parameter changes in chloroplast membrane lipid. Moreover, since NO is a relatively weak free radical, it probably does not bind co-valently to membrane constituents, but may loosely associate with the π bond in C=C sections which are comparatively prolific in MGDG fatty acyl side chains and may thus reduce the energy of *cis-trans* isomerisation of the C=C bond [8]. If so, this implies that NO should hinder LOX activity, a contention experimentally borne out [Table 1].

4. References

1. Bamberger, E.S., Alter, M., Landau, E.M., and Leshem, Y.Y. (1989) Biophysical effects of $\dot{O}_2$ on surface parameters of a model membrane, in O. Hayaishi, E. Niki, M. Kondo, and T. Yoshika (eds.), *Medical and Biochemical Aspects of Free Radicals*, Elsevier Publishers, Amsterdam, pp. 113-116.
2. Hery, P.J., Horowitz, J.D., and Louis, W.J. (1989) Nitroglycerin-induced tolerance affects multiple sites in the organic nitrate bioconversion cascade, *J. Pharmacol. Exp. Ther.* **248**, 762-768.
3. Leshem, Y.Y. and Haramaty, E. (1996) The characterization and contrasting effects of the nitric oxide free radical in vegetative stress and senescence of *Pisum sativum* Linn. foliage, *J. Plant Physiol.* **148**, 258-263.
4. Leshem, Y.Y., Rapaport, D., Frimer, A.A., Strul, G., Asaf, U., and Felner, I. (1993) Buckministerfullerene (C-60 carbon allotrope) inhibits ethylene evolution from ACC-treated shoots of pea, broad bean and flowers of carnation, *Ann. Bot.* **72**, 457-461.
5. Malik, Z., Cabib, D., Buchwald, R.N., Garmi, Y., and Soenkeson, D. (1994) A novel spectral imaging system combining spectroscopy with imaging, *Prog. Biomed. Optics* **2329**, 180-184.
6. Pinsky, A., Grossman, S., and Trop, M. (1971) Lipoxygenase content and anti-oxidant activity of some fruits and vegetables, *J. Food Sci.* **36**, 571-572.
7. Pryor, W. and Squadrito, G.L. (1995) The chemistry of peroxynitrite: a product from the reaction of NO with superoxide, *Am. J. Physiol.* **268**, L699-722.
8. Sprecher, M. (1996) personal communication.
9. Stober, F. and Lichtenthaler, H.K. (1993) Studies on the localization and spectral characteristics of the fluorescence emission of differently pigmented wheat leaves, *Bot. Acta* **106**, 365-370.

Acknowledgements: We thank Prof. M. Sprecher of the Department of Chemistry, Dr. Leonid Roitman and David Iluz of Bar-Ilan University, and Yehoram Leshem of the Faculty of Agriculture, Rehovoth, for their helpful aid and advice.

GLYCOSYLPHOSPHATIDYLINOSITOL-ANCHORED PROTEINS IN PLANTS

N. Morita[‡], H. Nakazato[†], H. Okuyama[†], Y. Kim[◊], and G. A. Thompson, Jr.[◊]
[‡]Department of Botany, Faculty of Science, Hokkaido University, Sapporo 060, Japan,
[†]Laboratory of Environmental Molecular Biology, Graduate School of Environmental Earth Science, Hokkaido University, Sapporo 060, Japan, and [◊] Department of Botany, University of Texas, Austin, Texas 78713

Introduction

Proteins containing glycosylphosphatidylinositol(GPI) anchors (Fig. 1) are widespread in animal cells [2], where they are highly localized on the outer face of the plasma membrane. There are over 150 different GPI-anchored proteins known, with alkaline phosphatase, acetylcholinesterase, and 5'-nucleotidase being noteworthy as extensively studied examples. Recently their occurrence has been reported in archaebacteria [3] and in the green alga *Chlorella saccharophila* [4]. In an effort to determine whether higher plants also contain GPI-anchored proteins, we examined the aquatic duckweed *Spirodela oligorrhiza* because it has been reported [5] to contain alkaline phosphatase, a protein typically GPI-anchored in animals [2]. Our work describing a GPI-anchored alkaline phosphatase in *S.oligorrhiza* [6] represents the first finding of a GPI-anchored protein in higher plants.

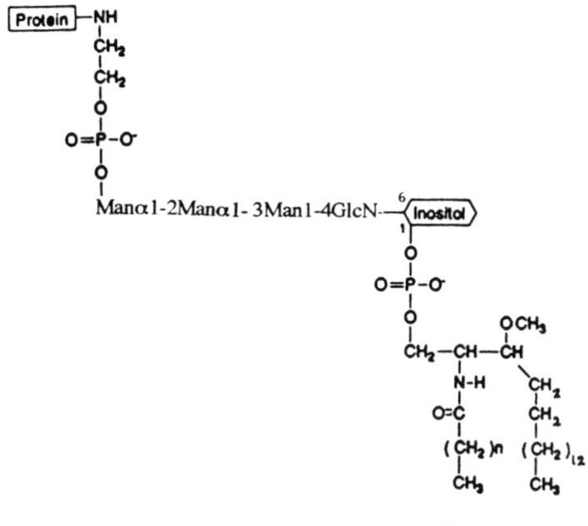

Fig. 1. Basic structure of a GPI anchor, exemplified by that found in *Tetrahymena mimbres* [1].

Results and Discussion

When *S. oligorrhiza* plants were placed in a phosphate-deficient medium (see ref 6 for experimental details), alkaline phosphatase activity increased within 20 days from a barely detectable level to a value some 13 fold higher. When the phosphate-deficient plants were incubated with [^{3}H]ethanolamine, radioactivity was incorporated into a protein migrating very slowly on non-denaturing polyacrylamide gels and staining positive for alkaline phosphatase. On denaturing SDS-PAGE gels enzymatic activity was lost, but the radioactive band now migrated with an apparent molecular weight of 50 kDa (Fig. 2).

Purification of a protein having high alkaline phosphatase activity was accomplished by the use of a CM-Toyopearl ion exchange column followed by a Protein-Pak SP 125 HR column and a Resource Q column. By using gel electrophoresis followed by Western blotting with polyclonal antibodies, this protein was shown to be the 50 kDa protein that was radiolabeled in Fig. 2.

Phosphate-deficient plants were also labeled with [^{3}H]myristic acid, and the radiolabeling pattern observed on SDS-PAGE gels again indicated a strongly radioactive protein migrating with an apparent molecular weight of 50 kDa (Fig 2).

Because of their hydrophobic anchor, GPI-anchored proteins can usually be extracted from tissues by detergents such as Triton X-100 or Triton X-114. However, partition of *S. oligorrhiza* between TX-114 and water resulted in more than 90% of the alkaline phosphatase activity being recovered in the aqueous phase. The properties of the small amount of activity associated with the detergent phase were distinctly different from that in the aqueous phase, suggesting that this less abundant form was GPI-anchored. For example, the detergent form, designated AP_{TX}, migrated slightly slower than the aqueous form, AP_{Aq}, on 5% acrylamide/0.1% Triton X-100 electrophoretic gels and slightly faster than AP_{Aq} on low temperature, non-denaturing LDS-PAGE gels [6]. This behavior is identical to that observed using bovine intestinal mucosa alkaline phosphatase, known to be a mixture of GPI-anchored and non-anchored forms [7].

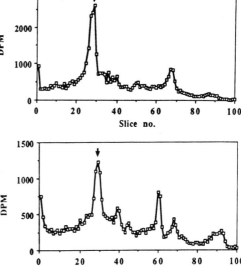

Fig. 2. Distribution of radioactivity in 1 mm slices of a denaturing LDS-PAGE gel containing proteins from [^{3}H]ethanolamine-labeled *Spirodela* (upper scan) and [^{3}H]myristic acid-labeled *Spirodela* (lower scan). Arrow indicates the 50 kDa band.

The unexpectedly large proportion of the total tissue alkaline phosphatase activity that appeared to be GPI anchor-free was puzzling and raised the possibility that *S. oligorriza* contains a very active phospholipase C or D capable of cleaving the phosphatidylinositol anchor moiety during extraction of the plant material. However, a number of control experiments, such as maintaining the tissue strictly at liquid nitrogen temperature while powdering and then adding it in small portions to hot SDS, indicated that the GPI anchor-free alkaline phosphatase was not an artifact of extraction [6]. These controls along with findings described below make it more likely that alkaline phosphatase is synthesized in the *S. oligorrhiza* endoplasmic reticulum as a GPI-anchored protein, targeted to the plasma membrane, and then gradually cleaved in vivo to a soluble form retaining only part of the GPI anchor.

The water-soluble, low phosphate-inducible alkaline phosphatase, because it was present in much larger quantities, was more amenable to detailed examination. It has been purified to homogeniety and shown to exist as a ~100 kDa dimer [H. Nakazato, T. Okamoto, K. Ishikawa, and H. Okuyama, submitted for publication]. Preliminary tests have shown that this phosphatase still retains covalently bound palmitate and stearate that can be released by mild alkaline hydrolysis, suggesting that the fatty acids are bound to the inositol ring.

Characterization of the limited amount of AP_{TX}, the putative GPI-anchored form of *S. oligorrhiza* alkaline phosphatase, showed that it resisted cleavage by a bacterial phosphatidylinositol-specific phospholipase C. A radioactive polar lipid was recovered after nitrous acid deamination of [^{3}H]myristate-labeled AP_{TX}. Hydrolysis of this deamination product or of the intact AP_{TX} in methanolic HCl yielded two radioactive products: fatty acid methyl esters and an apparent long chain base migrating on silica gel H TLC (solvent system: chloroform:methanol:2N NH$_4$OH, 40:10:1) at a position slightly above authentic sphingosine. In this and other TLC solvent systems the putative long chain base closely resembled the unusual 3-O-methylsphinganine recently shown to be a component of the *Tetrahymena mimbres* GPI-anchored proteins. Present data suggest a structure of the intact GPI anchor resembling that shown in Fig. 1, with the probable addition of a long chain fatty acid to one hydroxyl group of the inositol ring. The presence of a fatty acyl group on the inositol moiety ordinarily protects the anchor against hydrolysis by phospholipase C [2].

There are several hypotheses regarding the physiological function of protein GPI anchors. One is that attachment of a protein to the outside of the plasma membrane in this way provides a ready opportunity for a phospholipase-mediated release of the protein into the intracellular environment. Our work indicates that *S. oligorrhiza* cells may indeed release a sizeable proportion of their alkaline phosphatase into the cell wall, where it is well situated to hydrolyze exogenous organic phosphates to the more readily utilizable inorganic phosphate.

A more recently proposed function of the GPI anchor is in targeting the proteins to which they are linked to a final destination on the plasma membrane [8]. According to this hypothesis, which has considerable experimental support [9,10], newly synthesized GPI-anchored proteins gradually become associated with certain lipids, particularly sphingolipids and sterols, as they are processed in the Golgi apparatus and move outwards from the ER. These molecular associations, which are so strong as to resist TX-100 solubilization, are thought to be instrumental in targeting the entire protein/lipid domains to special vesicles destined for the plasma membrane.

Improved diagnostic tests, including several mentioned above, now make it much easier to detect the presence of a GPI-anchored protein in tissues. Because of their recognized role in nutrient uptake, cell-cell recognition, and metabolism in the extracellular compartment, an expanded search for GPI-anchored proteins in plants is in order.

References

1. Ko, Y.-G., Hung, C.-Y., and Thompson, G. A., Jr. (1995) Temperature regulation of the *Tetrahymena mimbres* glycosylphosphatidylinositol-anchored protein lipid composition, Biochem. J. 307, 115-121.
2. Low, M. G. (1989) The glycosyl-phosphatidylinositol anchor of membrane proteins, Biochim. Biophys. Acta 988, 427-454.
3. Kobayashi, T., Nishizaki, R., Maeda, T., and Ikezawa, H. (1996). Existence of GPI-anchored proteins in the archaebacterium *Sulfolobus acidocaldarius*, Abstracts of the Japanese Conference on the Biochemistry of Lipids, 38, 179-182.
4. Stöhr, C., Schuler, F. and Tischner, R. (1995) Glycosyl-phosphatidylinositol-anchored proteins exist in the plasma membrane of *Chlorella saccharophila* (Krüger) Nadson: Plasma-membrane-bound nitrate reductase as an example, Planta 196, 284-287.
5. Bieleski, R. L., and Johnson, P. N. (1972) The external location of phosphatase activity in phosphorus-deficient *Spirodela oligorrhiza*, Aust. J. Biol Sci 25, 707-720.
6. Morita, N., Nakazato, H., Okuyama, H., Kim, Y., and Thompson, G. A., Jr. (1996) Evidence for a glycosylinositolphospholipid-anchored alkaline phosphatase in the aquatic plant *Spirodela oligorrhiza*, Biochim. Biophys. Acta 1290, 53-62.
7. Ko, Y.-G., and Thompson, G. A., Jr. (1995) Purification of glycosylphosphatidylinositol-anchored proteins by modified Triton X-114 partitioning and preparative gel electrophoresis, Anal. Biochem. 224, 166-172.
8. Brown, D. A., and Rose, J. K. (1992) Sorting of GPI-anchored proteins to glycolipid-enriched membrane subdomains during transport to the apical cell surface, Cell 68, 533-544.
9. Brown, D. A. (1992) Interactions between GPI-anchored proteins and membrane lipids, Trends Cell Biol. 2, 338-343.
10. Futerman, A. H. (1995) Inhibition of sphingolipid synthesis: effects on glycosphingolipid-GPI-anchored protein microdomains, Trends Cell Biol. 5, 377-380.

EFFECTS OF EXOGENOUS FREE OLEIC ACID ON MEMBRANE FATTY ACID COMPOSITION AND PHYSIOLOGY OF *LEMNA MINOR* FRONDS

Gilles GRENIER, Alain MUANAMPUTU ZIMAFUALA and Jean-Pierre MARIER

Faculté des sciences, Département de biologie, Université de Sherbrooke, Sherbrooke, Québec, Canada, J1K 2R1.

INTRODUCTION

We report here the effects of exogenous free oleic acid, added to the growth medium of *Lemna minor*, on the membrane fatty acid composition, the chlorophyll content and the photosynthesis rate of this aquatic plant.

MATERIALS AND METHODS

After 5 d of growth, 0, 25, 50, 100, 200 or 300 μg mL^{-1} of free oleic acid was added to the culture medium of *Lemna minor* and the growth was continued for a subsequent 7 d. The temperature of the growth chamber was 20 °C (day) and 16 °C (night) with a photoperiod of 16 h and a photon flux density of 180 μmol m^{-2}s^{-1}. After washing the fronds copiously with distilled water, the plants were fixed in boiling water and total lipids were extracted according to the method of Bligh and Dyer (1959). Lipids were separated by TLC and the fatty acid methyl esters were analyzed by GLC.

Chlorophyll content was obtained according to Arnon (1949). O_2 released by fronds of *L. minor* was measured at a photon flux density of 360 μmol m^{-2}s^{-1} and 25 °C in a Gibson Differential Respirometer.

RESULTS AND DISCUSSION

Exogenous free 18:1 uptake by the cells of *L. minor* was high (Tab. 1). In function of increasing concentrations of exogenous 18:1, amounts of 18:2, 18:3 and 20:1 increased substantially as compared to control. Exogenous 18:1 induced at all concentrations an increase in NL, DL, DDG, DGG and SL content of *L. minor* fronds as compared to control (Tab. 2). Most exogenous 18:1 was incorporated into NL (mainly TAG), but substantial exogenous 18:1 was also incorporated into PL, DDG, DGG and SL.

These results are in accordance with data obtained by Williams *et al.* (1990) from *A. nidulans* and Grenier *et al.* (1991) from *C. reinhardtii*. The stimulation of the elongation of 18:1 to 20:1, by adding free 18:1 to culture medium, is reported, to our knowledge, for the first time. *L. minor* cells seem to regulate their membrane lipid composition quite efficiently by rapidly incorporating excess fatty acids into NL, and they attempt to maintain their fatty acid composition by increasing *de novo* synthesis of fatty acids in proportion to incorporated 18:1.

The decrease observed in chlorophyll a and chlorophyll b content (Fig. 1) and the evolution of O_2 (Fig. 2) suggests that the uptake of large quantities of 18:1 modifies the chloroplast ultrastructure and perturbs the functional organization of the photosynthetic membranes of *L. minor* fronds.

REFERENCES

Arnon, D.I. (1949) Copper enzymes in chloroplasts. Polyphenol exidase in *Beta Vulgaris*, *Plant Physiol.* **24**, 1-15.

Bligh, E.G. and Dyer, W.J. (1959) A rapid method for total lipid extraction and purification, *Can. J. Biochem. Physiol.* **37**, 911-917.

Grenier, G., Guyon, D., Roche, O., Dubertoret, G., and Trémolières, A. (1991) Modification of the membrane fatty acid composition of *Chlamydromonas reinhardtii* cultured in the presence of liposomes, *Plant Physiol. Biochem.* **29**, 429-440.

Williams, J.P., Moissan, E., Mitchell, I., and Khan, M.U. (1990) The manipulation of the fatty acid composition of glycerolipids in cyanobacteria using exogenous fatty acid, *Plant Cell Physiol.* **31**, 495-503.

ABBREVIATIONS

DDG, diacyldigalactosylglycerol; DGG, diacylgalactosylglycerol; NL, total neutral lipids, PL, total phospholipids; SL, sulfoquinovosyldiacylglycerol; TAG, triacylglycerol; 16:0, palmitic acid; 16:1-t, Δ_3-trans-hexadecenoic acid; 18:0, stearic acid; 18:1, Δ_9-cis-octadecenoic acid, oleic acid; 18:2, $\Delta_{9,12}$-cis-octadecadienoic acid, linoleic acid; 18:3, $\Delta_{9,12,15}$-cis-octadecatrienoic acid, α-linolenic acid; 20:1, Δ_{11}-cis-eicosenoic acid, gadoleic acid.

ACKNOWLEDGEMENT

This research was supported by a Natural Sciences and Engineering Research Council of Canada grant.

Table 1. Total fatty acid content and composition of *L. minor* fronds grown for 7 days with different concentrations of free 18:1.

Treatment (μg 18:1 mL^{-1})	Fatty acid content (μg g F W $^{-1}$)	Fatty acid composition (mole %)						
		16:0	16:1-t	18:0	18:1	18:2	18:3	20:1
0	3 864	20	2	1	9	16	49	3
25	6 196	15	2	1	25	15	39	3
50	10 361	9	1	1	47	13	27	2
100	21 445	4	1	tr	71	10	13	1
200	44 097	2	tr	tr	82	7	6	3
300	96 329	1	tr	tr	90	4	3	2

tr= trace

Table 2. NL, PL, DDG, DGG and SL content of *L.minor* fronds grown for 7 days with different concentrations of free 18:1.

Treatment (µg 18:1 mL^{-1})	Lipid class	Fatty acid content (µg g F W^{-1})	18:1 composition (mole %)
0	NL	208	51
	PL	1 045	4
	DDG	299	2
	DGG	588	1
	SL	73	10
100	NL	8 694	96
	PL	1 407	8
	DDG	427	8
	DGG	928	3
	SL	154	17
300	NL	58 579	98
	PL	1 631	15
	DDG	379	12
	DGG	997	12
	SL	215	22

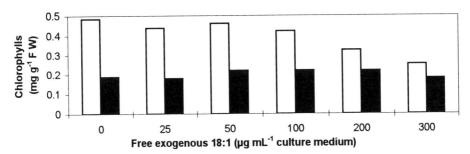

Figure 1. Chlorophyll a and chlorophyll b content of *L. minor* fronds after 7 days of culture in the presence of exogenous 18:1. White, chlorophyll a; Black, chlorophyll b. Means, n=8 for each treatment.

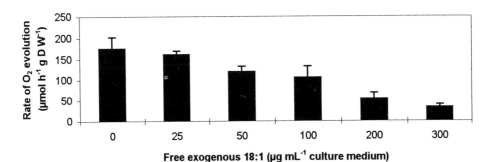

Figure 2. Rate of O_2 evolution of *L. minor* fronds after 7 days of culture in presence of exogenous 18:1. Means ± SD, n = 6 for each treatment.

A NEW *ARABIDOPSIS* MUTANT WITH REDUCED 16:3 LEVELS

M. Miquel[1] and J. Browse[2]
1 *Laboratoire Biogenèse Membranaire, CNRS UMR 5544, Univ. Bordeaux 2, 33076 Bordeaux Cedex, France.*
2 *Institute of Biological Chemistry, Washington State Univ., Pullman, WA 99164-6340, USA.*

Higher plants possess two distinct pathways for the synthesis of chloroplast glycerolipids in leaf cells [1]. The chloroplast is the sole site of *de novo* fatty acid synthesis. The final products of fatty acid synthesis and of the soluble 18:0-ACP desaturase are 16:0-ACP and 18:1-ACP. These either enter the prokaryotic pathway of the chloroplast inner envelope to produce chloroplastic lipids or they are exported as CoA thioesters to the ER to enter the eukaryotic pathway. Both pathways are initiated by the synthesis of phosphatidic acid (PA). Because of the specificities of the plastid acyltransferases, the PA made by the prokaryotic pathway has 16:0 at the sn-2 position and, in most cases, 18:1 at the sn-1 position. This PA is used for the synthesis of PG or is converted to DAG by a phosphatidic acid phosphatase. This DAG pool is the precursor for the synthesis of MGDG, DGDG, and SL, the major plastid membrane lipids. The PA synthesized in the ER by a different set of acyltransferases than the plastid isozymes is characteristically enriched in 18-carbon fatty acids at the sn-2 position, 16:0 when present, is confined to the sn-1. This PA is used to produce phospholipids characteristic of the various extrachloroplast membranes of the cell. In *Arabidopsis*, a portion of PC produced by the eukaryotic pathway is returned to the chloroplast and used in the production of chloroplast lipids. Consequently, both pathways contribute about equally to the synthesis of MGDG, DGDG, and SL, and the leaf lipids characteristically contain substantial amounts of 16:3 which is found only in MGDG and DGDG molecules produced by the prokaryotic pathway. Mutants of *Arabidopsis* with altered fatty acid composition have been isolated [2]. One of them, *act1*, is deficient in chloroplast acyl-ACP:*sn*--glycerol-3-phosphate acyltransferase activity, and its fatty acid composition is characterized by greatly reduced levels of 16:3 [3]. Here we show that another mutant with also reduced 16:3 levels is not allelic to *act1* suggesting that other mutations can (partly) block the prokaryotic pathway.

The mutant line we will describe here, JB19, was isolated during the screening of the M2 progeny of EMS-mutagenized seeds as its leaf lipids contained reduced levels of 16:3 compared to these of wild type (WT). The mode of inheritance of this altered fatty acid composition was determined by crossing a JB19 mutant with WT. All the plants of the F1 progeny exhibited a leaf fatty acid composition very similar to that of WT plants. The proportions of homozygote, heterozygote and wild types in the F2 progeny resulting from self-fertilisation of F1 plants were in agreement with the mendelian inheritance of a single nuclear mutation The progeny of a cross between JB19 and *act1* exhibited a complementation indicating that both mutants are not allelic.

As shown in figure 1, leaves of JB19 exhibit a nearly 3-fold decrease in 16:3 amounts compensated by an increase in 18-carbon fatty acids.

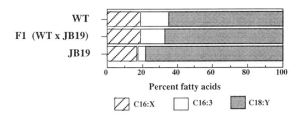

Figure 1. Overall leaf compositions of wild type (WT) and mutant *Arabidopsis* grown at 22°C under continuous illumination. C16:X, fatty acids containing 16 carbons and 0 to 2 double bonds; C18:Y, fatty acids containing 18 carbons and 0 to 3 double bonds.

To study the fatty acid fluxes through the prokaryotic and the eukaryotic pathways during lipid synthesis of JB19 and WT, positional analyses of individual leaf lipids were carried out. Indeed, it has been shown previously that the total proportion of C16 fatty acids at the sn-2 position of each lipid is an estimate of the proportion of that lipid that is synthesized via the prokaryotic pathway, with the remainder being formed by the eukaryotic pathway [4]. The results of these analyses are shown in table 1 and reveal that the decrease in 16:3 levels reflects a reduction of the amount of total fatty acids going through the prokaryotic pathway. This decreased lipid synthesis via the prokaryotic pathway in JB19 compared with WT plants is largely compensated by an increased synthesis of chloroplast lipids by the eukaryotic pathway so that the proportions of chloroplast lipids in the mutant is similar to that in WT.

TABLE 1. Mass and fatty acid composition of JB19 and WT leaf lipids and their lyso-derivatives. WT values are indicated between brackets. Cn:x, fatty acids at n carbons and 0 to 3 double bonds. Chloroplast lipids are designated by the shaded area.

Lipid Lyso-derivative	Mass of fatty acids (mol/1000 mol)	Fatty acid composition (mol%)	
		C16:X	C18:X
MGDG	348 (*390*)	20 (*37*)	80 (*63*)
sn-2		**34** (*70*)	66 (*30*)
DGDG	184 (*152*)	8 (*15*)	92 (*95*)
sn-2		**4** (*12*)	96 (*88*)
SL	23 (*28*)	30 (*43*)	70 (*57*)
sn-2		**39** (*63*)	61 (*37*)
PG	83 (*81*)	51 (*52*)	49 (*48*)
sn-2		**86** (*83*)	14 (*17*)
PC	219 (*218*)	22 (*22*)	78 (*78*)
sn-2		**2** (*1*)	98 (*99*)
PE	113 (*100*)	28 (*29*)	72 (*71*)
sn-2		**1** (<*1*)	99 (*99*)

It is not yet clear which biochemical step is affected by the mutation in JB19. The synthesis of chloroplast PG is apparently not affected in the mutant suggesting that JB19 could be a leaky mutation in the chloroplast phosphatidic acid phosphatase. Alternatively, preliminary results of glycerol feeding to whole JB19 and WT plants show an alleviation of the mutant phenotype (Figure 2). This would be then in agreement with the mutation being a defect in glycerol-3-phosphate supply.

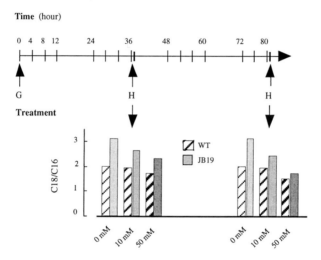

Figure 2. Effect of glycerol feeding to whole JB19 and WT plants on their leaf fatty acid composition.
Wild type and mutant plants were fed glycerol by spraying 0, 10 or 50 mM glycerol, respectively, at different times. Thirty minutes after both the 36-h and the 80-h feedings, 6-15 plants for each treatment and from each genotype were harvested and total leaf lipids were extracted from each sample. G, glycerol feeding; H, harvest; C18/C16, ratio of fatty acids with 18 carbons to fatty acids with 16 carbons in total leaf lipids.

The results of the first characterization steps concerning this new *Arabidopsis* mutant with reduced 16:3 levels demonstrate that other mutations than *act1* can (partly) block the prokaryotic pathway of lipid synthesis. We are currently investigating the two hypotheses indicated above in order to identify the biochemical step affected by the mutation in JB19.

References

1. Browse, J., and Somerville, C.: Glycerolipids synthesis: biochemistry and regulation, *Annu. Rev. Plant Physiol. Plant Mol. Biol.* **42** (1991), 467-506.
2. Browse, J., and Somerville, C.R.:Glycerolipids, in E.M. Meyerowitz and C.R. Somerville (eds.), *Arabidopsis*, Cold Spring Harbor Laboratory Press, Cold Spring Harbor, pp. 881-912, 1994.
3. Kunst, L., Browse, J., and Somerville, C.: Altered regulation of lipid biosynthesis in a mutant of *Arabidopsis* deficient in chloroplast glycerol-3-phosphate acyltransferase activity, *Proc. Natl. Acad. Sci. USA* **85** (1988), 4143-4147.
4. Browse, J., Warwick, N., Somerville, C.R., and Slack, C.R.: Fluxes through the prokaryotic and the eukaryotic pathways of lipid synthesis in the "16:3" plant *Arabidopsis thaliana*, *Biochem. J.* **235** (1986), 25-31.

STUDY ON THE FATTY ACIDS OF LIPIDS BOUND TO PEPTIDES OF PHOTOSYSTEM II OF *NICOTIANA TABACUM*

A. Gasser, S. Raddatz, A. Radunz and G.H. Schmid
University of Bielefeld, Faculty of Biology, Cell Physiology, P.O.Box 10 01 31, D-33501 Bielefeld, Federal Republic of Germany.

Introduction

The lipid composition of the thylakoid membrane and also that of functional regions such as photosystem I, photosystem II, cytochrome b_6 and the ATPase is mostly known [1-9]. Also known is the unsaturated character of the fatty acids of these lipids [1-3, 8-11]. However, insufficient is the information on lipids bound to membrane proteins as well as their fatty acid composition. In this publication we report on lipids bound to the LHCP-complex and to the D1-core peptide of photosystem II from *Nicotiana tabacum* as well as on the saturated character of the fatty acids of these lipids.

Results and Discussion

The LHCP-complex was prepared according to Ghanotakis and Yocum [12] from a PS II-complex preparation of wild type tobacco *Nicotiana tabacum* John William's Broadleaf by incubation with the detergent acetyl-β-D-glucopyranoside and subsequent centrifugation at 40.000 x g. According to SDS-PAGE-analyses the LHCP-complex prepared is composed of peptides with the molecular masses of 28, 26 and 24 kDa, with the 28 kDa peptide representing the major portion. Lipids were extracted with boiling ethanol followed by extraction with methanol and diethylether and compared via thin layer chromatography with the lipids of the PS II-preparation. It is seen that due to the strong increase of the chlorophyll content the two galactolipids decrease in the LHCP by approx. 30%, whereas the content of sulfolipids is even reduced to 1/3. The two phospholipids phosphatidylcholine and phosphatidylinositol are not detectable anymore in this complex. It is the content of phosphatidylglycerol which increases comparably 3-fold in the LHCP-complex. Whereas β-carotene is reduced to 1/3 of the original content in the LHCP-complex, the two xanthophylls violaxanthin and neoxanthin decrease slightly; only lutein appears in higher concentrations.

The D1-core peptide was isolated via SDS-PAGE from the PS II-complex of the *Nicotiana tabacum* mutant Su/su as a dimer with the molecular mass of 66 kDa. This isolation method presents the advantage that no impurities from LHCP-peptides, from the D2-peptide or from the extrinsic PS II-peptide with the molecular mass 33 kDa (OEC_1) can occur. The obtained D1-peptide reacted in the Western blot procedure exclusively with the D1-antiserum and gave no cross-reaction with antisera to the potential impurities namely the D2-, the LHCP- and the OEC_1-peptides. The D1-preparation did not react with the antiserum to the CP1-peptide of PS I. For their characterization the lipids of the D1-peptide were extracted with a mixture of boiling methanol/chloroforme (1:2, v/v) and analyzed by means of HPLC on a RP18-reversed-phase-column. The only identified lipids were monogalactosyldiglyceride and the two negatively charged lipids sulfoquinovosyldiglyceride and phosphatidylglycerol. A quantitative estimation shows that one D1-peptide contains (somehow bound) approx. 10 monogalactolipids, 30 phosphatidylglycerol and 170 sulfolipid molecules.

With immunological methods using the Western blot procedure and by means of lipid and carotenoid antisera [13] the binding of monogalactolipid and phosphatidylglycerol onto the isolated LHCP-complex was confirmed. The reaction with the digalactolipid and the sulfolipid antiserum came out negatively. Here, however, we were able to show that, if in the Western blot and the SDS-PAGE the original material to be analyzed was the PS II-preparation, the LHCP-complex gave with these antisera a positive reaction. From the carotenoid antisera only the antiserum to β-carotene and that to violaxanthin reacted positively with the LHCP. Reactions with the antisera to lutein, neoxanthin and zeaxanthin were negative. From these observations it is concluded that certain lipids such as the digalactolipid and the sulfolipid as well as the xanthophylls reacting negatively in the Western blot have, due to protein denaturation during the LHCP isolation or the SDS-gel electrophoresis, suffered such a rearrangement that antigenic determinants do not react anymore with antibodies (are not accessible) or that these lipids have been stripped off.

The reactions of the isolated D1-core peptide with the lipid antisera in the Western blot also came out differently. Thus, the D1-peptide isolated via SDS-PAGE reacted only with the antiserum to the monogalactolipid and with that of the sulfolipid and not with those to digalactolipid or phosphatidylglycerol. The dimer of the D1-peptide was only labelled with these antisera, if a PS II-preparation was used in the Western blot and not the isolated D1-peptide. Apparently, the monogalactolipid and the sulfolipid are tighter bound to the D1-protein. Moreover, as seen from the molar ratio of sulfolipid to D1-protein a large number of sulfolipid molecules seems to get unspecifically adsorbed onto this peptide during the isolation procedure.

The gaschromatographic analysis of fatty acids led to the result that the fatty acids of the lipids bound to the LHCP-complex as well as those bound to the D1-core peptide exhibit a higher degree of saturation than the fatty acids of the PS II-complex (Tab. I). Whereas the saturated fatty acids in the PS II-complex make up for 11%, they double in the LHCP-complex. The palmitic acid alone makes up for 20%. The two trienoic fatty acids, linolenic acid and hexadecatrienoic acid which are both the ester components of monogalactosyldiglyceride [14-17] appear reduced due to the decrease of this lipid in the LHCP. The ratio of C_{16}- and C_{18}-acids is shifted in favor of C_{16}-acids. As a characteristic feature of the LHCP it is seen that the hexadecenoic acid with trans-configuration increases 2-fold. As, however, phosphatidylglycerol, generally containing this trans fatty acid by 50% as an ester component [8, 18], increases 3-fold, it must be concluded that the LHCP also contains bound phosphatidylglycerol, which contains less trans-hexadecenoic acid. In the lipids of the D1-core peptide, in which the sulfolipid usually amounts to 4-fold the quantity of the other lipids monogalactosyldglyceride and phosphatidylglycerol, palmitic acid alone makes up for 74% of all fatty acids. Amongst the unsaturated acids oleic acid as monoenoic acid makes up for 14%, and a C_{16}-dienoic acid for 8% and the phosphatidylglycerol ester component trans-hexadecenoic acid for 4.5% of all fatty acids. It is surprising that for the detected lipid monogalactosyldiglyceride, the characteristic and highly unsaturated fatty acids linoleic and linolenic acid are not found [14-17]. Hence, the lipids bound to the D1-peptide also differ in their fatty acid composition from the monogalactolipid and the anionic lipids of the thylakoid membrane.

The comparison of the ratio of fatty acids with 16 and 18 carbon atoms in protein-bound lipids of the PS II-complex, in the LHCP-complex and in the D1-core peptide reveals an important increase of fatty acids with 16 carbon atoms and a corresponding decrease of fatty acids with 18 carbon atoms. Whereas the C_{16} fatty acids in PS II make up for 23.4%, they increase in the LHCP to 38% and make up for 86.5% of total fatty acids in the D1-core peptide.

This change in the fatty acid composition of the lipids bound to the protein, that is the increase of C_{16}-acids and of saturated fatty acids refers in the LHCP-complex to monogalactosyldiglyceride and phosphatidylglycerol and in the D1-peptide besides these lipids also to sulfoquinovosyldiglyceride. Whereas the anionic lipids of the thylakoid membrane contain up to 50% saturated fatty acids and as unsaturated acids linoleic and linolenic acid [8, 17-18], the monogalactolipid of the thylakoid membrane contains linolenic and hexadecatrienoic acid which makes up for 95% of total fatty acids [14-17]. For the increased occurrence of saturated fatty acids and the increase of carbonic acids with only 16 carbon atoms, sterical reaons and the function of these protein-bound lipids must play the decisive role. The function of these lipids bound to the core peptide and the LHCP-peptides and having a highly saturated character, obviously is the connection of these peptides amongst each other, and the anchoring of the complex in the bimolecular lipid film of the thylakoid membrane which consists of lipids with highly unsaturated fatty acids.

Fatty acids bound covalently to proteins do not seem to occur besides these lipids bound to the described proteins. An alcaline hydrolysis of the proteins already extracted with organic solvents and a subsequent extraction of the hydrolysis products with petrolether/diethylether (1:1 v/v) did not lead to the gaschromatographic detection of further fatty acids.

Tab. I: Fatty acid composition of the lipids of the LHCP II- and the PS II-complex as well as that of the core peptide D1 and of chloroplasts of *N. tabacum* var. JWB calculated as total fatty acids.

Fatty acids	D1-Peptide	LHCP II	PS II	Chloroplasts
$C_{12:0}$	-	-	2.5	0.6
$C_{14:0}$	-	0.9	3.0	0.3
$C_{16:0}$	74.0	19.7	5.2	11.4
$C_{16:1\,trans}$	4.5	6.9	3.2	4.3
$C_{16:2}$	8.0	-	1.5	1.3
$C_{16:3}$	-	11.5	13.5	9.8
$C_{18:1}$	13.5	2.0	1.4	3.0
$C_{18:2}$	-	2.4	4.1	6.9
$C_{18:3}$	-	56.7	65.6	63.3
C_{16}/C_{18}	9.1	0.62	0.33	0.37
saturated/unsaturated fatty acids	2.9	0.26	0.12	0.14

References

1. Koenig, F., Z. Naturforsch. 26b, (1971), 1180-1187.
2. Quinn P.J. and Williams W.P., Biochim. Biophys. Acta 737 (1983), 223-266.
3. Bednarz J., Radunz A. and Schmid G.H., Z. Naturforsch. 43c (1988), 423-430.
4. Voß R., Radunz A. and Schmid G.H., Z. Naturforsch. 47c (1992), 406-415.
5. Makewicz A., Radunz A. and Schmid G.H., Z. Naturforsch. 51c (1996), in press.
6. Makewicz A., Radunz A. and Schmid G.H., Z. Naturforsch. 49c (1994), 427-438.
7. Haase R., Unthan M., Couturier P., Radunz A. and Schmid G.H., Z. Naturforsch. 48c (1993), 623-631.
8. Kruse O., Radunz A. and Schmid G.H., Z. Naturforsch. 49c (1993), 115-124.
9. Murata N., Higeshi S.-I. and Fujimura Y., Biochim. Biophys. Acta 1019 (1990), 261-268.
10. Radunz A., Flora, Abt. A, 157 (1966), 131-160.
11. Radunz A., Hoppe-Seyler's Z. Physiol. Chem. 349 (1968), 303-309.
12. Ghanotakis D.F. and Yocum C.F., FEBS Lett. 197 (1986), 244-248.
13. Schmid, G.H., Radunz A. and Gröschel-Stewart U. (1993)
    Immunologie und ihre Anwendung in der Biologie, Georg Thieme Verlag, Stuttgart, New York.
14. Radunz A., Z. Naturforsch. 27b (1972), 822-826.
15. Radunz A., Z. Naturforsch. 31c (1976), 589-593.
16. Heinz E., Biochim. Biophys. Acta 144 (1967), 333-343.
17. Heinz E., Biochim. Biophys. Acta 144 (1967), 321-232.
18. Radunz A., Z. Naturforsch. 26b (1971), 916-919.
19. Radunz A., Hoppe-Seyler's Z. Physiol. Chem. 350 (1969), 411-417.

THE CYANOBACTERIUM *GLOEOBACTER VIOLACEUS* LACKS SULFOQUINOVOSYL DIACYLGLYCEROL

E. SELSTAM[1] and D. CAMPBELL[2]
[1]*Department of Plant Physiology, University of Umeå, S-901-87 Umeå, Sweden.* [2]*Department of Biology, Mount Allison University, Stockville, New Brunswick, Canada.*

1. Introduction

Gloeobacter violaceus is a unicellular cyanobacteria that besides the outer wall layer only has one membrane the cell membrane. It performs oxygenic photosynthesis and the photosynthetic complexes are localised in the cell membrane (Rippka *et al.*, 1974). Long phycobilisomes are oriented perpendicular to the cell membrane and the phycobiliproteins carry phycoerythrobilin, phycourobilin, phycocyanin and allophycocyanin (Guglielmi *et al.*, 1981, Bryant *et al.*, 1981)

2. Materials and Methods

Gloeobacter violaceus sp. PCC 7421 was grown under illumination of white fluorescent tubes (10 µmol photons m^{-2} s^{-1}) at 26 °C in inorganic BG 11 media (Rippka *et al.*, 1979), buffered to pH 7.5 with 10 mM Mops. The cells were harvested by centrifugation and cell lipids were extracted and analysed according to Wingsle *et al.* (1992).

3. Results and discussion

3.1. LIPID COMPOSITION

The lipid composition of *Gloeobacter violaceus* included the major cyanobacterial membrane lipids monogalactosyl diacylglycerol (MGDG), digalactosyl diacylglycerol (DGDG) and phosphatidyl glycerol (PG). No sulfoquinovosyl diacylglycerol (SQDG) was found (Table 1). Trace amounts of monoglucosyldiacylglycerol (MgluDG) (Feige *et al.*, 1980) as well as another lipid, probably phosphatidic acid (PA), was also present. The PA was tentatively identified from the mobility on two dimensional thin layer chromatography.

Table 1. Lipid composition of *Gloeobacter violaceus*.

Lipid	MGDG	DGDG	PG	PA?	MgluDG	SQDG
mol %	51	24	18	4	trace	0

The lack of SQDG in a photosynthetic membrane is exceptional. Although the presence of small traces of SQDG (< 1 %) is not definitively excluded, we conclude that *Gloeobacter violaceus* lacks significant amounts of this lipid and certainly contains far less than the typical cyanobacterial thylakoid complement of 11-14%, found in *Anacystis nidulans* (Murata et al., 1981, Omata and Murata, 1983). In the envelope and cell membrane the SQDG concentration is 5 and 7% respectively. SQDG is found in some purple nonsulfur bacteria, which contain a PS II-type reaction centre (Imhoff and Bias-Imhoff, 1995). In *Rhodobacter sphaeroides* this SQDG is not required for photosynthetic electron transport, and could be substituted by another negatively charged membrane lipid, phosphatidyl glycerol, during growth under phosphate replete conditions (Benning et al., 1993). In *Gloeobacter violaceus* the presence of PA(?) might substituted for the lack of SQDG. In contrast, a mutant strain of *Chlamydomonas reinhardtii* defective in SQDG synthesis had a slowed growth rate and impaired photosystem II function (Sato et al., 1995). Interestingly, the lack of SQDG in *Gloeobacter violaceus* also correlates with a slow growth rate.

3.2. ACYL GROUPS OF MEMBRANE LIPIDS

The acyl groups esterified to the membrane lipids are presented in Table 2. The major fatty acids are 16:0, 18:0, 18:1, 18:2 and α–18:3. Trace amounts of 14:0 and 16:1 was also found. The presence of 32 % 16-carbon fatty acids and 63 % 18-carbon fatty acids place *Gloeobacter violaceus* outside any of the typical cyanobacterial fatty acid groupings (Murata et al., 1992, Murata and Wada, 1995).

Table 2. Acyl groups (mol %) esterified to membrane lipids of *Gloeobacter violaceus*.

Lipid	mol % of acyl groups				
	16:0	18:0	18:1	18:2	18:3
MGDG	32	3	2	11	50
DGDG	33	4	3	11	48
PG	37	8	7	15	32
PA?	29	8	4	5	53
total membrane acyl groups	32	4	3	11	45

4. References

Benning, C., Beatty, J.T., Prince, R.C. and Somerville, C.R. (1993) The sulfolipd sulfoquinovosyl-diacyglycerol is not required for photosynthetic electron transport in *Rhodobacter sphaeroides* but enhances growth under phosphate limitation. *Proc. Natl. Acad. Sci. USA* **90**, 1561-1565.

Bryant, D.A., Cohen-Bazire, G. and Glazer, A.N. (1981) Characterization of the biliproteins of *Gloeobacter violaceus*. Chromophore content of a cyanobacterial phycoerythrin carrying phycouorbilin chromophore. Arch Microbiol **129**, 190-198.

Feige, G.B., Heinz, E., Wrage, K., Cochems, N. and Ponzelar, E. (1980) Discovery of a new glyceroglycolipid in blue-green algae and its role in galactolipid biosynthesis, in P. Mazliak, P. Benveniste, C. Costes and R. Douce (eds), *Biogenesis and Function of Plant Lipids*, Elsevier/North-Holland Biomedical Press, pp.135-140.

Guglielmi, G., Cohen-Bazire, G. and Bryant, D.A. (1981) The structure of *Gloeobacter violaceus* and its phycobilisomes. *Arch. Microbiol.* **129**, 181-189.

Imhoff, J.F. and Bias-Imhoff, U.(1995) Lipids, quinones and fatty acids of anoxygenic phototrophic bacteria, in R.E. Blankenship, M.T. Madigan and C.E. Bauer (eds.), *Anoxygenic Photosynthetic Bacteria*, Kluwer Academic Publishers, Dordrecht, pp.179-205.

Murata, N. and Wada, H. (1995) Acyl-lipid desaturases and their importance in the tolerance and acclimatization to cold of cyanobacteria. Biochem. J. **308**,1-8.

Murata, N., Sato, N., Omata, T. and Kuwabara, T. (1981) Separation and characterization of thylakoid and cell envelope of the blue-green alga (Cyanobacterium) *Anacystis nidulans*. *Plant Cell Physiol.* **22**, 855-866.

Murata, N., Wada, H. and Gombos, Z. (1992) Modes of fatty-acid desaturation in cyanobacteria. *Plant Cell Physiol.* **33**, 933-941.

Omata, T. and Murata, N. (1983) Isolation and characterization of cytoplasmic membranes from the blue-green alga (Cyanobacterium) *Anacystis nidulans*. *Plant Cell Physiol.* **24**, 1101-1112.

Rippka, R., Waterbury, J. and Cohen-Bazire, G. (1974) A cyanobacterium which lacks thylakoids. *Arch. Microbiol.* **100**, 419-436.

Rippka, R., Deruelles, J., Waterbury, J.B., Herdman, M. and Stanier, R.Y. (1979) Generic assignments, strain histories and properties of pure cultures of cyanobacteria. *J. Gen. Microbiol.* **111**, 1-61.

Sato, N., Sonoike, K., Tsuzuki, M. and Kawaguchi, A. (1995) Impaired photosystem II in a mutant of Chlamydomonas reinhardtii defective in sulfoquinovosyl diacylglycerol. *Eur. J. Biochem.* **234**,16-23.

Wingsle, G., Mattson, A., Ekblad, A., Hällgren, J.-E. and Selstam, E. 1992. Activities of glutathione reductase and superoxide dismutase in relation to changes of lipids and pigments due to ozone in seedlings of *Pinus sylvestris* (L.). *Plant Science* **82**, 167-178.

Section 4:

Sterols and Isoprenoids

A NOVEL MEVALONATE-INDEPENDENT PATHWAY FOR THE BIOSYNTHESIS OF CAROTENOIDS, PHYTOL AND PRENYL CHAIN OF PLASTOQUINONE-9 IN GREEN ALGAE AND HIGHER PLANTS

H. K. LICHTENTHALER[1], M. ROHMER[2], J. SCHWENDER[1], A. DISCH[2], M. SEEMANN[3],

[1] *Botanisches Institut II, Universität Karlsruhe, Kaiserstr. 12, D-76128 Karlsruhe, Germany*
[2] *Université Louis Pasteur, Institut Le Bel/CNRS, 4 rue Blaise Pascal, F-67070 Strasbourg Cedex, France*
[3] *Ecole Nationale Supérieure de Chimie de Mulhouse, 3 rue Alfred Werner, F-68093 Mulhouse Cedex, France*

1. Introduction

The acetate/mevalonate pathway, which provides IPP, had been investigated in the 50's and 60's in detail, and the universal occurrence of this pathway in animal and plant isoprenoid biosynthesis was generally accepted. However, some observations conflicted with this pathway. The biosynthesis of plastidic isoprenoids of higher plants (carotenoids, chlorophylls, plastoquinone-9) was not inhibited by mevinolin, a highly specific inhibitor of mevalonate formation (1, 2). In order to test if the plastidic isoprenoids of algae and higher plants are formed via the acetate/mevalonate pathway or via a different pathway some ^{13}C-labelling experiments were carried out with two green algae and the higher plant *Lemna*. Here we show that in all three cases carotenoids and the prenyl side chains of chlorophylls and plastoquinone-9 are labelled via a new, bacterial and mevalonate-independent pathway of IPP biosynthesis (3, 4, 5).

2. Procedure

Chlorella fusca was grown heterotrophically on [1-^{13}C]glucose with 10 % isotopic abundance (4). Axenic cultures of duckweed (*Lemna gibba* L.) were cultivated on a mineral medium containing 0.3% of labelled glucose. Low light (40 µmol photons m^{-2} s^{-1}) stimulated growth and greening. In several experiments, *Lemna* was grown on glucose with 10 % or 15 % abundance of [1-^{13}C]glucose. Extraction and purification of phytol, ß-carotene and plastoquinone-9, assignment of ^{13}C-NMR chemical shifts and evaluation of isotopic enrichments were performed as described in (4).

3. Results and Discussion

When plants are grown heterotrophically on [1-^{13}C]glucose, glycolysis and acetyl-CoA formation via pyruvate dehydrogenase yields [2-^{13}C]acetyl-CoA. The latter will label the isoprenic precursor IPP via mevalonic acid at carbon positions 2, 4 and 5 of IPP (Fig. 1). Indeed, sitosterol (*Lemna*) was essentially labelled according to this classical IPP-formation (6). In contrast, phytol as well as ß-carotene and plastoquinone-9 of *Lemna* were labelled in a different way via a new non-mevalonate pathway (Fig. 1, black dots) which had first been detected in some eubacteria and in *Scenedesmus* (3, 4, 5). C-1 and C-5 of the isoprenic units were labelled from [1-^{13}C]glucose (Fig. 1) in agreement with this new glyceraldehyde phosphate/pyruvate pathway (3-5). Our results show, that plastidic isoprenoid compounds of higher plants (phytol, ß-carotene, plastoquinone-9) are produced via this new pathway, whereas the cytoplasmic sterols of *Lemna* are formed via the classical acetate/mevalonate pathway (6).

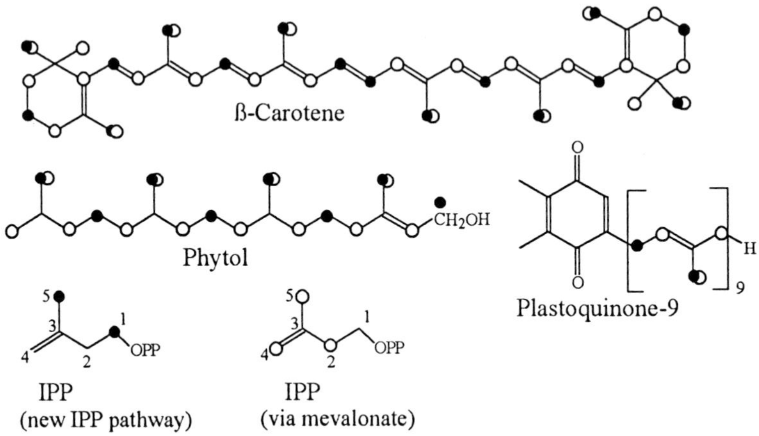

Figure 1. ^{13}C-labelling pattern of ß-carotene, phytol and nona-prenyl chain of plastoquinone-9 from [1-^{13}C]glucose as found in *Lemna gibba* and in two green algae (black circles: ^{13}C-enrichment e.g 3% versus 1% at unlabelled positions; new IPP pathway). Open circles: This labelling was expected if the compounds would have been formed via the classical acetate/mevalonate pathway of IPP formation.

The new ^{13}C-labelling pattern of plastidic isoprenoids of *Lemna* (C-1 and C-5 of IPP from [1-^{13}C]glucose) was found earlier in some eubacteria (triterpenic hopanoids, prenyl side chain of ubiquinone) and in *Scenedesmus obliquus* in the isoprenic units of phytol, ß-carotene, lutein, plastoquinone-9 and in the three main sterol components (4). We now found this labelling pattern also in another green alga, *Chlorella fusca*, grown on [1-^{13}C]glucose for the plastidic ß-carotene, phytol and for the sterols (see also ref. 6).

The new IPP biosynthesis pathway was investigated in great detail by performing various ^{13}C-labelling experiments in *Scenedesmus* (4) as well as with some eubacteria (3). This novel mechanism of IPP formation was also detected in *Ginkgo biloba* in the diterpenic ginkgolides (7). Our results demonstrate that in green algae and higher plants the plastidic prenyllipids are synthesized via a non-mevalonate pathway which had first been detected in some eubacteria (3, 5). The formation of IPP in plastids of higher plants via this novel pathway explains some older observations which could not fully be explained with the acetate/mevalonate pathway of IPP formation (8, 9, 10). We assume that this new pathway, which starts from glyceraldehyde 3-phosphate and pyruvate, is bound to the plastid compartment (s. Fig. 2), a hypothesis which is under investigation. In higher plants cytoplasmic sterol biosynthesis proceeds via the acetate/mevalonate pathway as also shown for *Lemna* (6). In contrast, the two green algae form sterols via the new IPP pathway (4). Thus it appears that higher plants possess two independent and different pathways for IPP and prenyllipid formation, a cytoplasmic and a plastidic one.

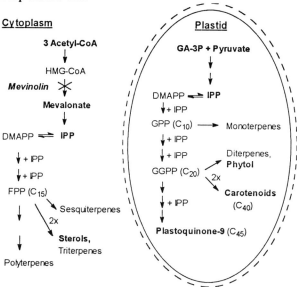

Figure 2. Scheme of the suggested compartimentation of IPP and isoprenoid lipid biosynthesis in higher plants based on ^{13}C-labelling studies. The inhibition site of sterol formation by mevinolin is indicated.

References

1 Bach, T. J. and Lichtenthaler, H.K. (1982) in Biochemistry and Metabolism of Plant Lipids (Wintermanns J. F. G. M. and Kuiper P. J. C., eds.), pp 515-521, Elsevier Biochemical Press, Amsterdam
2 Bach, T.J. and Lichtenthaler, H.K. (1983) Physiol. Plant. 59, 50-60
3 Rohmer, M., Knani, M., Simonin, P., Sutter, B.and Sahm, H. (1993) Biochem. J. 295, 517-524
4 Schwender, J., Lichtenthaler, H. K., Seemann, M., Rohmer, M. (1996) Biochem J. 316, 73-80
5 Rohmer, M., Seemann, M., Horbach, S., Bringer-Meyer, S. & Sahm, H. (1996) J. Am. Chem. Soc. 118, 2564-2566
6 Schwender, J., Lichtenthaler, H. K., Seemann, M., Rohmer, M. (1996) (this volume)
7 Schwarz, K. M. (1994), Doctoral Thesis, ETH Zürich
8 Goodwin, T.W. (1959) in Biosynthesis of Terpenes and Sterols (Wolstenholme, G.E.W. and O'Connor, M., eds.), pp. 279-291, Churchill, London
9 Griffith, W. T., Threlfall, D. R., Goodwin, T. W. (1968) Eur. J. Biochem. 5,124-132
10 Shah, S. P. J. & Rogers, L. J. (1969) Biochem. J. 114, 395-405

BIOSYNTHESIS OF STEROLS IN GREEN ALGAE (*SCENEDESMUS, CHLORELLA*) ACCORDING TO A NOVEL, MEVALONATE-INDEPENDENT PATHWAY

J. SCHWENDER[1], H. K. LICHTENTHALER[1], A. DISCH[2], M. ROHMER[2]

[1] *Botanisches Institut II, Universität Karlsruhe, Kaiserstr. 12, D-76128 Karlsruhe, Germany*
[2] *Université Louis Pasteur, Institut Le Bel/CNRS, 4 rue Blaise Pascal, F-67070 Strasbourg Cedex, France*

1. Introduction

The biosynthesis of isopentenyl pyrophosphate (IPP), the biological precursor of isoprenoid substances, is known as the acetate mevalonate pathway. In the latter pathway the condensation of three acetyl-CoA, followed by two reduction steps (HMG-CoA reductase), yields mevalonic acid which is converted to IPP. In higher plants, the biosynthesis of sterols occurs via mevalonic acid (1) and can be inhibited by mevinolin, a highly specific inhibitor of mevalonate formation (2). In contrast, in several green algae we could not find any inhibition effect of mevinolin on growth and multiplication of cells. Some ^{13}C-labelling experiments in the green alga *Scenedesmus* (3) showed that the main sterol components are synthesized via a novel mevalonate-independent glyceraldehyde phosphate/pyruvate pathway of IPP biosynthesis first found in some eubacteria (4, 5). Here we show that this also applies to *Chlorella*.

2. Procedure

Chlorella fusca (strain 211-8b; Algensammlung Göttingen, Germany) was grown heterotrophically on [1-^{13}C]glucose (10% abundance) as described for *Scenedesmus obliquus* (3). The main sterol components in both species were chondrillasterol, 22,23-dihydrochondrillasterol and ergost-7-enol. The isolation of the three sterol components and the ^{13}C-NMR analysis of the isolated substances was as previously described (3).

3. Results and Discussion

In *Chlorella* cells grown on [1-^{13}C]glucose the sterol carbon skeleton is not labelled according to the acetate/mevalonate pathway. Chondrillasterol, 22,23-dihydrochondrillasterol and ergost-7-enol of *Chlorella* were labelled in a different way than in higher plants. The label of [1-^{13}C]glucose was found in all carbon atoms which derive from C-1 and C-5 of IPP (Table 1) instead of labelling C-2, C-4 and C-5 of isoprenic units (Fig. 2).This labelling is in agreement with the novel IPP-pathway first found in some eubacteria (4) and in *Scenedesmus* (3). Figure 2 resumes all labelling experiments which have been performed with *Scenedesmus* and which give further evidence for a totally different non-mevalonate IPP biosynthetic pathway.

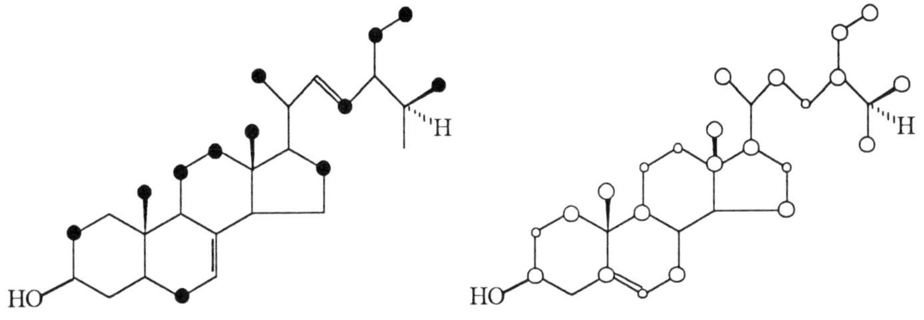

Chondrillasterol Sitosterol

(*Scenedesmus, Chlorella*) (*Lemna*)

Figure 1. ^{13}C-Labelling patterns of the algal sterol chondrillasterol (*Scenedesmus obliquus, Chlorella fusca*) and of sitosterol from duckweed (*Lemna gibba*) (NMR-data for sitosterol cf. 7). Each organism was grown heterotrophically on [1-^{13}C]glucose. Chondrillasterol of *Chlorella fusca* exhibited a labelling pattern according to the novel pyruvate/glyceraldehyde phosphate IPP biosynthesis pathway, while sitosterol of the higher plant was labelled via the acetate/mevalonate pathway. *Black circles*: ^{13}C-enrichment significantly higher than the 1.1% natural enrichment, *open circles*: significant ^{13}C-enrichment of 2.7-4.3 % (large circles) and 2.2-2.7 % (small circles).

In contrast to chondrillasterol, in the higher plant *Lemna gibba*, which was also grown on [1-^{13}C]glucose (6), sitosterol was labelled as expected via the acetate/mevalonate pathway (Fig. 1). [1-^{13}C]Glucose essentially labelled all carbon positions of sitosterol which correspond to C-2, C-4 and C-5 of IPP (Fig. 2). Some minor but significant ^{13}C-enrichment was also found at carbon positions which correspond to C-1 of IPP. This can be interpreted in the frame of the incorporation of ^{13}CO$_2$ (liberated from [1-^{13}C]glucose via the oxidative pentose phosphate cycle) into IPP possibly via the leucine bypass or the mevalonate shunt.

TABLE 1. ^{13}C-Isotopic abundances (%) in carbon atoms of isoprenoids of *Chlorella fusca*, grown on [1-^{13}C]glucose. The range of ^{13}C-abundances of all carbon atoms corresponding to one carbon of IPP is given. The numbering of carbon atoms of IPP is shown in Figure 2.

compound	C-1	C-2	C-3	C-4	C-5
chondrillasterol	4.0-4.6	0.8-1.2	0.9-1.3	1.1-1.7	3.6-4.9
22,23- dihydro-chondrillasterol	3.2-3.8	0.9-1.6	1.0-1.6	1.2-1.5	3.2-3.7
ergost-7-enol	2.5-5.4	1.3-1.5	1.1-2.0	1.2-2.1	4.0-4.6
phytol	3.2-3.6	1.0	1.0-1.3	1.0-1.3	3.2-3.7
ß-carotene	2.5-3.4	0.8-1.1	0.8-0.9	1.0-1.5	2.2-3.5

With our labelling experiments in two green algae we have shown for the first time that the carbon skeleton of a sterol is not built up from acetate units. While it was shown in several higher plants that mevalonic acid is incorporated into sterols at good rates, some unsuccessful attempts for such incorporation experiments in green algae had been reported (1). Since the cytoplasmic sterols as well as the plastidic isoprenoids (*e.g.* carotenoids) are synthesized in the green algae *Scenedesmus* and *Chlorella* via the new IPP pathway (Table 1; ref. 3 and 6), the question arises, if in green algae the acetate mevalonate pathway is fully absent. Also, it is not clear if green algae possess a cytoplasmic and a plastidic IPP pathway.

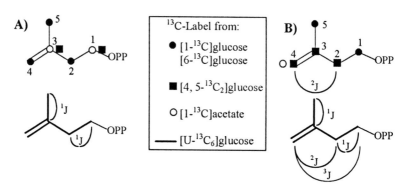

Figure 2. ^{13}C-Labelling of isoprenic units (represented by IPP) from various ^{13}C labelled precursors: A) As expected via the classical acetate/mevalonate pathway and observed for instance after feeding *L. gibba* with [1-^{13}C]glucose. The incorporation of acetate units is evident since glucose breakdown occurs via glycolysis and the pyruvate dehydrogenase complex yields acetyl-CoA.   B) As observed in bacteria (4, 5) and *S. obliquus* (3) via the novel IPP biosynthetic pathway. The indicated labelling pattern was observed *e. g.* for the isoprene units of phytol and chondrillasterol. ^{13}C/^{13}C coupling (^{1}J, ^{2}J, ^{3}J) was observed in ^{13}C-NMR spectra for carbon atoms of several isoprenic units indicating the incorporation of a C$_3$ unit in carbon positions 1, 2 and 4 of IPP (glyceraldehyde 3-phosphate) and a C$_2$ unit (coming from pyruvate) in carbon positions 3 and 5 of IPP. For further details see (3-5).

References

1. Goodwin, T. W. in Biosynthesis of Isoprenoid Compounds (1981) Vol. 1, (Porter, J.W. & Spurgeon, S.L., eds.), pp. 443-480, John Wiley and Sons, New York
2. Bach, T.J. and Lichtenthaler, H.K. (1983) Physiol. Plant. 59, 50-60
3. Schwender, J., Lichtenthaler, H. K., Seemann, M., Rohmer, M. (1996) Biochem J. 316, 73-80
4. Rohmer, M., Knani, M., Simonin, P., Sutter, B., Sahm, H. (1993) Biochem. J. 295, 517-5246
5. Rohmer, M., Seemann, M., Horbach, S., Bringer-Meyer, S., Sahm, H. (1996) J. Am. Chem. Soc. 118, 2564-2566.
6. Lichtenthaler, H. K., Schwender, J., Seemann, M., Rohmer, M. (this volume)
7. Akihisa, T., Thakur, S., Rosenstein, F. U., Matsumoto, T. (1986) Lipids 21, 39-47.

PLANT STEROL BIOSYNTHESIS: IDENTIFICATION AND CHARACTERIZATION OF Δ⁷-STEROL C5(6)-DESATURASE

M. TATON and A. RAHIER
Institut de Biologie Moléculaire des Plantes, CNRS-UPR 406
28 rue Goethe, 67083 Strasbourg Cédex, France

1. Introduction

In plants, the dominant sterols are 24-alkyl-Δ^5-sterols which play multiple roles in plant growth and development, i.e. membranes constituents and as precursors to steroid growth regulators such as brassinosteroïds.
The transformation of cycloartenol to Δ^5-sterols is a multistep process including the transposition of the 9β,19-cyclopropyl ring unsaturation to the double bond at the C5(6) position.
Introduction of a 5(6) double bond is a crucial step in higher plants except in some families like Cucurbitaceae, Theaceae etc... in which the dominant sterols are 24-alkyl-Δ^7-sterols.

The C5(6) desaturase is a membrane bound enzyme localized in the endoplasmic reticulum, and for the first time, we have developped an enzymatic bioassay. The enzymatic $\Delta^{5,7}$-sterol formed was detected using its typical UV absorbance at 281,5 nm. The properties of the microsomal system have been studied and the kinetics of the desaturation reaction has been established [1].

2. Detection of enzymatic C5(6) desaturation of cholest-7-en-3β-ol by maize microsomes.

Cholest-7-en-3β-ol was incubated aerobically with microsomes of *Zea mays* in the presence of NADH or NADPH and 5 μM fenpropimorph in order to block metabolism by $\Delta^{5,7}$-sterol-Δ^7-reductase. The metabolite was detected by H.P.L.C. analysis monitored at 281,5 nm.

3. Requirements for coenzymes and molecular oxygen for desaturation.

In the absence of any added coenzymes, the desaturation process was poorly but significantly effective. The addition of exogenous NADH strongly promoted the reaction. All these results (Table 1) indicate the requirement for a reducing equivalent in the Δ^5-desaturase step delivered by NADH or NADPH and the absolute involvment of molecular oxygen.

TABLE 1. Molecular oxygen and coenzymes requirements for C5(6) desaturation of Δ^7-sterols by maize microsomes

Additions to microsomes	Relative reaction rate %
none	26
NAD^+ regenerating system	7
NADH	88
NADPH	45
NADH + NADH regenerating system	29
NADH + NAD^+	100
NADPH + NAD^+	100
NAD^+	100
$NADP^+$	43
NADPH + NADPH regenerating system	60
NADH + NAD^+ + Argon + Glu + Glu 1-oxidase	0

The role of oxidized pyridine nucleotides NAD^+ or $NADP^+$ is not yet elucidated. However it has been suggested that NAD^+ could be involved. Nevertheless, studies performed in mammals to recover the 5α and 6α substrate hydrogen atoms in NADH have failed.

4. Investigations of the redox proteins and electron transport chain involved in the Δ^7-sterol C5(6) desaturase.

To obtain greater insight into the nature of the enzymatic systems involved in the C5(6) desaturation reaction, C5(6) desaturase was challenged with a variety of chemicals (Table 2). We can notice a strong sensitivity to cyanide (I_{50} = 15x10^{-6}M) and an absolute insensitivity to carbon monoxide strongly excluding the participation of a cytochrome P450 dependent monooxygenase.
The use of metal chelating agents like 1,10,phenantroline and SHA considerably inhibit the reaction process, indicating probably that plant desaturase involves a metal ion presumably iron in an enzyme-bound form. Recently a diiron-oxocluster has been identified as the active site component of the soluble castor bean ω^9-desaturase, thus assigning this desaturase to the class of O_2-activating proteins containing these clusters.

TABLE 2. Effects of some reagents on Δ^7-sterol C5(6)-desaturase

Reagent	Normalized C5(6)-desaturase activity
No addition	100
CN⁻	$I_{50} = 15$ μM
CO (saturated)	100
Menadione 5 μM	88
Menadione 50 μM	32
1,10-phenanthroline	$I_{50} = 2$ mM
Salicylhydroxamic acid	$I_{50} = 500$ μM
Cytochrome c	$I_{50} = 8$ μM

5. Kinetics and alternative substrate reactivities for Δ^7-sterol C5(6)-desaturase.

A series of different potential substrates was examined with regard to C5(6) desaturation. These results indicate that the presence of a 7(8) double bond is necessary for desaturation. This Δ^7-double bond may be required during catalysis for the activation of the allylic 6-hydrogen atom. C5(6)-desaturation occurs preferentially on sterols possessing a C24 unsaturation and without a Δ^{22}-unsaturation, in good accordance with the sequence of reaction resulting of *in vivo* studies.

6. Conclusion

C5(6) desaturase is an enzyme whose activity is dependent of cofactors such as NADH or NADPH requires absolutely molecular oxygen, and probably contains one or several metal ions in its active site. The data is in good accordance with the participation of cytochrome b5 as an intermediate electron carrier.

The mechanism by which the double bond is introduce has not been elucidated sofar, especially with regard to the question of weather C5(6) desaturation involves C5 or C6 hydroxylated intermediates or a direct dehydrogenation process, further investigation will be needed to clarify this alternative.

Cloning of *A. thaliana* and yeast desaturases have proven the existence of three histidine clusters and site-directed mutagenesis experiments, in order to show if these histidine are catalytically essential as already shown in the case of other desaturation process, are now in progress [2].

7. References

1. Taton, M. and Rahier, A. (1996) Plant sterol biosynthesis ; Identification and characterization of higher plant Δ^7-sterol C5(6)-desaturase. *Arch. Biochem. Biophys.* **325**, 279-288.
2. Shanklin, J., Whittle, E. and Fox, B.G. (1994) Eight histidine residues are catalytically essential in a membrane-associated iron enzyme, stearoyl-CoA desaturase and are conserved in alkane hydroxylase and xylene monooxygenase. *Biochemistry*, **33**, 12787-12794.

PHYTOSTEROL SYNTHESIS. IDENTIFICATION, ENZYMATIC CHARACTERIZATION AND DESIGN OF POTENT MECHANISM-BASED INHIBITORS OF $\Delta^{5,7}$-STEROL-Δ^7-REDUCTASE

A. RAHIER and M. TATON
Institut de Biologie Moléculaire des Plantes, CNRS-UPR 406
28 rue Goethe, 67083 Strasbourg Cédex, France

During sterol biosynthesis in higher plants, the 9β-19-cyclopropyl ring unsaturation of cycloartenol is transposed via a multi-enzyme cascade to the double bond at the C5(6) position. $\Delta^{5,7}$-sterols have been rarely isolated in higher plants and very little is known of the pathway involved in the reduction of the Δ^7-bond in these organisms. $\Delta^{5,7}$-sterols have been reported in a few species such as bryophytes, *Ochromonas danica* and lichens strongly suggesting that they are intermediates in sterol synthesis in plants. Inhibition of the Δ^7-reduction step is of interest since inhibitors of mammalian $\Delta^{5,7}$-sterol Δ^7-reductase (Δ^7-SR) have been tested as therapeutic drugs against hypercholesterolemia and it was recently reported that the Smith-Lemli-Optiz syndrome, a recessive autosomal disorder is associated with a deficient activity of Δ^7-SR. In addition, the inhibition of Δ^7-SR is of interest not only because inhibition of this enzyme could decrease the rate of end-pathway sterol formation in animals and plants but also from the aspect of the mechanism of enzyme reaction.

1. Identification and properties of maize $\Delta^{5,7}$-sterol-Δ^7-reductase

We have recently demonstrated the existence in a microsomal preparation from maize seedlings of an enzyme catalyzing the NADPH dependent reduction of the Δ^7-bond of $\Delta^{5,7}$-sterol precursors to produce the final Δ^5-sterols, i.e. Δ^7-SR, thus providing the first direct evidence for the participation of $\Delta^{5,7}$-sterols in Δ^5-phytosterols biosynthesis [1] (Figure 1).
The observed strict NADPH specificity of Δ^7-SR is shared with the two others ene-reductases involved in sterol synthesis i.e. $\Delta^{8,14}$-sterol Δ^{14}-reductase (Δ^{14}-SR) and $\Delta^{24(24')}$-sterol Δ^{24}-reductase (Δ^{24}-SR). In these lines, Δ^7-SR from *Arabidopsis thaliana* has been recently cloned and its amino acid sequence demonstrates that the enzyme belongs to the same sequence family as the Δ^{14}-SR and Δ^{24}-SR [2]. The kinetics of the maize Δ^7-SR and Δ^{14}-SR indicates that these reduction steps are not rate determining in the sterol biosynthetic flux, in agreement with the rare isolation of $\Delta^{8,14}$- and $\Delta^{5,7}$-sterols in plants [1].

2. Potent inhibition of maize $\Delta^{5,7}$-sterol Δ^7-reductase by novel 6-aza-B-homosteroids and other analogs of a presumptive carbocationic intermediate of the reduction reaction.

Little is known of the mechanism of Δ^7-SR. Data of previous isotopic labeling studies is consistent with a mechanism for Δ^7-reduction initiated by the electrophilic addition of a proton to the C7(8) double bond giving the carbocationic intermediate at C7 which is then neutralized by the delivery of an hydride ion from NADPH to yield the product (Figure 1).

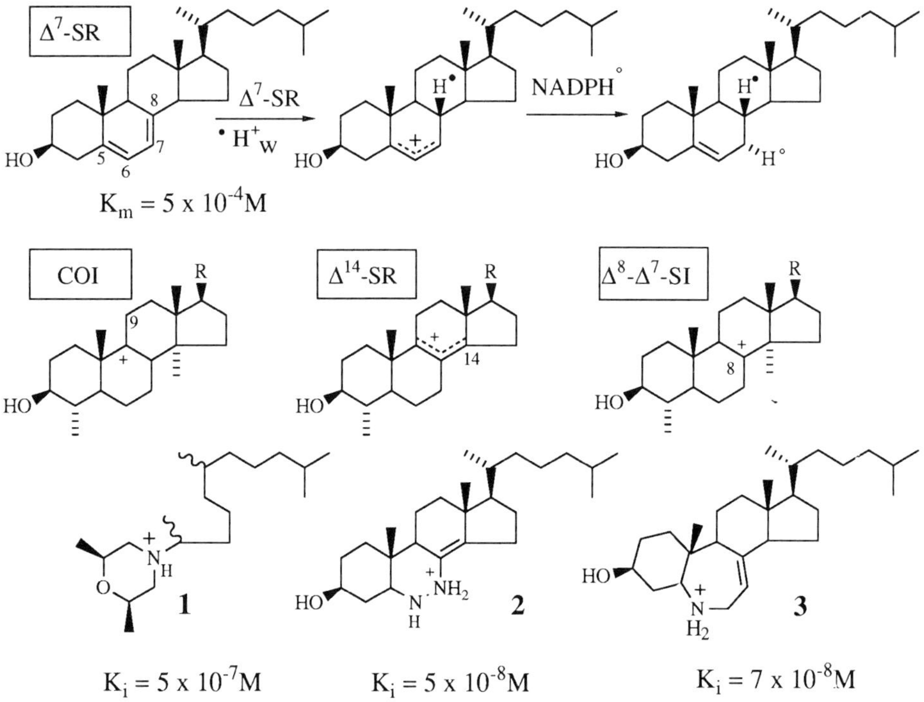

FIGURE 1. Proposed reaction pathway for $\Delta^{5,7}$-sterol-Δ^7-reductase. Structural and charge analogies between the cationic high energy intermediates involved in $\Delta^{5,7}$-sterol Δ^7-reductase (Δ^7-SR), cycloeucalenol isomerase (COI), $\Delta^{8,14}$-sterol Δ^{14}-reductase (Δ^{14}-SR), and Δ^8-Δ^7-sterol isomerase (Δ^8-Δ^7-SI) and the protonated forms of tridemorph **1**, 6,7-diazacholest-8(14)-en-3β-ol **2** and 6-aza-B-homocholest-7-en-3β-ol **3**.

In order to get experimental verification of this proposal and also to interfere with sterol synthesis, we designed, first synthesized and evaluated novel azasteroid analogs of the presumptive carbocationic intermediate involved in the Δ^7-SR mechanism, based on the transition-state analog concept. These novel compounds were shown to be very powerful inhibitors of Δ^7-SR *in vitro* in a maize microsomal preparation (Figure 1)

($K_{i,\,app}$ = 50-70 nM, $K_{i,app}/K_{m,app}$ = 1.0 x 10^{-4} to 1.3 x 10^{-4} [3]. The data provides strong suggestive evidence for the intermediacy of the predicted delocalized C5-C7 allylic carbonium ion intermediate and for the electrophilic nature and course of this reduction reaction [3].

Compound 3 in contrast to compound 2 displayed in the same microsomal preparation more than 50-fold selectivity for inhibition of the Δ^7-SR versus Δ^8-Δ^7-sterol isomerase (Δ^8-Δ^7-SI), cycloeucalenol isomerase (COI), and $\Delta^{8,14}$-sterol Δ^{14}-reductase (Δ^{14}-SR), the mechanism of these four enzymes involving presumptive cationic intermediates centered respectively at C7, C8, C9 and C14 (Figure 1) (Table I). These observations highlight the paramount importance of the location of the positively charged nitrogen atom(s) in the B-ring structure for selectivity among these enzymes involving structurally close cationic reaction intermediates. Efficient *in vivo* inhibition of sterol biosynthesis in bramble cell suspension cultures by a low concentration of compound 3 was demonstrated and confirmed the *in vitro* properties of this derivative [3].

Table I : Specificity of inhibition by azahomosteroid 3 and diazasteroid 2

enzyme	position of predicted carbocationic intermediate	K_i (μM)	
		6-aza-B-homocholest-7-en-3β-ol **3**	6,7-diaza-5α-cholest-8(14)-en-3β-ol **2**
COI	C9	3	100
Δ^{14}-SR	C14 (delocalized to C9)	30	nd[a]
Δ^8-SI	C8	4	0.1
Δ^7-SR	C7 (delocalized to C5)	0.07	0.05

[a] nd is not determined

Maize Δ^7-SR was also shown to be inhibited by ammonium-ion fungicides such as tridemorph (1). The data is in accordance with the abovementioned cationic mechanism. Moreover it indicates that plant Δ^7-SR might be a target of such fungicides.

Finally, this study yielded compounds which are powerful inhibitor of Δ^7-SR and of sterol synthesis, thus supporting the validity of our approach. In addition to their value to explore and manipulate sterol synthesis in plants, the novel compounds described herein could be of pharmacological interest as inhibitors of Δ^7-SR in cholesterol synthesis in animals.

3. References

1. Taton, M. and Rahier, A. (1991) Identification of $\Delta^{5,7}$-sterol-Δ^7-reductase in higher plant microsomes, *Biochem. Biophys. Res. Commun.* **181**, 465-473.
2. Lecain, E., Chenivesse, X., Spagnoli, R. and Pompon, D. (1996) Cloning by Metabolic interference in yeast and enzymatic characterization of *Arabidopsis thaliana* sterol Δ^7-reductase, *J. Biol. Chem.* **271**, 10866-10873.
3. Rahier, A. and Taton, M. (1996) Sterol biosynthesis : strong inhibition of maize $\Delta^{5,7}$-sterol-Δ^7-reductase by novel 6-aza-B-homosteroids and other analogs of a presumptive carbocationic intermediate of the reduction reaction, *Biochemistry* **35**, 7069-7076.

THE OCCURRENCE AND BIOLOGICAL ACTIVITY OF FERULATE-PHYTOSTEROL ESTERS IN CORN FIBER AND CORN FIBER OIL

ROBERT A. MOREAU, MICHAEL J. POWELL, AND
KEVIN B. HICKS
Eastern Regional Research Center
Agricultural Research Service
United States Department of Agriculture
600 East Mermaid Lane, Wyndmoor, PA 19038, USA

1. Summary

Corn fiber is a pericarp-rich fraction obtained during the processing of corn via "wet-milling." Wet milling of corn is used by all companies that produce corn starch and corn sweeteners, and by many companies that produce fuel ethanol from corn. All commercial "corn oil" is prepared by the extraction of only the germ fraction of the kernel. In contrast, extraction of corn fiber with hexane yielded an oil (comprising about 1.2 wt% of the fiber) which we termed "corn fiber oil." This oil contained ferulate-phytosterol esters, similar in structure to "oryzanol," a cholesterol-lowering substance found in rice bran and rice bran oil. The oil extracted from corn fiber contained high levels of ferulate-phytosterol esters (6.0 wt%), which is about 4-fold higher than their levels in rice bran oil. Corn fiber oil also contains free phytosterols (2.2 wt%) and phytosterol fatty acyl esters (6.8 wt%)

2. Introduction

Norton (1995) and Seitz (1989) reported that hexane extraction of corn bran produced an extract that contained high levels of ferulate-phytosterol esters, similar in composition to "oryzanol" found in rice bran and rice bran oil. Oryzanol has been shown to lower the levels of serum cholesterol in laboratory animals and man (Kahlon et al., 1992 and Nicolosi et al., 1991). We recently reported that an oil containing high levels of ferulate-phytosterol esters was extracted from corn fiber by hexane and supercritical CO_2 (Moreau et al, 1996). The present study was undertaken to investigate the quantities of oil and composition of oil obtained from "white fiber," a wet milling product that is partially dried before being mixed with steepwater and dried to yield corn gluten feed.

3. Materials and Methods

Corn fiber oil (crude) was prepared by shaking 4g common corn fiber ("white fiber" from yellow dent #2 corn supplied by Cargill Inc., Dayton, OH) ground to 20 mesh with a Wiley Mill (Thomas Scientific, Philadelphia, PA), with 40 ml hexane, in a 55ml screw-top tube for 1h. The lipid extract was then filtered through a 5.5 cm Whatman (Clinton, NJ) GF/A, glass microfiber filter and injected directly into the HPLC.

The HPLC system used for the separation of nonpolar lipid classes consisted of a Hewlett Packard (Avondale, PA, USA) Model 1050 ternary gradient system (HPLC pump, autosampler, and UV/visible detector), and an Alltech-Varex Mark III ELSD (Alltech Associates, Deerfield, IL). The column was a LiChrosorb DIOL, 5 µm, (3 x 100 mm from Chrompack, Inc., Raritan, NJ), and the flow rate was 0.5 ml/min. The solvents were: A, hexane/acetic acid, 1000/1, v/v; and B, hexane/isopropanol, 100/1, v/v, (Both were mixed fresh daily to eliminate variability caused by evaporation and/or absorption of moisture). The linear gradient timetable was: At 0 min, 100/0; at 8 min, 100/0; at 10 min, 75/25; at 40 min, 75/25; at 41 min, 100/0; at 60 min, 100/0; (%A/%B, respectively).

4. Results and Discussion

Hexane extraction of ground fiber yielded an oil that comprised about 1.2 wt% of oil from the fiber. The levels of hexane-extractable oil in samples of ground fiber from other wet milling plants ranged from 0.54 to 3.68 wt% (Moreau et al, 1996). HPLC-ELSD analysis of the corn fiber oil (Figure 1) revealed the following components: phytosterol-fatty acyl esters (6.75 wt%), triacylglycerols (79.14 wt%), free fatty acids (4.45 wt%), *p*-coumarate-phytosterol esters (<0.1 wt%), γ-tocopherol (<0.1 wt%), free phytosterols (2.21 wt%), 1,2-diacylglycerols (0.64 wt%), ferulate-phytosterol esters (6.04 wt%), 1,2-diacylglycerols (0.77 wt%).

Norton (1995) reported that the most abundant ferulate-phytosterol ester in corn pericarps is sitostanyl-ferulate (Figure 1). Free sitostanol is more effective at lowering serum cholesterol than other phytosterols (Vanhanen et al, 1993). In contrast oryzanol, the cholesterol-lowering ferulate-phytosterol ester fraction that comprises about 1.5% of rice bran oil is comprised mainly of cycloartenyl-ferulate (Kahlon et al, 1992). We have recently conducted preliminary hamster-feeding studies and verified that corn fiber oil is at least as effective as rice bran oil in lowering serum total and LDL cholesterol, and we now plan to conduct similar studies with purified ferulate-phytosterol ester from corn fiber oil.

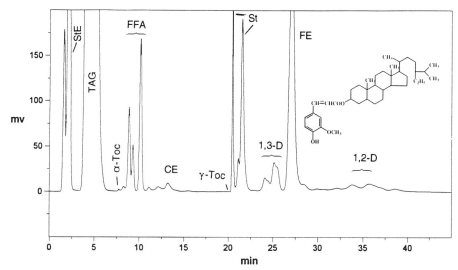

Figure 1. HPLC-ELSD chromatogram of corn fiber oil. Insert shows the structure of sitostanyl-ferulate, the major ferulate-phytosterol ester in corn fiber oil (Norton, 1995). Abbreviations: StE, phytosterol fatty acyl ester; TAG, triacylglycerol, α-Toc, α-tocopherol; FFA, free fatty acids; CE, *p*-coumarate-phytosterol esters; γ-Toc, γ-tocopherol; ST, free phytosterols; 1,3-D, 1,2-diacylglycerols; FE, ferulate-phytosterol esters; 1,2-D, 1,2-diacylglycerols.

5. References

Kahlon, T.S.; Saunders, R.M.; Sayre, R.N.; Chow, F.I.; Chiu, M.M.; Betschard, A.A. (1992) Cholesterol-lowering effects of rice bran and rice bran oil fractions in hypercholesterolemic hamsters. *Cereal Chem.* **69**, 485-489.

Moreau, R.A.; Asmann, P.T.; Norman, H.A. (1990) Quantitative analysis of major classes of plant lipids by high performance liquid chromatography and flame ionization detection (HPLC-FID). *Phytochemistry* **29**, 2461-2466.

Moreau, R.A; Powell, M.J.; Hicks, K.B. (1996) The extraction and quantitative analysis of oil from commercial corn fiber. *J. Ag. Fd. Chem.* (in press).

Nicolosi, R.J.; Ausman, L.M.; Hegsted, D.M. (1991) Rice bran oil lowers serum total and low density lipoprotein cholesterol and apo B levels in nonhuman primates. *Atherosclerosis* **88**, 133-138.

Norton, R.A. (1995) Quantitation of steryl ferulate and p-coumarate esters from corn and rice. *Lipids* **30**, 269-274.

Seitz, L.M. (1989) Stanol and sterol esters of ferulic and p-coumaric acids in wheat, corn, rye, and triticale. *J. Agric. Food Chem.* **37**, 662-667.

Vanhanen, H.T.; Blomqvist, S.; Ehnholm, C.; Hyvonen, M.; Jauhiainen, M.; Torstila, I.; Miettinen, T.A. (1993) Serum cholesterol, cholesterol precursors, and plant sterols in hypercholesterolemic subjects with different apo E phenotypes during dietary sitostanol ester treatment. *J. Lipid Res.* **34**, 1535-1544.

THE OCCURRENCE OF LONG CHAIN POLYPRENOLS IN LEAVES OF PLANTS

EWA SWIEZEWSKA, MALGORZATA SZYMANSKA, KAROLINA SKORUPINSKA AND TADEUSZ CHOJNACKI
Institute of Biochemistry and Biophysics, Polish Academy of Sciences
Pawinskiego St. 5a, 02-106 Warsaw, Poland

1. Mevalonate pathway lipids

Mevalonate pathway products are actively biosynthesized in all cell studied. In plants more than 20 000 end-products differing in structure and functions resulting from this pathway were described. Biological role of many of these compounds is well characterized (for example vitamins, hormones, phytoalexines, antioxidants). However in some cases the function of the compound has not yet been defined. Among the most abundant compounds biosynthesized from farnesyl diphosphate - a branch point product of mevalonate pathway, are sterols, quinones (ubiquinone and plastoquinone) and polyisoprenoid alcohols.

2. Polyprenols

Polyprenols are linear polymers of isoprenoid units (Fig.1.).

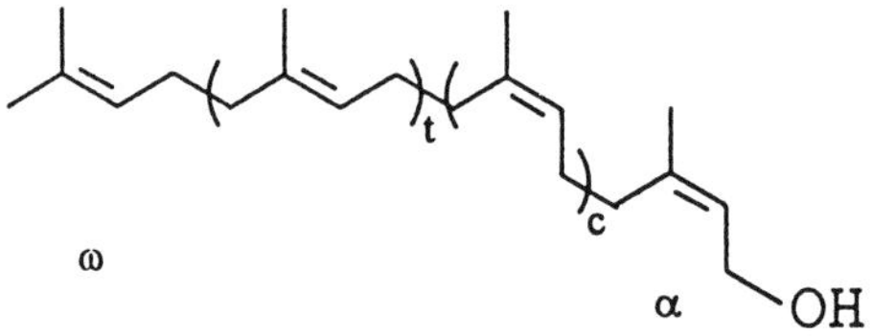

Figure 1. Structure of polyprenols. - isoprene residue, t - number of internal *trans* residues, c - number of internal *cis* residues, - isoprene residue

They are accumulated in leaves of various plants (up to ca 5% of wet weight). Size of polyprenol molecules varies from 9 to about 100 isoprene units. It corresponds to molecular weight range from 500 to 7000. Polyprenols are structural analogues of natural rubber, however those described so far are smaller than low weight low weight fraction of rubber polymer. In the cell polyprenols are found as the mixture of homologues and composition of this 'polyprenol family' is species-specific. It can serve

as a chemotaxonomic marker [1]. In large number of systematic families of angiosperms the size of dominating polyprenol ranges from 9 to 13 isoprene units. In most gymnosperm it ranges from 14 to 23 isoprene units. Interestingly in many representatives of *Rosaceae* family the type of polyprenol pattern
resembles that found in *Cycadopsida*. The occurrence of -dihydropolyprenols (dolichols) characteristic for animal tissues is unique in plants [2].
The longest polyprenols described so far were found in the representatives of *Combretaceae* family. HPLC records of polyprenols isolated from leaves of plants belonging to *Combretaceae* family are shown in Fig.2.

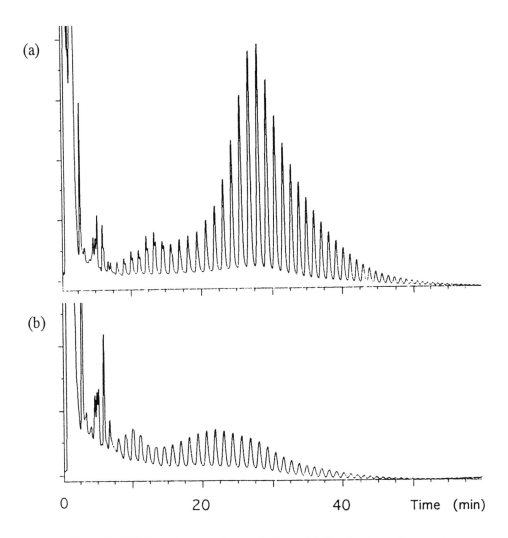

Figure 2. HPLC spectra of polyprenols from (a) *Combretum adenogonium*, dominating Prenol-31 and (b)*Terminalia laxiflora*, dominating Prenol-26.

3. Biological role of polyprenols

In plant tissues two major forms of polyprenols are found, e.g. free alcohols and esters with fatty acids. In some species one form is highly dominating. Similarly in animal tissues dolichols and dolichyl esters were described, accompanied by usually small amounts of dolichyl phosphate. Biological role of this latter compound is well established. Dolichyl phosphate is involved as cofactor in biosynthesis of glycoproteins and GPI-anchored proteins [3]. Biological role of free alcohol and its acyl ester has not been defined yet. As the components of biological membranes these compounds modulate membrane properties. Dolichol increases fluidity and permeability and decreases the stability of model membranes [4]. Experiments with polyprenols point to the same direction; polyprenols and polyprenyl phosphates increases the fluidity and permeability of model membranes [5]. Both polyprenols and dolichols could also serve as a reserve isoprenoid pool utilized by corresponding kinases.

Recent observations of prenylated proteins in plants suggest the new role for polyprenyl diphosphates. They could be utilized as prenyl donors for isoprenylation of proteins. Both *in vivo* [6] and *in vitro* [7] experiments show that besides farnesyl and geranylgeranyl group also long chain polyprenyl groups could be covalently attached to the proteins.

4. References

1. Swiezewska, E., Sasak, W., Mankowski, T., Jankowski, W., Vogtman, T., Krajewska, I., Hertel, J., Skoczylas, E. and Chojnacki, T (1994) The search for plant polyprenols, *Acta Biochim.Polon.* **41**, 221-260.
2. Jankowski, W., Swiezewska, E., Sasak, W. and Chojnacki, T. (1994) Occurrence of polyprenols and dolichols in plants, *J.Plant Physiol.* **143**, 448-452.
3. Hemming, F.W. (1985) Glycosyl phosphopolyprenols. In *Glycolipids*, Wiegandt ed., Elsevier Science Publishers B.V. (Biomedical Division), 261.
4. Chojnacki, T. and Dallner, G. (1988) The biological role of dolichol, *Biochem.J.* **251**, 1-9.
5. Janas, T., Chojnacki, T., Swiezewska, E. and Janas, T. (1994) The effect of undecaprenol on bilayer lipid membranes, *Acta Biochim.Polon.* **41**, 351-358.
6. Swiezewska, E., Thelin, A., Dallner, G., Andersson, B. and Ernster, L. (1993) Occurrence of prenylated proteins in plant cells, *Biochem.Biophys.Res.Commun.* **192**, 161-166.
7. Skoczylas, E. and Swiezewska, E. (1996) Protein farnesyltransferase in plants, *Biochimie*, in press

TAXONOMIC ASPECTS OF THE STEROL AND Δ11-HEXADECENOIC ACID (C16:1, Δ11) DISTRIBUTION IN ARBUSCULAR MYCORRHIZAL SPORES

Grandmougin-Ferjani, A.[1], Dalpé, Y.[2], Hartmann, M-A.[3], Laruelle, F.[1], Couturier, D.[4],& Sancholle, M.[1]. [1] Université du Littoral, BP 699, 62228 Calais cedex, France [2] ECORC, Agriculture & Agrifood Canada Ottawa K1A OC6 Canada [3] Dept. Enzymologie cellulaire et moléculaire, Institut de Botanique, 67083 Strasbourg cedex, France [4] : Chimie organique et Environnement USTL, 59655 Villeneuve d'Ascq cedex France.

1. Abstract

The sterol and fatty acid contents of spores of 16 species of arbuscular mycorrhizal fungi were analyzed. Results indicated that ergosterol, the most predominant sterol in fungi,was absent. The major sterols found belonged to the Δ-5-sterols, mainly cholesterol, 24-methylcholesterol and 24-ethylcholesterol. One of the major sporal fatty acid was an unusual monounsaturated fatty acid, Δ11-hexadecenoic acid, never previously found in fungal material. It occurred in all taxa studied. Its presence can be directly correlated with the fatty acid metabolism of these fungi which is essentially a C16 pathway. Results are discussed with reference to both taxonomic and evolutionary aspects. Comparative analysis of spore sterols and fatty acids from different species may help to clarify some aspects of the systematics and evolution of these organisms since the classification of the arbuscular mycorrhizal fungi has been based almost entirely on spore morphology,.

2. Introduction

Arbuscular mycorrhizal fungi(AMF) are soil-borne organisms living in obligate mutualistic association with terrestrial plants. Known from the Devonian era (Taylor et al. 1995) AMF actually comprised three families (Glomaceae, Gigasporaceae and Acaulosporaceae grouped in the order Glomales (Zygomycetes). Recent investigations on sterols and fatty acids composition have been done on these organisms (Sancholle & Dalpé, 1993, Gaspar & Pollero, 1994,Grandmougin et al, 1995), pointing out the abundance of Δ11-hexadecenoic acid and of cholesterol and the absence of ergosterol. A variety of additional species representing taxa of the taxonomical families were analyzed and results presented.

3. Materials and methods

3.1. FUNGAL MATERIAL

Arbuscular mycorrhizal fungi were propagated in pot-culture over leek plants (*Allium porrum* L.). Spores were isolated by wet-sieving and sucrose gradient techniques and preserved by lyophilization (Dalpé, 1991). The 16 studied species are listed in Table 1.

3.2. LIPID ANALYSIS

A mean of 100 freeze-dried spores were extracted in CH_2CL_2 + methanol (2:1;v:v). Residue was saponified in KOH (60g/L in methanol). The unsaponifiable fraction was extracted 3 times with hexane and purified on silica gel TLC plates. The saponifiable fraction was adjusted to pH 1.0 with HCL before similar extraction. Fatty acids and sterols were analyzed by gas chromatography in the form of methyl ester and acetate derivatives respectively. Identifications were made by comparing retention time relative to methyl heptanoate (C17:0) and cholesterol as internal standards.

4. Results and discussion

24-ethylcholesterol is the major sterol found in all the studied Glomales species (Table 1). The presence of this sterol segregates the Glomales from the Mucorales species, the latter being characterized by the presence of ergosterol (24β-methylcholesta(5,7,22) trienol). No ergosterol was detected in Glomales spores although it is considered the main sterol for most of the fungi (Weete,1989; Lösel,1988). Considerable amounts of α-amyrin, a triterpen classically found in higher plants were detected. Representatives of the three studied genera presented rather similar sterol profiles. This supports, based on biochemical data,the monophyletic origin of these obligate mutualistic symbionts. Complementary experiments are required to determine the 24 chirality of the 24 ethylsterols of arbuscular mycorrhizal fungi. The study of the 24-stereochemistry of 24-ethylsterols coupled with incorporation of $2\text{-}^{14}C$ acetate would allow us to verify whether these fungi are able to synthetize their own sterols or whether they depend on plant partners for providing them.

TABLE 1 Percent sterols of arbusuclar mycorrhizal fungi

	1	2	3	4	5	6	7	8		1	2	3	4	5	6	7	8
GLOMACEAE									Glomus macrocarpum								
Glomus aggregatum									mature spores			tr.	8	75	16	tr.	
spores & hypha	10	tr.	tr.	59	30	tr.			Glomus mosseae								
Glomus borealis									young spores	5	11	11	70	14			
young spores	tr.			79	21	tr.			mature spores	58		tr.	24	7			
mature spores				88	12				sp. and peridium	13	15		56	16	tr.	tr.	
Glomus caledonium									Glomus pustulatum								
mature spores	tr.	tr.		80	20	tr.			mature spores			15	80	5			
Glomus clarum									Glomus tortuosum								
mature spores	tr.	tr.	tr.	79	21	tr.	tr.		young spores			10	70	20			
Glomus etunicatum									mature spores			tr.	80	20			
mature spores	9	tr.	tr.	74	17	tr.	tr.		Glomus versiforme								
Glomus fasciculatum									mature spores	tr.	tr.	tr.	67	33	tr.	tr.	
mature spores	5	tr.	5	71	10		9		Glomus vesiculiferum								
Glomus intraradices									mature spores	tr.	9	74	9	tr.	tr.	8	
young spores	7			76	11				GIGASPORACEAE AND ACAULOSPORACEAE								
mature spores	4			81	15				Gigaspora margarita								
hypha	14			68	18				mature spores	tr.	tr.	tr.	77	22	tr.	tr.	
Glomus lamellosum									Acaulospora scrobiculata								
mature spores	5	tr.		65	30		tr.		mature spores	5	tr.	tr.	84	10	tr.		

1: cholesterol; 2: 24-methylcholesterol; 3: 24-ethylcholesta-5,22-dien-3ß-ol; 4:24-ethylcholesterol; 5: α-amyin ; 6: ß-amyrin; 7: 24-ethylcholesta-5,24(24^1)-dien-3ßol; 8:24-methylenecycloartanol

Major fatty acids found in Glomales spores were palmitic acid (C16:0), stearic acid (C18:0) and oleic acid (C18:1)(Table 2). Δ11-16:1 hexadecenoic acid appeared to be present in the majority of Glomales spores analyzed. Found abundantly in 9 of the 12 studied species, it remained absent from *Gigaspora margarita,* the sole representative of the studied Gigasporaceae. Further investigations with GC-MS would confirm the Δ11 position of the double bond as the specificity of this fatty acid to the AMF.

TABLE 2. Percent fatty acids of spores of arbuscular mycorrhizal fungi

	14:0	14:1	16:0	16:1 cis	18:0/18:1 cis	18:2 cis	18:3 trans	18:3 cis	18:3 γ	20:4	22:?	22:6	U
GLOMACEAE													
Glomus aggregatum	2		42	21	23	7		3			2		
Glomus borealis			46		22	23	4			2	2		
Glomus clarum	2	1	34	12	15	28	5	2			1	tr.	
Glomus etunicatum	2		37	8	15	19	12				7		
Glomus fasciculatum	1		50		6	25	13		3		1	1	1
Glomus intraradices													
young spores			32	47	14	7							
mature spores	1	1	30	11	13	38	4	1			tr.		
Glomus lamellosum			50	8	16	19					tr.		
Glomus macrocarpum	2		37	6	17	25	9		2	2			
Glomus mosseae													
young spores			8	11	2	17	20	16		14	1	11	
spores and peridium	tr.		9	27	4	55	2	tr.			2		
Glomus tortuosum													
young spores	2	1	39	3	16	30	5	2			1		
mature spores	1		43		20	25	6	1			1	1	1
GIGASPORACEAE AND ACAULOSPORACEAE													
Gigaspora margarita	1	5	33		14	37	4			4			2
Acaulospora scrobiculata	tr.		29	12	18	23	9		2			6	

tr. : traces; U: unidentified

5. Conclusion

The analysis performed on the spores of sixteen Glomales species pointed out the uniqueness of sterol, triterpene and fatty acid profiles of AMF. Different from other Zygomycetes by the total absence of ergosterol, the sterol profiles of AMF spores lays closer to the one of higher plants than to the one of fungi. The 400 millions years of symbiosis with higher plants (Pirozynski & Dalpé, 1989; Taylor et al. 1995) might have strongly determined these patterns. Up to now, C16:1-Δ11 fatty acid has been reported only into Glomales species. Our results suggest that this very specific fatty acid could be considered as a potential taxonomic marker for AMF. The lipid analysis of additional species of *Gigaspora* and *Scutellospora* species (Gigasporaceae) would provide confirmation of specificity for AMF.

6. References

Dalpé, Y. 1991. Vesicular-arbuscular mycorrhizae. in M.R. Carter (ed.), *Manual of Soil Sampling and Methods of Analysis*. Canadian Society of Soil Science Lewis Pub. of CRC Press, pp. 287-301.
Gaspar, M.L. & Pollero, R.J. 1994. *Glomus antarcticum*: Lipids and fatty acid composition. *Mycotaxon* 51,129-136.
Grandmougin, A., Dalpé,Y., Veignie, E., Rafin, C. & Sancholle, M. 1995. Infection by arbuscular mycorrhizal fungus *Glomus mosseae* of leek plants (*Allium porrum* L.): effect on lipids, in Kader, J.C. & Mazliak, P.(eds), *Plant Lipid Metabolism*, Kluwer Academic Publishers, pp. 444-446.
Lösel, D.M. 1988. Fungal lipids, in Ratledge, C., Wilkinson, S.G. (eds) *Microbial Lipids* Vol. 1. Academic Press London, pp. 669-806.
Pirozynski, K. and Dalpé, Y. 1989. The geological history of the Glomaceae with particular reference to mycorrhizal symbiosis. *Symbiosis* 7,1-36.
Sancholle, M. & Dalpé, Y. 1993. Taxonomic relevance of fatty acids of arbuscular mycorrhizal fungi and related species. *Mycotaxon* 59,187-193.
Taylor, T.N., Remy, W., Haas, H., and Kerp, H. 1995. Fossil arbuscular mycorrhizae from the Early Devonian. *Mycologia* 87, 560-573
Weete, J.D. 1989. Structure and function of sterol in fungi. *Advances inLipid Research*. Vol 23. Academic Press, London pp. 115-167

Section 5:

Environmental Effects on Lipids

LACK OF TRIENOIC FATTY ACIDS IN AN ARABIDOPSIS MUTANT INCREASES TOLERANCE OF PHOTOSYNHESIS TO HIGH TEMPERATURE

Jean-Marc Routaboul, P. Vijayan and John Browse
Institute of Biological Chemistry, Washington State University,
Pullman, WA 99164-6340, USA.

1. Introduction

Oxygenic photosynthesis occurs in a uniquely constructed bilayer membrane known as thylakoid. The glycerolipid molecules that form the thylakoid membranes are characterized by their sugar head groups and a very high level of unsaturation in the fatty acid chains. In *Arabidopsis*, trienoic fatty acids - α-linolenic (18:3) and hexadecatrienoic acids (16:3) - account for more than 70% of the total fatty acids in thylakoid lipids. Therefore, this fatty acids might have some crucial role in maintaining photosynthetic functions.

Acclimation of higher plants to high or low temperatures is often accompanied by changes in fatty acid saturation. This observation has lead to the proposal that the photosynthetic apparatus adapts to temperature changes by altering the fatty acid composition of the membrane lipids (1, 2). Several groups have modified the degree of saturation of fatty acids of thylakoid lipids by changing the growth temperature of the plants, treating with chemicals or using genetic manipulations. However, the conclusions reached were often antagonistic.

We previously showed that trienoic fatty acid have a crucial role in maintaining photosynthetic function at low temperatures (3). Conversely, we also tested the hypothesis that trienoic fatty acids influence the thermotolerance of the photosynthetic apparatus using a triple mutant *fad3fad7-2fad8* of *Arabidopsis thaliana* that lacks 18:3 and 16:3 fatty acids in its membrane lipids (4). This trienoic fatty acid deficient mutant carried mutations in the three genes FAD3, FAD7 and FAD8 that mediate the synthesis of trienoic fatty acids from 18:2 and 16:2. The FAD3 gene product is the endoplasmic reticulum desaturase and the FAD7 and FAD8 genes encode chloroplast isozymes (4).

2. Results

Fad3fad7-2fad8 **photosystem II is thermotolerant.**
The most thermosensitive complex of photosynthesis is the photosystem II (PSII). The heat-induced damages to PSII was followed by measuring the ratio of variable to maximal chlorophyll *a* fluorescence yield (Fv/Fm, 6). The data presented in figure 1 show the gradual heat inactivation of PSII in wild-type and *fad3fad7-2fad8* when detached leaves were slowly heated in the dark from 30°C to 50°C ($1°C.min.^{-1}$).

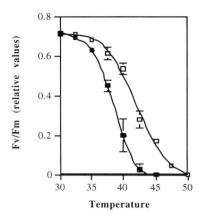

Figure 1: Changes in the quantum yield of PSII photochemistry, Fv/Fm, when wild-type (closed symbols) and mutant (open symbols) leaves are heated from 30°C to 50°C (1°C/min.).

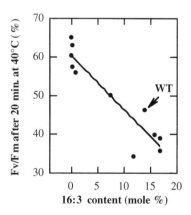

Figure 2: Relationship between the quantum yield of PSII photochemistry (Fv/Fm) measured on detached leaves exposed for 20 min. at 40°C and the 16:3 content in leaves for different lipid mutants. Measurements were made at 25°C under 100μEm^{-2}s^{-1} light intensity. The data points represent from top to bottom: *fad3fad7-2 fad8, fad5, fad7-2 fad8, fad6, fab2,* wild-type, *fad3, fad2, fab1* and *fad4*.

The temperatures of half inactivation of PSII were 38°C and 42°C for wild-type and mutant respectively. This indicated that the mutant is 4°C more thermotolerant that wild-type.

The thermostability of PSII were also measured by to other widely used methods: the fluorescence rise under low exciting light, Fo (5) and the PSII dependent oxygen evolution of isolated thylakoids (data not shown). Isolated thylakoid were incubated in darkness for 10min. at various temperatures from 25°C to 45°C and then we measured the PSII dependent oxygen evolution from water to DCPIP at 25°C. Both measurements, Fo rise and PSII dependent oxygen evolution were less sensitive to temperature in the mutant *fad3fad7-2fad8*.

Lack of 16:3 confers the thermotolerance to the mutant.

To differentiate which of the two trienoic fatty acid is conferring the PSII thermotolerance, we dark incubated excised leaves of various fatty acid desaturase mutants at 40°C for 20 min. and then we measured Fv/Fm at 25°C (Figure 2)(for characterization and total leaf fatty acid contents see reference 7). All thermotolerant mutants, namely *fad3fad7-2fad8, fad7fad8, fad5, fad6, act1,* and fab2 contained less than 7% of 16:3. By contrast, they contained varying proportions in 18:3: from 0 to 54%. Thermosensitive mutants, namely *fad2, fad3, fad4,* and *fab1,* contained more than 14% 16:3 and more than 39% 18:3. The coefficient of the linear regression between the 16:3 content and thermotolerance was, r^2=0.86 whereas the same coefficient for the linear regression between 18:3 content and the thermotolerance was 0.24. No other fatty acid content showed a high relationship to thermotolerance (data not shown). Hence the lack of 16:3 appears as the main factor contributing to the thermotolerance of these mutants and not 18:3.

***Fad3fad7-2fad8* mutant died when grown at 33°C.**
To find if the short-term PSII thermotolerance result in a better growth of the plant at high temperature we transferred *fad3fad7-2fad8* triple mutant and wild-type plants at 33°C after 13 days initial growth at 22°C. After 6 days, growth of the mutant decreased, the quantum yield of electron transfer was inhibited and plant turn chlorotic. Finally, after 15 days all mutant plants died whereas wild-type plants continued to grow. Preliminary results showed that the polypeptide profile of the mutant thylakoids were not different from the wild-type indicating that these effects were not due to any major disruption of existing photosystems.

5. Discussion

During acclimation of *Arabidopsis* to high temperature, the amount of 16:3 decrease from 14% to less than 5% after 13 days at 33°C (Routaboul et al., unpublished). In some desert plant species, like *Atriplex lentiformis,* the hexadecatrienoic acid can totally disappear after prolonged exposure to high temperature (12). We show that the lack of trienoic fatty acid in the *fad3fad7-2fad8* mutant results in increased thermotolerance of photosynthesis compared to the wild-type. Among the two trienoic fatty acids, the lack of 16:3 appears as the main factor contributing to the thermotolerance of these mutants and not 18:3. Hence, changes in the 16:3 content may provide a very flexible means for the long-term regulation of the thermal properties of the thylakoid membrane in response to temperature stress.

It has been postulated that MGD and DGD may be involved in stabilizing the polypeptides of the photosynthetic apparatus (8-11). The physical properties of the glycerolipids depend on the degree of unsaturation of the fatty acids. As all the 16:3 fatty acids are specifically esterified to these two galactolipids, our results raise the possibility that galactolipids modulate PSII thermotolerance by decreasing their 16:3 content in their acyl moiety.

However, *fad3fad7-2fad8* plants died after 15 days at 33°C showing that trienoic fatty acids are critical for high temperature growth of *Arabidopsis.* In the light of recent results showing that this *fad3fad7-2fad8* mutant do not synthesize a critical hormone, the jasmonic acid, the observed long-term heat phenotype could be linked to a deficiency in hormonal regulation in these plants (4).

6. References

1. J. K. Raison, J. K. M. Roberts, J. A. Berry, *Biochim. Biophys. Acta* **688**, 218-228 (1982).
2. D. Lynch, G. A. Thompson, *Plant Physiol.* **74**, 198-203 (1984).
3. J.M. Routaboul and J. Browse, in *Progress in Photosynthesis Research* P. Mathis Eds. (Kluwer Academic Publishers, Dordretch, Boston, London, (1995).
4. M. McConn, J. Browse, *Plant Cell* **8**, 403-416 (1996).
5. U. Schreiber, J. A. Berry, *Planta* **136**, 233-238 (1977).
6. M. Kitajima, W. L. Butler, *Biochim. Biophys. Acta* **376**, 105-115 (1975).
7. J. Browse, C. Somerville, in *Arabidopsis* E. Meyerowitz, C. Somerville, Eds. (Cold Spring Harbor Press, New York, 1995) pp. 881-912.
8. P. A. Siegenthaler, et al., in *Plant lipid metabolism* J.-C. Kader, P. Mazliak, Eds. (Kluwer Academic Publishers, Dordretch, Boston, London, 1995) pp. 170-172.
9. P. Dörmann, S. Hoffmann-Benning, I. Balbo, C. Benning, *Plant Cell* (1996).
10. A. Tremolieres, et al., *Eur. J. Biochem.* **221**, 721-730 (1994).
11. N. Murata, et al., *Biochim. Biophys. Acta* **1019**, 261-268 (1990).
12. R. W. Pearcy, *Plant Physiol* **61**, 484-486 (1978).

A TRIENOIC FATTY ACID DEFICIENT MUTANT OF ARABIDOPSIS IS DEFECTIVE IN RECOVERY FROM PHOTOINHIBITION AT LOW TEMPERATURES

Perumal Vijayan Jean-Marc Routaboul and John Browse
Institute of Biological Chemistry, Washington State University,
Pullman, WA 99164-6340, USA.

1. Introduction

Photosystem II reaction center is susceptible to light induced damage and steady state photosynthesis is maintained by continual repair of the photo-inactivated photosystems. An increase in the rate of inactivation (by over excitation of PS II) or a decrease in the rate of repair of PSII centers can result in accumulation of damaged PS II and consequent photoinhibition of photosynthesis. Since the assembly and stability of PS II complex involves close interaction with membrane lipids, photoinhibition and recovery of PS II centers are likely to be affected by the lipid composition of the thylakoid membrane.
A triple mutant *fad3 fad7-2fad8* of *Arabidopsis*, which is unable to make trienoic fatty acids (TFA) has enabled us to investigate the role of this specific fatty acid in the structure and function of the photosynthetic machinery. Remarkably enough, the mutant is indistinguishable from the wild type by its vegetative features and the photosynthetic functions of the mutant plants are normal (1). However, the ability of *fad3 fad7-2 fad8* to adapt to low temperatures is severely impaired as revealed by chlorosis and decreased quantum yields of photosynthesis when grown at 5°C (2, data not shown) . Since it has been shown that low temperature stress can actually mimic high light stress and adaptation to low temperature involves development of resistance to photoinhibtion of photosynthesis (3), we investigated the phenomenon of photoinhibition in the TFA deficient triple mutant *fad3 fad7-2 fad8* .

2. Results

Photoinhibition
Excised leaves of wild type and *fad3 fad7-2 fad8* plants grown at 22°C and 120 $\mu E/m^2/sec$ white light were photo inhibited under 1200 $\mu E/ m^2/sec$ white light at 22°C. The decrease in the activity of PSII units was followed by measuring the ratio of variable to maximal chlorophyll a fluorescence yield (Fv/Fm). The inhibition of PS II activity was 40% in wild type leaves as compared to 73% in *fad3fad7-2fad8* leaves after photoinhibitory treatment for 6 hours. Thus the leaves of *fad3fad7-2fad8* were photoinhibited at a higher rate at 22°C (Fig1)
When high light irradiation of the leaves was carried out at 3°C, the rate of photoinhibition increased both in the wild type and *fad3 fad7-2 fad8* . The quantum yield of PS II in both

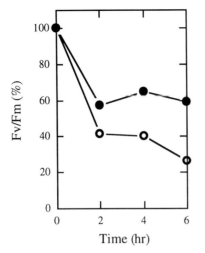

Fig.1 Photoinhibition of excised leaves at 22°C measured as decrease in quantum yield of PS II (Fv/Fm)

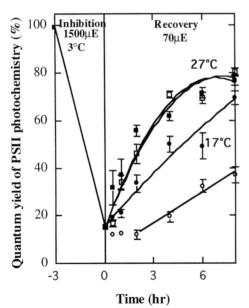

Fig.2 Light dependent recovery of mutant (open symbols) and wild type (closed symbols) leaves after photo-inhibition at 3°C.

decreased by about 80% in 3 hours. However the rate of photoinhibtion was equal in mutant and wild type leaves at this temperature. Thylakoids isolated from wild type and mutant leaves were also photoinhibited at 2000 $\mu E/m^2/sec$ at 22°C and the rate of PSII activity measured as DCIP dependent O_2 evolution subsequently. Both the wild type and mutant thylakoids were photoinhibited at similar rates (data not shown).

Repair of PS II in highly photoinhibited samples would be mainly mediated by light dependent synthesis and assembly of new reaction center proteins. Since no protein synthesis is expected to take place in leaves at 3°C and thylakoids at 22°C, the net rates of photoinhibtion observed under these conditions in the mutant and the wild type samples reflect the rate of light induced damage alone. In contrast, the net rate photoinhibition observed in leaves at 22°C is the resultant of both photoinactivation as well as simultaneous repair of PS II. Since the mutant leaves are more susceptible to photoinhibition at 22°C only, the results demonstrated that rates of primary inactivation of PS II units were similar in the wild type and mutant leaves while the recovery of photoinactivated PS II centers is likely to be inhibited in the mutant *fad3 fad7-2 fad8*.

Recovery from photoinhibition

Excised leaves were photoinhibited for 3 hours at 3°C to prevent any recovery. They were subsequently warmed to 22°C and allowed to recover in complete darkness. PS II activity measured fluorimetrically as Fv/Fm was used to follow the kinetics of the recovery process. Both the wild type and mutant leaves recovered about 9% of the lost activity in 6 hours of dark incubation . However the PS II activity in the mutant leaves further decreased by about 3% in the first two hours of dark incubation before recovering to the same level as the wild type in four hours (data not shown).

In highly photoinhibited samples, recovery of the photo-inactivated PSII centers is predominantly light dependent. This light dependent recovery has two components: 1) The light dependent re-ligation of ligands released during photoinactivation, and 2) The proteolytic degradation of photo-damaged D1 protein of inactivated PS II reaction centers followed by synthesis of new D1 and its re-assembly into active PS II centers.

Excised leaves of wild type and mutant plants were photoinhibited at 3°C to about 15% of original PS II activity and allowed to recover under 70 µE/m^2/sec illumination at different ambient temperatures. Fig. 5 shows the kinetics of light recovery at 27°C and 17°C. The rates of recovery of photoinhibited PS II was identical in the wild type and mutant at 27°C. The rate of recovery decreased with temperature in both samples. However at all temperatures below 27°C the kinetics of PS II recovery was slower in the mutant compared to the wild type. The difference was most clearly visible at 17°C where the rate of recovery in the mutant was about 20% less than the wild type. Additionally an actual decrease of about 3% to 4% was observed in the PS II activity of the mutant during the first 2 hours of the recovery phase (Fig.2)

3. Discussion

The higher rate of photoinhibition observed in *fad3fad7-2fad8* leaves appears to be primarily due to a slower rate of light dependent recovery of photoinactivated PS II centers and not due to a higher rate of photoinactivation.

This deficiency of the mutant leaves in recovering from photoinhibition was apparent only at temperatures below 27°C and became more severe with decreasing temperature.

At 17°C, the rate of light dependent recovery from photoinhibition in the mutant was only 80% that of the wild type in a time scale of 8 hours.

Light dependent recovery from photoinhibition involves proteolytic degradation of the reaction center protein D1 in photoinactivated PS II centers, and its replacement with newly synthesized D1 polypeptide (3). The absence of trienoic fatty acids in the thylakoid lipids of the mutant *fad3 fad7 -2fad8* probably inhibits one or more of the steps involved in this process in a temperature dependent manner. A similar temperature dependent deficiency in the recovery from photoinhibition has also been reported in a cyanobacterial mutant and transgenic tobacco deficient in other fatty acids (4,5)

A small but significant decrease in PS II activity could be observed during the initial phase of light and dark recovery in mutant leaves and not in the wild type leaves. The relative contribution of this component to the overall inhibition of PS II recovery requires further investigation.

5. References

1. M. McConn and J Browse, Plant Cell **8**, 403-406 (1996)
2. J.M. Routaboul and J. Browse, in P. Mathis Eds.*Progress in Photosynthesis Research* (Kluwer Academic Publishers, Dordretch, Boston, London, (1995).
3. van Wijk KJ, Nilsson LO, & Styring S, J Biol Chem **269**, 28382-28392(1994)
4. E. Kanervo ,E-M. Aro &N. Murata , FEBS Letters **364**, 239-242 (1995)
5. Y.M. MooN, S-I. Higashi , Z. Gombos and N.Murata, PNAS **92**, 6219-6223 (1995)

GROWTH TEMPERATURE AND IRRADIANCE MODULATE TRANS-Δ^3-HEXADECENOIC ACID CONTENT AND PHOTOSYNTHETIC LIGHT-HARVESTING COMPLEX ORGANIZATION

GORDON R. GRAY [1], MARIANNA KROL [1], MOBASHSHER U. KHAN [2], JOHN P. WILLIAMS [2] & NORMAN P.A. HUNER [1]
[1]*Department of Plant Sciences, The University of Western Ontario, London, Ontario, CANADA N6A 5B7 and* [2]*Department of Botany, University of Toronto, Toronto, Ontario, CANADA M5S 3B2*

1. Introduction

Photosynthetic acclimation is an essential physiological process allowing plants to tolerate environmental stress conditions, such as low temperature [5]. Chloroplast thylakoid membranes represent a unique combination of pigments, proteins, and lipids whose organization allows for light harvesting, electron transport, and ultimately, the fixation of carbon. The major chlorophyll (Chl) *a/b*-binding protein associated with photosystem II (PSII), light-harvesting complex II (LHCII), is thought to be a trimer composed of Lhcb1 and Lhcb2 polypeptides and plays a fundamental role in photosynthetic light-harvesting [6]. In addition, phosphatidylglycerol (PG) is the major phospholipid present in thylakoid membranes and characterized by the presence of a novel fatty acid, *trans*-Δ^3-hexadecenoic acid (*trans*-16:1)[3]. Previously, it has been established that growth at cold-hardening temperatures (5°C) modulates LHCII organization such that the oligomeric form (LHCII $_1$) predominates at 20°C and monomeric (LHCII $_3$) and/or intermediate forms (LHCII $_2$) predominate at 5°C [3, 4, 7, 8, 9]. However, low temperature also results in a specific decrease in *trans*-16:1 content associated with PG, both *in vivo* and *in situ*, with minimal changes in lipid and pigment profiles [3, 4, 7]. Thus, the modulation of the supramolecular organization of LHCII may be achieved by specifically decreasing the *trans*-16:1 content associated with thylakoid PG, which has been shown to be important for the stabilization of oligomeric LHCII [3, 4, 7, 8, 9].

It has recently been demonstrated in cereal crops, such as winter rye, that tolerance to high-light stress (photoinhibition) is attributable to excitation pressure on PSII, reflecting chloroplastic redox poise, which can be modulated by either growth at high-light or low temperature [2]. Since low temperature growth can be utilized to modulate chloroplastic redox poise and *trans*-16:1 content, we have examined the contribution of both growth temperature and irradiance on the ability to alter the lipid composition of the thylakoid membrane. The capacity to modulate *trans*-16:1 may reflect a mechanism for the regulation of energy distribution in the photosynthetic apparatus, allowing for photosynthesis to occur efficiently under temperature and irradiance stress conditions [3, 5].

2. Materials and Methods

2.1 PLANT MATERIAL AND GROWTH CONDITIONS

Winter rye (*Secale cereale* L cv Musketeer) was germinated from seed at 20°C, with an irradiance of 50, 250 or 800 μmol m^{-2} s^{-1}, or at 5°C with an irradiance of 50 or 250 μmol m^{-2} s^{-1} as previously outlined [2]. Fully expanded leaves from plants grown at the various temperature/irradiance regimes were used for all experiments.

2.2 LIPID AND FATTY ACID ANALYSES

Uppermost fully expanded leaves were extracted and analyzed for their lipid content and fatty acid compositions as described in detail by Williams et al. [12].

3. Results and Discussion

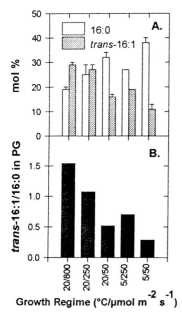

Figure 1. A, Effect of growth under various temperature/irradiance regimes on the fatty acids 16:0 and trans-16:1 in phosphatidylglycerol (PG) as indicated. B, Ratio of trans-16:1 to 16:0 in PG as a result of development at the temperature/irradiance regimes indicated. All values represent means ± SD, n = 3.

The results presented in Figure 1A demonstrate that as growth irradiance is decreased, the amount of palmitic acid (16:0) in PG increases, irrespective of the growth temperature (20 or 5°C). This increase in 16:0 was accompanied by a concomitant decrease in trans-16:1. Absolute amounts of trans-16:1 present a 43% decrease upon growth at 20/50 in comparison to the other 20°C growth regimes. Growth at cold-hardening temperatures (5/250) results in a 30% decrease of trans-16:1 in PG in comparison to typical non-hardened controls (20/250), confirming previous results [4, 7]. In addition, growth at 5/50 resulted in a 42% reduction of trans-16:1 in comparison to cold-hardened controls (5/250) simply as a result of decreasing the growth irradiance.

Ultimately, it is the ratio of trans-16:1 to 16:0 in PG which is important for stabilization of the oligomeric form of LHCII [3, 8, 9]. These data are presented in Figure 1B and demonstrate a clear irradiance dependence of this ratio at either 20 or 5°C. Decreasing the growth irradiance from 800 to 50 μmol m^{-2} s^{-1} at 20°C results in a 3-fold reduction in the trans-16:1 to 16:0 ratio. Similarly, decreasing growth irradiance at 5°C from 250 to 50 μmol m^{-2} s^{-1} results in a 2.3-fold reduction in the trans-16:1 to 16:0 ratio. Low temperature growth at 5/250 also results in a 36% decrease in this ratio when compared to typical non-hardened controls (20/250), consistent with previous reports [4, 7]. Thus, the ratio of trans-16:1 to 16:0 present in PG is dependent not only on growth temperature, but also on growth irradiance. This occurs irrespective of the growth temperature, as this light-dependence is observed at 20 and 5°C.

TABLE 1. Fatty acid composition of phosphatidylglycerol (PG) from total leaf extracts of winter rye developed at the temperature/irradiance growth regimes indicated. Values represent means ± SD, n = 3.

Growth Regime (°C/μmol m^{-2} s^{-1})	Fatty Acid Profile in PG (mol %)					
	16:0	trans-16:1	18:0	18:1	18:2	18:3
20/800	19 ± 1	29 ± 1	1 ± 0	1 ± 0	9 ± 1	41 ± 1
20/250	25 ± 4	27 ± 2	2 ± 1	2 ± 1	8 ± 1	36 ± 2
20/50	32 ± 2	16 ± 1	1 ± 0	2 ± 0	10 ± 1	39 ± 2
5/250	27 ± 0	19 ± 0	1 ± 0	1 ± 0	7 ± 1	45 ± 0
5/50	38 ± 2	11 ± 2	1 ± 1	1 ± 1	8 ± 0	41 ± 1

The temperature and irradiance effects observed are specific for the 16:0 and *trans*-16:1 fatty acids, as no significant changes were observed in any of the other major fatty acids present in PG (Table 1). In addition, no major changes were observed in the fatty acid profiles of the other phospholipids (phosphatidylcholine and phosphatidylethanolamine), galactolipids (monogalactosyldiacylglycerol and digalactosyldiacylglycerol), or sulfolipid (sulfoquinovosyldiacylglycerol) (data not shown). This is consistent with previous studies [4, 7].

Separation of the chlorophyll-protein complexes with a deoxycholate/SDS detergent system [7] indicated a 18% decrease in the ratio of oligomeric:monomeric LHCII with a decrease in irradiance from 800 to 50 µmol m^{-2} s^{1} at 20°C. This was even more pronounced at 5°C with a 39% decrease in the oligomeric:monomeric ratio when growth irradiance was decreased from 250 to 50 µmol m^{-2} s^{-1} (data not shown). This occurs with only slight changes to the absolute amount of LHCII when examined through immunoblotting [2]. The Lhcb1 and Lhcb2 polypeptides, thought to compose the LHCII trimer [6], show minimal change as a result of growth under any of the temperature/irradiance regimes examined (data not shown).

The role of thylakoid lipids in the stabilization of LHCII remains controversial [10, 11], although several reports have implicated a functional association between PG and the stabilization of oligomeric LHCII [1, 7, 8, 9]. Dubacq and Trémolières [1] were the first to present evidence that the *trans*-16:1 of PG is correlated to LHCII stabilization. This was verified by Huner and co-workers who demonstrated that it is the capacity to change the *trans*-16:1 content of PG that is correlated with the capacity to change the supramolecular organization of LHCII [3, 4, 7, 8, 9]. We further extend this in this report, confirming previous results that the reduction of *trans*-16:1 is temperature-dependent, but also demonstrating a light-dependence at both 5 and 20°C. In addition, this cannot be explained by chloroplastic redox poise, which may also be modulated by temperature and irradiance [2]. The changes which we observe in the ratio of *trans*-16:1 to 16:0 are also correlated to an organization of LHCII. The capacity to modulate *trans*-16:1 may reflect an acclimatory mechanism within the photosynthetic apparatus under conditions of environmental stress [3, 5], however, the physiological significance of this change in LHCII organization has yet to be fully elucidated.

4. Acknowledgements

This research was supported by research grants from the Natural Sciences and Engineering Research Council of Canada (NSERCC) to NPAH and JPW. GRG was supported, in part, by a NSERCC Postgraduate Scholarship and an Ontario Graduate Scholarship.

5. References

[1] Dubacq JP, Trémolières A (1983) Physiol Veg **21**: 293-312
[2] Gray GR, Savitch LV, Ivanov AG, Huner NPA (1996) Plant Physiol **110**: 61-71
[3] Huner NPA (1988) CRC Crit Rev Plant Sci **7**: 257-278
[4] Huner NPA, Krol M, Williams JP, Maissan E, Low PS, Roberts D, Thompson JE (1987) Plant Physiol **84**: 12-18
[5] Huner NPA, Öquist G, Hurry VM, Krol M, Falk S, Griffith M (1993) Photosynth Res **37**: 19-39
[6] Jansson S (1995) Biochim Biophys Acta **1184**: 1-19
[7] Krol M, Huner NPA, Williams JP, Maissan EE (1989) J Plant Physiol **135**: 75-80
[8] Krupa Z, Huner NPA, Williams JP, Maissan E, James DR (1987) Plant Physiol **84**: 19-24
[9] Krupa Z, Williams JP, Khan MU, Huner NPA (1992) Plant Physiol **100**: 931-938
[10] McCourt P, Browse J, Watson J, Arntzen CJ, Somerville CR (1985) Plant Physiol **78**: 853-858
[11] Plumley FG, Schmidt GW (1987) Proc Natl Acad Sci USA **84**: 146-150
[12] Williams JP, Khan MU, Mitchell K, Johnson G (1988) Plant Physiol **87**: 904-910

RESPONSES TO COLD IN TRANSGENIC RUSSET BURBANK AND BOLIVIAN POTATO

DAN GUERRA, KASIA DZIEWANOWSKA AND JIM WALLIS
Department of Molecular Biology and Biochemistry
University of Idaho
Moscow, ID. 83844-3052

1. Introduction

Proteins with specific antifreeze properties were first characterized in winter flounder (*Pseudopleuronectes americanus*) and shorthorn sculpin (Yang et al., 1988). The simplest of these proteins, Type I from the winter flounder, are alanine-rich, amphiphilic, α-helices about 3-5 kD in size (Davies and Hew, 1990).
We synthesized, cloned and expressed in potato an AFP-I fused to the signal peptide coding region of phytohaemogluttinin, PHA (gift from M. Chrispeels) for delivery of the protein to the extra-cytoplasmic space where ice first forms. This report describes responses of these transgenic plants and of cold-tolerant Bolivian potato leaf tissue to low temperature.

2. Materials and Methods

The strain *Agrobacterium tumefaciens* LBA4404 (Ooms et al., 1982) was used for all plant transformations.

2.1 CONSTRUCTION OF THE PHA-AFP SYNTHETIC GENE

The synthetic AFP gene was constructed based on the amino acid sequence of mature Type I AFP of winter flounder (Davies et al., 1982), with codon usage frequencies observed for higher plants (Beremand et al., 1987). The synthetic AFP gene was fused to the sequence representing the PHA signal sequence. For introduction into plants, a 240 bp fragment representing the complete PHA-AFP was cloned into pKYLX35S^2 (Schardl et al., 1987).

2.2 POTATO CULTURE AND TRANSFORMATION

Tissue culture and *Agrobacterium*-mediated transformation of potato microtuber discs was conducted as described (Ishida et al., 1989). Transgenic shoots which rooted vigorously on 75 μg ml^{-1} kanamycin were transferred to microtuber induction medium. Microtubers which formed from this tissue were regenerated into shoots on medium containing kanamycin at 100 μg ml^{-1}. Following this second selection, shoots were transferred to soil in a growth chamber. These vegetatively propagated plants were subjected to molecular and biochemical analyses. The Bolivian potato Ch'aska (*Solanum gonoicalaix*) was a gift of Chuck Brown, USDA.

2.3 FROST TOLERANCE ASSAY

To determine the phenotypic effect of expression of the PHA-AFP gene, leaves of selected plants expressing the fusion protein were analyzed by an electrolyte release assay (Sukumaran and Weiser, 1972a). Young but fully expanded leaves were excised from the plants and placed in sample tubes closed with a rubber stopper through which a glass tube had been inserted. The samples were incubated for one hour in a controlled temperature bath at 0 °C to equilibrate the tissue. In experiments measuring the effect of slowly lowering the temperature, the bath temperature was lowered in 1 degree increments at each hour, stepwise to -4 °C. Samples were taken every hour. In experiments to measure effects of cold duration, the bath temperature was lowered to -2 °C and maintained for up to three hours, with samples taken every hour. Ice crystallization was initiated by insertion of wooden sticks dipped in liquid nitrogen through the glass tube. Each time and temperature point were analyzed in triplicate. After exposure to cold, each sample was incubated at +4 °C for one hour, then transferred to an Erlenmeyer flask containing 15 ml of double deionized water. After a 1 hour incubation at 30 °C on a shaker, during which electrolytes released from the tissue are leached into the distilled water, a meter (Corning M90) was used to measure the conductivity of the solution. Each sample was then frozen in liquid nitrogen and its electrolyte release measured again. The ratio of the conductivity measurement taken after the experimental treatment to total conductivity after liquid nitrogen freezing is the measure of electrolyte release.

3. Results and Discussion

Plant lines with consistent levels of AFP expression (according to Western blot) were analyzed by electrolyte release tests. These tests measure the release of ions, primarily K^+, from leaf tissue that is damaged during a freeze-thaw cycle.

The effect on leaf tissue of prolonged exposure to a constant temperature of -2 °C was examined. The Bolivian potato Ch'aska, known to be highly resistant to episodic frost, served as a positive biological control.

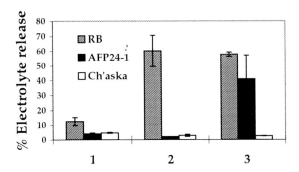

Figure 1. Electrolyte leakage of potato leaf tissue as a percent of total electrolytes in the leaf. Leaf tissue was chilled to -2 °C and incubated for 1, 2 or 3 hours. RB, untransformed Russet Burbank. AFP24-1; second generation PHA-AFP transgenic Russet Burbank line. Ch'aska, frost-resistant Bolivian potato variety.

Leaf tissue from untransformed Russet Burbank, transgenic plant line AFP24-1 and Ch'aska where brought from 22 to -2 °C in one hour and then held at that temperature for up to 3 additional hours. Russet Burbank leaf tissue released over 60% of total electrolyte after 2 hours at the sub-zero temperature (*Figure 1*). The frost-resistant Ch'aska leaf tissue showed no sign of frost-induced electrolyte release even up to 3 hours at -2 °C. There was no detectable electrolyte release from the PHA-AFP expressing plant line AFP24-1 after 2 hours at -2 °C, and only after 3 hours did an increase in electrolyte release occur. The amount of electrolyte release was only about two-thirds of that observed in non-transformed Russet Burbank leaves (40% vs. 60%). Expression of the fusion AFP was sufficient to retard frost damage for at least 2 hours.

We also examined the fatty acid composition of several lipid classes in potato leaf tissue. Phosphatidylcholine fatty acid profiles from leaf tissue grown either at 21° C or 6° C from plants in soilless mix are shown in *Figure 2*.

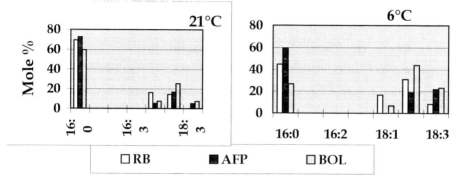

Figure2. Fatty acid mole% in PC lipid fractions of three potato lines. RB= nontransgenic Russet Burbank. AFP=24-1 and BOL=Ch'aska.TLC and GLC were performed using standard techniques.

The data suggest that Ch'aska possesses less 16:0 than both the transgenic and parental RB and that growth at 6° C for a prolonged period results in an increase in 18:3 for all potato lines with Ch'aska containing the highest level (over 60 mole%) and the AFP transgenic leaf is intermediate. It is too early to suggest that AFP expression has an effect on fatty acid composition but it is clear that the highly frost resistant Bolivian potato has higher levels of polyunsaturated fatty acids in both high and low growth temperature environments suggesting its frost resistant phenotype may correlate with the degree of unsaturation. It is currently not known if Ch'aska also expresses anti-freeze type proteins. This will be the subject of future studies.

4. References

Beremand PD, Hannapel DJ, Guerra DJ, Kuhn DN, Ohlrogge JB (1987) Arch Biochem Biophys 256: 90-100
Davies PL, Hew CL (1990) FASEB J 4: 2460-2468
Davies PL, Roach AH, Hew, CL (1982) Proc Natl Acad Sci USA 79: 335-339
Ishida BK, Snyder GW, Belknap WR (1989) Plant Cell Reports 8: 325-328
Ooms G, Hooykaas PJJ, Van Veen RJM, Van Beelen P, Regensburg-Tuink TJG, Schilperoort RA (1982) Plasmid 7: 15-29
Schardl CL, Byrd AD, Benzion G, Altshuler MA, Hildebrand DF, Hunt AG (1987) Gene 61: 1-11
Sukumaran MP, Weiser CJ (1972a) HortScience 7: 467-468
Yang DSC, Sax M, Chakrabartty A, Hew CL (1988) Nature 333: 232-237

LOW-TEMPERATURE RESISTANCE OF HIGHER PLANTS IS SIGNIFICANTLY ENHANCED BY A NON-SPECIFIC Δ9 DESATURASE FROM CYANOBACTERIA

OSAMU ISHIZAKI-NISHIZAWA, TOSHIO FUJII,
TAKESHI OHTANI and TOSHIHIRO TOGURI
Central Laboratories for Key Technology, Kirin Brewery Co.
1-13-5 Fukuura, Kanazawa, Yokohama, Kanagawa 236, Japan

Many important crops are injured or killed by exposure to low nonfreezing temperatures. The presence of only one centrally positioned *cis*-double bond in the fatty acids in polar lipids effectively decreases the phase transition temperature [1]. There has been sustained interest in the possible role of the saturation level of membrane lipids as a determinant factor of chilling sensitivity of higher plants. However, it has been difficult to establish correlations between the saturation level of total membrane lipids of different higher plant species and chilling sensitivity [2].

Three enzymatic activities are known to be involved in desaturation of saturated fatty acids in higher plants [3]. The first desaturation reaction from 18:0-ACP to 18:1c9-ACP is accomplished by stearoyl-ACP desaturase. The second enzyme desaturates 16:0 at *sn*-2 in PG to 16:1*t*3. Because physical properties of 16:1*t*3 are very similar to those of a saturated fatty acid, 16:1*t*3 is often considered equivalent to a saturated fatty acid. The third enzyme desaturates 16:0 at *sn*-2 in MGDG to 16:1*c*7. Thus, saturated fatty acids esterified to other lipids or other positions on the glycerol backbone are not desaturated at all in higher plants. By contrast, in cyanobacteria, desaturation of saturated fatty acids is only accomplished by a Δ9 desaturase which introduces a *cis*-double bond at the Δ9 position of saturated fatty acids linked to any position of all membrane lipids [4]. Therefore, we predicted that the saturation level of total membrane lipids in higher plants might be highly reduced by expressing the Δ9 desaturase from cyanobacteria. We cloned the Δ9 desaturase gene from *Anacystis nidulans* R-2 and introduced it into tobacco plants. We demonstrated that the achieved reduction of saturated fatty acids in most membrane lipids can significantly enhance the chilling resistance of higher plants.

Cloning and identification of Δ9 desaturase gene from *Anacystis nidulans* R-2.

Fatty acid composition analyses showed that the Δ9 desaturases in *Anacystis nidulans* R-2 have a high substrate specificity for lipid-linked 16:0 [5]. Because higher plants contain a high amount of 16:0 and a very low amount of 18:0 in membrane lipids [6], we isolated a Δ9 desaturase gene (EMBL accession No. X77367) designated as des9 from *A. nidulans* R-2. We obtained enzymatic activity of the des9 gene product expressed in *E. coli* and analyzed the total fatty acid composition. The level of 16:1c9 was increased 48% relative to that in control cells after-induction of the des9 gene.

Furthermore, no 18:1c9 was detected in control cells, whereas 1.4% of 18:1c9 was detected in des9 cells after-induction. These results signify that the des9 gene encodes a Δ9 desaturase which introduces a *cis*-double bond at the Δ9 position of saturated fatty acids esterified to lipids; the enzyme shows higher specificity for 16:0 but also utilizes 18:0 as substrate. This indicates that the substrate specificity of the des9 gene product from *A. nidulans* R-2 is significantly different from that of the desC gene product from *A. variabilis* or *Synechocystis* PCC6803 [7].

Expression of Δ9 desaturase in tobacco plants.

The desaturases of cyanobacteria utilize ferredoxin as an intermediate electron donor which is indispensable for expression of the enzymatic activity. In higher plants, lipid biosynthetic activities are mainly confined to the plastid and ER. For fatty acid desaturation, ferredoxin and cytochrome b_5 is utilized in the plastid and ER, respectively [8,9]. For this reason, expression of a functional *A. nidulans* Δ9 desaturase in higher plants requires that the gene product must be transported to plastids. The transit peptide of the pea RuBisCO small subunit was placed at the 5' end of the des9 protein coding region to produce a chimeric gene Rbcs-des9. Thirty transgenic tobacco plants expressing the Rbcs-des9 gene were analyzed. Two independent transgenic lines having a single copy of the Rbcs-des9 gene and a medium amount of the mRNA were subjected to a series of evaluations. Because the two independent lines showed almost the same phenotype in a series of analyses, the results from only one line, D9-1, is reported here. In the following, WT indicates a transgenic tobacco plant containing the T-DNA region of pBI121, which was phenotypically similar to the parental wild-type line, was used as the control for all analyses.

Fatty acid composition of membrane lipids in transgenic tobacco plants.

The level of saturated fatty acids (16:0 + 16:1t3 + 18:0) in leaves and roots of D9-1 was reduced to 72% and 61% of that of WT, respectively. By contrast, the level of monounsaturated fatty acids (16:1c9 plus 18:1) in leaves and roots of D9-1 was 16.6-fold and 8.6-fold increased relative to that of WT, respectively. This result indicates that the Δ9 desaturase of *A. nidulans* localized in plastids significantly reduces the saturation level of membrane lipids in photosynthetic and non-photosynthetic tissues in higher plants.

The level of saturated fatty acids in individual membrane lipids in D9-1 was reduced to 48 - 86% of that of WT. By contrast, the level of monounsaturated fatty acids in individual membrane lipids in D9-1 was 11 - 25% higher than that of WT. Although the Δ9 desaturase of *A. nidulans* is thought to be localized in plastids in D9-1, a high amount of monounsaturated fatty acids were also detected in membrane lipids, PC, PE and PI, which are synthesized only in the ER. This suggests that desaturation occurs on many saturated fatty acid residues linked to membrane lipids in plastids by the Δ9 desaturase of *A. nidulans* and then a high amount of Δ9-desaturated acyl groups appear as ER lipids by an unknown mechanism. Transport of acyl groups from plastids to ER is most likely. The form of acyl groups transported may be free fatty acid, acyl-CoA or lipid esterified to acyl groups, however, this is still unresolved.

Chilling resistance of transgenic tobacco plants.

We analyzed plants grown on MS agar media in plastic boxes to maintain constant humidity in all experiments, because the chilling sensitivity of higher plants exhibits humidity dependency. When plants were exposed to 1 °C under constant light for 11 days, WT exhibited chlorosis; however D9-1 showed no injury. This indicates that D9-1 is resistant to rapid chilling injury due to short-period exposure to low chilling temperature. Seeds were germinated at 10 °C under constant light for 52 days. Although WT seedlings had very pale green cotyledons, D9-1 seedlings had the same green color cotyledons as those of seedlings germinated at normal temperature (25 °C). This indicates that D9-1 is able to develop chloroplasts normally at the germination stage under moderate chilling temperature. Seeds of WT and D9-1 were germinated and grown at 10 °C under constant light for 70 days. Chlorosis developed on almost the whole region of the leaves of WT. By contrast, the primary and secondary leaves of D9-1 mostly showed no chlorosis. These results indicate that D9-1 is able to grow like naturally chilling-resistant plant species under chilling temperatures in the range of 0 - 15 °C. The significant decrease of the saturated fatty acid level in most membrane lipids, as opposed to a particular membrane lipid, is thought to be important in enhancing chilling-resistance of higher plants. This suggests that chilling resistance of many kinds of chilling-sensitive higher plants may be significantly enhanced by expressing the Δ9 desaturase of *A. nidulans* in plastids.

References

1. Silvius, J.R.: Thermotropic phase transitions of pure lipids in model membranes and their modification by membrane proteins, in P.C. Jost and O.H. Griffith (eds.), *Lipid-Protein Interactions*. Vol. 2, Wiley, New York, (1982), pp. 239-281.
2. Somerville, C.: Direct tests of the role of membrane lipid composition in low-temperature-induced photoinhibition and chilling sensitivity in plants and cyanobacteria. *Proc. Natl. Acad. Sci. U.S.A.* **92** (1995), 6215-6218.
3. Ohlrogge, J. and Browse, J.: Lipid biosynthesis. *Plant Cell* **7** (1995), 957-970.
4. Lem, N.W. and Stumpf, P.K.: In vitro fatty acid synthesis and complex lipid metabolism in the cyanobacterium *Anabaena variabilis*. *Plant Physiol*. **74** (1984), 134-138.
5. Bishop, D.G., Kenrick, J.R., Kondo, T. and Murata, N.: Thermal properties of membrane lipids from two cyanobacteria, *Anacystis nidulans* and *Synechococcus* sp.. *Plant Cell Physiol*. **27** (1986), 1593-1598.
6. Browse, J., Warwick, N., Somerville, C.R. and Slack, C.R.: Fluxes through the prokaryotic and eukaryotic pathways of lipid synthesis in the '16:3' plant *Arabidopsis thaliana*. *Biochem. J.* **235** (1986), 25-31.
7. Sakamoto, T., Wada, H., Nishida, I., Ohmori, M. and Murata, N.: Δ9 acyl-lipid desaturases of cyanobacteria. *J. Biol. Chem.* **269** (1994), 25576-25580.
8. McKeon, T.A. and Stumpf, P.K.: Purification and characterization of the stearoyl-acyl carrier protein desaturase and the acyl-acyl carrier protein thioesterase from maturing seeds of safflower. *J. Biol. Chem.* **257** (1982), 12141-12147.
9. Smith, M.A., Cross, A.R., Jones, O.T.G., Griffiths, W.T., Stymne, S. and Stobart, K.: Electron-transport components of the 1-acyl-2-oleoyl-*sn*-glycero-3-phosphocholine Δ12-desaturase in microsomal preparations from developing safflower (*Carthamus tinctorius* L.) cotyledons. *Biochem. J.* **272** (1990), 23-29.

BIOCHEMICAL ASPECT ON THE EFFECT OF TITAVIT TREATMENT ON CAROTENOIDS, LIPIDS AND ANTIOXIDANTS IN SPICE RED PEPPER

PÉTER A. BIACS, HUSSEIN G. DAOOD AND ÁRON KERESZTES*
Central Food Research Institute (KÉKI), H-1022 Herman Ottó u. 15. Budapest,
*Eötvös Loránd University-Department of Plant Physiology, Budapest, Hungary

Introduction

Titanium is a widely distributed metal and ranks tenth among the most common elements in the earth crust. In the soils, titanium is generally found in the form of titanium dioxide or titanium silicates, and due to its low solubility is practically unavailable for plants. The application of soluble forms of titanium, Ti-chloride, Ti-sulphate or Ti-ascorbate (Titavit), has been proved to be beneficial to the growth and quality of some important agricultural crops [1, 2].
The aim of this work is to study in detail the effect of two types of foliar spray of titanium ascorbate (Titavit) on the yield and quality of red pepper fruits using as control both untreated plants and plants treated with potassium ascorbate. This will allow us to differenciate between the effect of titanium itself and that of other components of the Titavit solution. Determination of yield, carotenoids, organic acids, tocopherols, fatty acid composition of chloro and chromoplast and LOX have been carried out at two stages of development (green and red) of the fruits.

Materials and Methods

Seeds of red pepper (Capsicum annuum L., c.v. Mihályteleki) were sown in small pots. Two months after sowing, the seedlings were transpalnted to the experimental field.
Nine plots, 36 m^2 each, were set in the sandy soil experimental field of the University of Horticulture of Budapest in Soroksár, Hungary.
The plants in three of the plots were sprayed with 10 l/plot of a titanium-ascorbate solution (Titavit, 10 mg/l). In other three plots the plants were sprayed with 10 l/plot of a kalium-ascorbate solution (with the same ascorbate concentration as in the Titavit solution). The remaining three plots were used as a control and were spray with water. A group of ten plants from every plot received a second shock-treatment, with the same solutions and concentrations, two weeks before sampling.
20 plant where randomly selected from each plot. All their fruits were harvested. The number of fruits/plant and the total weight harvested were recorded. In the case of the plants which received a second treatment, only ten plants were harvested, and the same parameters were recorded.
Quality components such carotenoids, ascorbic acid and tocopherol were analysed by high performance liquid chromatographic (HPLC) methods [3, 4, 5].
Lipids were extracted by 3:2 hexane-isopropanol, saponified and transmethylated. The methyl esters were analysed by GLC.
The chloroplasts and chromoplasts were isolated after immersion of fruit pericarp in sucrose solution containing β-mecraptoethanol, EDTA, Tris and ascorbic acid.

Result and Discussion

As shown in Table 1 Titavit treatment stimultated to a high extent carotenoid formation, specially β-carotene and red coloured xanthophylls (capsorubin and capsanthin). Esterification of capsanthin with fatty acids increased 1.4 times as a function of Titavit treatment. This was accompanied by structural change on the chromoplast. The fatty bodies disappeared and the fibriles became much thicker in chromoplast from Titavit-treated fruits.

TABLE 1. Carotenoid content (g/kg raw) of red pepper from plants teated with Titavit. Treatment I, 10 ppm; treaatment II, 10 + 5 ppm

Carotenoids	Control	K-Ascorb I	Titavit I	K-Ascorb II	Titavit II
Free xanthophylls					
Capsorubin	0.2 ± 0.3	-	5.4 ± 4.4	-	0.3 ± 0.3
Violaxanthin	11.3 ± 2.6	12.2 ± 3.0	21.1 ± 2.6	14.1 ± 2.9	18.0 ± 1.0
Capsanthin	31.1 ± 3.4	35.4 ± 6.3	41.5 ± 1.2	35.8 ± 9.0	37.2 ± 5.0
Mutatoxanthine	4.2 ± 0.3	4.2 ± 1.1	6.3 ± 0.7	5.4 ± 1.7	5.7 ± 0.4
Lutein	15.2 ± 1.3	12.0 ± 1.2	16.6 ± 3.0	17.4 ± 3.5	16.4 ± 1.7
Zeaxanthin	2.7 ± 0.2	1.5 ± 0.8	4.0 ± 0.5	2.4 ± 1.7	4.0 ± 0.5
cis- Zeaxanthin	2.9 ± 0.4	1.9 ± 1.6	5.1 ± 0.7	5.3 ± 0.7	5.1 ± 0.3
Neolutein	5.2 ± 1.4	4.5 ± 1.7	9.7 ± 1.6	6.7 ± 2.6	9.8 ± 0.5
Cryptocapsin	9.7 ± 0.9	8.0 ± 3.7	15.4 ± 3.7	11.9 ± 2.0	16.0 ± 1.5
β-Cryptoxanthin	18.7 ± 1.3	13.7 ± 1.3	21.6 ± 3.4	19.7 ± 2.0	28.6 ± 2.3
Monoesters					
Capsorubin ME	15.9 ± 0.4	14.0 ± 3.9	26.2 ± 5.4	19.3 ± 4.0	24.9 ± 2.0
Violaxanthin ME	17.8 ± 3.4	19.1 ± 3.1	29.9 ± 4.7	26.3 ± 3.2	34.7 ± 1.5
Capsanthin ME	39.7 ± 4.9	32.6 ± 7.4	55.1 ± 11.1	50.0 ± 6.9	61.6 ± 5.3
Cryptocapsin ME	25.0 ± 6.5	15.2 ± 5.1	39.6 ± 5.1	33.9 ± 2.6	42.3 ± 4.0
AntheroxanthinME	33.0 ± 3.5	20.0 ± 4.6	39.2 ± 7.1	37.2 ± 5.3	50.5 ± 6.3
β-Cryptoxanthin ME	7.8 ± 2.8	4.8 ± 1.8	8.3 ± 2.7	6.4 ± 1.7	9.9 ± 2.4
β-carotene	65.6 ± 9.3	51.5 ± 2.0	79.8 ± 13.7	67.1 ± 11.9	104.3 ± 13.7
Diesters					
Capsorubin DE	24.2 ± 1.6	18.2 ± 1.1	35.8 ± 6.5	27.4 ± 5.4	34.9 ± 3.2
Violaxanthine DE	17.8 ± 3.9	10.2 ± 1.2	16.0 ± 3.0	18.4 ± 2.6	24.2 ± 2.8
Capsanthin DE	104.2 ± 5.6	78.1 ± 6.5	126.6 ± 11.9	99.9 ± 6.9	142.7 ± 6.2
Lutein + Zeax. DE	17.0 ± 4.0	10.7 ± 0.8	14.0 ± 4.4	11.8 ± 1.5	19.2 ± 6.1
Unidentified	18.4 ± 3.2	12.8 ± 3.2	31.5 ± 8.5	25.9 ± 14.6	27.4 ± 5.0
TOTAL	496.9 ± 11.9	391.5 ± 51.3	643.6 ± 92.8	540.7 ± 85.3	715.5 ± 52.8

Illustrated in Table 2 is the change in the ascorbic acid content of red pepper fruit as a function of different treatments. It was found that Titavit treatment consed significant decrease in ascorbic acid fruits level in unripe green fruits, whereas no variation was obtained between treated and untreated ripe fruits in their ascorbic acid content.

The tocopherol content of fruits as affected by Titavit treatments is also shown in Table 2. Two-spray treatment of Titavit resulted in a significant increase in both α- and β-tocopherol and marked decrease in the content of γ-analogue.

TABLE 2. Tocopherol and ascorbic acid content of red pepper from plants treated with titavit and K-ascorbate. Treatment I, 10 ppm; Treatment II, 10 ppm + 5 ppm.

Treatment	Tocopherol (µg/g) in red fruit			Ascorbic acid (mg/g)	
	α-tocopherol	β-tocopherol	γ-tocopherol	green	red
Control	52.8 ± 5.0	1.36 ± 0.34	1.14 ± 0.56	1.32 ± 0.06	1.46 ± 0.07
K-Ascorbate I	54.2 ± 9.2	1.19 ± 0.15	1.28 ± 0.22	1.30 ± 0.03	1.42 ± 0.04
Titavit I	62.1 ± 1.4	1.33 ± 0.23	0.69 ± 0.22	1.14 ± 0.05	1.46 ± 0.04
K-Ascorbate II	58.7 ± 8.5	1.48 ± 0.49	0.77 ± 0.10	1.07 ± 0.01	1.43 ± 0.06
Titavit II	70.5 ± 9.5	2.67 ± 0.31	0.99 ± 0.16	0.30 ± 0.01	1.47 ± 0.07

Fatty acid composition was found to be affected only by the highest doses of both K-ascorbate and Titavit. These treatments distributed ,in part, further desaturation of linoleic acid. This was clear from the increase of C18:2 at the account of C18:3.

TABLE 3. Fatty acid composition (mol %) of the cholorplast and chromoplast of green and red paprika from plants treated with titavit and K-ascorbate. Treatment I, 10 ppm; treatment II, 10 + 5 ppm. results are the average ± sd (three replicates).

CHLOROPLAST	14:0	16:0	16:1	18:0	18:1	18:2	18:3
Control	1.2 ± 0.2	19.5 ± 0.5	0.4 ± 0.1	8.4 ± 1.0	4.6 ± 1.4	35.2 ± 7.1	30.2 ± 2.9
K-Ascorbate I	0.8 ± 0.4	16.9 ± 1.3	0.9 ± 0.2	7.0 ± 0.7	4.2 ± 2.7	36.1 ± 4.4	33.8 ± 4.6
Titavit I	0.8 ± 0.1	17.0 ± 0.4	0.6 ± 0.1	6.4 ± 0.1	2.4 ± 0.4	42.6 ± 4.9	30.2 ± 4.4
K-Ascorbate II	0.7 ± 0.1	15.7 ± 1.2	1.1 ± 0.8	5.2 ± 0.3	3.6 ± 1.3	51.9 ± 4.7	21.6 ± 2.6
Titavit II	0.7 ± 0.3	18.4 ± 2.4	0.9 ± 0.8	4.6 ± 0.7	3.8 ± 1.1	50.6 ± 4.1	20.8 ± 3.0
CHROMOPLAST							
Control	0.4 ± 0.1	15.7 ± 0.7	0.9 ± 0.2	4.5 ± 0.8	10.3 ± 1.3	52.2 ± 3.1	15.6 ± 2.4
K-Ascorbate I	0.3 ± 0.1	16.0 ± 0.4	1.1 ± 0.2	4.0 ± 0.6	10.8 ± 1.6	51.8 ± 2.3	15.6 ± 2.4
Titavit I	0.5 ± 0.3	16.9 ± 1.7	0.8 ± 0.1	4.2 ± 0.4	10.2 ± 1.2	52.2 ± 0.7	14.6 ± 0.3
K-Ascorbate II	0.3 ± 0.1	17.4 ± 1.2	1.3 ± 0.8	3.8 ± 0.4	10.4 ± 1.4	51.5 ± 2.0	14.7 ± 2.4
Titavit II	0.2 ± 0.1	17.8 ± 1.3	1.5 ± 0.6	4.7 ± 0.5	8.8 ± 1.0	48.8 ± 0.1	18.1 ± 1.4

Acknowledgement

This work is financially supported by the Hungarian National Committee of Technical Development (OMFB) (Hungarian-Spanish Intergovermental Scientific and Technical Cooperation).

References

[1] Pais, I. (1983) The biological importance of titanium, *J. Plant Nutr.* **6**, 3-131.
[2] Dummon, J.C. and Ernst, W.H.O. (1988) Titanium in plants, *J. Plant Physiol.* **133**, 203-209.
[3] Biacs, P. and Daood H.G. (1994) HPLC and photodiode-array detection of carotenoids and carotenoid esters, *J. Plant Physiol.* **143**, 520-525.
[4] Daood, H.G., Biacs, P.A., Dakar, M. and Hajdú, F. (1994) Ion-pair chromatography and photodiode-array detection of vitamin C and organic acids, *J. Chromatogr. Sci.* **32**, 481-487
[5] Speek, A.J., Schrijver, F. and Shreurs, H.P. (1985) Vitamin E composition of some oils as determined by fluorimetric detection, *J. Food Sci.* **50**, 121-124.

EFFECT ON ENVIRONMENTAL CONDITIONS ON THE MOLECULAR SPECIES COMPOSITION OF GALACTOLIPIDS IN THE ALGA *PORPHYRIDIUM CRUENTUM*

Daniel Adlerstein, Inna Khozin, Chiara Bigogno and Zvi Cohen
The Laboratory for Microalgal Biotechnology, Jacob Blaustein Institute for Desert Research, Ben-Gurion University of the Negev, Sede-Boker Campus 84990, Israel

INTRODUCTION

The lipid composition in higher plants and algae can be modulated by changing growth conditions such as temperature and light intensity. One of the major effects is on the level of fatty acid desaturation. Polyunsaturated fatty acids (PUFAs), especially those of the ω3 family appear to be essential for survival at low temperatures (Wada and Murata, 1989; Browse and Somerville, 1991). While there is an abundance of information pertaining to the effect of environmental conditions on the fatty acid composition of algae (Cobelas, 1989), very little is known concerning the effects on the molecular species composition of the individual lipids. The lipids of higher plants and algae are further divided according to their molecular species composition. The diacylglycerol (DAG) moieties of the prokaryotic species of galactolipids are assembled in the chloroplast and are characterized by the presence of C_{16} acyl groups at their *sn*-2 position, while the DAGs of eukaryotic species originate in the cytoplasm and contain C_{18} acyl groups in their *sn*-2 position. The aim of the present work was to study the effect of cultivation temperature and biomass density on the molecular species composition of the major galactolipids, MGDG and DGDG, in *Porphyridium cruentum*.

RESULTS

Effect of biomass density
The effect of biomass density on the fatty acid and the molecular species composition of the major galactolipids, MGDG and DGDG of *P. cruentum* was studied using daily diluted cultures at 3 different growth temperatures, 20 ^{0}C, 25 ^{0}C, and 30 ^{0}C. The cultures were kept at biomass densities corresponding to chlorophyll concentration ranges of 2-6 and 8-16 mg l^{-1}, respectively. In keeping with previous reports (Klyachko-Gurvich et al., 1985; Cohen et al., 1988), the proportion of eicosapentaenoic acid (20:5ω3, EPA) in the total lipids was lower in the denser cultures while that of arachidonic acid (20:4ω6, AA) was higher (Table 1). Following separation of MGDG and DGDG by reverse-phase HPLC to their constituent molecular species, we found that EPA decreased in both prokaryotic and eukaryotic molecular species (Table 2). Similar results were obtained for DGDG, which in *P. cruentum* is entirely prokaryotic (Table 2). E.g., the ratio of 20:5/16:0 to 20:4/16:0 increased in the less dense cultures, at 25 ^{0}C, from 1.3 and 2 to 27 and 91 in MGDG and DGDG, respectively. Similarly, the ratio of 20:5/20:5 to 20:4/20:4 in MGDG increased from 1.9 to 26.

Effect of growth temperature

The proportion of EPA in MGDG and DGDG increased with decreasing growth temperature while that of AA decreased (Table 2). This phenomenon was observed in both the prokaryotic and eukaryotic molecular species of MGDG as well as in DGDG (Table 2). Furthermore, comparison of the molecular species composition at different temperatures have shown that the share of the eukaryotic components was inversely related to the growth temperature. At 20°C, almost 60% of the molecular species were eukaryotic in comparison to only 20% at 30°C.

Table 1. Fatty acids composition of *P. cruentum* biomass cultivated at different temperatures and biomass densities.

Temp (°C)	Biomass density	fatty acids (% of total)		
		16:0	20:4	20:5
20	HD	27.3	18.4	39.1
20	LD	28.2	19.8	38.0
25	HD	30.3	28.6	26.5
25	LD	32.5	16.9	36.9
30	HD	35.1	32.0	12.3
30	LD	34.7	23.8	24.6

HD, high density (cultures were daily diluted to 10 mg chl./ml); LD, low density (cultures were daily diluted to 2.5 mg chl./ml).

Table 2. Molecular species composition of MGDG and DGDG of *P. cruentum* cultivated at different temperatures and biomass densities.

Lipid	molecular species	(% of total)					
		20 °C		25 °C		30 °C	
		HD	LD	HD	LD	HD	LD
MGDG	20:5/20:5	40	46	19	26	6	11
	20:4/20:5	14	11	17	3	8	7
	20:4/20:4	5	1	10	1	8	2
	20:5/16:0	27	20	26	54	23	53
	20:4/16:0	5	2	20	2	45	18
	18:2/16:0	9	20	8	14	10	9
Total	eukaryotic	59	58	46	30	22	20
Total	prokaryotic	41	42	54	70	78	80
DGDG	20:5/16:0	82	90	57	92	42	76
	20:4/16:0	7	2	28	1	47	14
	18:2/16:0	11	8	15	7	11	10

HD, LD (see Tab. 1).
The molecular species composition was determined by HPLC using light scattering detector.

DISCUSSION

The proportion of 20:5ω3 in galactolipids of *P. cruentum* increased with decreasing biomass concentration (Table 2) or decreasing light intensity (Cohen et al., 1988). More recent reports, seemingly in contradiction, suggest an opposite effect in other EPA producing algae. In both *Phaeodactilum tricornutum* (Arao et al., 1994) and *Nannochloropsis* (Sukenik et al., 1989) enhancing light intensity reduced the proportion of EPA.

Recently, Khozin and Cohen (1996) suggested that the biosynthetic pathways of *P. cruentum* include a chloroplastic ω3 desaturation step. In contrast, in both *P. tricornutum* (Arao et al., 1994) and *Nannochloropsis* (Schneider et al., 1995), ω3 desaturation is entirely cytoplasmic. We thus suggest that enhancing the light intensity per cell results in an increase in the ω3 chloroplastic desaturation of 20:4ω6 to 20:5ω3 in *P. cruentum*, in an apparent similarity to chloroplastic desaturation of 18:2ω6 to 18:3ω3 in higher plants.

Our data indicate that the modulation of EPA levels in the galactolipids of *P. cruentum* in response to changes in environmental conditions follows two patterns:

1) Increased light intensity promotes the chloroplastic ω3 desaturation of AA to EPA.
2) Decreased temperatures enhance the synthesis of the eukaryotic components of MGDG.

In higher plants the difference between the prokaryotic and the eukaryotic molecular species is rather subtle and is expressed in a slightly shorter acyl group at the *sn*-2 position of the prokaryotic species. In *P. cruentum* however, the ultimate eukaryotic species (20:5/20:5) contains ten double bonds in contrast to only five in the prokaryotic one (20:5/16:0). Possibly, one of the roles of the eukaryotic components is the accommodation with changes in membrane fluidity inflicted by variations in the ambient temperature. Indeed, we have recently isolated a mutant of this alga selected for its cold sensitivity which was shown to be lacking in its ability to produce eukaryotic molecular species of MGDG (Khozin et al., 1996).

REFERENCES

Arao T, Sakaki T, Yamada M (1994) Biosynthesis of polyunsaturated lipids in the diatom *Phaeodactilum tricornutum*. Phytochemistry 36: 629-635.

Browse J, Somerville C (1991) Glycerolipid synthesis: Biochemistry and Regulation. Annu. Rev. Plant Physiol. Plant Mol. Biol. **42**:467-506

Cobelas MA (1989) Lipids in microalgae. A review II. Environment. Grasas y Acietes **40**: 213-223

Cohen Z, Vonshak A, Richmond A (1988) Effect of environmental conditions of fatty acid composition of the red alga *Porphyridium cruentum*: correlation to growth rate. J Phycol **24**: 328-332

Khozin I, Zheng Yu H, Adlerstein D and Cohen Z(1996) Triacylglycerols partecipate in the eukaryotic pathway of PUFAs biosynthesis in the red microalga *Porphyridium cruentum*. 12th International Symposium on Plant Lipids.; 1996 July 7-12, Toronto.

Klyachko-Gurvich GL, Yureva MI, Semenenko VE (1985) Specificity of the fatty acid composition of acyl-containing lipids in the unicellular red alga *Porphyridium cruentum*. Fiziolgiya Rastenii **32**: 115-123

Schneider JC, Livine A, Sukenik A, Roessler PG(1995) A mutant of *Nannochloropsis* deficient in eicosapentaenoic acid production. Phytochemistry (in press).

Sukenik A, Yamaguchi Y and Livne A (1993) Alterations in lipid molecular species of the marine eustigmatophyte *Nannochloropsis* sp. J. Phycol. **29**: 620-626

CHILLING INJURY AND LIPID BIOSYNTHESIS IN TOMATO PERICARP

H. YU and C. WILLEMOT

Department of Plant Science, Université Laval, Québec, QC G1K 7P4 and Food Research and Development Center, Agriculture and Agri-food Canada, Saint-Hyacinthe, QC J2S 8E3.

Introduction

Lyons (1973) proposed that the primary event leading to chilling injury was a phase transition in membrane lipids. Genetic engineering showed that high-melting point phosphatidylglycerol (hmp-PG) is an important factor (Murata et al., 1992; Wolter et al., 1992), although other factors are also involved (Wu and Browse, 1995).

However, Hmp-PG is not a factor in chilling injury of solanaceae, including the tomato. Nguyen and Mazliak (1990) reported significant loss of galactolipids from tomato fruits during chilling. L'Heureux et al. (1994) showed the accumulation of polyunsaturated phosphatidylcholine (PC) under the same conditions. Detailed lipid analysis (Yu et al., 1996), including positional analysis of galactolipid fatty acids (Yu and Willemot, 1996) suggested inhibition of eukaryotic galactolipid biosynthesis. We bring further information, by ^{14}C labelling of lipids in pericarp disks, on the involvement of lipid metabolism in the development of chilling injury in tomato fruits.

Materials and Methods

Pericarp disks of tomato cvs 'Early Cherry' ('EC', chilling-sensitive) and 'New York 280' ('NY', chilling-tolerant) were fed ^{14}C-acetate at 20 or 4 °C. The disks were prepared from unchilled control fruits and from fruits chilled for 6 h or 8 d. Lipids were separated by TLC. Positional analysis of the fatty acids of monogalactosyldiacylglycerol (MGDG) was carried out by reverse phase TLC after *sn*-1 specific lipase hydrolysis.

Results

In the fruits fed at 20°C and after 6 h pre-chilling then fed at 4 °C, ^{14}C-acetate was readily incorporated into FFA and then into DAG (Fig. 1, a and b). In contrast, in the fruits after

8 d pre-chilling then fed at 4 °C, this pattern was strikingly altered: the label accumulated in FFA to a much higher level than in DAG (Fig. 1c).

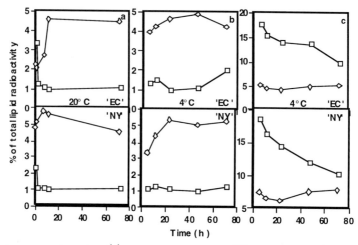

Figure 1. Incorporation of [1-^{14}C] acetate into FFA (□) and DAG (◊, % of total lipid radioactivity) at 20° C (unchilled control) (a) or after pre-chilling at 4° C for 6 h (b) or 8 d (c) pre-chilled fruits.

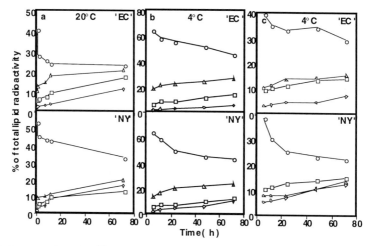

Figure 2. Incorporation of [1-^{14}C]-acetate (% of total lipid radioactivity) into PC, MGDG, DGDG and PE of tomato pericarp disks from mature-green fruits of chilling-sensitive cultivar 'EC' and tolerant 'NY'. Incorporation was carried out at 20° C (unchilled control) (a) or at 4° C for 6 h (b) or 8 d (c) pre-chilled fruits.   O : PC; ◊ : DGDG; □ : MGDG; Δ : PE.

PC showed as a precursor both at 20°C and at 4 °C (Fig. 2). In 6 h pre-chilling fruits (Fig.2b), PC accumulated label to a higher level in both cvs by feeding at 4 °C.

Autoradiographs of 2-D TLC of the labelled polar lipids (not shown) after 48 h feeding, PC was more heavily labelled and MGDG less labelled in the 'EC' than in the 'NY' in the disks prepared from the 8 d pre-chilling fruits.

Autoradiography of the labelled fatty acids in position sn-2 of MGDG (not shown) indicated only traces of label in the area covered by 16:0 (prokaryotic galactolipid) in 'EC' while this area was well labelled in 'NY'; 18:2 and 18:3 were well labelled in both cultivars (eukaryotic pathway).

Conclusion

Our data suggest inhibition of galactolipid biosynthesis during chilling, first at the step of glycerol-3-phosphate acyl transfer (prokaryotic pathway), and later at the transfer of polyunsaturated DAG from PC to MGDG, to a much greater extent in chilling-sensitive 'EC' than in tolerant 'NY'.

References

L'Heureux, G.P., Bergevin, M., Thompson, J.E. and Willemot, C. (1994) Molecular species of phosphatidylcholine, phosphatidylethanolamine and diglycerides of tomato pericarp during ripening, chilling and subsequent storage at 20 °C. *J. Plant Physiol.* **143**, 143:699.

Lyons, J.M. (1973) Chilling injury in plants. *Annu. Rev. Plant Physiol.* **24**, 445-466.

Murata, N., Ishizakinishizawa, Q., Higashi, S., Hayashi, H., Tasaka, Y. and Nishida, I. (1992) Genetically engineered alteration oin the chilling sensitivity of plants. *Nature*, **356**, 710-713.

Nguyen, X.V. and Mazliak, P. (1990) Chilling injury ninduction is accompanied by galactolipid degradation in tomato pericarp. *Plant Physiol. Biochem.* **28**, 283-291.

Wolter, F.P., Schmidt, R. and Heinz, E. (1992) Chilling sensitivity of *Arabidopsis thaliana* with genetically engineered membrane lipids. *EMBO J.* **11**, 4685-4692.

Wu, J. and Browse, (1995) Elevated levels of high-melting-point phosphatidylglycerols do not induce chilling sensitivity in an *Arabidopsis* mutant. *J. Plant Cell*, **7**, 17-27.

Yu, H., Thompson, J. E., Yelle, S. and Willemot, C. (1996) Impairment of galactolipid biosynthesis in tomato pericarp at chilling temperature. *J. Plant Physiol.*, in press.

Yu, H. and Willemot, C. (1996) Inhibition of eukaryotic galactolipid biosynthesis in mature-green tomato fruits at chilling tempeature. *Plant Science*, **113**, 33-41.

CEREBROSIDES IN SEED-PLANT LEAVES: COMPOSITION OF FATTY ACIDS AND SPHINGOID BASES

HIROYUKI IMAI[1], MASAO OHNISHI[2], MICHIYUKI KOJIMA[2] AND SEISUKE ITO[2]
[1]Department of Biology, Konan University, 8-9-1 Okamoto, Higashinada, Kobe 658, Japan, [2]Department of Bioresource Chemistry, Obihiro University of Agriculture and Veterinary Medicine, Inada, Obihiro 080, Japan

Introduction

Cerebroside (monoglycosylceramide) is a major lipid class in plasma membranes and tonoplasts of several plant species [1,2]. It has been reported that cerebrosides would set the phase transition temperature of lipids in plasma membranes and tonoplasts, which has proposed a possible mechanism of chilling injury of plants [3]. In addition, changes of cerebroside content in plasma membranes after acclimation to cold stress [2] and water-deficit stress [4] were reported. These findings have suggested that cerebroside may play physiological roles in plasma membranes and tonoplasts. In this communication, we report the composition of fatty acids and sphingoid bases in cerebrosides from leaves of various seed-plants for primarily discussing whether the apparent correlation of cerebroside composition with chilling sensitivity.

Distribution of unsaturated hydroxy fatty acids

The component fatty acids of cerebrosides from leaves were almost exclusively 2-hydroxy fatty acids [5]. These results were similar to the results as reported previously [6,7]. Hydroxytetracosenic acids (24h:1) were found only in 3 species of chilling-resistant plants (broccoli, chrysanthemum and dandelion) among the 21 plant species analyzed. Interestingly, any species of chilling-sensitive plants analyzed included no significant amounts of cerebroside species containing these fatty acids. Fig. 1 shows 2-hydroxy fatty acid composition of cerebrosides from leaves of five species of Gramineae. Unsaturated hydroxy fatty acids were found in chilling-resistant oats, wheat and rye, while these fatty acids were not found in chilling-sensitive rice and maize. In particular, it would be worth noting that high proportions of cerebroside species having unsaturated hydroxy fatty acids are detected in very cold-hardy plants such as rye and wheat. All of unsaturated hydroxy fatty acids in plant cerebrosides showed very long chains with more than C_{22}. These fatty acids combined mainly with trihydroxy bases such as 4-hydroxysphingenine (t18:1). We have previously observed that phase transition onset temperatures of cerebroside species containing t18:1(8Z)-24h:1, t18:1(8Z)-24h:0, t18:1(8E)-24h:1 and t18:1(8E)-24h:0 were 20 °C, 39 °C, 31 °C and 58 °C, respectively. These results have suggested that cerebroside species having unsaturated hydroxy fatty acids exhibited much lower phase transition temperatures than those having saturated hydroxy fatty acids. In addition, some cerebroside species containing trihydroxy bases and very long chain hydroxy fatty acid with more than C_{22} would appear to determine the phase transition temperature in the micelles system including unsaturated phospholipids [3]. It would therefore be assumed that introduction of the Z-double bond into the hydroxy fatty acid moiety of cerebroside can result in a lowering of the temperature for the gel to liquid phase transition of membrane lipids.

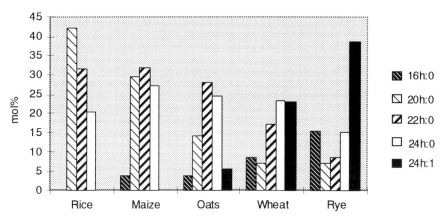

Fig. 1 2-Hydroxy fatty acid composition of cerebrosides from leaves of Gramineous plants

The characteristics of several biological families on sphingoid bases

The phase transition temperatures of the cerebroside species having 8-Z -unsaturated sphingoid bases showed lower phase transition temperatures than those having 8-E- forms, e.g. 39 °C for t18:1(8Z)-24h:0-Glc; 58 °C for t18:1(8E)-24h:0-Glc [8]. During the initial stages of the present study, we originally anticipated that the degree of 8-Z- unsaturation of the sphingoid bases would appear to be different between chilling- resistant and chilling-sensitive plants: the cerebroside species having 8-Z -unsaturated sphingoid bases would be enriched in chilling-resistant plants. Thus sphingoid bases of cerebrosides from leaves of 22 plants were analyzed by capillary gas chromatography after conversion to the corresponding fatty aldehydes. The results, however, indicated no demonstrable differences on the levels of 8Z- and 8E-isomers between chilling-sensitive and chilling-resistant plants.

Plots of the %t18:1(8Z) of total t18:1 versus the percentage of d18:1(8Z) plus d18:2(4E,8Z) in the sum of d18:1 and d18:2 indicated a simple linear regression with positive trend with the exception of Chenopodiaceae (spinach and beet) (Fig. 2). It is possible, therefore, that the molecular mechanism involved in the desaturation and/or isomeration at the C-8 position of plant sphingoid bases would coordinately operated between the dihydroxy and trihydroxy bases. Gaining insight into the mechanism responsible for the 8-unsaturated sphingoid base formation in plants could be important for discussing the possible roles of plant cerebroside in connection with membrane function in plasma membranes and tonoplasts. Additionally, plotting the proportions of the 8-unsaturated base isomers in plant cerebrosides may be applicable for characterizing some biological families, since its distribution seemed to be peculiar to each family.

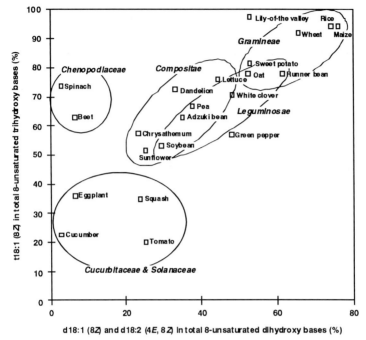

Fig. 2 Proportion of 8-Z-sphingoid bases between component dihydroxy and trihydroxy bases in leaf cerebrosides

References

1. Yoshida, S. and Uemura, M.: Lipid composition of plasma membrane and tonoplasts isolated from etiolated seedlings of mung bean (*Vigna radiata* L.), *Plant Physiol.* 82 (1986), 807-812.
2. Lynch, D.V. and Steponkus, P.L.: Plasma membrane lipid alterations associated with cold acclimation of winter rye seedlings (*Secale cereale* L. cv Puma), *Plant Physiol.* 83 (1987), 761-767.
3. Yoshida, S., Washio, K., Kenrick, J. and Orr, G.: Thermotropic properties of lipids extracted from plasma membrane and tonoplast isolated from chilling-sensitive mung bean, *Plant Cell Physiol.* 29 (1988), 1411-1416.
4. Norberg, P. and Liljenberg, C.: Lipids of plasma membranes prepared from oat root cells, *Plant Physiol.* 96 (1991), 1136-1141.
5. Imai, H., Ohnishi, M., Kinoshita, M., Kojima, M. and Ito, S.: Structure and distribution of cerebroside containing unsaturated hydroxy fatty acids in plant leaves, *Biosci. Biotech. Biochem.* 59 (1995), 1309-1313.
6. Ohnishi, M., Ito, S. and Fujino, Y.: Characterization of sphingolipids in spinach leaves, *Biochim. Biophys. Acta* 752 (1983), 416-422.
7. Cahoon, E.B. and Lynch, D.V.: Analysis of glucocerebrosides of rye (*Secale cereale* L. cv Puma) leaf and plasma membrane, *Plant Physiol.* 95 (1991), 58-68.

CHANGES IN PHOSPHATIDYLINOSITOL METABOLISM IN SUSPENSION CULTURED TOBACCO CELLS SUBMITTED TO STRESS

P. NORBERG, D. CHERVIN, M. GAWER, N. GUERN, Z. YANIV*AND P. MAZLIAK
Université Pierre et Marie Curie (Paris 6), Laboratoire de Physiologie Cellulaire et Moléculaire, CNRS - URA 2135, case 154, 4 place Jussieu, 75252 Paris Cedex 05, France. *The Volcani Center, A.R.O, Bet-Dagan 50250, Israel.

1. Introduction.

Changes in inositol phospholipids had been shown to occur in plant cells as a result of different stress (1). We decided to study the inositol phospholipid metabolism in suspension cultured Tobacco cells because they provide a relatively uniform system to which a stress might be easily applied.

2. Material and methods.

Cells (approximately 0.25g in 7ml of medium) were separately cultivated (2). For measuring the time course of [^{33}P] incorporation into phospholipids, 10μCi of [^{33}P] orthophosphate was provided to each flask. Incorporation was stopped at specified intervals by adding ice-cold extracting solvent to each culture. Acidic solvent for lipid extraction was chloroform: methanol: conc. HCl (100:100:1.5 v/v/v). After centrifugation, determination of radioactivity was performed on aliquots of the aqueous and the chloroform phase. Lipids were separated by TLC on plates presoaked in 1% potassium oxalate. The TLC solvent system composed of chloroform: methanol: ammoniac: water (90:70:5.5:15 v/v/v/v), routinely used, provided excellent and reliable separation of PIP2, LPIP, PIP and PC (3). To resolve PI - PA and PE - PG a TLC solvent system according to (4) was used containing chloroform: acetone: methanol: acetic acid: water (50:20:10:10:5 v/v/v/v/v). Phospholipids were located with authentic standards spotted on the TLC plate with the extract. Radiolabelled spots were visualized via autoradiography and scraped from TLC into scintillation vials for radioactivity determination.

Abbreviations; DAG, diacylglycerol; LPA, lyso phosphatidic acid; LPIP, lyso phosphatidyl inositol 4-phosphate; PA, phosphatidic acid; PC, phosphatidyl choline; PE, phosphatidyl ethanolamine; PG, phosphatidyl glycerol; PI, phosphatidyl inositol; PIP, phosphatidyl inositol 4-phosphate; PIP$_2$, phosphatidyl inositol 4,5-bisphosphate;

The phosphoinositides PIP and PIP2 were further identified by hydrolysis with phospholipase A2, giving products which co-migrated with respective standards. Results are means of 2 determinations from 3 separate experiments.

Choline chloride (10mM), ABA (10μM) and cold stress were imposed on cell cultures without stopping agitation. Cold temperature (1.5°C) was obtained within 3 min by transferring the samples to an ice-water bath. Other parameters were as described in (2).

3. Results

3.1. LABELLING OF PHOSPHOLIPIDS FROM [^{33}P] ORTHOPHOSPHATE IN 8 DAY-OLD CELLS

The time course of [^{33}P] incorporation into total lipids is shown in fig 1a. The labelling was linear between 5 and 60 min. The phosphoinositides were the first radiolabelled lipids. After 2 min incorporation PIP was by far the most heavily labelled lipid, and its labelling remained at a stable level during the first hour (fig 1b), LPIP and PIP2 were weakly represented (~2%). The labelling of PI increased steadily up to 4 hours incubation, with an incorporation rate comparable to that of the other structural phospholipids. The fast labelling of PIP, PIP2 and PA indicates high activities of PI, PIP and DAG kinases.

The absolute uptake and incorporation of [^{33}P] by tobacco cells depended on the stage of development, but the labelling of individual lipids followed the same course as described above for 8 day-old cultures.

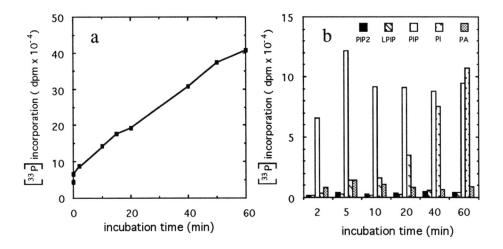

FIGURE 1. Time course of [^{33}P] orthophosphate incorporation, (a) into total lipids of tobacco cell cultures, (b) into the various phosphoinositides of tobacco cells.

3.2. STRESS TREATMENTS OF TOBACCO CELL CULTURES

Addition of 10 mM choline chloride to tobacco cells, 10 min before the end of a short term (10 min) or a long term (20 h) incubation with [^{33}P] markedly increased the labelling of PC but did not induce changes in the phosphoinositide pool labelling.

Addition of 10 µM ABA in the culture medium for 2 or 10 min during a 10 min incubation in the presence of [^{33}P] reduced significantly the labelling of total phospholipids. Moreover in cells cultivated during three days in the presence of 10 µM ABA before a 10 min [^{33}P] incubation, the incorporation was very low, and the only labelled phospholipids detected were PIP and PA.

Following a 10 min [^{33}P] incubation at 25°C an exposure to a 10 min cold stress (1-2°C) greatly enhanced PA labelling at the expense of the other phospholipids but did not change the total radioactivity incorporation. However when the cold treatment preceeded a [^{33}P] incubation, performed at 1-2°C, the incorporation in the lipids was reduced and only two labelled metabolites, PA and PIP were present, PIP being predominant. When the radiolabelled precursor was given at the onset of the cold treatment, the total [^{33}P] incorporation was strongly reduced and the only detected labelled metabolites were PA » PIP > LPA.

4. Conclusion

Our results show that the components of the phosphatidyl inositol cycle are present in suspension cultured tobacco cells, which is in agreement with previous report (5). The incorporation of [^{33}P] into phosphoinositides was rather fast, preferentially into PIP which peaked at 5 min (70% of tot. inc.) and then levelled (fig 1b). The fast incorporation of [^{33}P] into PIP, PIP2 and PA is considered as a the result of rapid phosphoinositide turnover (3). Moreover, changes in the [^{33}P] PA to PI ratio are good indicators of the rapid biosynthesis of the major membrane phospholipids. In tobacco cells we noted a lack of response of PIP2 to cold stress or to added substances as choline chloride or ABA to the culture medium. Cold temperature resulted in an enhanced level of PA and PIP suggesting that PI-kinase and the enzymes involved in PA formation (e.g. DAG-kinase and phospholipases C and D) remained active in cold.

So, our results, as those of others (6), do not allow to decide whether PIP and PA might be involved in signal transduction in plants.

5. References

1. Coté, G.G. and Crain, R.C. (1993) Biochemistry of phosphoinositides, *Ann. Rev. Plant Phys. Plant Mol. Biol.* **44**, 333-356.
2. Gawer, M., Guern, N., Chervin, D. and Mazliak, P. (1988) Comparison between a choline tolerant Nicotiana tabacum cell line and the corresponding wild type, *Plant Physiol. Biochem.* **26**, 323-331.
3. Munnik, T., Musgrave, A. and de Vrije, T. (1994) Rapid turnover of polyphosphoinositides in carnation flower petals. *Planta* **193**, 89-98.
4. Lepage, M. (1967) Identification and composition of turnip root lipids, *Lipids* **2**, 244-250.
5. Kamada, Y. and Muto, S. (1990) Ca^{2+} regulation of phosphatidylinositol turnover in the plasma membrane of tobacco suspension culture cells, *Biochim. Biophys. Acta* **1093**, 72-79.
6. Harryson, P. (1995) Lipid transfer and signal transduction in plants: A study of inositol phospholipids, Thesis, Göteborg University, Sweden.

INHIBITION OF POLYUNSATURATED FATTY ACID SYNTHESIS BY SALICYLIC ACID AND SALICYLHYDROXAMIC ACID AND THEIR MODES OF ACTION

A, BANAS[1], G. STENLID[1], M. LENMAN[2], F. SITBON[1] and S. STYMNE[2]

[1]*Department of Plant Physiology, Swedish University of Agricultural Sciences, Box 7047, S-750 07 Uppsala, Sweden,* [2]*Department of Plant Breeding Research, Swedish University of Agricultural Sciences, Herman Ehles v. 2-4, S-268 31 Svalöv, Sweden*

1. Introduction

Inhibition of root growth in wheat caused by the grass herbicide haloxyfop [2-(4-[(3-chloro-5-trifluoromethyl)2-pyridinyl)oxo]phenoxy)propanoic acid) is accompanied by a relative increase in linolenate in the membrane lipids of the root tips (Banas *et al.*, 1993a). A number of substances which counteract the growth inhibition also counteract the increase in linolenate (Banas *et al.*, 1993b). In order to elucidate possible relationship between these antagonistic compounds and their effect on polyunsaturated fatty acid content, we report in this communication the effect of two of these substances, salicylic acid (SAL) and salicylhydroxamic acid (SHAM), on the production of polyunsaturated fatty acids *in vivo* and *in vitro*.

2. Materials and Methods

Wheat seedlings (*Tricium aestivum*, cv. Diamant II) with central roots of 22-26 mm were grown in nutrient solution for 24h at 20°C in the dark with or without the addition of SAL or SHAM. In order to establish the *in vivo* effects of SAL and SHAM on linseed (*Linum usitatissimum*, cv. MacGregor) membrane lipids, cotyledons at the mid-stage of seed development (20-25 days after flowering) were detached from their seed coats and placed in petri dishes in medium (pH 5.8) according to Nitsch and Nitsch (1967) containing 0.33M sucrose, vitamins, macro- and micronutrients and SAL or SHAM. Cotyledons were incubated in light for 24h at 25°C before extraction and lipid analysis. Microsomes were prepared from root tips (the outermost 10 mm of the central root) and cotyledons of linseed according to Bafor *et al.* (1991) except that microsomes used in Δ15 desaturase assays were prepared from tissues homogenized in medium containing 4mM of NADH. Microsomal desaturase assays were performed with NADH and [^{14}C]oleoyl-CoA or [^{14}C]linoleoyl-CoA or *in situ* ^{14}C-labelled microsomal lipids according to methods described by Bafor *et al.* (1993) . Lipid extraction and analysis are described elsewhere (Bafor *et al.*, 1991)

3. Results and Discussion

Wheat plants supplied with SAL and SHAM in the absence of haloxyfop showed a decrease in linolenate and increase in oleate in root tips lipids compared to plants grown in only nutrient solution (Table 1). Unsaturated fatty acids in phosphatidylcholine of detached linseed cotyledons showed similar changes as seen in the lipids in wheat root tips after 24h incubations with SAL or SHAM (Table 1).

It is likely that the changes in oleate and linolenate content are caused by effects of SAL and SHAM on $\Delta12$ and $\Delta15$ desaturase activities. Therefore, the $\Delta12$ desaturase activity was measured in microsomal preparations of root tips from wheat plants treated with SAL or SHAM as well as in preparations from non-treated roots. The microsomes from the SAL treated roots had 23% lower $\Delta12$ desaturase activity than microsomes from control roots whereas microsomes from roots treated with SHAM had substantially higher $\Delta12$ desaturase activity than the control (Table 2).

TABLE 1. The effect of salicylic acid (SAL) and salicylhydroxamic acid (SHAM) on fatty acid composition in root tips of wheat seedlings and detached linseed cotyledons

Plant	Treatment[a]	Fatty acid composition[b]		
		18:1	18:2	18:3
		mol%		
Wheat	Control	5.3	44.8	25.5
	SAL (3×10^{-5}M)	14.1	47.6	14.9
	SHAM (3×10^{-4} M)	56.7	14.7	13.7
Linseed	Control	15.9	23.3	39.5
	SAL (10^{-4}M)	27.9	29.8	24.3
	SHAM (3×10^{-4}M)	33.3	28.7	19.7

[a] Treatment for 24 h.
[b] Fatty acid composition given for total fatty acids in root tips and in phosphatidylcholine in detached linseed cotyledons

Microsomes prepared from wheat roots from control plants were assayed for $\Delta12$ desaturase activity with the presence of SAL or SHAM in the assay mixture. SAL had no effect on desaturase activity whereas SHAM significantly inhibited the activity (Table 2). The wheat root microsomes did not show any $\Delta15$ desaturase activity. However, linseed microsomes were effective in desaturating both oleate and linoleate. The $\Delta12$ desaturase and $\Delta15$ desaturase activities in linseed microsomes were hardly effected by SAL but substantially inhibited by SHAM (Table 2).

The level of expression of the FAD2 and FAD3 genes were not significantly altered in linseed cotyledons treated with SHAM or SAL compared to non-treated cotyledons, as judged by northern blots (data not shown).

Although SAL and SHAM have a similar effect on membrane fatty acid composition, their mode of actions are obviously different. SAL has no direct effect on the $\Delta12$ desaturase (and $\Delta15$ desaturase) activity *in vitro*, but membranes prepared from SAL treated plants had somewhat depressed $\Delta12$ desaturase activity; possibly caused by a post-transcriptional mechanism. SHAM, on the other hand, has a direct inhibitory effect on the $\Delta 12$ desaturase (and $\Delta 15$ desaturase) activity but does not lower the amount of activity recovered in the membranes of the plants.

TABLE 2. The effect of salicylic acid (SAL) and salicylhydroxamic acid (SHAM) on desaturase activities in microsomes prepared from root tips of wheat and detached linseed cotyledons.

Plant	Treatment	Microsomal desaturase activities	
		$\Delta 12$ desaturase	$\Delta 15$ desaturase
		(% of control)	
Wheat	Control	100	[c]
	Plants grown in SAL (3×10^{-5}M)[a]	77[b]	[c]
	Plants grown in SHAM (3×10^{-4}M)[a]	143[b]	[c]
	Plants grown in control solution, microsomes incubated with SAL (5×10^{-4}M)	100[d]	[c]
	Plants grown in control solution, microsomes incubated with SHAM (5×10^{-4}M)	51[d]	[c]
Linseed	Control microsomes	100[d]	100[d]
	Microsomes incubated with SAL (5×10^{-4}M)	82[d]	92[d]
	Microsomes incubated with SHAM (5×10^{-4}M)	45[d]	24[d]

[a] Treatment of plants with SAL or SHAM and was performed for 24h before preparation of microsomal membranes.

[b] Desaturase activity calculated from the amount endogenous oleoylgroups + [^{14}C] oleoyl groups desaturated in microsomal membranes.

[c] Wheat microsomes did not show any $\Delta 15$ desaturase activity.

[d] Desaturase activity was calculated from the amount of [^{14}C]oleoyl substrate or [^{14}C]linoleoyl groups desaturated in microsomes with and without SAL or SHAM.

That SAL and SHAM have different modes of action on desaturase activities strengthens the hypothesis that the antagonistic effects of these compounds against haloxyfop are due to their common abillity to conteract the increase in linolenate caused by the herbicide (Banas *et al.*, 1993b).

REFERENCES

Bafor, M., Smith, M., Jonsson, L., Stobart, K. and Stymne, S. (1991) Ricinoleic acid biosynthesis and triacylglycerol assembly in microsomal preparations from developing castor bean endosperm, *Biochem. J.* **280**, 507-514.

Bafor, M., Smith, M.A., Jonsson, L., Stobart, K. and Stymne, S. (1993) Biosynthesis of vernoleate (cis-12-epoxyoctadeca-cis-9-enoate) in microsomal preparations from developing endosperm of *Euphorbia lagascae*, *Arch. Biochem. Biophys.* **303**, 145-151.

Banas, A., Johansson, I., Stenlid, S. and Stymne, S. (1993a) The effect of haloxyfop and alloxydim on growth and fatty acid composition of wheat roots, *Swedish J. Agric. Res.* **23**, 55-65.

Banas, A., Johansson, I., Stenlid, S. and Stymne, S. (1993b) Free radical scavengers and inhibitors of lipoxygenases as antagonists against the herbicides haloxyfop and alloxydim, *Swedish J. Agric. Res.* **23**, 67-75.

Nitsch, C. and Nitsch, J.P. (1967) The induction of flowering *in vitro* in stem segment of *Plumago indica*, *Planta* **72**, 355-370.

IDIOBLAST AS A MODEL SYSTEM FOR THE STUDY OF BIOSYNTHESIS OF LIPIDS WITH ANTIFUNGAL PROPERTIES IN AVOCADO FRUITS

A. I. LEIKIN-FRENKEL and D. PRUSKY.
Department of Post-Harvest Science of Fresh Produce.
Agricultural Research Organization,
Volcani Center, Bet Dagan 50250. Israel

1. Introduction

The most important pathological factor limiting life after harvest in subtropical fruits are quiescent infections caused by *Colletrotrichum gloeosporioides* (4). Prusky and Keen determined that the resistance of avocado to the fungi involves the action of a preformed biocide 1-acetoxy-2-hydroxy-4-oxo-heneicosa-13,15 diene (3). This compound, found in avocado pericarp inhibits the growth of *C. gloeosporioides* in unripe fruits and it's level decreases to sub-fungitoxic concentration during fruit ripening in parallel with the initiation of fungal development. *C. gloeosporioides* also decays the mesocarp of the fruits despite fungitoxic concentrations of antifungal compounds (1). This may be explained by the compartamentation of the antifungal diene in idioblast oil cells (1). The catabolism of the diene in pericarp after harvest is attributed to the oxidation by lipoxygenase whose activity is in turn regulated by epicatcchin, a non specific inhibitor present in the pericarp of unripe, but not ripe fruit (5). The biosynthetic process of this antifungal diene is, however, still unknown.
The objectives of the present work were: a-to localize the site of synthesis of antifungal lipids in avocado fruit, b- to determine the relationship between the synthesis and the activity of the compound and c- to develop a model system for these studies.

2. Experimental Procedures

Avocado fruits (cv. Fuerte) were employed within 24h of harvest. Pericarp and mesocarp slices (1-2 mm thick) were cut and their diene content was extracted with ethanol (20 ml/g fr.weight) (4). Idioblasts were isolated from avocado mesocarp by enzymatic digestion essentially according to Platt et al. (2) with slight modifications. The cells were obtained together with lipids from the mesocarp. The antifungal diene was extracted as above after boiling the cells for 3 min (1). The extracted diene was fractionated in dichloromethane-water and solubilized in ethylacetate . It was then analyzed and quantified by HPLC and TLC (4). I_2 detection and scintillation counting of the scraped spots were applied when appropiate. Ethylene treatment (40 µl/l) was applied to whole fruits or to fruit slices, in flasks sealed with a rubber septum placed under a humidified atmosphere. The same treatment was applied to idioblasts .

3. Results

The levels of antifungal diene in the pericarp of whole avocado fruits could be increased by exposure to ethylene (40 µl/l) during 3 h. The diene content of the extracted pericarp increased by 50% 6 h after ethylene removal. Similar results were obtained at 35°C.

Avocado pericarp and mesocarp slices or whole fruits exposed to ethylene for 3 h at 35°C showed that the diene content increased in the mesocarp as well as in the pericarp, when whole fruits were exposed. But when slices were exposed, the diene content increased only in the mesocarp. Pericarp's slices isolated from the fruit did not respond to ethylene. This suggests that avocado's mesocarp respond to ethylene treatment as the whole fruit does.

Idioblasts separated from avocado mesocarp by enzymatic digestion overnight and further centrifugation were also responsive to stimulation by ethylene. The cells responded to ethylene as a function of temperature; maximal increase in the diene content was observed at 35°C. At this temperature, a time course study showed that a minimum of 1 h of treatment is required for the cells to respond to the stimulus by increasing their diene content. This behaviour continued for at least 3h.

Since no increase of diene was induced in the pericarp alone, we hypothesized that the compound may be produced in the mesocarp and transported towards the pericarp. To test that hypotesis, the cells were exposed to ethylene before and after separation from their lipid environment by filtration. In both cases, the diene content was measured in the cells and in the medium. When the whole system was stimulated together, the diene content increased in the cells as a function of time, as well as in the surrounding medium. However when the separation was performed before the exposure, only the cells were able to increase their diene content 3h after exposure to ethylene. No response was found in the lipid environment thus suggesting that the increase of diene in the avocado masocarp's lipids after exposure of the whole system to ethylene is due to idioblasts stimulation and release of the formed diene towards the medium.

As an initial stage in the study of the metabolic pathway we observed that idioblasts were able to incorporate ^{14}C-2-malonyl CoA into a lipid that co-chromatographed with the antifungal diene both in TLC and in HPLC.

4. Conclusions

The level of antifungal lipids in avocado pericarp and mesocarp can be induced by ethylene stimulus.

The lack of response by pericarp alone suggests that avocado mesocarp may be responsible for the changes in the diene levels in the pericarp. Idioblast oil cells from avocado mesocarp are also susceptible to stimulation: the effect of ethylene on their antifungal diene induction is time and temperature dependent.

Idioblasts but not their lipid environment are responsible for the increased production of antifungal diene. They seem to export the diene towards their surrounding medium.

Idioblasts are metabolically active and able to incorporate labelled precursors into a compound putative as antifungal diene.

5. References

1. Kobiler, I., Prusky, D., Midlans, S., Sims, J. J. and Keen, N.T. (1993) . Compartmentation of antifungal compounds in oil cells of avocado fruit mesocarp and it's effect on susceptibility to *Colletotrichum gloeosporioides*. Physiol Mol. Pl. Path. 43, 319-328.

2. Platt, K.A. and Thomson,W.W. (1992). Idioblast oil cells of avocado: distribution, isolation, ultrastructure, histochemistry and biochemistry. Int. J. Plant. Sci. 153, 301-310.

3. Prusky, D. and Keen, N. (1993). Involvement of preformed antifungal compounds in the resistance of subtropical fruits to fungal decay. Plant Disease 77, 114-119.

4. Prusky, D. , Keen, N., Sims, J. J. and Midland, S. L. (1982). Possible involvement of an antifungal diene in the latency of *Colletotrichum gloeosporioides* in unripe avocado fruits. Phytopatology 72, 1578-1582.

5. Prusky, D., Jacoby, B., Kobiler,I., Sims, J. J. and Midland, S. L. (1985). Effect of inhibitors of lipoxygenase activity and it's possible relation to latency of *Colletotrichum gloeosporioides* on avocado fruits. Physiol. Mol. Pl. Path. 27, 269-279.

INHIBITION OF THREE SOYBEAN FUNGAL PLANT PATHOGENS BY LIPID DERIVATIVES AND NATURAL COMPOUNDS

Brian J. Barnes [1], Dept. of Biology, Pittsburg State University, Pittsburg, Kansas 66762
Helen A. Norman, Weed Science Laboratory, Plant Science Institute, U.S. Dept. of Agriculture, Beltsville Agricultural Research Center, Beltsville, Maryland 20705
Nancy L. Brooker, Dept. of Biology, Pittsburg State University Pittsburg, Kansas 66762

1.1 SYNOPSIS

The use of naturally occurring plant derived compounds, including lipids, fatty acids and triterpenoids to inhibit fungal plant pathogens represents a new approach to controlling agriculturally important fungal diseases. New evidence for the use of sesamol and glycerrhizic acid and their inhibition of three fungal plant pathogens of soybean (*Glycine max* (L.) Merrill) are collectively reviewed. In addition, the application and effectiveness of these alternative natural compounds for biological control of soybean fungal pathogens will be discussed.

1.2 INTRODUCTION

United States production of abundant food and fiber would not be possible without crop protection. Historically, this crop protection has relied on synthetic pesticides, with 85% of the total 1990 US sales of pesticides being used for agricultural applications.[1] Currently, the general public's concern over the use of chemical pesticides and their impact on environment, food safety and water quality has created a significant mandate for exploring alternative strategies for controlling crop pests.[2,3] Reductions in chemical use may be achieved through the use of resistant varieties and natural fungicidal compounds. One suggested approach to controlling fungal diseases is through the use of naturally occurring plant derived compounds including lipids, fatty acids and triterpenoids. Several of these natural compounds have been identified as having antifungal or antimicrobial activity.[4,5,6,7] However, no evaluation of these natural compounds for antifungal activity of soybean pathogens have been performed. In this study eight natural compounds were

---

[1] Currently Located at The University of Kansas, School of Pharmacy
1301 W. 24th C18
Lawrence, Kansas 66064

evaluated for antifungal activity *in vitro* to determine their potential as natural pesticides on soybeans (*Glycine max* (L.) Merrill). This report describes the results of this *in vitro* screening study and phytotoxicity tests on soybeans.

1.3 METHODS AND MATERIALS

1.3.1 *Research Procedure 1, Screening of Lipid, Fatty Acid and Triterpenoid Compounds for Antifungal Activity Against Economically Important Soybean Pathogens*

Strains and Media. Stock cultures of *Diaporthe phaseolorum* (Cke. and Ell.) var. sojae Wehm. (syn. D. sojae Leh.), *Cercospora sojina* Hara (syn. *C. daizu*), and *Septoria glycines* Hemmi were maintained on Potato Dextrose Agar at 25 C.

Lipids, Fatty Acids and Triterpenoids for Screening. Eight compounds were screened for antifungal activity; Linoleic Acid, Palmitic Acid, Arachidonic Acid, Trans-Linoelaidic Acid, Sesamol, Carbenoxolone, Glycyrrhizic Acid, Glycerrhetinic Acid.

Screening For Antifungal Activity. Two approaches to screening for antifungal activity were utilized. Direct incorporation of the compound in a minimal screening media measured inhibition as the amount of growth from the inoculation point (mm of mycelial growth) as compared to a non-amended control culture. The second screening approach was to use a disc assay with the dissolved compounds applied to inert paper discs. These discs were dried and placed on a plate with the culture of fungus 4 cm. apart. Inhibition was measured in mm zones of inhibition around the disc, as compared to nontreated paper disc and solvent only treated disc controls.

1.3.2 *Research Procedure 2, Plant Tests to Determine the Feasibility of Using Compounds to Control Soybean Diseases*

Phytotoxicity Assays. Compounds identified with antifungal activity were used for plant tests to determine disease resistance potentials. Soybean varieties Essex, KS5292, Crawford and Hutchenson at 2-4 true leaf stage, two plants per pot, two pots per treatment were used for the test. Antifungal compounds were incorporated into a spray formulation for application onto soybean plants. Two ml volumes of 0, 10, 100, 250 and 500 µg/ml concentrations were applied to the plants followed by incubation on a greenhouse bench at 30 C. Phytotoxicity was assessed daily over a 14 day period and the test was repeated.

1.4 RESULTS

Sesamol and Carbenoxolone possessed significant inhibitory characteristics after 4 days with mean fungal inhibitions of 100% and 69% respectively (Table 1). From the two screening methods which were performed, direct incorporation of the compounds in screening media displayed the highest levels of fungal inhibition. Soybean phytotoxicity tests on sesamol and carbenoxolone resulted in no visible toxicity symptoms on any of the soybean varieities after 14 days.

TABLE 1. Percent Inhibition of Three Fungal Plant Pathogens by Three Naturally Derived Compounds Using Media And Disc Assays

	Media Assay	Disc Assay	Media Assay	Disc Assay	Media Assay	Disc Assay
C. sojina	Sesamol		Carbenoxolone		Glycerrihizic Acid	
Day 1	0	0	0	0	0	0
Day 4	100	24±0.5	0	24±0.5	0	0
Day 7	100	33	60	11±0.5	60	23±0.5
S. glycines						
Day 1	0	0	0	0	0	0
Day 4	100	41	0	41	0	41
Day 7	100	23	0	33	0	23
D. phaseolorum						
Day 1	0	0	0	0	0	0
Day 4	100	89±1.9	66	69±1.2	23	12±4.6
Day 7	100	38±6.1	43	24±3.3	0	0

1.5 CONCLUSION

The results of this research yields a very promising outlook for the use of natural compounds to inhibit fungal pathogens on soybean plants. Sesamol's high level of fungal inhibition and stability over time makes it the most attractive compound for further *in plantae* tests. This pest control approach is expected to be of considerable interest to industry and use to the farmer, and will benefit U.S. soybean production and the environment. The significant fungal inhibition of Sesamol and Carbenoxolone along with their lack of phytotoxic symptoms makes them an ideal candidate for such an agricultural application.

1.6 REFERENCES

1. National Agricultural Chemicals Association. 1991. Annual Report 1991. Washington, D.C.

2. Ragsdale, N.N., Henry, M.J. and Sisler. 1993. Minimizing Nontarget Effects of Fungicides. National Agricultural Pesticide Impact Assessment Program. American Chemical Society 524:332-341.

3. National Resource Council. 1987. Regulating Pesticides in Food. National Academy Press, Washington, D.C.

4. Trione, E.J. and Ross, W.D. 1988. Lipids as Bioregulators of Teliospore Germination and Sporidial Formation in the Wheat Bunt Fungi *Tilletia* sp. Mycologia,38-45.

5. Cohen, Y., Gisi, U. and Mosinger, E. 1991. Systemic Resistance of Potato Plants Against Phytophthora infestans Induced by Unsaturated Fatty Acids. Physiological and Molecular Plant Pathology. 38:255-263.

6. Kabara, J.J. 1986. Fatty Acids and Esters Antimicrobial/Insecticidal Agents. ACS Symposium Series. 325:220-238

7. Branen, A.L., Davidson, P.M. and Katz, B. 1980. Antimicrobial Properties of Phenolic Antioxidants and Lipids. Food Technology. 5:42-63.

CAN LIPIDS FROM THE FRESHWATER ALGA *SELENASTRUM CAPRICORNUTUM* (CCAP 278/4) SERVE AS AN INDICATOR OF HEAVY METAL POLLUTION IN FRESHWATER ENVIRONMENTS?

Christian RICHES, Carole ROLPH, David GREENWAY and Peter ROBINSON. *Department of Applied Biology, University of Central Lancashire, Preston, Lancashire, PR1 2HE, U.K.*

SYNOPSIS

The effects of heavy metals, illumination and temperature on the lipid composition of *S. capricornutum* have been determined. Results suggest that metal exposure leads to characteristic changes in specific acyl-lipid and sterol components, which may enable the identification of heavy metal pollution in freshwater environments.

INTRODUCTION

Environmental factors are well known to influence lipid metabolism in both plants and algae. Previous studies have demonstrated that heavy metal pollution affects algal acyl-lipid composition with respect to fatty acid elongation and desaturation [1]. Changes in algal lipid composition from metal exposure could therefore serve as indicators of heavy metal pollution. However, the effects of other environmental factors on algal metabolism must be considered since their lipid composition has also been shown to respond to factors such as light and temperature (reviewed [2]). The aim of this study, therefore, is to investigate the effects of illumination, temperature and heavy metals on the fatty acid and sterol composition of the freshwater microalga *Selenastrum capricornutum*. When these effects are determined, the potential use of lipid profiles as a biochemical tool to identify heavy metal pollution in freshwater environments can be assessed.

MATERIALS AND METHODS

In order to investigate the effects of heavy metals on lipids from *Selenastrum capricornutum* (CCAP 278/4) cells were grown in batch culture to late-exponential phase in the presence or absence of copper, with continuous illumination (70 µmol $m^{-2} s^{-1}$ photosynthetically active radiation) and at 25°C. Cells were treated with 124µM Cu which inhibited the specific growth rate by 50 %. To determine the effects of illumination and temperature, cells were grown to mid-exponential phase under light and temperature regimes as described above, then subjected to either a dark period or a decrease in temperature from 25°C to 10°C for up to one week. Cells were harvested by centrifugation and freeze dried. Lipids were then extracted from freeze dried material [3] and fatty acid methyl esters prepared from a total lipid aliquot. Sterol components were separated from other lipids by TLC, and sterol acetates prepared. Acyl-lipid and sterol compositions were analysed by capillary GC using a Unicam 610 gas chromatograph operating in a splitless injection mode. Sterol identity was confirmed by GC-MS.

RESULTS AND DISCUSSION

Effect of Cu on lipid profiles from S. capricornutum. Typical acyl-lipid and sterol compositions from control and metal-treated algal cells are shown in *Figure 1a*. These results indicate that metal exposure had significant effects on acyl-lipid metabolism, leading to increased levels of palmitate (16:0) and oleate (18:1), consistent with previous research on marine diatoms [4]. A similar increase in oleate levels has been observed in marine brown algae with Cu exposure, although these authors found the levels of palmitate to be significantly reduced [2]. Changes observed in our study may be accounted for by inhibition of desaturase enzymes, as suggested by Ros and co-workers [5]. Exposure to Cu also resulted in changes in the percentage composition of individual sterols (*Figure 1b*), such that there was an observed increase in 24-ethylcholesta-7,22-dien-3-β-ol (EC-7,22) and a decrease in 24-ethylcholest-7-en-3-β-ol (EC-7) when compared with control cells. Consequently, this led to an increase in the EC-7,22 to EC-7 ratio from 0.28 (control) to 0.66 (Cu treated). This observed increase in Δ22 desaturation with metal treatment may represent an attempt by the cell to increase plasma membrane fluidity by the introduction of desaturated sterols into the membrane.

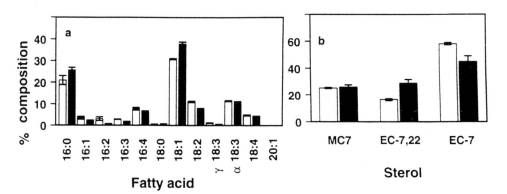

Figure 1. Effect of Cu on acyl-lipid and sterol composition. (Mean n=3 ± SEM) □ control, ■ 124 µM Cu.

Effect of illumination/temperature on acyl-lipid and sterol profiles. Algal cells subjected to dark treatment had a significantly reduced total lipid content when compared with cells grown under continuous illumination, which is consistent with previous research [6]. A period of dark treatment decreased slightly the levels of oleate and palmitate, whilst the levels of linoleate (18:2) and linolenate (18:3) increased slightly. These findings are broadly consistent with studies on marine algae which have demonstrated that low light intensity increased the levels of unsaturated fatty acids [7]. One novel feature of acyl-lipid metabolism in the dark period was the production of a 20 carbon fatty acid, tentatively identified as gadoleic acid (20:1), accounting for 2% of the total lipid composition.

When fatty acid profiles of algal cells subjected to a one day and seven day dark period were compared, no significant differences were observed, suggesting that changes arising from lack of illumination are rapid and stable. The sterol composition was also affected by dark treatment resulting in increased EC-7 levels, such that the ratio EC-7,22 to EC-7 decreased slightly. However, in contrast to the fatty acid profile, this change in sterol composition became more apparent with a longer dark period.

In response to a decrease in temperature algal cells increased their total lipid content. Low temperature led to a 51 % increase in the level of oleate whilst levels of linoleate and parinate (18:4) decreased, by 81 % and 43 % respectively, when compared to control cells. Low temperature also significantly affects the sterol profile, leading to a 48 % increase in 24-methylcholest-7-en-3β-ol (MC-7), and a decrease in the ratio of EC-7,22 to EC-7 from 0.41 to 0.28.

This study has therefore shown that exposure of cells to Cu, dark and low temperature significantly affects both the acyl-lipid and sterol profiles of *S. capricornutum*, leading to characteristic lipid profiles. With respect to acyl-lipid composition, when *S. capricornutum* was treated with Cu the levels of palmitate and oleate increased, whilst dark treatment resulted in decreased levels of palmitate and oleate, and low temperature led to little change in palmitate and an increase in oleate. Characteristic changes in the sterol profile were associated with the sterol ratio EC-7,22 to EC-7. Cu exposure led to an increase in this sterol ratio (from 0.28 to 0.66), dark treatment resulted in a slight decrease, and low temperature led to a significant decrease. Lipid profiles from *S. capricornutum* may therefore serve as a useful indicator of heavy metal pollution in freshwater environments.

REFERENCES

(1) Jones, L.A. and Harwood, J.L. (1993) J. Exp. Bot. **44**, 1203-1210.
(2) Harwood, J.L. and Jones, A.L. (1989) *In* Advances in Botanical Research (Callow, J.A. ed.) p.1. Academic Press, London.
(3) Moreton, R.S. (1985) Appl. Microbiol. Biotechnol. **22**, 41-45.
(4) Jones, G.J., Nichols, P.D., Johns, R.B. and Smith, J.D. (1987) Phytochemistry. **26**, 1343-1348.
(5) Ros, C., Cooke, D.T., Burden, R.S. and Jones, C.S. (1990) J. Exp. Bot. **41**, 457-462.
(6) Koskimies-Soininen, K. and Nyberg, H. (1991) Phytochemistry. **30**, 2529-2536.
(7) Floreto, E.A.T., Hirata, H. Ando, S. and Yamasaki, S. (1993) Botanica Marina. **36**, 149-158.

Section 6:

Lipid Degradation

PURIFICATION AND CHARACTERIZATION OF A MICROSOMAL PHOSPHOLIPASE A_2 FROM DEVELOPING ELM SEEDS

U. STÅHL[1], B. EK[2], A. BANAS[1], M. LENMAN[3], S. SJÖDAHL[3] and S. STYMNE[3].

[1]*Department of Plant Physiology, Swedish University of Agricultural Sciences, Box 7047,* [2]*Department of Cell Research, Swedish University of Agricultural Sciences, Box 7055, S-750 07 Uppsala, Sweden.* [3]*Department of Plant Breeding Research, Swedish University of Agricultural Sciences, Herman Ehles v. 2-4, S-268 31 Svalöv, Sweden.*

1. Introduction

The phospholipases A_2 (PLA_2s) hydrolyse specifically the *sn*-2-fatty acyl ester bond of phosphoglycerides (Waite, 1987). PLA_2s in animal systems are involved in many important processes, such as signal transduction, eicosanoid synthesis and inflammation. The available information about PLA2 from plant tissues is, however, very limited.

Recently we have shown that developing castor bean (*Ricinus communis*) seeds have microsomal PLA_2 activity which selectively removes newly synthesized ricinoleate (12-hydroxy-oleate) from phosphatidylcholine (PC) (Bafor et al., 1992). An acyl specific phospholipase was also found in microsomal preparations from *Euphorbia lagascae* seeds where vernoleoyl(12-epoxy-oleoyl)-PC was hydrolyzed to free vernolic acid (Bafor *et al.*, 1994). Furthermore, developing seeds from elm (*Ulmus glabra*) and *Cuphea procumbens* (which has triacylglycerols rich in caproyl groups) have microsomal PLA_2 activities which selectively hydrolyse medium chain fatty acids from the membrane lipids (Ståhl et al., 1995). These results led us to propose a role for microsomal PLA_2 activities in removing uncommon acyl groups from membrane lipids in developing oil seeds.

Here we report the purification of a PLA_2 from developing elm seed microsomes with properties similar to the low Mw, secretory PLA_2s found in various animal tissues.

2. Materials and Methods

Developing elm (*Ulmus glabra*) seeds were harvested from local trees, seed coats removed and frozen immediately in liquid nitrogen and stored at - 80° C.

Microsomes were prepared from frozen embryos as described previously (Ståhl *et al.*, 1995).

[14C]labeled substrates were synthesized according to the trifluoroacetic anhydride method described by Kanada and Wells (1981).

Solubilized PLA_2 was assayed in eppendorf tubes with up to 150 µl of enzyme fraction, if less the volume was compensated with assay buffer (20 mM of Tris-HCl buffer, pH 8.0 containing 0.06% (w/v) Lubrol PX and 5 mM $CaCl_2$). The assays were started by the addition of 10 nmol of sn-2-[^{14}C]PC, (spec. activity 10 000 dpm/nmol) solubilized in 50 µl of assay buffer and the mixtures were incubated at 30° C for 30 min. The assays were stopped by adding 400 µl of $CHCl_3$/MeOH/Hac, 50:50:1 (w/w) followed by shaking and centrifugation. The $CHCl_3$ phases were removed and fractionated using small silica gel columns. Hydrolysed fatty acids were eluted with 400 µl of $CHCl_3$ into a scintillation vial and radioactivity measured by liquid scintillation counting.

SDS-PAGE gels (Excel 8-18% gradient and Excel 15% homogeneous gels) were purchased from Pharmacia and electrophoresis was performed on a horizontal Pharmacia Multiphore II electrophoresis unit. The gels were either silver or colloidal Coomassie stained.

Protein concentration was measured with the BCA protein reagent (Pierce) and with BSA as standard.

2. Results and Discussion

The PLA_2 activity in microsomal preparations from developing elm seed embryos was solubilized with the non-ionic detergent Lubrol PX at 0.3% (w/v) in 0.15 M potassium phosphate buffer, pH 7.2. About 85% of the microsomal PLA_2 activity and 80-90% of the microsomal proteins were solubilized. Solubilization was verified by high speed centrifugation (the activity was not pelleted by 100 000 x g for 60 min) and by gel filtration on a Superose 6 column, where all activity was recovered in the included volume.

The solubilized PLA_2 was purified at least 30 000 fold using the following five column chromatographic steps: anion exchange (Q-Sepharose), cation exchange (SP-Sepharose), gel filtration (Superose 12), anion exchange (Mono Q) and finally a second gel filtration on a Superose 12 column (Table 1). An apparent molecular weight of 17 kDa

TABLE 1. Purification of microsomal phospholipase A_2 from developing elm seeds

	Protein	Tot. activity	Spec. activity	Purification	Yield
	mg	nmol/min	nmol/min/mg	-fold	%
Microsomes	307	88	0.29	1	100
Solubilized sup.	286	73	0.26	0.9	88
Q-Sepharose	4.10	43	10.5	36	49
Sp-Sepharose	0.37	41	111	383	47
Superose 12	a	25			28
Mono Q	a	17			19
Superose 12	<0.001[b]	8.7	8,700	30,000	10

[a] Protein concentration not measured due to very low levels.
[b] Protein content as estimated from colloidal Coomassie stained SDS-gel.

for the enzyme could be established after SDS-PAGE and recovery of PLA_2 activity (Fig. 1). The purification was not to homogeneity, as the most purified extract showed at least six protein bands when separated on a homogeneous 15% SDS-gel and stained with colloidal Coomassie. Recovered PLA_2 activity from this gel coincided with two close running major protein bands each having the same staining intensity and the same amount of PLA_2 activity.

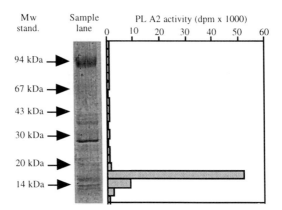

Figure 1. Apparent Mw of purified PL A2. The pooled peak fractions from the first gel filtration column were separated on a gradient 8-18% SDS -gel. The whole lane was sliced up in 5 mm pieces and proteins were eluted in a 0.2% SDS buffer over night. After removal of SDS, the proteins were assayed for phospholipase A2 activity.

The purified PLA_2 was specific for the *sn*-2 position in PC, and had no activity towards diacylglycerols (DAG) or LPC. The pH-optimum was between 7.5 and 8.5, and there was no activity below pH 5.5. It was absolutely dependent on Ca^{2+} and needed mM concentrations for maximum activity. Incubation (5min) of the purified enzyme at 95°C only reduced the activity by 4%. However, it was sensitive to reducing agents and 1% DTT completely abolished the activity. These properties of the purified elm seed microsomal PLA_2 are similar to the properties of the secretory PLA_2s found in animal tissues. The apparent molecular weight of 17 kDa is, however, somewhat higher than the 14 kDa reported for most of the secretory PLA_2s. The stability to heat and sensitivity to reducing agents of elm seed PLA_2 is probably due to the presence of several disulphide bridges. The secretory PLA_2s contain six or seven highly conserved disulphide bridges.

Contrary to the PLA_2 activity found in the intact microsomes, the PLA_2 in the purified fraction did not show any pronounced acyl specificity. The intact microsomal preparation had 50 times more activity with *sn*-2-caproyl-PC than with *sn*-2-oleoyl-PC; the purified enzyme showed only a two fold higher activity. A protein with several disulphide bridges, as the elm PLA_2 appears to have, is not likely to be localized in the cytosol due to the reducing environment in this cell compartment. It would appear from these facts that it is improbable that the purified elm PLA_2 has the role of removing medium-chain fatty acyl groups from membrane lipids, as proposed for the acyl specific PLA_2 activity in developing elm seed microsomes.

3. References

Bafor, M., Smith, M., Jonsson, L., Stobart, K. and Stymne, S. (1991) Ricinoleic acid biosynthesis and triacylglycerol assembly in microsomal preparations from developing castor bean endosperm. *Biochem. J.* **280**, 507-514.

Bafor, M., Smith, M., Jonsson, L., Stobart, K. and Stymne, S. (1993) Biosynthesis of vernoleate (cis-12-epoxyoctadeca-cis-9-enoate) in microsomal preparations from developing endosperm of Euphorbia lagascae, *Arch. Biochem. Biophys.* **303**, 145-151.

Kanda, P. and Wells, M.A. (1981) Facile acylation of glycerophosphocholine catalyzed by trifluoroacetic anhydrid. *J Lipid Res.* **22**, 877-879.

Ståhl, U., Banas, A. and Sten Stymne (1995) Plant microsomal phospholipid acyl hydrolases have selectivities for uncommon fatty acids *Plant Physiol.* **107**, 953-962.

Waite, M. (1987) *The Phospholipases*, Plenum Publishing Corp., New York.

BIOGENERATION OF GREEN ODOR EMITTED BY GREEN LEAVES – ON HPO LYASE AND RELATIONSHIP OF LOX-HPO LYASE ACTIVITIES TO ENVIRONMENTAL STIMULI

Kenji Matsui, Tadahiko Kajiwara and *Akikazu Hatanaka
Department of Biological Chemistry, Faculty of Agriculture, Yamaguchi University, Yamaguchi 753, JAPAN
*Department of Plant Life Science, Graduate School of Integrated Art and Science, University of East Asia, Shimonoseki 751, JAPAN.

What is the so called green odor? The so called green odor distributed in fresh leaves in forest, vegetables and fruits involved from eight volatile C_6 components such as (3Z)-, (3E)- and (2E)-hexenols and these corresponding aldehydes, added n-hexanal and n-hexanol. Since 1957, we have been conducting studies on these compounds found in tea leaves, using inter-disciplinary approaches including synthetic chemistry, natural products chemistry, physiological biochemistry and molecular biology.

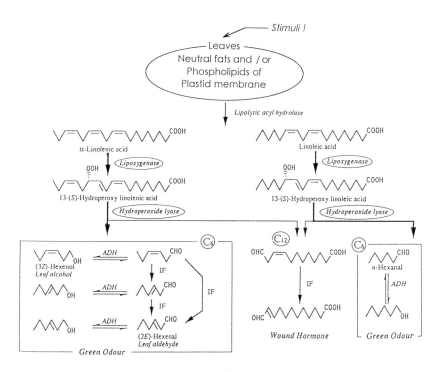

The biosynthetic route ! We found that the precursors of green odor are α-linolenic and linoleic acids as confirmed by ^{14}C-labelling experiments with tea chloroplasts. The 13-(S)-hydroperoxide was isolated as an intermediate in the formation of (3Z)-hexenal, key compound, in green odor (see scheme, ADH; Alcohol dehydrognase, IF; Isomerization factor). In this chapter, the subject shall be forcused on the cleavage reaction of 13-(S)-hydroperoxide (ω6-hydroperoxide) to (3Z)-hexenal or n-hexanal.

The purification of HPO lyase from tea leaves and it's properties !
The hydroperoxide(HPO) lyase was at first prepared from fresh tea leaves follwing steps; fresh tea leaves were macerated with phosphate buffer containing polyvinylpyrrolidone along with reducing agents such as gultathione and ascorobic acid in order to avoid non specific aggregation of proteins by polyphenol oxidase. And then it was filtered through 8 layers of cheesecloth and centrifuged at 100,000 x g for 60 min and membrane fraction was obtained. It was solubilized with 0.15% Triton X-100 then following this table, purified at the first. It was estimated as MW.; 55kD by SDS.

The synthesis of substrate for studies on substrate specificities of HPO lyase. Using on the substrate specificities for HPO lyase activities, ω6-hydroperoxides, synthetic ω6-hydroperoxy C_{14}~C_{24} fatty acids fixed from ω1 to ω11 containning ω7E, ω9Z- (A group) or ω3Z,ω7E, ω9Z-(B group), were synthesized by Jone's oxidation of the corresponding alcohols prepared via Wittig or Grignard coupling reaction of two counter parts using acetylen chemistry under reaction condition at -50°C in liquid ammonium and these fatty acids were peroxigenated by soybean lipoxygenase. These ω6-hydroperoxides were incubated with purified HPO lyase for 10 min at 25°C and then, prepared to hydrazon of (3Z)-hexenal or n-hexanal by 0.1% 2,4 DNPH solution. This graph of figure at next page correlates the relative conversion rate into aldehyde products with the chain length of substrate hydroperoxides;

	Total Protein (mg)	Total Activity (μmol)	Specific Activity (μmol/mg)	Yield (%)	Purification (fold)
crude	9515	1444	0.15	100	1
membrane	1332	720	0.54	49.9	3.61
solubilized	925	1383	1.5	95.8	9.97
PEG 6000	73.5	799	10.9	55.4	72.5
DEAE-Cellulofine	2.49	366	147	25.4	984
DEAE-Toyopearl	nd*	160	-	11.1	-
Hydroxy Apatite Fraction I	0.055	44.7	872	3.10	5816
Fraction II	0.010	9.84	879	0.68	5860

* nd ; not determined.

total carbon number 18 showing α-linolenic (●) or linoleic (○) acids. For both the tri- and di-enoic acid hydroperoxides, the product specificities of tea leaves HPO lyase is broad and elongation between the terminal carboxyl group and the hydroper-oxy group from an over length of C18 to C22 caused enhancement of the

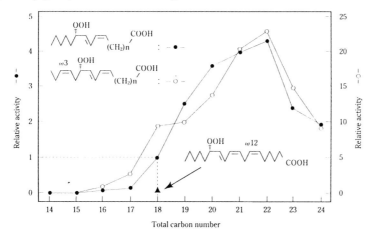

activity and more elongation beyond C22 decreased the activity. It should be noted that reactivities of the trienoic acid hydroperoxides were always four to seven times higher than those of the dienoic acid having the same carbon number. This indicates that introduction of a (Z)-double bond between $\omega 3$ and $\omega 4$ carbon positions is very effective in increasing the activity. Decomposition of γ-linolenic acid hydroperoxide was catalyzed at a rate of only about 2% of that α-linolenic acid. It is assumed that at α-linolenic acid, the compact turning of the side arm at the ω-terminal end caused by a (Z)-double bound facilitate recognition by HPO lyase and on the other hand, at γ-linolenic acid introduction of a (Z)-double into the carbonyl side arm between $\omega 12$ and $\omega 13$ carbon positions, decreases the activity strikingly.

On relationship Lipoxygenase (LOX)-HPO lyase activities to environmental stimuli,   The seasonal changes in (3Z)- hexenal formation activity in tea leaves, the activity begins to increase in April and in August it reaches a maximum, then gradually decreases and disappeares completely in December, below 10°C and it's changes were pallallel to the temperature and solar radiation. But LOX activity shows maximal value in summer leaves, on the other hand, HPO lyase activity in summer leaves was quite substantial but higher still in winter. This high HPO lyase activity was found throughout the year, and this activity does not disappear such as LOX activity in winter. So, the overall C_6-aldehyde-forming activity which signifies a sequential reaction of LOX-HPO lyase showes a seasonal changes similar to that of LOX, namely the step determiniing the seasonable changes activity is in LOX rather than HPO lyase.

References
1. Matsui, K., Toyota, H., Kajiwara, T. and Hatanaka, A. (1991) Fatty acid hydroperoxide cleaving enzyme, hydroperoxide lyase, from tea leaves, Phytochem., 30, 2109~2113.
2. Hatanaka, A. (1993) The biogeneration of green odour by green leaves, Phytochem., 34(5),1201~1218.

DO LIPOXYGENASES INITIATE ß-OXIDATION?

The reductase pathway

IVO FEUSSNER[1], HARTMUT KÜHN[2] AND CLAUS WASTERNACK[1]
[1]*Institute of Plant Biochemistry, Weinberg 3, D-06120 Halle, Germany;*
[2]*Institute of Biochemistry, Hessische Str. 3-4, Universitätsklinikum Charité, Humboldt-Universität- Berlin, D-10115 Berlin, Germany.*

Introduction

The etiolated germination process of oilseed plants is characterized by the mobilization of storage lipids which serve as a major carbon source for the seedlings growth. During this stage the lipid storing organelles, the lipid bodies, are degraded and a new set of proteins, including a specific form of lipoxygenase (LOX), is detectable at their membranes in different plants [1,2]. LOXs are widely distributed in plants and animals and catalyze the regio- and stereo-specific oxygenation of poly-unsaturated fatty acids [3]. The enzymatic transformations of the resulting fatty acid hydroperoxides have been extensively studied [4]. Three well characterized enzymes, a lyase, an allene oxide synthase, and a peroxygenase, were shown to degrade hydroperoxides into compounds of physiological importance, such as odors, oxylipins, and jasmonates. We have recently reported a new LOX reaction in plants where a specific LOX, the lipid body LOX, metabolizes esterified fatty acids. This reaction resulted in the formation of 13(S)-hydroxy-linoleic acid (13-HODE) and lead us to propose an additional branch of the LOX pathway: the reductase pathway. Besides a specific LOX form we suggest two additional enzyme activities, a lipid hydroperoxide reductase and a lipid hydroxide-specific lipase which lead to the formation of 13-HODE. 13-HODE might be the endogenous substrate for β-oxidation in the glyoxysomes during germination of oilseeds containing high amounts of polyunsaturated fatty acids.

Properties of the lipid body lipoxygenase

Lipid body lipoxygenase is high level expressed during the germination process [5]. It is bound at the lipid body membrane and exhibits an unusual pH-optimum of 8.5 and positional specificity which has not been described so far for any other lipoxygenase [6,7]. It is a 13-LOX, but converts arachidonic acid not only to the 15(S)-hydroperoxy-eicosatetraenoic acid, but also to 12(S)-, and 8(S)-hydroperoxy derivatives. This might be indicative for a different active site compared to other LOXs. Binding to the lipid body membrane enhances its enzymatic activity about 4-fold and alters the positional specificity towards the 8(S)-isomer. Additionally, we found for the first time in plants that this LOX form is capable of oxygenating the esterified linoleic acids which are located in the lipid bodies [8]. Because of these characteristic features this enzyme is clearly distinguishable from other plant LOXs. This might be indicative for a so far not identified physiological role of plant LOXs.

Physiological role of the lipid body LOX: the reductase pathway

Lipid fractionation studies of lipid bodies indicated a stepwise metabolization of their constituents. In the phospholipid fraction LOX products were found at first during the time course of germination. The amount of oxygenated fatty acids in the storage lipids increased drastically during germination [8]. When the monolayer membrane structure is hampered due to a high amount of oxygenated fatty acids in the phospholipids, the triacylglycerols became available for the LOX, too and LOX products were also found in the triacylglycerol fraction. Besides large amounts of esterified 13(S)-hydroperoxy-linoleic acids minor amounts of esterified 13-HODE were found. Interestingly, in the cytosol only the amount of 13-HODE and not for linoleic acid was increased during germination and *in vitro* experiments showed that lipid bodies prepared from this stage preferentially released 13-HODE. In order to test whether this reaction was restricted to cucumber we analyzed lipid bodies from other oilseeds. Large amounts of oxygenated fatty acids which are due to an *in vivo* action of a LOX were detected in germinating soybean, tobacco, and rape seedlings, either.

These observations lead us to propose an alternative model for the degradation of polyunsaturated fatty acids during germination (*Figure 1, right hand*) compared to the peroxisomal degradation of linoleic acid suggested by Gerhardt and Kleiter (*Figure 1, left hand*) [9]. We suggest as initial step a LOX-catalyzed oxygenation of polyunsaturated fatty acids located within the phospholipids and triacylglycerols (TG). This may exhibit a dual function. (i) The surrounding monolayer became disrupted and (ii) the storage lipids became accessible to this specific LOX and the following metabolic enzymes. The resulting lipid hydroperoxides were reduced by a yet not identified enzymatic reaction and the hydroxy fatty acids (OH-FA) were subsequently cleaved from the lipids by a highly specific lipase. We suggest that the formation of hydroxy fatty acids is a specific signal for lipid degradation of polyunsaturated fatty acids. By this way plants could discriminate between this final catabolic reaction and currently occurring lipid turnover. Additionally, in the case of 13-HODE the C12=C13 cis double bond of linoleic acid is changed towards a C11=C12 trans double bond. This might be an advantage for the plant in metabolizing fatty acids: the so-called dehydratase pathway during the β-oxidation is by-passed since it is not needed for metabolizing this derivative in contrast to linoleic acid *(Figure 1, left hand)* [9].

Interestingly, parallels of this reaction pathway between plants and mammalians exist [10]. In mammalians a similar process has been proposed for the degradation of mitochondria during the maturation of the erythrocytes.

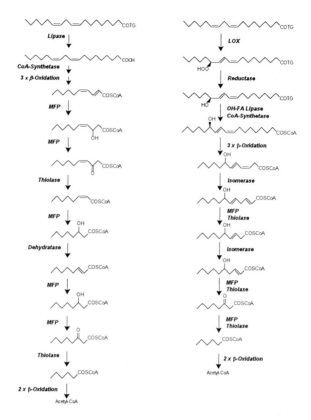

Figure 1. Proposed pathways for the degradation polyunsaturated fatty acids in plants. The left hand side displays the peroxisomal degradation of linoleic acid according to Gerhardt and Kleiter [9]. The left hand side shows the proposed reductase pathway.

References

1. Feussner, I. and Kindl, H. (1992) A lipoxygenase is the main lipid body protein in cucumber and soybean cotyledons during the stage of triglyceride mobilization, FEBS Lett. **298**, 223-225.
2. Radetzky, R., Feussner, I., Theimer, R.R., Kindl, H. (1993) Transient occurrence of lipoxygenase and glycoprotein gp49 in lipid bodies during fat mobilization in anise seedlings, Planta **191**, 166-172.
3. Siedow, J. N. (1991) Plant lipoxygenase - structure and function, Annu. Rev. Plant Physiol. **42**, 145-188.
4. Hamberg, M. (1993) Pathways in the biosynthesis of oxylipins in plants, J. Lipid Med. **6**, 375-384.
5. Feussner, I., Hause, B., Nellen, A., Wasternack, C. and Kindl, H. (1996) Lipid-body lipoxygenase is expressed in cotyledons during germination prior to other lipoxygenase forms, Planta **198**, 288-293.
6. Feussner, I. and Kindl, H. (1994) Particulate and soluble lipoxygenase isoenzymes - comparison of molecular and enzymatic properties, Planta **194**, 22-28.
7. Feussner, I. and Kühn, H. (1995) The lipid body lipoxygenase from cucumber seedlings exhibits unusual reaction specificity, FEBS Lett. **367**, 12-14.
8. Feussner, I., Wasternack, C., Kindl, H. and Kühn, H. (1995) Lipoxygenase-catalyzed oxygenation of storage lipids is implicated in lipid mobilization during germination, Proc. Natl. Acad. Sci. USA **92**, 11849-11853.
9. Gerhardt, B. and Kleiter, A. (1995) Peroxisomal catabolism of linoleic acid, in J.-C. Kader and P. Mazliak (eds.), *Plant Lipid Metabolism,* Kluwer Academic Publishers, Dordrecht, pp. 265-267.
10. Schewe, T. and Kühn, H. (1991) Do 15-lipoxygenases have a common biological role?, Trends Biochem. Sci. **16**, 369-373.

DEGRADATION OF ACETYLENIC TRIACYLGLYCEROLES AND THE INACTIVATION OF MEMBRANE PREPARATIONS FROM MOSS PROTONEMA CELLS

Peter Beutelmann and **Karola Menzel**
Institut für Allgemeine Botanik, Universität Mainz,
Müller Weg 6, D-55099 Mainz, Germany

Introduction

Protonema cells of the moss *Ceratodon purpureus* accumulate triacylglycerols with two acetylenic fatty acids, 9,12-octadecadien-6-ynoic acid (18:2A) and 9,12,15-octadecatrien-6-ynoic acid (18:3A), as main components. By following the incorporation of the [^{14}C]-precursors acetate, linoleate, α-linolenate, γ-linolenate, stearidonate and 18:2A, into 18:3A in triacylglycerol accumulating cells, the pathway for acetylenic acids could be established. 18:2A and 18:3A could be synthesized by a second desaturation of the ∆6 double bond of γ-linolenate and stearidonate, respectively. However the major pathway for 18:3A synthesis was via a ∆15 desaturation of 18:2A. Since 18:2A was found exclusively in the triacylglycerols of the cell, there is evidence that the triacylglycerols in *Ceratodon purpureus* can act as a direct substrate for the ∆15 desaturation [1]. Our next goal is to confirm these *in-vivo* results by demonstrating the pathways *in vitro* with subcellular fractions of the moss cells. In the following we will report mainly on the obstacles we met on our way to successful *in-vitro* experiments. Cell fractionations, in-vitro measurements and analyses were carried out according to established methods [1], [3], [4].

Results and Discussion

The triacylglycerols (TAG), the main point of our interest, are deposited in ball shaped lipid bodies with a diameter of 0.1-0.3 µm. As electron micrographs show, these lipid bodies are obviously not surrounded by membrane structures, although they are in close connection to membrane systems which are very likely to belong to the ER. In old cells the lipid bodies flow together into large lipid drops which can be seen in the light microscope.

At least for small lipid bodies a tight association to ER membranes seems to exist, since it is possible to isolate a kind of a "liposomal membrane fraction" from the moss cells by saccharose density gradient centrifugation. On the gradient this fraction is located between 15 and 20 % saccharose, which indicates a lower density than normal ER membranes and a higher density than common liposomes . Besides TAG and phospholipid, NADH-cytochrom C reductase was found to be associated with this fraction, which means evidence for the presence of ER (Fig.1)

Both, preparations of liposomal membranes and whole microsmal fractions of the moss protonema cells were incubated with radioactive oleoyl-CoA and glycerol-3-phosphate, in order to study incorporation into different lipid fractions. But so far practically no incorporation of the precursors into any lipid fraction could be demonstrated, even though the cells, used for the membrane preparations, were in a stage of high lipid accumulation. Only degradation and oxydation products of the fatty acid (FA) could be detected.

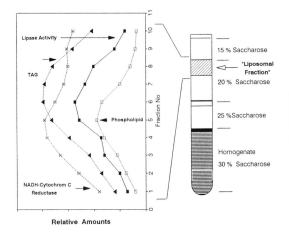

Fig. 1: Analysis of the "liposomal fraction" from protonema cells, isolated by flotation in saccharose density step gradient

One hint for an explanation of the situation seemed to us to be the presence of high amounts of free fatty acids in homogenates and membrane preparations, and also the fact that liposomal membrane preparations were only possible from cells which had already accumulated extremly high amounts of TAG. These findings led us to study the lipolytic activity in the moss cell homogenates, which showed that homogenization of the cells activates an extremly high lipase activity which degrades more than half of the TAG within 1 min, even at 0° C. The activity is however inhibited to almost zero within 5 min (Fig.2). The lipase activity seems to be associated with the liposomal membrane fraction. The hydrolytic activity does obviously not affect phospholipids and glycolipids. Among the fatty acids, liberated by the lipase, the acetylenic FA, which make up about 80 % of the total FA, are further degraded or modified (Fig.2) to less than half of the original amount within 10 min.

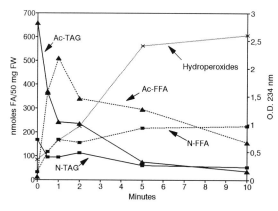

Fig.2: TAG and fatty acid degradation in a homogenate of Ceratodon protonema cells at 0° C.

Degradation of acetylenic FA is indicated by the production of ethane in homogenates of moss cells. The ethane evolution takes place for about 15 min after breaking up the cells. The amount of ethane produced is dependent on the amount of acetylenic FA present in the cells. Also malonaldialdehyd, an indicator of fatty acid degradation, is formed in homogenates. A marked increase of the absorbance at 234 nm in the fatty acid fraction of homogenates during the first 10 min indicates the formation of FA hydroperoxides from acetylenic FA (Fig.2).

We conclude that in homogenates of protonema cells of *Ceratodon* an active lipase which cannot be inhibited by low temperatures, produces high amounts of particularly acetylenic fatty acids, which are further degraded or modified into components which are probably potent enzyme inhibitors. In order to avoid oxydative degradation of fatty acids we added various antioxidants to the extraction media. Ascorbate, Sodiummetabisulfite, DTT, Diethyldithiocarbamate were used in various concentrations, but none of them did improve the result. We also used a method to isolate a microsmal fraction by polyethyleneglycol precipitation [2] by which it was possible to carry out the isolation procedure in a nitrogen atmosphere. Even this method did not yield membrane preparations which incorporated precursors in anabolic lipid pathways.

Acknowledgement: The work was supported by the Deutsche Forschungsgemeinschaft.

References:

1. Kohn G, Hartmann E, Stymne S, Beutelmann P. Biosynthesis of acetylenic fatty acids in the moss *Ceratodon purpureus* (Hedw.) Brid. J. Plant Physiol. 144, 265-271 (1994)

2. van der Hoeven TA. Isolation of hepatic microsomes by polyethyleneglycol 6000 fractionation of the postmitochondrial fraction. Anal. Biochem. 115, 398-404 (1981)

3. Lord JM, Kagawa T, Moore T, Beevers H. Endoplasmic reticulum as the site of lecithin formation in castor bean endosperm. J. of Cell Biol.57, 659-667 (1973)

4. Bafor M, Smith M, Jonsson L, Stobart K, Stymne S. Ricinoleic acid biosynthesis and triacylglycerol assembly in microsomal preparations from developing castor bean (*Ricinus communis*) endosperm. Biochem J. 280, 507-514 (1991)

STORAGE LIPID MOBILIZATION DURING NITROGEN ASSIMILATION IN A MARINE DIATOM

TONY LARSON[1] AND PAUL J. HARRISON[1,2]
Departments of Botany[1] and Oceanography[2]
University of British Columbia
6270 University Blvd.
Vancouver, B.C. V6T 1Z4
Canada

1. Introduction

Marine diatoms are major contributors to coastal and oceanic primary productivity. Assimilation of nitrate, the predominant limiting nutrient, is usually tightly coupled to photosynthetic carbon fixation; however, during darkness or periods of rapid nitrate assimilation, stored carbon reserves are respired for the energetic requirements of the cell and for protein synthesis (Turpin and Weger 1990).

In general, algae possess peroxisomes and the enzymes of the glyoxylate cycle, which is known to be important for gluconeogenesis in higher plants. However, it may be possible that diatoms can directly respire lipids through the tricarboxylic acid (TCA) cycle, because B-oxidation enzymes have been found in the mitochondria (Winkler and Stabenau 1995). Therefore, it is unclear if the major metabolic fate of storage lipid in diatoms is in gluconeogenesis or respiration.

This study investigates how lipid metabolism is affected by nitrate assimilation in *Phaeodactylum tricornutum.*

2. Results

Cultures of the marine diatom *P. tricornutum* were grown in an enriched artificial seawater medium Cultures were left to naturally deplete the nitrate from the growth medium. Recovery from nitrogen starvation was achieved by diluting the cultures with fresh medium.

Intracellular carbohydrate was degraded within 24 h of addingnitrate to the medium. Fatty acids associated with storage lipid only started to be degraded when carbohydrate had been depleted to a quota of approximately 2 pg.cell^{-1} (Fig.1). NR activity peaked within 6 to 12 h of nitrate addition. ICL activity peaked at 24 h (Fig. 2). The peak in ICL activity coincided with the beginning of storage lipid degradation.

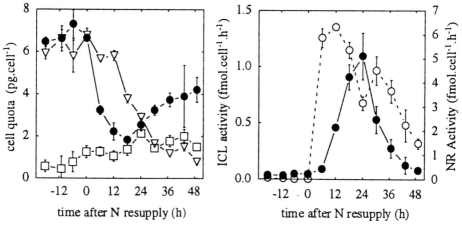

Figure 1. Mobilization of stored carbon reserves following nitrate addition to N-starved cultures of *P. tricornutum*. Cultures were N-starved for 3 d prior to dilution with fresh medium containing nitrate. Plots show changes in the intracellular concentrations of storage (▽) and structural (□) fatty acids (FA), and carbohydrates (●).

Figure 2. *In vitro* enzyme activities for nitrate reductase (NR) and isocitrate lyase (ICL) following nitrate addition to N-starved cultures of *P. tricornutum*. Culture conditions were as for Fig. 1. NR (○) activity was measured at the culture incubation temperature of 18 °C. ICL (●) activity was measured at 30 °C and normalized to 18 °C.

3. Discussion

There are several possible reasons why storage lipid is mobilized and the glyoxylate cycle stimulated following nitrate addition to N-starved cultures. These are listed as follows: 1.) to replenish, *via* gluconeogenesis, carbohydrates exhausted during nitrate assimilation; 2.) to replenish, anaplerotically, TCA cycle intermediates drawn off for protein synthesis during rapid nitrate assimilation; 3.) to increase, *via* malate glycolysis, the amount of cytosolic NADH available for NR. These possibilities are detailed in Fig.3.

There is obviously an intimate relationship between carbohydrate and storage lipid metabolism in *P. tricornutum*. Researchers have shown that endogenous or exogenous carbohydrates can inhibit ICL activity and acetate metabolism in a variety of plants and microorganisms (e.g. McLaughlin and Smith 1994). However, the role of the glyoxylate cycle may go beyond gluconeogenesis; it may be important for supplying reductant and providing for anaplerotic replenishment of TCA cycle intermediates used during biosynthetic processes.

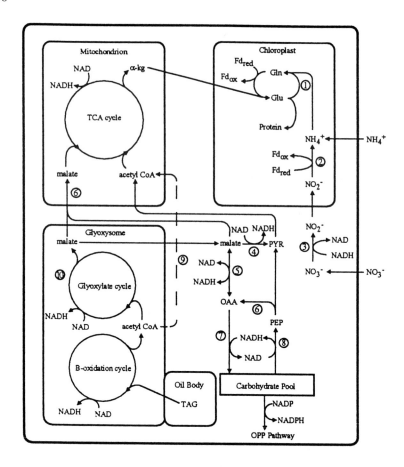

Figure 3. Model showing possible fates for lipid and carbohydrate carbon during nitrogen assimilation in *P. tricornutum*. Fluxes shown do not represent actual stoichiometries. (1) glutamine synthetase; (2) nitrite reductase; (3) nitrate reductase; (4) malate glycolysis; (5) cytosolic malate dehydrogenase; (6) anaplerotic carbon flux; (7) gluconeogenesis; (8) glycolysis; (9) carnitine acyltransferase; (10) isocitrate lyase.

3. References

McLaughlin, J.C., Smith, S.M. (1994) Metabolic regulation of glyoxylate-cycle enzyme synthesis in detached cucumber cotyledons and protoplasts. Planta **195**, 22-28

Turpin, D.H., Weger, H.G. (1990) Interactions between photosynthesis, respiration and nitrogen assimilation. In: Plant physiology, biochemistry and molecular biology. pp 422-433, Dennis, D.T., Turpin, D.H., eds, Longman Scientific and Technical, Essex

Winkler, U., Stabenau, H. (1995) Isolation and characterization of peroxisomes from diatoms. Planta **195**, 403-407

BIOCATALYTIC OXIDATION OF ACYLGLYCEROL AND METHYL ESTER

GEORGE J. PIAZZA, THOMAS A. FOGLIA, AND ALBERTO NUÑEZ
Eastern Regional Research Center, ARS, USDA,
600 East Mermaid Lane
Wyndmoor, PA 19038, USA

1. Introduction

Soybean lipoxygenase (LOX; linoleate, oxygen oxidoreductase, EC 1.13.11.12) generates fatty acid hydroperoxides from polyunsaturated fatty acids and oxygen. In earlier work the high pH form of soybean LOX was shown to strongly prefer free fatty acid as a substrate, rather than esterified acid (1). However experimental evidence with structural analogues suggests that 15(S)-lipoxygenases, such as soybean LOX, recognize and bind their substrates through the methyl end, and therefore modification to the carboxylic end of the substrate should have a minimal impact upon substrate reactivity (2). In support of this contention, recently it was shown that LOX acted upon esterified fatty acids in phosphoglycerides when a bile salt surfactant was present (3-5). Phosphoglyceride oxidation was highly positionally and stereochemically specific and therefore was not merely an enzyme promoted radical-mediated autoxidation (4,5). There is a single report demonstrating the action of LOX upon fatty amide (6).

That phosphoglyceride and fatty amide can be good substrates for LOX raises the question as to whether neutral fatty esters can also be oxidized by LOX. Therefore the susceptibility of acylglycerol and methyl ester toward oxidation by LOX in the presence of the bile salt deoxycholate was measured. These studies are important prerequisites to possible industrial scale processes in which lipoxygenase and other enzymes are used to introduce readily derivatizable oxygen functionality into fats and oils.

2. Results and Discussion

The pH and deoxycholate concentration were varied to determine the optimal oxidation rate of dilinolein (7,8). It was found that the fastest rate of oxidation occurred at pH 8-9 in the presence of 10 mM deoxycholate; critical micelle concentration of deoxycholate: 0.25 - 2.5 mM (9). Having established optimal reaction conditions for dilinolein, other esters of linoleate were tested (Figure 1). Monolinolein was oxidized

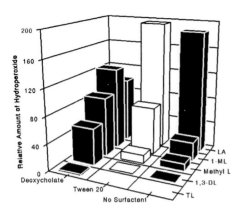

Figure 1. Relative Amount of Oxidized Linoleate Formed by Lipoxygenase in 15 min. LA, linoleic acid; 1-ML, monolinolein; Methyl L, methyl linoleate; 1,3-DL, 1,3-dilinolein; TL, trilinolein.

TABLE 1
Relative Rate of Oxidation of Mixtures of Substrates by LOX in Deoxycholate

Substrate Mixture	Relative Oxidation Rate[a]
1. Linoleic acid	100[b]
1,3-Dilinolein	36 ± 2
2. Linoleic acid	100[b]
Methyl linoleate	39 ± 3
3. Linoleic acid	100[b]
1-Monolinolein	75 ± 2
4. Linoleic acid	100[c]
1,3-Dilinolein	10 ± 2
Methyl Linoleate	18 ± 2
5. Linoleic acid	100[c]
1,3-Dilinolein	36 ± 9
1-Monolinolein	65 ± 4

[a]Relative oxidation rates were calculated by the method of Schellenberger et al (10).
[b]Each time point sample contained 75 µg LOX.
[c]Each time point sample contained 113 µg LOX.

more than two times faster than dilinolein and slightly faster than linoleic acid. The methyl ester of linoleic acid was oxidized at approximately the same rate as linoleic acid. In contrast, earlier work that was performed with Tween 20 (polyoxyethylene-sorbitan monolaurate), but without deoxycholate showed that the methyl ester of linoleic acid was a poor substrate for LOX (1). Since the results obtained in deoxycholate were at such variance with these earlier results, assays were repeated in the presence of no added surfactant and in Tween 20. Fig. 1 shows that without surfactant or in the presence of Tween 20 the oxidation of linoleic acid was accelerated, but monolinolein and methyl linoleate were oxidized at rates that were significantly reduced compared to those obtained in deoxycholate. No oxidation of dilinolein could be detected in the presence of Tween 20 or without surfactant.

The data in Table 1 show that when the oxidation of dilinolein, methyl linoleate, and monolinolein was followed in the presence of linoleic acid, their rates of oxidation were reduced relative to that of linoleic acid (Table 1, entries 1-3). However, even in the presence of linoleic acid, monolinolein was still the most rapidly oxidized ester substrate. When 1,3-dilinolein and methyl linoleate were oxidized together in the presence of linoleic acid, their relative rates of oxidation were reduced further, although the amount of LOX was increased proportionately (Table 1, entry 4). The combination of 1,3-dilinolein and monolinolein was also oxidized in the presence of linoleic acid (Table 1, entry 5). Each substrate was oxidized at approximately the same relative rate as before when assayed separately against linoleic acid.

The data found in Fig. 1 and Table 1 show that the presence of a free carboxyl group on linoleic acid influences the magnitude of its interaction with LOX. In the presence of deoxycholate linoleic acid is oxidized most rapidly in the competition experiments (Table 1), but monolinolein is oxidized most rapidly when the substrates are assayed individually (Fig. 1). If the order of reactivity of the substrates were determined solely by the ability of deoxycholate to solubilize the substrates, then a change in the reaction order would not be expected depending on whether the substrates were assayed separately or together. That linoleic acid is always oxidized most rapidly in the competition experiments must be because of preferential binding of linoleic acid by LOX.

3. Conclusion

In conclusion the results demonstrate that methyl linoleate as well as mono- and diacylglycerol that contain linoleoyl residues are rapidly oxidized by LOX in buffers that contain deoxycholate surfactant. The oxidized products are useful chemical intermediates for further transformation to emulsifiers, coatings, and lubricants with enhanced hydrophilic properties.

4. References

1. Christopher, J., Pistorius, E., and Axelrod, B. (1970) Isolation of an isozyme of soybean lipoxygenase, *Biochim. Biophys. Acta* **198**, 12-19.
2. Lehmann, W.D. (1994) Regio- and stereochemistry of the dioxygenation reaction catalyzed by (S)-type lipoxygenases or by the cyclooxygenase activity of prostaglandin H synthases, *Free Radical Biol. Med.* **16**, 241-253.
3. Eskola, J. and Laakso, S. (1983) bile salt-dependent oxygenation of polyunsaturated phosphatidylcholines by soybean lipoxygenase-1, *Biochim. Biophys. Acta* **751**, 305-311.
4. Brash, A.R., Ingram, C.D., and Harris, T.M. (1987) Analysis of a specific oxygenation reaction of soybean lipoxygenase-1 with fatty acids esterified in phospholipids, *Biochemistry* **26**, 5465-5471.
5. Therond, P., Couturier, M., Demelier, J.-F., and Lemonnier, F. (1993) Simultaneous determination of the main molecular species of soybean phosphatidylcholine or phosphatidylethanolamine and their corresponding hydroperoxides obtained by lipoxygenase treatment, *Lipids* **28**, 245-249.
6. Ueda, N., Yamamoto, K., Yamamoto, S., Tokunaga, T., Shirakawa, E., Shinkai, H., Ogawa, M., Sato, T., Kudo, I., Inoue, K., Takizawa, H., Nagano, T., Hirobe, M., Matsuki, N., and Saito, H. (1995) Lipoxygenase-catalyzed oxygenation of arachidonylethanolamide, a cannabinoid receptor agonist, *Biochim. Biophys. Acta* **1254**, 127-134.
7. Piazza, G. J. and Nuñez, A. (1995) Oxidation of acylglycerols and phosphoglycerides by soybean lipoxygenase, *J. Am. Oil Chem. Soc.* **72**, 463-466.
8. Nuñez, A. and Piazza, G.J. (1995) Analysis of lipoxygenase kinetics by high-performance liquid chromatography with a polymer column, *Lipids* **30**, 129-133.
9. Murada, Y., Sugihara, G., Nishikido, N., and Tanaka, M. (1982) Study of the micelle formation of sodium deoxycholate, in K.L. Mittal and E.J. Fendler (eds.), *Solution Behavior of Surfactants*, Plenum Press, New York, pp. 611-627.
10. Schellenberger, V., Siegel, R.A., and Rutter, W.J. (1993) Analysis of enzyme specificity by multiple substrate kinetics, *Biochemistry* **32**, 4344-4348.

VOLATILE PRODUCTION BY THE LIPOXYGENASE PATHWAY IN OLIVE CALLUS CULTURES

M. WILLIAMS and J.L. HARWOOD
School of Molecular and Medical Biosciences
University of Wales Cardiff, P.O. Box 911, Cardiff CF1 3US, Wales,
U.K.

Introduction

The aroma of an oil is derived from the production of the oil's volatile constituents. Some of the major volatiles of olive oil include the derivatives of the C_6 aliphatic hexanal (1). Of the large number of volatiles present, only a few have organoleptic properties which are of any importance. Therefore, the emphasis of analysis has been on the identifying only those components with organoleptic properties. We have chosen to use olive cultures as a model to investigate olive volatile production. Callus cultures have already been found to be reliable material to study lipid synthesis as they have an acyl composition compatible with olive fruit (2). The use of cultures is also most useful because it is possible to control strictly the growth conditions of the cultures and hence limit biological variation, while investigating different cultivars.

Materials and Methods

Callus cultures were established with three olive (*Olea europaea* L.) varieties using a method reported previously (2). Volatiles were collected from the olive cultures (20g) using dynamic headspace sampling techniques described previously (3). Analysis of volatiles was performed using an Automatic Thermal Desorption System linked to a Perkin Elmer Autosystem GC. Lipoxygenase (LOX) activity of the cultures was assayed routinely using an oxygen electrode.

Table 1 ANOVA of some important volatile compounds of three different varieties of olive callus

	Peak-area ratio (×100)[a]			
	Picual	Koroneiki	Coratina	F probability
Aldehydes				
Acetaldehyde	0.262	0.388	0.861	0.006*
3-Methyl butanal	5.418	4.520	3.790	0.499
Hexanal	0.687	0.104	0.104	<0.001*
Trans-2-hexenal	0.607	0.141	0.181	0.001*
Octanal	0.358	0.165	0.038	0.002*
Nonanal	0.059	0.123	0.042	0.269
Alcohols				
1-Penten-3-ol	1.058	0.320	0.429	0.039*
3-Methyl butanol	0.203	0.101	0.057	0.048*
Hexan-1-ol	0.276	0.175	0.050	0.004*
Cis-3-hexen-1-ol	0.068	0.072	0.023	0.063
Trans-3-hexen-1-ol	0.168	0.077	0.061	0.191
Trans-2-hexen-1-ol	0.426	0.342	0.171	0.367
Octan-2-ol	0.084	0.033	0.059	0.258
Cis-2-hexen-1-ol	0.072	0.315	0.075	0.018*
Hydrocarbons				
n-Hexane	0.045	0.092	0.097	0.612
Octane	0.149	0.027	0.119	0.057
Nonane	0.256	0.412	0.694	0.052
Methylbenzene	1.742	0.335	0.427	<0.001*
Esters				
Ethyl propionate	0.216	0.062	0.117	0.006*
Hexyl acetate	0.683	0.081	0.078	<0.001*
Cis-3-hexenyl acetate	0.858	0.880	0.272	0.128
Methyl nonanoate	0.629	0.233	0.057	0.009*
Ketones				
Acetone	0.195	0.161	0.118	0.644
3-Pentanone	2.257	0.056	0.683	<0.001*

Analysis of variance (ANOVA) was used to study the reproducibility of the samples, which were analysed in triplicate. The three cultivars were cultured with the same growth conditions so that the only variable factor was the variety. Hence, the experiment structure consisted of a 2^1 factorial set. Data show means (n=3). Abreviations: [a] =Ratio of the compound peak area to the internal standard area; * =Statistically significant (5% level).

Results and Discussion

Volatile analysis of olive oil has been well reported (1, 3). Environmental factors as well as cultivar are a major influence on volatile production in the oil (1). In order to study the effect of cultivar we have used callus cultures which have previously proved to possess an acyl composition compatable with the developing olive fruit (2). Table 1 shows some of the major volatile constituents of the callus cultures which have also been identified in virgin olive oil (3). They included aldehydes, alcohols, esters, hydrocarbons and ketones. The presence of aldehydes, such as hexanal and *trans*-2-hexenal, as well as their alcohol derivatives (commonly known as the leaf alcohols) were indicative of an active LOX pathway and alcohol dehydrogenase activity. The aromatic properties of the oil are attributed to just a few of these constitutents. Hexanal, *trans*-2-hexenal, 3-methylbutan-1-ol, nonanal and *trans*-2-hexen-1-ol are particularly important volatile components of olive oil (1). Morales *et al.* (3) have also found that butyl acetate, 3-methylbutanal, 1-penten-3-ol, 3-hexyl acetate, *trans*-3-hexen-1-ol and methyl nonanoate are prominent volatile constituents.

The data also indicate that there were quantitative differences in the volatile constituents of the three varieties. ANOVA indicated that of these differences, 13 were statistically significant. The aldehydes were by far the most prominent of the volatiles, amounting to 50% of the total volatile composition, which is in good agreement with data published previously for olive oil (1). Hexanal, *trans*-2-hexenal and 3-methyl butanal were the most prominent aldehydes, the former two volatiles being major products of hydroperoxide lyase activity. In most instances Picual produced the greatest amount of volatiles, particularly of the aldehydes. The identification of these volatile constituents in the callus cultures verifies the use of callus cultures as a model system for the study of the LOX pathway in olive.

References

1. Montedorro, G., Bertuccioli, M., Anichini, F. (1978). Aroma analysis of virgin oil by head space (volatiles) and extraction (polyphenols) technique, in G. Charalambous and G.E. Inglett (eds.), *Flavor of Foods and Beveridges Chemistry and Technology,* Academic Press, New York, pp. 247-281.
2. Williams, M., Sanchez, J., Hann, A.C. Harwood, J.L. (1993). Lipid biosynthesis in olive cultures. *J. Exp. Bot.*, **44**: 1717-1723.
3. Morales, M.T., Aparicio, R., Rios, R.R. (1994). Dynamic headspace gas chromatographic method for determining volatiles in virgin olive oil. *J. Chromatog.* **668**: 455-462.

IMMOBILIZATION OF LIPOXYGENASE IN AN ALGINATE-SILICATE SOLGEL MATRIX: FORMATION OF FATTY ACID HYDROPEROXIDES

An-Fei Hsu, Thomas A. Foglia and George J Piazza
Eastern Regional Research Center, ARS, USDA
600 E. Mermaid Lane, Wyndmoor, PA 19038

1. Introduction

Lipoxygenase (EC1.13.11.12) (LOX) catalyzes the positionally specific dioxygenation of polyunsaturated fatty acids to their corresponding hydroperoxy derivatives. Reduction of the latter compounds gives the corresponding hydroxy acids which are potential substitutes for ricinoleic acid, an industrially important fatty acid. LOX has been conventionally immobilized by adsorption to glutenin, glass wool, talc, polymer beads and ion exchange supports or by covalent linkage to matrices such as CNBr-activated Sepharose, agarose, or oxirane acrylic beads (1,2) or a carbonyl diimidazole activated polymer (3,4). In this study, a novel method for immobilizing LOX entrapped in an alginate-silicate sol-gel matrix was developed. The stability and reusability of LOX immobilized by this improved method was investigated.

2. Materials and Methods

Sodium alginate (4%, w/v) in 0.2 M sodium borate buffer (pH 9.0) was mixed well with an equal volume of soybean LOX (5 mg/ml in 0.2 M sodium borate buffer, pH 9.0). Droplets of this mixture were dispensed through a pasteur pipette into a cold solution of 0.2 M $CaCl_2$ (100 ml) to form single beads. The beads were collected within 10 min. and transferred to a beaker containing enough hexane to cover the beads. Tetramethoxy-ortho-silicate (TMOS, 1-1.5 vol/vol of the beads) was added and the mixture was left overnight at room temperature to complete the polymerization process. The beads were left to dry overnight under vacuum at room temperature. The beads then were either soaked in 0.2 M borate buffer (sodium pH 9.0), 75 % glycerol in water, or borate buffer saturated with isooctane for 6 h. Finally, the beads were removed from the solution, air dried and assayed for lipoxygenase activity by measuring the hydroperoxide formation from linoleic acid described previously (5). The level of hydroperoxide was determined spectrophotometrically by xylenol orange method (6).

3. Results and Discussions

The formation of the product of the linoleic acid oxidation, 13-hydroperoxyoctadecadienoic acid (HPOD), was maximal at pH 9.0. The enzymatic activity of immobilized LOX in calcium alginate beads was affected by drying treatment. Vacuum drying was most efficient in removing water from samples while retaining the highest LOX activity. The average entrapped activity relative to that of free LOX varied from 15 to 30%.

The enzymatic activity of entrapped LOX after immobilization could be enhanced by soaking the beads in 75% glycerol or in borate buffer or borate buffer saturated with isooctane. For long term storage, samples were soaked in 75% glycerol because samples that were not stored in glycerol lost more than 50% of their activity after one week at room temperature. The entrapped LOX was stable at room temperature with no loss in activity after 25 days of storage (Fig. 1). In contrast, free LOX in solution lost most of its activity after 24 hours at room temperature. Other studies on immobilized β-glucosidase in alginate silicate gel (7) also showed that the enzymatic activity was stable at ambient temperature for at least several months. In contrast, LOX covalent bound to a carbonyl-imidazole activated support (3) had a half-life of only 75 hours at 15°C. Other conventionally immobilized LOX preparations had half lives on the order of only several hours (1,2) at room temperature.

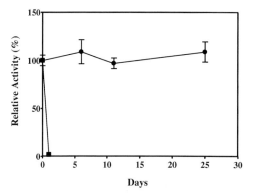 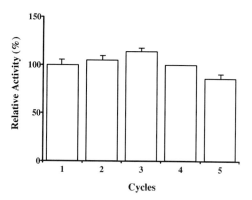

Fig. 1. Comparison of the immobilized (●—●) and free (■—■) lipoxygenase at room temperature. At the times indicated, the samples were assayed to measure the lipoxygenase. Zero day of free enzyme or immobilized lipoxygenase is expressed as 100%

Fig. 2. Reusability of the immobilized lipoxygenase. The immobilized beads were incubated with linoleic acid for five cycles. The activity of lipoxygenare during the first cycle is expressed as 100%.

In our study, the reusability of LOX immobilized in calcium-alginate beads was determined by repeatedly using the same beads for five successive reactions with linoleic acid (LA) (Fig. 2). The LOX activity was measured after each cycle, and the beads were recovered and washed with sodium borate buffer before reuse. To initiate the next cycle of oxidation, LA was added to the incubation mixture containing the recovered beads. The data (Fig. 2) demonstrates that LOX immobilized in beads can be reused at least five times without substantial loss in enzyme activity. In contrast, free lipoxygenase is typically inactivated by hydroperoxide accumulation and the partial anaerobic condition that develops in the reaction mixture (8).

5. Conclusion

In this study, lipoxygenase was successfully entrapped by a novel immobilization method within a calcium-alginate silicate gel matrix. The entrapped lipoxygenase had stability better than LOX immobilized by other methods. The immobilized preparation also was reusable. This procedure should enhance the potential of LOX as a biocatalysts in the preparation of oxygenated fatty acid derivatives that have unique chemical and physical properties.

References

1. Cuperus, F.P., Kramer, G.F.H., Derksen, J.T.P. and Bouwer, S.T. (1995) Activity of immobilized lipoxygenase used for the formation of perhydroxyacids, *Catalysis Today* **25**, 441-445.
2. Battu, S., Rabinovitch-Chable, H. and Beneytout, J.-L. (1994) Effectiveness of talc as adsorbent for purification and immobilization of plant lipoxygenase, *J. Agric. Food Chem.* **42**, 2115-2119.
3. Piazza, G.J., Brower, D.P. and Parra-Diaz, D. (1994) Synthesis of fatty acid hydroperoxide in the presence of organic solvent using immobilized lipoxygenase, *Biotechnol. Appl. Biochem.* **19**, 243-252.
4. Parra-Diaz, D., Brower, D.P., Medina, M.P. and Piazza, G.J. (1993) A method for immobilization of lipoxygenase, *Biotech. Appl. Biochem.* **18,** 359-367.
5. Piazza, G.J. and Nunez, A. (1995) Oxidation of acylglycerols and phosphoglycerides by soybean lipoxygenase, *J. Am. Oil Chem. Soc.* **72**, 463-466.
6. Jiang, T.Y., Wollard, C.S. and Wolff, S.P. (1991) Lipid hydroperoxide measurement by Fe^{+2} in the presence of xylenol orange. Comparison with TBA assay and an iodometric method, *Lipids* **26**, 853-856.
7. Heichal-Segal. O., Rappoport, S. and Braun, S. (1995) Immobilization in alginate-silicate sol-gel matrix protects -glucosidase against thermal and chemical denaturation, *Biotechnology* **13**, 798-800.
8. Seidow, J.N. (1991) Plant lipoxygenase: Structure and function, *Annu. Rev. Plant Physiol. Plant Mol. Biol.* **42**, 145-188.

EXPRESSION AND LOCATION OF LIPOXYGENASES IN SOYBEAN SEEDS AND SEEDLINGS

CUNXI WANG, K. P.C. CROFT, D. HAPSORO & D. F. HILDEBRAND
Department of Agronomy, University of Kentucky, Lexington, KY 40546

Introduction

Expression of LOX is differentially regulated during soybean development and germination. It was suggested that the plant hormones may affect the expression of LOXs in soybean (Liu *et al.* 1991). The fact that pre-existing LOX-1, -2 and -3 disappear and new LOXs (LOX-4, -5 and -6) appear in germinating soybean cotyledons suggests that these newly synthesized LOX may play an important role in some aspect of seedling growth. Mobilization of lipid body triglyceride starts after 3 days of germination, and at this stage LOX-4, -5 and -6 reach high levels in cotyledons. These observations raise the obvious question of whether LOX (seedling or embryo LOX) were associated with lipid body and also where LOX was localized in germinating soybean seedlings. In this report, the effects of plant growth regulators on LOX expression and localization of LOX were thoroughly investigated in soybean.

Materials and methods

Wild type soybean (*Glycine max* (L.) Merr) cv. Century or Fayette, four mutant lipoxygenase-lacking lines (L0, L1, L2 and L3) and one cucumber (*Cucumis sativa*) cultivar (Boston Pickling) were used in this study. Soybean and cucumber seeds were germinated as described by Park and Polacco (1989). Imbibition of soybean seed was in deionized water at 4° C for 12 h. Lipid bodies were purified by the discontinuous sucrose gradient method described by Sturm *et al.* (1985) with minor modifications. Fifteen mL of the homogenate was layered onto 15 mL of 20% sucrose (w/v) in 20 mM Tricine buffer at pH 7.5. Five mL of 8% sucrose (w/v) in 20 mM Tricine buffer at pH 7.5 was layered onto the homogenate in an Oakridge centrifuge tube.

IEF- and SDS-PAGE and Western blotting were performed as described by Liu *et al.* (1991). LOX activity was determined by the oxygen electrode polarography method (Siedow and Girvin, 1980).

The seedling cotyledons and imbibed soybean seed were cut with a razor blade into approximately 1 mm^3 blocks, fixed in 0.1 M phosphate buffer, pH 7.3, 2% (v/v) paraformaldehyde, 1% (v/v) glutaraldehyde, dehydrated through an ethanol series and infiltrated with L. R. White (The London Resin Co. Ltd.). Sections, 60-90 nm thick on nickel grids, were blocked with 0.1% (w/v) BSA and 5% (v/v) heat denatured normal goat serum in PBS buffer (0.148% Na$_2$HPO$_4$, 0.043% KH$_2$PO4, 0.72% NaCl, 0.13% NaN$_3$, pH 7.3) for 1 h and incubated in a mixture of soybean embryo LOX1+2+3 antibody. After being washed with PBS plus 0.5% (v/v) Tween20 four times for 10 min., the sections were incubated with goat anti-rabbit immunoglobulin-gold conjugate (15 nm, Pelco, Ted Pella, inc, Redding CA; 1:40 in PBS) for 1 h. The sections were then rinsed with PBS plus 0.1% (w/v) BSA and 0.5% (w/v) Tween20 three times for 5 min. each time, and three times with deionized water, dried, stained with uranyl acetate for 4 min. and then with lead citrate for 4 min..

Results and discussion

NAA (54 µM), ABA (0-10 µM), TIBA (0-20 µM), Jasmonic acid (JA) (0-1 µM) and Methyl jasmonate (0-1 µM) were used in immature embryo culture. The results showed that expression of LOX4 in immature embryos was mainly induced by the NAA treatment. Expression of LOX4 was significantly increased if NAA was combined with JA. During soybean germination, ABA at 10 µM resulted in a decrease in the level of LOX4 at 1 day after germination and an increase in the expression of LOX4 at 2 or more days of germination. Seedling LOX were not expressed when germination was suppressed by 100 µM ABA.

All the mutants examined including dry seeds and germinated cotyledons except for the triple null had LOX present in purified lipid bodies washed *via* discontinuous sucrose gradient centrifugation. LOX protein cross-reacting with soybean seed LOX1+2+3 antibodies were found in germinated cucumber lipid bodies, but not in those of dry seed. That purified lipid body from seedling cotyledons of the triple null line had no LOX indicates that seedling LOX (LOXs 4-6) do not co-purify along with the lipid bodies. Most of the LOX activity remained in the 15% sucrose gradient fraction after centrifugation. No activity for germinating Century and triple null mutant was measurable in purified lipid bodies at pH 6.83, indicating that germination-associated LOXs and type II LOX (LOX2+3) were not present in the lipid bodies. LOX activity at pH 9.0 was found in purified lipid bodies from Century. The amount of LOX activity at pH 9.0 which finally appeared in the washed lipid bodies of germinating Century represented less than 0.01% of the total activity found in the crude extract. Purified lipid bodies from 4-day germinated seedling cotyledons or dry seed extracts of the triple null showed no detectable LOX. The lipid bodies from the LOX triple null mixed with Century soluble LOX showed that LOX can stick to lipid bodies.

In 12 h imbibed Century cotyledon in which LOX1, 2 & 3 are present, the lipid bodies and protein bodies fill most of the cytoplasm in the cell. Using LOX antibody, immunogold label mostly appeared in the cytoplasm of parenchyma cells, with some present in protein bodies. No specific label was found in lipid bodies, mitochondria or cell walls. In four day germinated seedling cotyledons of Century, lipid bodies and protein bodies were less predominant in storage parenchyma cells. Vacuoles appeared,

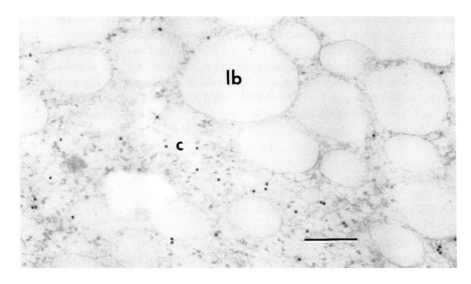

Fig. 1: Electron micrographs of portions of parenchyma cells. Sections from 4-day germinated cotyledon of Century cross-reacting with soybean LOX1-3 antibodies. X 580,000, bar: 0.25 μm. Abbreviations: c: cytoplasm; lb: lipid body.

especially in epidermal cells. Immunogold label was found in cytoplasm and protein bodies of parenchyma cells, but no specific label appeared in lipid bodies, mitochondria, plasma membranes or cell walls. In parenchyma cells of 4-day triple null germinated seedling cotyledons, LOX was observed in cytoplasm at a lower concentration compared to Century. In epidermal cells of germinated Century and the triple null, LOX appeared both in cytoplasm and vacuoles. These results suggest that LOX-1, -2, -3, -4, -5 and -6 may not be directly involved in reserve lipid mobilization of soybean seeds and seedlings. The LOX localization in vacuoles supports a hypothesis that when cells are damaged, LOX may be released from vacuoles and play a role in fatty acid peroxidation. The changing pattern of LOX expression in soybean seeds and seedlings may be regulated by ratios of auxin and various other hormones.

References

Liu, W., Hildebrand, D. F., Grayburn, W. S., Phillips, G. C. and Collins, G. B. (1991). Effects of exogenous auxins on expression of lipoxygenases in cultured soybean embryos. Plant Physiol. 97, 969-976
Park, T. K. and Polacco, J. C. (1989). Distinct lipoxygenase species appear in the hypocotyl/radical of germinating
soybean. Plant Physiol. 90, 295-290.
Sturm, A., Schwennesen, K. and Kindl, H. (1985). Isolation of proteins assembled in lipid body membranes during
fat mobilization in cucumber cotyledons. Eur. J. Biochem. 150, 461-468.
Siedow, J. N. and Girvin, M. E. (1980). Alternative respiratory pathway: its role in seed respiration and its inhibition by propyl gallate. Plant Physiol. 65, 669-674.

RELATIONSHIP BETWEEN LIPOXYGENASE – CATALYZED FORMATION OF DIHYDROXYLATED UNSATURATED FATTY ACIDS AND ITS AUTO-INACTIVATION

MEE REE KIM and DAI-EUN SOK
Department of Food & Nutrition and Department of Pharmaceutics
Chungnam National University, Yuseng-Ku, Gung-dong, Taejon, 305-764, KOREA

INTRODUCTION

Earlier, soybean lipoxygenase had been reported to undergo inactivation during the incubation with polyunsaturated fatty acids, and the inactivation was greater using polyunsaturated fatty acids containing a cis,cis-1,4-pentadiene system into hydro peroxy dienoic acids (Smith & Lands, 1972). Soybean lipoxygenase was found to convert polyunsaturated fatty acids such as α- linolenic acid, γ-linolenic acid, dihomo-γ-linolenic acid or arachidonic acid into the respective dioxygenation products (Bild, Ramadas & Axelrod, 1977 ; Sok & Kim, 1994). Subsequently, only the conjugated hydroperoxy acids containing cis, cis-1,4-pentadiene group correspond to effective inactivators of lipoxygenase (Kim & Sok, 1991).

However, the relationship between lipoxygenase- catalyzed formation of soybean lipoxygenase was not clarified.

MATERIALS AND METHODS

1. MATERIALS.

Soybean lipoxygenase (type 1), α-linolenic acid (99%), γ-linolenic acid (99%), dihomo-γ-linolenic acid, arachidonic acid (99 %), hemoglobin (type 1, bovine 75%), sodium borohydride, sodium cholate, Detapac, Brij 56 and DEAE-Sephacel were products of Sigma Chemical Co.

Preparation of hydroperoxy acids and chemical analyses of reaction products were performed as described before (Kim & Sok, 1991 ; Sok & Kim, 1994).

2. ASSAY OF LIPOXYGENASE ACTIVITY.

Lipoxygenase acitivy was determined at 235nm using a spectrophotometer (Gilford Model 250), as previously described (Schewe et al., 1981 ; Sok et. al., 1990). One unit is expressed as the activity of enzyme forming 1 μmol of oxygenation products / min.

3. PREPARATION OF SOYBEAN LIPOXYGENASE-1.

Soybean lipoxygenase-1 was prepared by applying 50mg of soybean lipoxygenase-1 (type I, Sigma) on DE-AE-Sephacel column (Sigma, 1×20cm) and eluting the column with a gradient concentration (20~220mM) of phosphate buffer, pH 6.8, as described (Axelrod et. al., 1981).

4. INCUBATION OF FATTY ACIDS OR HYDROPEROXY ACIDS WITH SOYBEAN LIPOXYGENASE-1.

The incubation was started by adding soybean lipoxygenase-1 (14 units) to the 0.1 M phosphated buffer, pH 6.5 (50 ml) containing 5 mg of fatty acids or hydroperoxy acid (1 mg) and contained for 20 min at 20℃ as described before (Sok & Kim, 1994).

5. INACTIVATION OF SOYBEAN LIPOXYGENASE-1 BY UNSATURATED FATTY ACIDS OR HYDRO(PER)OXY ACIDS.

Fatty acids dissolved in ammonium hydroxide or hydroperoxy acids were preincubated with 3 units of soybean lipoxygenase-1 in 200 μl of 0.1 N borate buffer at 25℃. The aliquot (20 μl) was taken, and used for the assay of remaning activity of lipoxygenase-1 as described before (Kim & Sok, 1991). Additionally, 5,15-HPETE (6 nmol), 8,15-diHPETE (6 nmol), 15(S)-HPETE (12 nmol) or 15-KETE (6 nmol) was preincubated with lipoxygenase-1, and the remaning enzyme acitivity was measured as described.

6. UV AND GC/MS SPECTROMETRY.

The acidified samples were extracted with ethyl ether twice. The extract was dried by azeotropic evaporation with ethanol. HPLC and spectrometric analyses were performed as reported previously (Sok & Kim, 1990).

RESULTS AND DISCUSSION

When the respective unsaturated fatty acid was incubated with soybean lipoxygenase-1, the spectral change, indicative of the conversion of the unsaturated acids into multiple oxygenation products, was observed. Noteworthy, arachidonic acid containing three 1,4-pentadiene structure showed a further conversion, suggestive of a triple oxygenation (Fig. 1). the pH for the conversion of monooxygenated aicd inoto di- or tri-oxygenation products, different from that for primary lipoxygenation, differed according to the position for secondary oxygenation (Fig. 2). Thus, the formation of dihydroxylated fatty acid seems to depend on the kind of unsaturated acids and the pH of the enzyme reaction.

Next, the inactivation of lipoxygenase-1 acitivity during the incubation with the respective unsaturated fatty acid was examined. Time-dependent inhibition of soybean lipoxygenase-1 was found during the preincubation with arachidonic acid but neither linoleic acid nor 11,14,17-eicosatrienoic acid (Fig. 3). Thus, the inactivation was feasible during the incubation of lipoxygenase-1 with unsaturated fatty acids, monohydroperoxy acids of which can be further subjected to a further oxygenation. As demonstrated in Fig. 4 as a representative example, the inactivation during the incubation with monooxygenated acids followed the pseudo first order kinetics. Table 1 shows the K_i value, k_3 value and inactivation potency (K_3 / k_i) of the respective peroxy acid. It is noteworthy to observed that while hydroperoxy acids possessing 1,4-pentadiene moiety closer to carboxyl terminal show higher binding affinity, those with 1,4-pentadiene closer to ω carbon terminal express higher inactivation rate.

In further studies, the inactivation of lipoxygenase-1 by 15(S)-hydroperoxyeicosatetraenoic acid (15-HPETE) was examined in more detail. Further conversion of 15-HPETE could result in the formation of 5,15-dihydroxyeicosatetraenoic acid and 8,15-dihydroperoxyeicosatetraenoic acid. However, these two dihydroperoxy acid did not inactivate the enzyme remarkably (Table 2).

Therefore, it was strongly suggested that the binding of hydroperoxy acid to active site of lipoxgyenase-1 is required for the potent inactivation. This might explain why 15-HPETE with a higher number of double bonds was the most potent inactivator, consistent with the previous observation that unsaturated fatty acids with more double bonds express a higher binding affinity.

However, the k_i value (56 μM) of 15-HPETE in the enzyme inactivation was much different from the Km value (440 μM) of 15-HPETE in the second lipoxygenation, suggesting that the site of enzyme molecule involved in the inactivation by hydroperoxy acids may be different from the catalytic site.

Noteworthy, the effective inactivation of lipoxygenase-1 requires the homolytic cleavage of 15-HPETE in the active site. In support of this, the addition of hemoglobin (Fe^{2+}) enhanced the inactivation, and moreover, the timely addition of hemoglobin was also important (Fig. 5). Next, the effect of pH on the inactivation of soybean lipoxygenase-1 by 15-HPETE was examined (Fig. 6). The pH optimum was found to be around pH 10, much different from the optimum pH (pH 8.7) for secondary lipoxygenation. All the observations support the previous observation that the enzyme inactivation by 15-HPETE and the second lipoxygenation of 15-HPETE may occur at different subsites in the active site.

CONCLUSION

1) Soybean lipoxygenase-1 converts polyunsaturated fatty acids containing a cis,cis-1,4-pentadiene system into dioxygenated acids via monohydroperoxy acids.
2) The transformation of monohydroperoxy acids into dihydroperoxy acids shows a pH dependence and positional specificity.
3) The effect inactivation of lipoxygenase-1 by hydroperoxy acids requires the structural requirement as a substrate for lipoxygenase-1.
4) The target site involve in the inactivation by hydroperoxy acids seems to be separate from the catalytic site in the consideration of optimum pH and binding affinity.
5) There is a good structure-activity relationship between lipoxygenase-1- catalyzed conversion of monohydroperoxy acids into dioxygenated acids and auto-inactivation of lipoxygenase-1 by monohydroperoxy acids ; hydroperoxy acids possessing 1,4-pentadiene moiety closer to carboxyl group rather than ω carbon terminal correspond to more effective inactivator.
6) Thus, it is proposed that the activity of soybean lipoxygenase-1 may be regulated during secondary oxygenation and / or auto-inactivation, differing according to pH of medium, concentration of Fe^{2+}, etc.

REFERENCES

Axelrod, B., Cheesebrough, T.T. & Laakso, S. (1981) Methods Enzymol. 71, 441.
Bild, G. S., Ramados, C. S. & Axelrod, B. (1977) Arch. Biochim. Biophys., 184, 36.
Borgeat, P. & Samuelsson, B. (1979), Proc. Natl. Acad. Sci., 76, 3213.
Galliard, T. & Philips, D. R. (1971) Biochem. J., 124, 431.
Hamberg, M. & Samuelsson, B. (1967) J. Biol. Chem., 242, 5329.
Kim, Mee Ree & Sok, Dai-Eun. (1991) Arch. Biochim. Biophys., 288, 270.
Schewe, T., Wiesner, R. & Rapoport, S. M. (1981) Methods in Enzymol. vol. 72, pp. 403~441.
Smith, W. L. & Lands, W. E. M. (1972) J. Biol. Chem. 247, 1038.
Sok, Dai-Eun & Kim, Mee Ree (1990) Arch. Biochim. Biophys. 277, 86.
Sok, Dai-Eun & Kim, Mee Ree (1994) J. Agric. Food Chem. 42, 2703.
Van Os, C. P. A., Rijk-Schilder, G. P. M., Halbeek, H. V., Verhagen, J. & Vliegenthart, J. F. G. (1981) Biochem. Biophys Acta., 663, 177.

Table 1. Kinetic constants of the respective hydroperoxy acids in inactivating lipoxygenase-1.

	K_i (μM)	K_3 (sec^{-1})	Inactivation potency (K_3 / K_i) μM^{-1}sec^{-1}
15-hydroperoxy-eicosatetraenoic acid	100	6.6×10^{-3}	6.6×10^{-5}
15-hydroperoxy-eicosatrienoic acid	600	8.3×10^{-3}	1.4×10^{-5}
9-hydroperoxy-octadecatrienoic acid	2,530	1.2×10^{-2}	4.7×10^{-6}
13-hydroperoxy-octadecatrienoic acid	720	6.0×10^{-3}	8.3×10^{-6}

Table 2. Inactivation of soybean lipoxygenase by various hydro(pero)xy acids. The incubation was carried out for 10 min in 200 μl of 0.01 N borate buffer, pH 9.0 as described in Fig. 1.

Hydro(pero)xy Acids	Inactivation, %
15(S)-HPETE (60 μM)	70
15(S)-HETE (60 μM)	< 5
15-KETE (60 μM)	< 5
5,15-diHPETE (30 μM)	< 5
8,15-diHPETE (30 μM)	< 5

CHANGES OF PHOSPHOLIPASE D ACTIVITY DURING RAPE SEED DEVELOPMENT AND PROCESSING

O. VALENTOVÁ, Z. NOVOTNÁ, J.-C. KADER[1] and J. KÁŠ

Institute of Chemical Technology - Department of Biochemistry and Microbiology, Technická 3, 166 28 Prague 6, Czech Republic, [1]Laboratoire de physiologie cellulaire et moléculaire des plantes, U.R.A. - C.N.R.S. 1180, 4, place Jussieu, 75 252 Paris Cedex 05, France

Introduction

Rape seed represents one of the most important oil seeds in the countries of Central Europe. The annual crop in the Czech Republic is about 1/2 milion of tons and it is the main source for the plant oil production. Rape seed oil contains approx. 0.8 - 3.5% of lecithin (1), the main components of which are phosphatidylcholine (PC), phosphatidylethanolamine (PE) and phosphatidylinositol (PI). All phospholipids mentioned above are potential substrates for phospholipase D (E.C.3.1.4.4). The product of this enzyme reaction is phosphatidic acid ("nonhydratable" phospholipid), which is not readily removed during degumming process and affects flavour of the final product.

Experimental

Winter rape seed variety Lirajet was used in all experiments. Extracts were prepared by the procedure described recently (2). Protein content in the extracts was determined by the Bradford method in the format adapted for microtitre plate reader. PLD activity was determined by choline biosensor (3) using egg L-α-phosphatidylcholine as substrate.

Phospholipids were extracted from grinded rape seeds by chloroform/methanol (2/1).and separated on silica column 250×4 mm, LiChrospher Si60 (5 µm). Column was equilibrated with solvent A (2-propanol/hexane, 4/3) and elution performed for 20 min by linear gradient to 100% of solvent B (2-propanol/hexane/water, 8/6/1.5).

Results and Discussion

PLD ACTIVITY IN DIFFERENT STAGES OF SEED MATURITY

Samples of rape seeds variety Lirajet were gradually harvested during seed development at six different planting localities in week intervals from green to fully matured seeds. Owing to the significant changes of PLD activity during seed maturation in all tested samples(2), more detailed study of the PLD activity distribution of soluble (cytosolic) and bound (microsomal) fraction has been provided. Results summarized in TABLE 1 show a significant increase of the total activity within the ripening process which is caused by the activity rise in soluble fraction. Specific activity of the soluble PLD in green and brown seeds is almost the same and increases rapidly in the fully matured seeds.

TABLE 1. Subcellular distribution of rape seed PLD activity during the seed development

fraction of PLD	14 days before full maturity (green seeds)		7 days before full maturity (brown seeds)		matured seeds (black seeds)	
	Total activity (U)	Specific activity (U/mg)	Total activity (U)	Specific activity (U/mg)	Total activity (U)	Specific activity (U/mg)
crude extract	7.9	0.3	18.5	0.3	95.0	1.1
cytosolic PLD	7.1	0.3	18.1	0.3	79.2	1.1
microsomal PLD	0.5	0.3	0.5	0.3	0.6	0.2

CHANGES OF PLD ACTIVITY AND PHOSPHOLIPID CONTENT DURING STORAGE

PLD activity and content of PC, PE and PI in the seeds dried to the different moisture were determined. Values of the activity were measured in August, October 1995 and March 1996 (TABLE 2), the amount of phospholipids in March 1996 (TABLE 3). The lowest values of PLD activities were found in the seeds with 6% of water content and the activity even decreases during storage, while the activity extracted from the seeds containing 10% resp. 13% (non-dried) of water doesn´t change during the 8 months storage. The highest amount of remaining PC was detected in the seeds with 6% moisture and lowest PLD activity. The results of these experiments indicate that PLD activity is significantly influenced by the dewatering of seeds and are in agreement with findings of List and Mounts (4) published recently that the content of phosphatidic acid increases in soybeans during storage of seeds with different moisture content.

TABLE 2 PLD activity changes during storage of rape seeds, var. Lirajet with different content of water

	PLD activity (U/g of seeds)			
water content	6%	8%	10%	13%
Aug 95	15.1	17.5	18.2	18.2
Oct 95	18.6	17.1	17.1	17.1
Mar 95	10.4	13.0	18.4	18.4

TABLE 3 Content of PC, PE and PI in the rape seeds, variety Lirajet with different water content (March 1996)

	Content of phospholipid (mg/g of seeds)			
water content	6%	8%	10%	13%
PC	1.56	1.22	1.39	0.88
PE	1.25	0.94	1.18	0.68
PI	0.41	0.35	0.58	0.29

MONITORING OF PLD ACTIVITY IN DIFFERENT STEPS OF OIL PROCESSING

The PLD activity was measured during the industrial oil processing at 7 stages of the process (*Figure 1*). The serie of samples was collected within 24 hours in four hours intervals.The enzyme remains active up to fifth step of the process despite of the high temperature (approx. 100°C) and pressure treatment of the material.

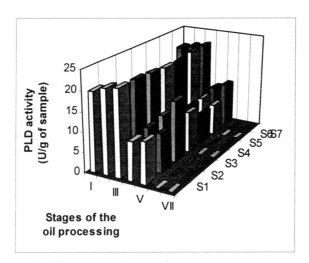

Figure 1 Changes of PLD activity during oil processing: S1 - S7 samples collected within 24 h in the intervals of four hours. I - seeds entering the process (15 - 25 °C), II - seeds after steaming (temperature increases by 10 - 15 °C), III - seeds are grinded - flakes (35 °C), IV - flakes after conditioning (85 - 90 °C), V - press cake (high pressure within hundreds of magnitude of atp.) VI - crush after extraction of the press cake by isohexane (110 - 130 °C) VII - step VI. treated with lecithin slurry.

The enzyme in its natural "environment" is obviously more stable under the process conditions comparing to termostability of partially purified preparation which is inactivated after 5 min at 60 °C (unpublished results).

References

1. Hougen, W.F., Thompson, V.J. and Daun, J.K.: Rape seed lecithin. In: *Lecithins*, Am.Oil Chem.Soc. 1985, pp.79 - 95.
2. Novotná Z., Valentová O., Svoboda Z., Schwarz W., Kader J.-C. and Káš J.(1996) Changes of phospholipase D activity during rape seed development and processing, *Potrav. Vědy* **14**, 63-70.
3. Vrbová E., Kroupová I., Valentová O., Novotná Z., Káš J. and Thévenot C. (1993) Determination of phospholipase D activity with choline biosensor, *Anal. Chim. Acta* **280**, 43 - 48.
4. List G.R. and Mounts T.L. (1993) Origin of the nonhydratable soybean phosphatides: whole beans or extraction? *J. Am. Oil Chem. Soc.* **70**, 639 - 641.

LIPASE ACTIVITY IN GERMINATING OIL PALM SEEDS

ABIGOR, D.R. and D.A. OKIY
BIOCHEMISTRY DIVISION, NIGERIAN INSTITUTE FOR OIL
PALM RESEARCH (NIFOR), P.M.B. 1030, BENIN CITY,
NIGERIA.

SYNOPSIS:

Lipase (Glycerol ester hydrolase, E.C. 3.1.1.3) Isolated from dormant and germinating seeds of the tenera fruit type showed activity in the extracts of whole sprouted seed and the various parts studies during germination.
The Lipase was active in the endosperm from weeks 1-8, in the plumule from weeks 4-10, in the haustorium from weeks 5-10, and in the radicle from weeks 5-10.

INTRODUCTION:

Lipases are used to catalyze hydrolytic, esterification and transesterification reactions. These reactions alter the physical properties of fats and oils and thereby produce a wide range of products (Mukherjee, 1990). Lipases have also been used for the kinetic resolution of Isomers of alcohols or fatty acids (Hills, et al. 1990). The lipases for these purposes have been mostly obtained from microorganisms. However, another source of Lipases not considered until recently is that from the seedlings of young oilseed plants. During the early growth of oilseed plants, Lipases are produced in large amounts to hydrolyze triacylglycerols to fatty acids and glycerol which go to support growth of young plants (Hills et al. 1990).
In this study, Lipase from the various parts of germinating oil plam seeds, normally the endosperm, haustroium, plumule and radicle was isolated wtih the aim of determining the enzyme activity and also the optimal period of production.

METHODS:

Seeds of the tenera variety fo the oil palm obtained from the Nigerian Institute for Oil Palm Research (NIFOR), Benin City, Nigeria, were germinated and the Acetone poweders of the crude lipase prepared from the germinating seeds from weeks one to ten at weekly intervals according to the method of Wetter (1957). Lipase (Glycerol eater hydrolase, E.C. 3.1.1.3) activity was determined by the titrimetric method according to Khor et al, (1986).

RESULTS AND DISCUSSION

The endosperm showed Lipase activity from weeks 1-8 of germination with the highest activity at the 6th week. The haustorium showed Lipase activity from weeks 5-10 and above with the highest activity at the 6th week. The plumule showed lipase activity from weeks 4-10, while the radicle showed lipase activity from weeks 5-10. Both showed increasing lipase activity even at week 10 of germination. (Fig 1).

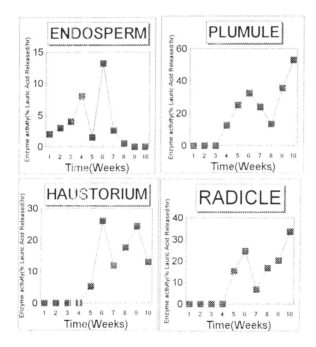

Fig. 1: Effect of length of germination on Lipase activity.

Lipase activity in germinating oil palm seed follows the pattern of development of the various parts, i.e. endosperm, haustorium, plumule and radicle. At the initial stage of germination, the bulk of the seed was endosperm and lipase activity is detected almost immediately. The activity continues to increase until it attains a maximum at the 6th week after which it begins to decline. As germination proceeds the constituents of the endosperm decrease until they are finally absorbed by the hautorium after 12 weeks (Boatman & Crombie, 1958) It is obvious therefore that the decrease in Lipase activity is due to the decrease in the endosperm constituents. As the Constituents of the edosperm are being absorbed by the haustorium, Lipase activity in the germinating oil plam seed becomes restricted to the hautorium, plumule and radicle. The activity in the plumule and radicle increase with increase in these parts while it declines in the haustroium.

REFERENCES:

Boatman, S.G, and W.M. Crombie (1958).
J. Exp. Bot. $\underline{96}$ (5) 52 - 74.
Hills, M.J; I. Kiewitt, and K.D. Mukherjee (1990).
The proceedings of the 9th Int. Symposium on Plants Lipids held at Wye 1990 (pp. 474 - 476).
Khor, H.T; Tan, N.N. and Chua, C.L. (1986).
J. Am. Oil Chem. Soc. $\underline{63}$: 360.369.
Mukherjee, K.D. (1990).
Biocatalysis, $\underline{3}$:277 - 293.
Wetter, L.R. (1957).
J. AM. Oil. Chem. Soc. $\underline{34}$: 66 - 70.

EFFECT OF γ-IRRADIATION ON TOMATO FRUIT RIPENESS WITH SPECIAL REGARD TO LIPOXYGENASE ACTIVITY, CAROTENOID CONTENT AND ANTIOXIDATION POTENCY

ABDULNABI ABUSHITA, HUSSEIN G. DAOOD, E HEBSHI AND PÉTER A. BIACS
Central Food Research Institute (KÉKI), H-1022 Herman Ottó u. 15. Budapest, Hungary

Introduction

Several studies on irradiation of fruits and vegetables have been conducted to delay fruit ripeness, thus, their shelf life might be extended by few days to few weeks [1]. However in some horticultural crops gamma irradiation treatments initiated the climacteric ripeness sequences by inducing the preclimacteric fruits to produce stimulatory amount of ethylene [2]. Factors most likely to affect response of fruit and vegetables to irradiation may include type of fruit (climacteric or nonclimacteric), ripeness stage at expossure, dose of irradiation and post-treatment storage conditions.
The purpose of the present study was to investigate the effects of low and relatively high doses of γ-irradiation on some physiological orders of tomato fruit ripeness (pigmentation, bioantioxidant content and lipoxygenase activity).

Materials and Methods

Immature green fruits of tomato (Lycopersicon esculentum var. Floriset) were obtained from experimental farms of the Faculty of Horticulture, Kecskemét, Hungary).
The ^{60}CO-radiation source of ^{65}kCi capacity of the Central Food Research Institute was used. The applied dose level were 0, 0.05, 0.1, 0.5, 1 and 2 kGy. The irradiated fruits were stored at ambient conditions for weeks to ripen.
High performance liquid chromatographic methods (HPLC) were applied to analyse carotenoid [3], organic acid [4] and tocopherols [5]. The activity of lipoxygenase was spectrophotometrically determined by measuring Δ absorbance at 234 nm using linoleic acid as the substrate [6].

Result and Discussion

One week after irradiation, some fruits in most of the treated batchs turned red in colour. These fruits were taken as 1st the harvest for different analysed. The remaining fruit were slowly ripening, therefore, the 2nd harvest was made at the end of 4th week of storage.
Shown in Table 1 is the change in major carotenoid components of tomato (1st harvest) as a function of irradiation dose irradiation level of 0.5 and 2 kGy resulted in a rapid accumulation of carotenoids mainly lycopene. In fruits treated with 0.5 kGy the whole carotenoid profile was improved. Whereas conversion of lycopene to β-carotene and lutein via cyclization and hydroxylation reaction respectively was partially distributed in fruit irradiated with a dose of 2 kGy.

TABLE 1 Carotenoid content (μg/g raw) tomato fruit ripened after γ-irradiation treatment (1st harvest)

Carotenoids	γ-irradiation dose (kGy)			
	0	0.1	0.5	2
Lutein	0.36	0.68	1.87	0.19
Lycoxanthin	0.29	0.21	0.44	0.81
Lycopene	22.50	27.90	48.90	72.00
Neolycopene	2.25	2.70	5.1.	8.00
β-carotene	1.50	3.00	6.90	2.30
Total carotenoids	27.40	35.00	63.75	84.00

The data in Table 2 implied that the fruits of the 2nd harvest responded by different manner to irradiation revealing that these fruits were at different stage of immaturity at exposure. The lowest dose (0.05) activated carotenoid biosynthesis but irradiation doses higher than 0.1 kGy exhibited marked inhibition of color development.

TABLE 2 Carotenoid content(μg/g fresh) of tomatoes irradiated at green stage ripeness (1st harvest)

Carotenoid	Irradiation dose (kGy)					
	0	0.05	0.1	0.5	1	2
Lutein	1.135 ± 0.05	0.985 ± 019	1.11 ± 0.34	1.165 ± 0.43	1.05 ± 0.07	0.71 ± 0.15
Lycoxantin	0.435 ± 0.02	1.92 ± 0.04	0.635 ± 0.02	0.61 ± 0.09	0.23 ± 0.04	0.155 ± 0.05
Lycopene	65.33 ± 5.38	93 ± 8.76	71.35 ± 2.33	40.5 ± 1.83	26.38 ± 2.42	22.66 ± 3.16
Neolycopene	3.34 ± 0.31	6.0 ± 1.02	4.45 ± 0.07	3.3 ± 0.42	2.4 ± 0.14	1.46 ± 0.22
β-caroten	4.1 ± 0.42	1.785 ± 0.26	2.9 ± 0.14	1.985 ± 019	3.35 ± 0.07	2.665 ± 0.05
Total carotenoid	74.0 ± 3.88	103.76 ± 9.80	80.5 ± 2.82	47.7 ± 2.96	24.54 ± 15.2	27.95 ± 3.04

As shown in Table 3 promotion of carotenoid formation was accompained by remarkable increase in LOX activity in fruits of 1st harvest. This held true with fruits of 2nd harvest in which distributance of colour development was associated with reduced LOX activity. These results led to the conclusion that LOX pathway takepart in the onset of ripeness in tomatoes by producing ethylene via oxidation of unsaturated fatty acids.

TABLE 3 Lipoxigenase activity as affected by irradiation treatment

Irradiation dose	LOX activity $\Delta A_{234\,nm}$	
	1st harvest	2nd harvest
0	0.10	0.18
0.1	0.22	0.15
0.5	0.44	0.07
1	nd	0.03
2	0.60	0.018

nd = not determined

As for natural antioxidants, irradiation treatments caused proportional decrease in the tocopherol and ascorbic acid content of tomato at both 1st and 2nd harvest (Table 4). In exception was that a treatment of 0.05 slightly improved ascorbic acid of 2nd harvest fruits.

TABLE 4 Tocopherol and ascorbic acid content (μg/g fresh) of tomato irradiated at green stage ripeness

Irradiation dose (kGy)	α-tocopherol	γ-tocopherol	Ascorbic acid
		1st harvest	
0	0.625 ± 0.22	1.53 ± 0.15	230.5 ± 13.44
0.1	0.33 ± 0.08	1.165 ± 0.09	166 ± 14.14
0.5	0.475 ± 0.09	1.28 ± 0.18	158 ± 28.3
1	0.175 ± 0.05	0.705 ± 0.07	154 ± 19.8
2	0.115 ± 0.02	0.27 ± 0.08	181 ± 18.3
		2nd harvest	
0	2.2 ± 0.52	1.96 ± 0.22	144.4 ± 8.48
0.05	0.345 ± 0.19	1.365 ± 0.02	170.35 ± 10.96
0.1	0.21 ± 0.014	1.14 ± 0.028	150.45 ± 4.03
0.5	0.099 ± 0.015	1.01 ± 0.13	120.5 ± 12.02
1	0.079 ± 0.0014	0.995 ± 0.1	121.5 ± 17.11
2	0.03 ± 0.014	0.46 ± 0.056	130.15 ± 1.06

Acknowledgement

This work is financially supported by the Hungarian National Committee of Technical Development (OMFB) (Hungarian-Germany Intergovermental Scientific and Technical Cooperation).

References

[1] Akamine, E.K. and Moy, J.H. (1983) Delay in postharvest ripening and senescence of fruits. In preservation of food by ionizin radiation (Josephson and Peterson eds.) CRC press, Boca Raton, Florida
[2] Lee, T.H., McGlasson, W.B. and Edwards, R.A. (1968) Effect of gamma radiation on tomato fruit picked at four stages of development, *Radiat. Bot.* **8**, 259-264.
[3] Biacs, P. and Daood H.G. (1994) HPLC and photodiode-array detection of carotenoids and carotenoid esters, *J. Plant Physiol.* **143**, 520-525.
[4] Daood, H.G., Biacs, P.A., Dakar, M. and Hajdú, F. (1994) Ion-pair chromatography and photodiode-array detection of vitamin C and organic acids, *J. Chromatogr. Sci.* **32**, 481-487
[5] Speek, A.J., Schrijver, F. and Shreurs, H.P. (1985) Vitamin E composition of some oils as determined by fluorimetric detection, *J. Food Sci.* **50**, 121-124.
[6] Daood, H.G. and Biacs, P.A. (1988) Some properties of tomato fruit lipoxygenase, *Acta Aliment.* **17**, 53-65.

EXPRESSION AND SUBSTRATE SPECIFICITY OF LIPOXYGENASE ISOENZYMES EMBRYOS OF GERMINATING BARLEY

WESSEL L. HOLTMAN, GERT VAN DUIJN, JAN R. VAN MECHELEN, NORBERT J.A. SEDEE, ANNEKE C. DOUMA AND NATHALIE SCHMITT*
*Center for Phytotechnology RUL-TNO, Department of Plant Biotechnology TNO, Wassenaarseweg 64, 2333 AL, Leiden, The Netherlands. * TEPRAL, Centre de Recherche et de developpement du Groupe Danone, Brasseries Kronenbourg, 68 Route d'Oberhausbergen. 67200 Strasbourg*

Introduction

Plant lipoxygenases (lox) catalyze the oxygenation of polyunsaturated fatty acids with a 1,4-*cis,cis*-pentadiene structure to form conjugated diene hydroperoxides. These in turn can be converted into a variety of secondary metabolites by an array of enzymes. The physiological function of lox is only poorly understood. In barley two isoenzymes have been described, namely lox-1 and lox-2, which clearly differ in properties, e.g. products formed from linoleic acid [1,3,4]. This suggests that lox-1 and lox-2 each give rise to different lipoxygenase pathway end-products and fulfil distinct physiological roles in the germinating kernel. To get more insight into these roles we have performed a detailed study on the expression of lox-1 and lox-2 during germination at the activity, protein and mRNA level. Additionally, kinetic data are presented concerning the specificity of lox-1 and lox-2 towards different substrates.

Material and Methods

Barley grains (*Hordeum vulgare* L. cv Caruso, harvest 1992) were germinated between 2 x 3 layers of moist filter paper at 25 °C in the dark.
  Lipoxygenase activity was measured spectrophotometrically as described by Doderer *et al.* [1] or, in the case of substrate specificity studies, polarographically by following the oxygen consumption using a Clark-type electrode.
  Northern blotting, as well as SDS-PAGE, Western blotting, and subsequent detection of lox-1 and lox-2 protein by specific monoclonal antibodies, was performed as described by Holtman *et al.* [2].
  Identification of 9- and 13-hydroperoxide of linoleic acid (9-HPOD and 13-HPOD) was performed by reverse-phase HPLC as described by Holtman *et al.* [2].
  Lox-1 and lox-2 isoenzymes were separated after hydroxyapatite chromatography, as described by Yang *et al.* [4], using embryos from 4 day-old seedlings as a starting material. Substrate specificity of lipoxygenase 1 and 2 was studied using a range of concentrations of linoleic- and linolenic acid, methyl linoleate, dilinolein and trilinolein. Reaction velocity was plotted as a function of the substrate concentration using the Lineweaver-burk equation and K_m and V_{max} values were determined.

Results and Discussion

EXPRESSION OF LIPOXYGENASE ISOENZYMES IN EMBRYOS OF GERMINATING BARLEY

Lipoxygenase activity

Figure 1 shows the total lipoxygenase activity in barley embryos, as well as the ratio of 9-HPOD and 13-HPOD formed from linoleic by extracts from barley embryos at different stages of germination, which is an indication of the contribution of lox-1 and lox-2 respectively to total activity. Upon germination the ratio of 9- to 13-HPOD decreased from 96:4 in embryos of quiescent grains to a ratio of 62:38 in embryos of 2-day-old seedlings. These data suggest that in embryos of quiescent grains almost exclusively lox-1 contributes to total activity, and that the sharp decrease during the first day of germination is the result of decreasing lox-1 activity. Furthermore, these data suggest that lox-2 activity increases during germination making up one-third of total activity from day 2 onwards.

Lipoxygenase isoenzyme protein and mRNA levels

Levels of lox-1 and lox-2 protein in embryos of germinating barley were followed by Western blotting experiments. Lox-1 is already present in embryos of quiescent grains (data not shown). On the first day of germination the level of lox-1 protein shows a sharp decrease, suggesting that activity decrease at this stage is caused by decreasing lox-1 protein levels. From day 1 onwards levels of lox-1 protein increase again, accompanied by increasing levels of loxA mRNA, which probably encodes lox-1 (Fig. 2a). This suggests that the expression of lox-1 is principally regulated at the pre-translational level. The expression of lox-2 may be under pre-translational control as well, as increasing lox-2 protein levels during the first days of germination are accompanied by a sharp increase of mRNA levels of loxC, which probably encodes lox-2 (Fig. 2b).

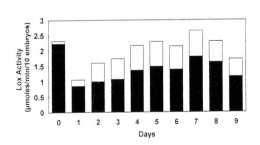

Figure 1. Time course of lipoxygenase activity in the embryo of barley upon germination. The ratio of 9- and 13-HPOD formed from linoleic acid by extracts from barley embryos is indicated within the bars.

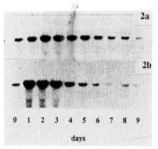

Figure 2. Patterns of loxA (Fig. 2a), and loxC mRNA (Fig. 2b) in barley embryos in the course of germination as determined by Northern blotting. Each lane contains 7.5 μg of total RNA.

SUBSTRATE SPECIFICITY OF BARLEY LIPOXYGENASE ISOENZYMES

Purification of lox-1 and lox-2
Lox-1 and lox-2 were separated after hydroxyapatite chromatography. Two activity peaks were detected, the first eluting at 79 mM potassium phosphate, the second after 102 mM. Product analysis showed that the first activity peak almost exclusively contained lox-1, as it mainly formed 9-HPOD after incubation with linoleic acid. The second peak mainly contained lox-2. Western blotting experiments confirmed the separation of both isoenzymes by this single column step (data not shown).

Substrate specificity of lox-1 and lox-2
K_m and V_{max} values from lox-1 and lox-2 for 5 different substrates are depicted in Table 1, which shows that lox-1 as well as lox-2 can efficiently oxidize esterified lipids beside FFA. However, K_m values from both lox-1 and lox-2 for free fatty acids (FFA) are much lower than those for esterified lipids. For lox-1 as well as lox-2 highest V_{max} values were calculated with methyl linoleate and dilinolein. Also, lox-1 had significantly higher K_m values for esterified lipids than lox-2.

TABLE 1 : V_{max}, K_m, and V_{max}/K_m from lox-1 and lox-2 for 5 different substrates. K_m was expressed in μM and V_{max} in units/mg protein.

	LOX-1 Vmax	LOX-1 Km	LOX-1 Vmax/Km	LOX-2 Vmax	LOX-2 Km	LOX-2 Vmax/Km
linoleic acid	2.1	26	0.08	4.8	38	0.13
linolenic acid	2.3	50	0.05	7.3	72	0.10
methyl linoleate	5.4	909	0.006	12.7	279	0.04
dilinolein	5.2	537	0.01	18.6	304	0.06
trilinolein	0.95	370	0.002	4.1	174	0.02

Literature

1 Doderer A., Kokkelink I., van der Veen S., Valk B.E., Schram A.W. and Douma A.C. (1992) Purification and characterization of two lipoxygenase isoenzymes from germinating barley, *Biochim. Biophys. Acta* **1120**, 97-104.
2 Holtman W.L., van Duijn G., Sedee N.J.A. and Douma A.C. (1996) Differential expression of lipoxygenases isoenzymes in embryos of germinating barley, *Plant Physiol.* **111**, 569-576.
3 Hugues M., Boivin P., Gauillard F., Nicolas J., Thiry J-M and Richard-Forget F. (1994) Two lipoxygenases from germinating barley-heat and kilning stability, *J. Food Sci.* **59** (4) 885-889.
4 Yang G., Schwarz P.B. and Vick B.A. (1993) Purification and characterization of lipoxygenase isoenzymes in germinating barley, *Am. Assoc. Cereal Chem.* **70** (5) 589-595.

This work was supported by EUREKA grant no. EU270 and is ABIN publication no. 152

Section 7:

Oil Seeds and Fruits

OLEOSINS: THEIR SUBCELLULAR TARGETING AND ROLE IN OIL-BODY ONTOGENY

D.J. MURPHY[1], C. SARMIENTO[1], J.H.E. ROSS[1], E. HERMAN[2]
1 Department of Brassica & Oilseeds Research
John Innes Centre, Norwich, NR4 7UH, United Kingdom
2 Plant Molecular Biology Laboratory, USDA-ARS
Beltsville, Maryland 20705, USA

Seed oil bodies in plants are surrounded by a monolayer of highly unusual and abundant proteins, the oleosins, which can comprise as much as 8-15% total seed protein [1,2]. However, relatively little is known about the subcellular targeting and processing of oleosins, and their role in oil-body formation and maturation. We have expressed a soybean 24 kDa oleosin in transgenic rapeseed, in order to elucidate these issues. Detailed experimental data will be published elsewhere [3], but in this report, we discuss the implications of our findings for the mechanism of oil-body ontogeny in plants.

Five independent T_1 transgenic rapeseed lines were produced, each containing 1-2 copies of a soybean genomic clone encoding a 24 kDa oleosin. In all five transgenic lines, the soybean oleosins accumulated specifically at the mid-late stages of cotyledon development, similarly to the endogenous rapeseed oleosins. The ratio of rapeseed:soybean oleosin in the transgenic plants varied between 4:1 and 6:1, as determined by both EM-immunogold and SDS-PAGE methods. The apparent molecular weight of the recombinant soybean oleosin changed from 23 to 24 kDa during seed development, indicating a possible conformational change in the protein. Up to 5% of the recombinant soybean and endogenous rapeseed oleosins were associated with ER-like membrane fractions of 1.10-1.13g.ml^{-1}, as determined by both sucrose density gradient centrifugation and confirmed by EM immunogold labelling. Oleosins were only found on ER membranes in the vicinity of oil bodies, and none were detected on the bulk ER cisternae. Oleosins attached to oil bodies, could be immunoprecipitated by the appropriate antiserum. However, oleosins associated with the ER-like membrane fractions were unavailable for immunoprecipitation, which is consistent with a conformational difference between the two pools of oleosin.

In the light of the above results and evidence from the literature, we propose the following mechanism for oleosin targeting and oil-body ontogeny in plants as shown in Figure 1.

A. Triacylglycerol (TAG) biosynthesis and oil accumulation often occurs in the absence of oleosin biosynthesis, most notably in oily mesocarp tissues, such as olive, avocado and oil palm [4, 2]. TAGs are probably deposited between the leaflets of the phospholipid bilayer of ER, within specialised regions enriched in enzymes of the Kennedy pathway [5, 6]. TAG accumulation causes a localised swelling of the ER membrane, which then buds off to form a nascent oil-body consisting of a relatively small TAG droplet enclosed by phospholipid monolayer [7]. Very small oil droplets will tend to coalesce in order to minimise their surface area:volume ratios - a process that is accentuated by dehydration and/or high salt conditions. Therefore, in the absence of oleosins or other emulsifying agents, oil bodies in mesocarp cells and possibly in some desiccation-intolerant tropical species, will tend to undergo repeated fusions until they are as much as 10-25 µm in diameter, as shown in Figure 1A.

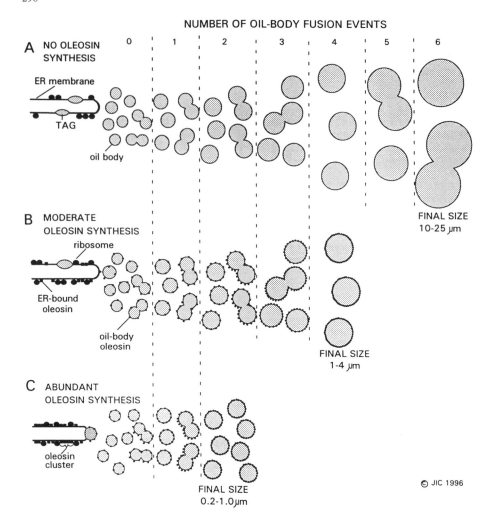

Figure 1: Role of oleosins in oil-body ontogeny

Oil bodies probably bud off from the ER as small TAG droplets of 60-100 nm diameter. Such small oil droplets are unstable in aqueous media and would tend to coalesce, rapidly at first and then more slowly, until reaching a stable size as determined by their surrounding environment. **A,** In the absence of oleosins and in non-desiccating tissues like palm mesocarp or cocoa seeds, the final oil-body size is about 10-25 μm. **B,** Directly after their synthesis, oleosins are inserted onto the ER membrane from where they may diffuse onto nascent oil bodies, probably undergoing a conformational change. Alternatively, oleosin-poor oil bodies may re-fuse with the ER to undergo TAG turnover and acquire more oleosins. If the ratio of oleosin:TAG synthesis is only low to moderate, the nascent oil bodies will need to undergo many fusion events before acquiring a complete oleosin monolayer. In oilseeds which have oil bodies of 1-4 μm, approximately 9-14 fusions will be required to get from a diameter of 100 nm to 1-4 μm. **C,** With much higher ratios of oleosin:TAG synthesis, fewer fusions are required before the oleosin monolayer is complete, which results in smaller oil bodies. Therefore, in rapeseed, mature oil-body sizes of 0.2-1.0 μm can be achieved by only 3-8 fusion events from 100 nm oil bodies.

B. In several oilseed species, the rate of oleosin biosynthesis early in seed development is relatively low compared to that of oil accumulation, leading to a much lower oleosin:oil ratio in such oil bodies compared to those produced at later developmental stages [8,9]. We propose that oleosins are inserted into the ER membrane, either during or shortly after translation in a conformation that may be different from that adopted when it finally reaches the surface of oil bodies, as originally proposed by Loer & Herman [10], and shown here by the differential availability of the two oleosin pools for immunoprecipitation. Oleosin molecules readily self-associate to form dimers and higher-order oligomers, even under denaturing conditions on SDS-PAGE. Following tissue homogenisation, the ER will fragment into a heterogenous population of vesicles. Those vesicles enriched in clusters of oleosins will have a slightly higher buoyant density than other ER regions containing marker proteins such as BiP and choline phosphotransferase, as found in the present study. Nascent oil bodies budding off from the ER at this stage of seed development would contain relatively few oleosin proteins. These oleosin-poor oil bodies would tend to fuse with other oil bodies, although perhaps at a slower rate than in the complete absence of oleosins. The presence of an increasing surface density of oleosins following each fusion event would limit the eventual size of the oil bodies, as shown in Figure 1B.

C. During the mid-late stages of seed development, the rate of oleosin biosynthesis increases substantially and even newly formed oil bodies will contain a relatively dense coating of oleosins. Such oil bodies undergo fewer fusion events before becoming completely enclosed by a monolayer of oleosins, which would then stop further fusions. Therefore, the final diameters of oil bodies produced later in seed development would tend to be somewhat smaller than those produced at earlier stages, as has been observed experimentally [11,12]. It should be noted that TAG, and therefore oil-body biosynthesis, can still occur even after net oil accumulation has ceased. For example, TAGs from oil bodies are still available for ER-mediated desaturation, implying that at least part of the oil-body TAG pool is still exchangeable with the ER [13]. Such oil bodies may not be totally enclosed by oleosins, so as to facilitate their refusion with ER membranes. The complete coating of the entire oil-body population with oleosins is probably driven by the increase in oleosin biosynthesis and the onset of desiccation that both occur at the latter stages of seed development.

According to this model, oleosins are not required for TAG formation and oil-body biogenesis, but the relative rate of oleosin biosynthesis and accumulation on oil-body surfaces during the entire period of seed development will be the major determinant of the final size of oil bodies in mature seeds. This model is consistent with all published data relating to oleosin biosynthesis and oil-body ontogeny.

References:
1. Murphy, D.J. (1993) Structure, function and biogenesis of storage lipid bodies and oleosins in plants. *Prog. Lipid Res.* **32**,:247-280.
2. Huang, A.H.C. (1996) Oleosins and oil bodies in seeds and other organs. *Plant Physiology* **110**, 1055-1061.
3. Sarmiento, C. et al (1996) Expression and subcellular targeting of a soybean oleosin in transgenic rapeseed. Implications for the mechanism of oil body formation in seeds. *Submitted for publication.*
4. Ross, J.H.E., Sanchez, J., Millan, F. and Murphy, D.J. (1993) Differential presence of oleosins in oleogenic seed and mesocarp tissues in olive (*Olea europea*) and avocado (*Persea americana*). *Plant Sci.* **93**, 203-210.
5. Herman, E.M. (1995) Cell and molecular biology of seed oil bodies, in J. Kiegel and G. Galili (eds.) *Seed development and germination.* Marcel Dekker, New York, Basel, Hong Kong, pp 195-214.
6. Vogel, G. and Browse, J. (1996) Cholinephosphotransferase and diacylglycerol acyltransferase. *Plant Physiol.* **110**, 923-931
7. Wanner, G. et al (1981) The ontogeny of lipid bodies (spherosomes) in plant cells. *Planta* **151**, 109-123.
8. Cummins, I., et al. (1993) Differential, temporal and spatial expression of genes involved in storage oil and oleosin accumulation in developing rapeseed embryos. *Plant Mol. Biol.* **23**. 1015-1027.
9. Aalen, R.B. (1995) The transcripts encoding two oleosin isoforms are both present in the aleurone and in the embryo of barley (*Hordeum vulgare* L.) seeds. *Plant Mol.Biol.* **28**, 583-588.
10. Loer, D.S. and Herman, E.M. (1993) Cotranslational integration of soybean (*Glycine max*) oil body membrane protein oleosin into microsomal membranes. *Plant Physiol.* **101**, 993-998.
11. Rest, J.A. and Vaughan, J.G. (1972) The development of protein & oil bodies in the seed of *Sinapis alba* L. *Planta* **105**, 245-62.
12. Cummins, I. and Murphy, D.J. (1990) Mechanism of oil body synthesis and maturation in developing seeds, in J.L. Harwood and P.J. Quinn (eds.), *Plant Lipid Biochemistry, Structure and Function*, Portland Press, London. pp. 231-233.
13. Garces, R. et al (1992) Temperature regulation of oleate desaturase in sunflower seeds. *Planta* **186**, 461-465.

EVOLUTION OF OLEOSINS

ANTHONY H.C. HUANG
Department of Botany and Plant Sciences
University of California, Riverside, CA 92521–0124, USA

The seeds of most plant species store triacylglycerols (TAGs) as food reserves for germination. The TAGs are present in small subcellular spherical oil bodies of about 0.6–2 µm in diameter. Each oil body has a matrix of TAGs surrounded by a layer of phospholipids (PL) embedded with structural proteins called oleosins (Huang 1996). The small entities of oil bodies provide a large surface area per unit TAG that would facilitate lipase binding and lipolysis during germination. Oleosins are also present on the storage oil bodies in pollens and in lipid-containing organelles in the tapetum cells of anthers (Roberts et al. 1993; Lee et al. 1994; Robert et al. 1994).

Oleosins have unique secondary structures that enable the proteins to interact with other molecules on the surface of oil bodies. Each oleosin molecule has three structural domains: an N-terminal amphipathic domain, a central antiparallel β-stranded hydrophobic domain, and a C-terminal amphipathic domain. The central domain anchors the protein onto the oil body or in the lipid-containing organelles, whereas the N-terminal and C-terminal domains are exposed to the exterior. Among oleosins of different organs and plant species, only the central hydrophobic stretches (72 residues) are highly conserved, whereas the N-terminal and C-terminal sequences are quite dissimilar.

Prokaryotes, in general, do not store TAGs as food reserves. A minor exception occurs in *Actinomyces*, which produces TAGs under certain nutritional conditions. TAGs likely evolved as efficient food reserves in early eukaryotes by the addition of one enzyme, diacylglycerol acyltransferase, which diverted diacylglycerols from the ubiquitous PL metabolic pathway to TAGs. Initially, the hydrophobic TAGs were stored between the two PL layers of the endoplasmic reticulum (ER) membrane, where the acyltransferase was. Today, seeds of some species on occasion still have some TAGs present in the ER membrane. The presence of excess TAGs in the ER membrane would interfere with the normal functioning of the ER. This problem was overcome with the evolutionary appearance of oleosins, which had a long hydrophobic stretch. Oleosins were synthesized on bound ER, and from there they extracted the TAGs to produce solitary oil bodies. It is assumed that algae and primitive plants have oleosins associated with the storage TAGs. It is less certain if *Euglena*, yeast, and animals also have oleosins.

The hydrophobic stretch of 72 residues in the oleosin is the longest one found in all prokaryotic and eukaryotic proteins. The mechanism by which it has evolved is intriguing. A postulation can be made partly based on the length of 72 residues being roughly 4 times that of a transmembrane polypeptide and on the occurrence of several relatively hydrophilic residues at the center of the stretch. Initially, a short hydrophilic polypeptide joining two transmembrane hydrophobic polypeptides, of possibly an ER membrane protein, became hydrophobic through DNA sequence mutation. A continuous hydrophobic stretch resulted, which consisted of about half of the final 72 residues. This primitive hydrophobic stretch could stabilize an oil body, though not efficiently. Its length was doubled by one of the following DNA sequence mutations (Fig. 1): (a) duplication of the DNA segment encoding the primitive hydrophobic stretch, (b) mutation of an adjacent pair of transmembrane polypeptides similar to that which produced the primitive hydrophobic stretch, and (c) continuous mutation of the cytosol-exposed hydrophilic residues flanking the primitive hydrophobic stretch to hydrophobic residues. Once evolved, the 72-residue hydrophobic stretch has been preserved because of structural constraints. More modifications have occurred in the regions of oleosin exposed to the cytosol. The resulting oleosins are functional in seeds, pollen, tapetum, and other organs.

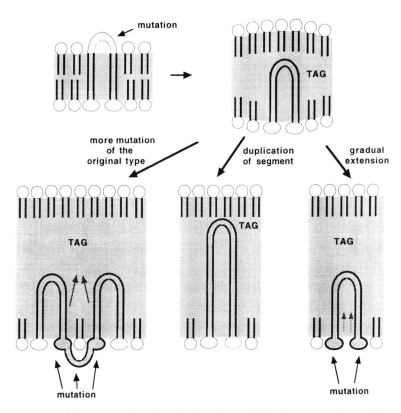

Figure 1. A proposal of the evolution of oleosins. The shaded area and thick lines represent hydrophobic regions; they include the acyl moieties of PL (two lines joining a circle), TAGs and the hydrophobic portion of an evolving oleosin polypeptide (enclosed column). Open circles and columns with thin lines depict hydrophilic portions of PL and proteins. Oleosin evolved via mutations of a transmembrane protein. See text for explanation.

A phylogenetic tree of the 35–40 reported oleosins is constructed based on analyses of the amino acid sequences of the highly conserved 72 residues of the central hydrophobic stretch and its adjacent moderately conserved residues. The tree is unrooted, since no single oleosin can be assigned to the basal lineage (Huang 1996). Two major seed oleosin lineages representing the two isoforms are apparent. Each lineage has both monocot and dicot members. The Brassicaceae tapetum oleosins form another lineage. The two *Brassica* pollen oleosins make up a rather distinct lineage; they are of male gametophytic origin, which is different from the sporophytic origin of the seed and tapetum oleosins. Similarly, pine female gametophytic oleosin represents a separate lineage.

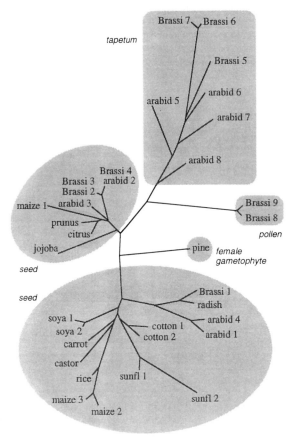

Figure 2. An unrooted phylogenetic tree of oleosins. The oleosin numbers after the names of plants follow roughly the chronological appearance of their genes in the GCG data bank. The large and small shaded ovals denote the two seed oleosin isoforms. The large and small shaded rectangles represent the tapetum and pollen (male gametophytic) oleosins, respectively. The shaded circle indicates the female gametophytic oleosin.

References

Huang, A.H.C. (1996) Oleosins in seeds and other organs, *Plant Physiology* **110**, 1055–1061.
Lee, K., Bih, F.Y., Learn, G.H., Ting, J.T.L., Sellers, C., and Huang, A.H.C. (1994) Oleosins in the gametophytes of *Pinus* and *Brassica* and their phylogenetic relationship with those in the sporophytes of various species, *Planta* **193**, 461–469.
Robert, L.S., Gerster, J., Allard, S., Cass, L., and Simmonds, J. (1994) Molecular characterization of two *Brassica napus* genes related to oleosins which are highly expressed in the tapetum, *Plant J.* **6**, 927–933.
Roberts, M.R., Hodge, R., Ross, J.H.E., Soremsen, A., Murphy, D.S., Draper, J., Scott, R. (1993) Characterization of a new class of oleosins suggests a male gametophyte-specific lipid storage pathway, *Plant J.* **3**, 629–636.

OILS-TO-OLEOSINS RATIO DETERMINES THE SIZE AND SHAPE OF OIL BODIES IN MAIZE KERNELS

JULIE T.L. TING, ANTHONY H.C. HUANG
Department of Botany and Plant Sciences
University of California, Riverside, CA 92521–0124, USA

Introduction

The seeds of many plant species store triacylglycerols as food reserves in small subcellular spherical organelles called oil bodies. Each oil body contains a triacylglycerol matrix surrounded by a monolayer of phospholipids embedded with abundant structural proteins called oleosins (Huang 1992).

The size of the spherical oil bodies in situ and in isolated preparations is apparently species-specific, and the average diameter ranges from 0.6 to 2.0 μm (Tzen et al. 1993). The mechanism which controls the size of the oil bodies in a species-specific manner has not been known previously. The controlling factors may be of physicochemical and/or biological nature. The physicochemical factor would reside in the species-specific structures of the oleosins, which could dictate the surface curvature and thus the size of the oil bodies. The biological factor would rest on the relative amounts of the matrix oils and the surface oleosins/phospholipids synthesized in the cell during seed maturation. When the proportion of oils to oleosins is high, the oil bodies would become larger, and vice versa. This proportion of oils to oleosins is controlled by the genetic background of the species, such as the relative activities of the oil- and oleosin-synthesizing machinery and the availability of their respective substrates, and the environmental parameters, such as the growth temperature and the availability of nitrogen for oleosin synthesis.

We used two maize (*Zea mays* L.) strains having diverse kernel (seed) oil contents to study the effects of varying the oil and oleosin contents on the structure of the oil bodies. Illinois High Oils (IHO, 15% w/w oils) and Illinois Low Oils (ILO, 0.5%) maize kernels were the products of breeding for diverse oil contents for about 100 generations (Dudley and Lambert 1992).

Results

Southern analyses were performed to determine if the extensive breeding of maize for kernel oil contents had altered the oleosin genes. There are three oleosin genes, *ole16*, *ole17*, and *ole18*, which are expressed in the maturing embryos, encoding oleosins OLE16, OLE17, and OLE18, respectively. In Southern blot analyses, we observed identical EcoRI restriction fragments of each of the three genes from the genomic DNA of both IHO and ILO.

The proteins and oleosins in IHO and ILO embryos were analyzed by SDS-PAGE and immunoblotting. When the crude extracts obtained from an equal weight of embryos were separated by SDS-PAGE, both IHO and ILO samples were found to possess the three oleosins, OLE16, OLE17, and OLE18 in the proportion of about 2:1:1. Similarly, oil bodies isolated from an equal weight of IHO and ILO embryos contained the same amount and proportion of the three oleosins.

When the crude extracts containing equal amounts of oils from IHO and ILO were resolved by SDS-PAGE, IHO contained much less oleosins than ILO. Similarly, oil bodies containing equal amounts of oils from IHO had substantially less oleosins than those of ILO.

Electron microscopy of the embryonic cells in the maturing kernels was performed. The oil bodies in situ in IHO embryos were substantially larger than those in ILO embryos, even though other subcellular structures such as the mitochondria appeared to be quite similar in size (Fig. 1). The oil bodies in IHO were essentially spherical, unless they were pressing against other subcellular structures and thus would assume irregular shapes. In contrast, a large percentage of the oil bodies in ILO had irregularly contoured surfaces. In addition, the oil bodies in ILO were more heterogenous in size.

The size and shape of the oil bodies of IHO and ILO, as observed in situ, were preserved after the organelles had been isolated in vitro (Fig. 1). In isolated preparations, the oil bodies of IHO were totally spherical, whereas those of ILO still retained their irregular shapes.

Overall, we interpret the occurrence of larger and spherical oil bodies in IHO and the smaller and irregularly shaped oil bodies in ILO to reflect the difference in the ratio of matrix oils to surface oleosins. A high oils/oleosins ratio enables the oil bodies to assume a large size and a spherical shape, whereas a low ratio dictates the oil bodies to be of a small size with irregularly contoured surface which has more surface area per unit matrix volume.

Discussion

The current findings show that the size of an oil body in a plant species is not dictated, at least not strongly, by the specific molecular configuration of the oleosins on the organelle surface. Rather, the size can be highly variable, depending on the ratio of oils to oleosins synthesized during seed maturation.

Expression of the genes for the synthesis of oils and oleosins during seed maturation are uncoupled in maize strains having diverse kernel oil contents. This finding has at least two significant implications. First, the occurrence of the oleosin gene loci at the vicinity of the quantitative trait loci linked to high oil contents (Lee and Huang 1994) appears to be fortuitous. Therefore, using oleosin loci as markers for high oil traits in breeding is unwarranted. Also, using genetic engineering to increase the seed oleosin contents in the hope of enhancing the oil contents will not be successful. Second, it is inappropriate to use the oleosin gene promoter as a model, albeit its great strength, to study the general pattern of expression of genes encoding oil-synthesizing enzymes.

References

Dudley, J.W., and Lambert, R.J. (1992) 90-Generations of selection for oil and protein in maize, *Maydica* **37,** 81–87.

Huang, A.H.C. (1992) Oil bodies and oleosins in seeds, *Annual Review of Plant Physiology Plant Molecular Biology* **43,** 177–200.

Lee, K., and Huang, A.H.C. (1994) Genes encoding oleosins in maize kernels of inbreds Mo17 and B73, *Plant Mol. Biol.,* **26,** 1981–1987.

Tzen, J.T.C., Cao, Y.Z., Laurent, P., Ratnayake, C., and Huang, A.H.C. (1993) Lipids, proteins, and structure of seed oil bodies from diverse species, *Plant Physiology* **101,** 267–276.

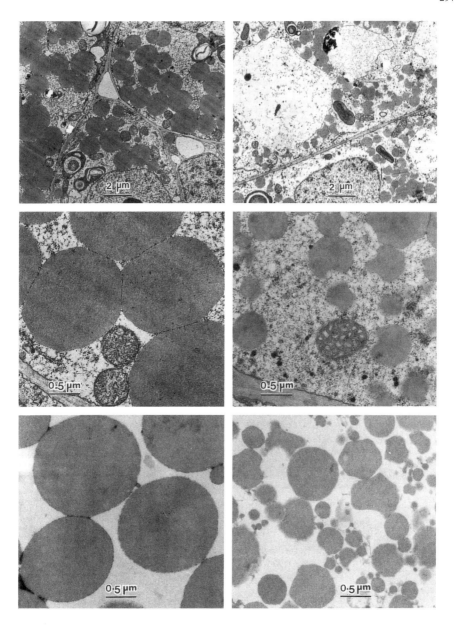

Figure 1. Electron micrographs of oil bodies of Illinois High Oils (IHO) and Illinois Low Oils (ILO). Inside the embryonic scutellar cells (*upper and middle panels*), oil bodies of IHO (*left*) were substantially larger and generally spherical, whereas those of ILO (*right*) were smaller and more heterogeneous in size with irregularly contoured surfaces. Note that the mitochondria with well-recognized double membranes were of a similar size in both IHO and ILO. In isolated preparations (*lower panels*), the oil bodies of IHO (*left*) and ILO (*right*) resembled those in situ in both size and shape.

DEVELOPMENT OF GENETICALLY ENGINEERED OILSEEDS

From Molecular Biology to Agronomics

ANTHONY J. KINNEY
DuPont Experimental Station
P.O. Box 80402, Wilmington, Delaware 19880-0402 USA

Although the relative fatty acid composition of the oil varies between species, almost all of the important commodity vegetable oils are rich in 18-carbon, polyunsaturated fatty acids (linolenic and/or linoleic acids). Polyunsaturated fatty acids are prone to oxidation during storage (linolenic acid, for example, is over 50 times more chemically reactive than oleic acid) and chemical changes during cooking, which result in off-flavors. Oils rich in polyunsaturated and monounsaturated fatty acids are liquid at room temperatures which makes them, in their unmodified form, unsuitable for applications such as butter and lard substitutes . Thus most edible vegetable oils used in commercial frying, margarines, baking shortenings and hydrophobic coatings have been modified by chemical hydrogenation after refining.

Hydrogenation results in the formation of monounsaturated fatty acids with a double bond in the *trans* configuration. These so-called *trans* fatty acids have the physical properties of a fully saturated fat rather than those of a monounsaturated fatty acid with a *cis* double bond and their consumption has been correlated with unfavorable changes in blood cholesterol. Therefore, value may be added to generic vegetable oils by eliminating hydrogenation and producing *trans* free oils with a fatty acid composition directly suited to the intended food application. In addition, because palmitic acid consumption has also been correlated with unfavorable changes in blood cholesterol, table oils with lower saturated fats may also command a premium over unmodified oils.

We have cloned many of the genes which encode fatty acid biosynthetic enzymes in soybeans, corn and canola. Using these genes in conjunction with seed-specific promoters we have been able to increase or decrease the activity of individual fatty acid biosynthetic enzymes in oilseeds and thus produce edible vegetable oils with modified fatty acid profiles. For example, by suppressing a gene which encodes a palmitoyl-ACP thioesterase (GmFatB1) in soybean we have been able to produce an oil with less than 4% total saturates, compared with about 15% in unmodified soy oil. We have also produced oxidatively stable soybean oil by reducing the total polyunsaturated fatty acid content to less than 5% and increasing the oleic acid content to about 85%. This was achieved by suppressing a microsomal omega-6 desaturase gene and this phenotype is described in more detail below.

To illustrate the potential for the commercial production of transgenic oilseed crops with modified fatty acid contents, I will present high oleic soybean oil as a case study. The aim of this project was to produce a soybean oil which, in its unmodified form, would have the equivalent properties of hydrogenated vegetable oils without any of the costs or disadvantages associated with the hydrogenation process. Initial experiments had confirmed that the major components of soybean oil affecting the oxidative stability were the polyunsaturated fatty acids, linoleic and linolenic acids. Thus the desired fatty acid phenotype of the genetically modified oil was low polyunsaturated fatty acids (5% or less) and high oleic acid (80% or greater).

The initial step in the synthesis of polyunsaturated fatty acids in developing soybeans is the conversion of oleic acid to linoleic acid by the action of a microsomal omega-6 desaturase. Almost all of the linoleic and linolenic acids in soybean oil are derived from this initial reaction. The microsomal omega-6 desaturase is encoded by one of two genes, termed Fad 2-1 and Fad 2-2. The Fad 2-2 gene is expressed in all tissues of the soybean including the leaves, stems and roots. In the developing seed it is expressed at a constant, relatively low level. The Fad 2-1 gene is expressed only in the developing seed and its expression level greatly increases when oil biosynthesis is induced, at about 15 to 20 days after pollination. The Fad 2-2 gene product appears to be a "housekeeping" enzyme, responsible for maintaining the polyunsaturated content of soybean membranes whereas the polyunsaturates of soybean oil are the result of the action of the Fad 2-1 gene product. It follows, therefore, that the most effective way of reducing the polyunsaturate content of soybean oil would be to completely suppress the expression of the Fad 2-1 gene.

One of the methods we have used to suppress the activity of endogenous genes is gene-transgene cosuppression, or Transwitch. In this method all or part of the coding region of the target gene, in a sense orientation behind a plant promoter, is reintroduced into a plant. As a result the expression of both the transgene and any homologous, endogenous genes are suppressed.

We have produced high oleate (85%), transgenic soybeans which were derived from a soybean transformed with a Fad 2-1 Transwitch construct. The construct contained 100% of the coding sequence of the soybean Fad 2-1 cDNA under the control of a beta-conglycinin (seed-specific) promoter. Based on Southern analysis, there appeared to be two *loci* of transgene integration. One *locus* contained only a Fad 2-1 cDNA and seeds segregating with only this *locus* had an oleic acid content of 80-85%. The other *locus* also had a sense Fad 2-1 cDNA but this was overexpressing, resulting in a seed oleic acid content of about 2-3%. When both Fad 2-1 *loci* were present in the same seed the resulting oleic acid content of the oil was about 75%. Both of these *loci* segregated in a Mendelian fashion in the R1:2 and R2:3 seeds.

Northern analysis of high oleic acid plants indicated that all of the detectable, endogenous Fad 2-1 mRNA was suppressed by the Transwitch *locus*. The mRNA of the constitutive Fad 2-2 gene was also reduced, although not completely suppressed, in Transwitched seeds but not in the leaves of the same plants.

Since these transgenic soybeans were producing seeds with the desired phenotype our next step was to investigate the stability of this transgenic event in the field as well as the effect of different growing environments on the oleic acid content of the seeds and the effect of the phenotype on the yield of the transgenic lines.

Plants (R2:3) which were homozygous for the Transwitch locus were planted in the summer of 1995 in field plots in Delaware and at a number of other US sites, including sites in Iowa and Michigan. The bulk oleic acid content of seeds from plants grown at each site was about 85% (range 84-88% at each site). Thus the high oleate phenotype was stable over a number of different environments during a single growing season. During the same season the oleate content of normal soybeans and of high oleate mutants varied considerably across the different sites. The highest oleic content we observed in seeds from a single field-grown plant was 88%. The linolenic acid content of this very high oleic plant was 2%, compared with about 10% in normal soybeans and the linolenic content was less than 1%, compared with over 50% in commodity soybean oil. The total oil content of the transgenic seeds was similar to those of normal seeds.

We also observed a novel fatty acid in some of the high oleic lines. This fatty acid was an isomer of linoleic acid, 9,15 octadecanoic acid, and was probably the result of omega-3 desaturase (Fad 3) activity on oleic acid, since the normal substrate for this enzyme was not present in these plants (this isomer can be reduced or eliminated by crossing the Fad 2-1 plants with plants containing a suppressed Fad 3 gene). The 9,15 octadecanoic acid has not be detected in soybeans with a normal oleic acid content but it is found in many other foodstuffs (from about 5% in mango pulp to about 0.5% in human milk) and thus is not of a concern in terms of food safety. It is also found in partially hydrogenated oils, animal fats and buttermilk at concentrations similar to that of the high oleic soybean oil.

In addition to growing R2:3 seeds in field plots we also wanted to test the stability of the Transwitch under the same conditions. Three generations of seeds (R0:1, R1:2 and R2:3), all containing the same Transwitch *locus*, were planted in parallel plots in the field at our Delaware location. The bulk oleic acid content of homozygous plants (85%) was similar in all three generations. The Transwitch *locus* appeared to be a stable, dominant Mendelian gene both in controlled environments and in the field. We have subsequently grown further generations (R4:5 and R5:6) of these transgenic lines in Puerto Rico, during the Winter of 1995/1996. Again, the seeds of homozygous Transwitch plants grown at this site had an oleic acid content of about 84%.

Extensive yield trials of the transgenic, high-oleate lines are iplanned for the summer of 1996. Preliminary data from the plants grown in the US during the Summer of 1995, and from those grown in Puerto Rico this Winter, indicated that the transgenic plants were indistinguishable from the Elite parental lines. This means that neither the transformation process nor the fatty acid phenotype has had a detrimental effect on the high yield of the Elite line used in the transformation. The high oleic soybeans appeared to be competitive in terms of yield with all commercial lines of soybeans currently being grown in the US.

An important consideration prior to commercialization of this new product was to determine if the modification of the relative fatty acid content significantly improved the cooking and storage properties of the oil. In laboratory tests the high oleic oil from our transgenic soybean lines performed remarkably well. The oxidative stability of the high oleic oil was nearly ten times that of unmodified soybean oil, with an AOM induction time of around 150 hours compared with less than 15 hours for soybean salad oil. In fact the induction time of the high oleic oil was very similar to that of fully hydrogenated, heavy-duty frying shortening. The high oleic oil also performed well in temperature stability tests, having a very low rate of polymer and polar compound formation even after 40 to 50 hours of elevated temperature. This indicated that the high oleic oil would have an extended storage life even without antioxidant additives.

We were surprised to find that the high oleic soybean oil had an oxidative stability three to four times that of high oleic canola and high oleic sunflower oils, despite the fact that these latter oils have a fatty acid composition very similar to that of the high oleic soybean oil. Thus it is clear that there are other components of plant oils which affect their oxidative stability, but which only become apparent when the predominant effect of the polyunsaturated fatty acids is eliminated. These other components might include well known antioxidants such as tocopherols and perhaps other, unknown, antioxidative compounds.

We are currently in the process of producing larger quantities of the high oleic soybean oil for sensory analysis and for customer evaluation. However our experience with these transgenic soybean lines is very encouraging. We have demonstrated that is possible to drastically alter the relative fatty acid composition of an oilseed without affecting germination rates or seed yield under normal seed production conditions. We have also demonstrated that it is possible to produce transgenic oilseeds with modified fatty acid compositions that are stable over a number of different field environments. Thus it now seems realistic to anticipate the commercialization of soybeans, canola and other oilseed plants with fatty acid profiles modified for many edible and industrial uses.

I would like to acknowledge all those who have contributed to the work described above including my research assistants, Kevin Stecca, Russ Booth, Bruce Schweiger and my DuPont colleagues Susan Knowlton, Bill Hitz, Guo-Hua Miao, Rich Broglie, Ted Klein, Chris Kostow and Scott Sebastian. I would also like to acknowledge the collaboration of Allen Leroy and his colleagues at the Asgrow Seed Co.

REGULATION OF OIL COMPOSITION IN OILSEED RAPE (*B. NAPUS L*): EFFECTS OF ABSCISIC ACID AND TEMPERATURE

J. A. WILMER[1], J. P.F.G. HELSPER[1] & L. H.W. van der PLAS[2]
[1] *DLO-Research Institute for Agrobiology and Soil Fertility (AB-DLO), PO Box 14, 6700 AA Wageningen, NL,* [2] *Department of Plant Physiology, Wageningen Agricultural University, Arboretumlaan 4, 6703 BD Wageningen, NL*

Introduction

The use of vegetable oils increases over the years and demands for specific oil composition are currently creating new uses for temperate oilseeds. Research concentrates on identifying and cloning enzymes in lipid metabolism, in part aimed at the introduction of new fatty acids in existing oilseed crops. An area that has gained little interest is the regulation of fatty acid composition by external factors. Here we present some data on the effects of two such factors, temperature and abscisic acid (ABA), and a model for regulation of fatty acid composition in oilseed rape, which focuses on the elongase activity involved in erucic acid (22:1) production.

Experimental

Plants of a high erucic acid rape cultivar, Reston, were grown in the greenhouse. In *in vivo* studies plants were transferred to growth chambers at 15°C or 25°C at the onset of flowering. Seeds were harvested at different timepoints during development. In a second experiment plants were transferred from 15°C to 25°C or *vice versa* at different developmental stages during seed development and mature seeds were harvested.

For *in vitro* studies microspore-derived embryos (MDEs) were initiated from flower buds of 3.4 to 3.6 mm. The MDEs were grown at 15 or 25°C as previously described (Wilmer et al., 1996). ABA was added at 350 °Cday together with fresh medium, while control treatments received fresh medium with a trace of methanol, the solvent for ABA.

Lipids were extracted from MDEs and seeds the using a Bligh & Dyer type extraction. Fatty acid composition of the triacylglycerol (TAG) fraction was determined using capillary GC after methylation of the fatty acids. Levels of ABA were determined using an indirect immunoassay with oxime-coupled ABA as plate coating as described elsewhere (Wilmer et al., submitted).

Results & discussion

We compared the oil composition in seeds grown on plants kept at 15 or 25 °C. At 15°C the amount of oil, containing 40 mol% of 22:1, accumulated was more than twice the amount found in seeds grown at 25 °C, which contained 30 mol% of 22:1 (data not shown). However, the time course of oil accumulation was similar at both temperatures, as is shown for total oil and 22:1 at 15 °C (fig 1A). In plants transferred from 15 °C to 25 or *vice versa* changes could be observed in oil composition at maturity. Plants exposed to 15 °C at the time of maximum oil synthesis (temperature switch before 400 °Cday at a starting temperature of 25 °C, or later than 600 °Cday at the lower starting temperature) contained more 22:1 in the oil than seeds exposed to 25°C at that time (Fig 1B).

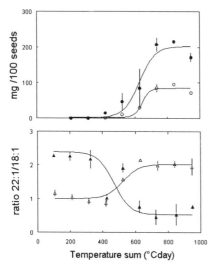

Figure 1: Oil accumulation and effects of temperature switch in developing seeds of Reston.
Top panel: accumulation of total oil (closed symbols) and erucic acid (open symbols) in Reston seeds grown at 15 °C.
Bottom panel: ratio between 22:1 and 18:1 in oil from mature seeds after a temperature switch from 15 to 25 °C or *vice versa* at the indicated timepoint. Open symbols: initial temperature 15 °C, closed symbols: initial temperature 25 °C.

In studies with MDEs similar effects were observed, but the absolute levels of 22:1 were lower. Addition of exogenous ABA increased the level of 22:1 to that found in seeds (compare mol% 22:1 in table 1 with values mentioned above). However, ABA is not an intermediate in the effect of temperature as its levels in seeds, 5 to 10 µM, much higher than those for maximum sensitivity determined in MDEs, around 0.3 µM. Moreover no changes in sensitivity with changes in temperature could be found (Wilmer et al., submitted).

Thus we identified 2 mutually independent factors that influence the level of 22:1 in oilseed rape. Do both just regulate the elongase directly producing 22:1, or is a more intricate control to be found in this system? A first answer to this question can be found in the absolute levels of 22:1 (table 1). The increase in the level of 22:1 is about 10% when the level at 25 °C is compared with either the level obtained with addition of ABA or the level obtained with culture at 15 °C. However, the absolute amounts of 22:1 are very different with the latter two treatments. Addition of ABA results in both a higher production of total oil and a higher level of 22:1 implicating a large increase in the production of 22:1, whereas a change in temperature only results in a shift in fatty acid composition without an increase in total oil content. These results suggest that temperature and ABA do not affect the same regulatory mechanism of oil biosynthesis.

This observation has led us to the hypothesis that regulation of 22:1 levels is not only a matter of regulation of elongase activity producing 22:1, but also involves the balance between fatty acid elongation and oil synthesis (see also fig. 2). Cao and Huang (1987) showed that the diacylglycerol acyltransferase (DAGAT) of oilseed rape has a slight preference for 22:1, but effectively incorporates fatty acids proportional to their abundance at the site of lipid synthesis and that its activity is saturated at 5 or 10 µM of acyl-CoA.

Table 1: Effects of temperature and adding 5 µM ABA on the level erucic acid in the oil and elongase activity of MDEs of *B. napus* cv. Reston.

		mol% of FAs	mg/100 emb	elongase*
25°C	- ABA	17.0 (51)	1.96 (30)	6.32 (28)
	+ ABA	26.1 (78)	4.70 (71)	11.31 (51)
15°C	- ABA	23.7 (71)	2.14 (32)	6.52 (29)
	+ ABA	33.5 (100)	6.59 (100)	22.40 (100)

*: elongase activity in nmol/g FW, determined after 10' incubation with 10 µM 18:1-CoA

Recent results show that ABA influences the total level of elongase changes (table 1), temperature may also influence other properties of the enzyme. The activity is maximum at similar concentrations of acyl-CoA as for the DAGAT (data not shown). This results in the model of regulation in fig. 2, showing that the level of 22:1 in oilseed rape is regulated by the balance (or competition) between levels, as expressed by the apparent V_{max} for the various enzyme systems, of the enzymes utilising acyl-CoAs.

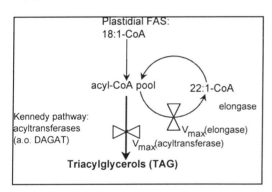

Figure 2: Schematic representation of regulation of lipid production in oilseed rape with apparent V_{max} of fatty acid incorporation and apparent V_{max} of elongation. as the main control points for oil quantity and very long chain fatty acid content.

References

Cao YZ & AHC Huang (1987): Plant Physiol. **84**: 762-765.

J.A. Wilmer, J.P.F.G. Helsper & L.H.W. van der Plas (1996): J.Plant Physiol. **147**: 486-492.

J.A. Wilmer, J.P.F.G. Helsper & L.H.W. van der Plas: Effects of abscisic acid and temperature on erucic acid accumulation in oilseed rape (*Brassica napus* L.). J.Plant Physiol. (submitted).

BROAD-RANGE AND BINARY-RANGE ACYL-ACP THIOESTERASES FROM MEDIUM-CHAIN PRODUCING SEEDS

TONI VOELKER, AUBREY JONES, ANN CRANMER, MAELOR DAVIES, DEBORAH KNUTZON.
Calgene Inc., 1920 Fifth Street, Davis, CA 95616, USA

Medium-chain (C_8-C_{14}) fatty acids are found in the seed oils of certain species or genera of very divergent angiosperms families (1). In all cases published to date, medium-chain acyl-ACP thioesterases with specificities closely reflecting the seed-oil composition were detected. Furthermore, when such specialized thioesterases were expressed in transgenic long-chain plants, a large proportion of the storage oil was directed to the respective medium chains (2, 3). This evidence resolved the long-standing mystery of medium-chain genesis in plants. In the current model a medium-chain acyl-ACP thioesterase "prematurely" hydrolyzes an elongation intermediate of the plastidial fatty acid synthase and the resulting free fatty acid is subsequently deposited in triglycerides. Phylogenetic sequence analysis showed that such specialized enzymes evolved independently several times from the common long-chain saturated acyl-ACP thioesterase, FatB (4, 5).

The seed acyl groups of elm (*Ulmus americana*) consist almost exclusively of medium chains, predominantly 10:0 and lesser amounts of 8:0. Nutmeg (*Myristica fragrans*) accumulates almost exclusively 14:0. We used evolutionary PCR and interspecific hybridization in order to obtain FatB cDNAs. A total of 29 independent nutmeg *FatB* clones were obtained. About 400-600 bp DNA sequence at the 3' end of each clone was obtained, and all were identical in the coding area covered. The nutmeg thioesterase gene was named *Mf FatB1*. In order to determine the hydrolytic specificity of the nutmeg thioesterase, the respective mature-enzyme coding area was expressed as lacZ fusion. Figure 1 shows the results obtained with *Mf FatB1*. When compared to the control, all substrates of equal or longer chain length than 14:0 were hydrolyzed at much higher rates and about equally. The hydrolysis of 12:0-ACP was only slightly elevated relative to the control. Clearly, *Mf* FatB1 is a rather nonspecific long-chain ACP thioesterase, and equally active at 14:0-ACP hydrolysis. It is very similar to the specificity of the "housekeeping" FatB, which is not restricted to medium-chain producing tissues and species (4). A cDNA fragment containing the preprotein reading frame was inserted into a maturing-seed-specific expression cassette. The seeds of most transformants accumulated large quantities of myristate, between 8 and 20 mol%, up from about 0.1 % for the untransformed plants. Laurate proportion increased from 0.02 mol% in the control to 0.34 mol%, 16:0 increased from 5 mol% to 31 mol%, stearate (18:0) increased from 2 mol% to 7.5 mol%. Even the very long-chain saturates 20:0 and 22:0 increased from 0.8 mol% to 2.1 mol% and from 0.4 mol% to 0.8 mol%, respectively. In summary, the transgenic phenotype does reflect the specificity of the enzyme *in vitro* very well, but is rather different from the original nutmeg oil composition, in which 14:0 alone represents 80 mol%.

We applied a evolutionary PCR strategy for the cloning of elm FatB cDNAs and we obtained nine positives clones, all apparently derived from the same transcript, now designated *Ua FatB1*. We expressed *Ua* FatB1 as a His-tagged polypeptide in *E. coli*. The His-tagged *Ua* FatB1 was affinity purified and Fig. 1 shows the results of *in vitro* assays. Clearly, as was expected from the acyl composition of elm seeds, *Ua* FatB1 acts on 10:0-ACP. *Ua* FatB1's activity on 12:0-ACP is low, but it is very active on 14:0-ACP and it prefers 16:0-ACP over any other substrate tested. Even 18:0-ACP and 18:1-ACP are hydrolyzed well. Restricting oneself to the series of saturated substrates tested, it is obvious that this enzyme's specificity has two peaks, at C10 and at C16. With its novel dual specificity, *Ua* FatB1 can be described as the common long-chain FatB housekeeping enzyme with an additional activity, not unlike the nutmeg thioesterase of this paper, except that the new preference is shifted further down to shorter chain lengths. The lipid compositions of mature seeds from the elm thioesterase transgenic plants were determined. Most saturates were elevated in seeds of all *Ua FatB1* transformants, especially 16:0, which ranged from 15 to 33 mol%. In a high expressor, proportion of 16:0 is 30%, 14:0 is 13 mol%, 10:0 is 4 mol%, also 12:0 increases to 1.5%. The proportion of 18:0 is 4 mol%. Only a trace-production of 8:0 was seen.

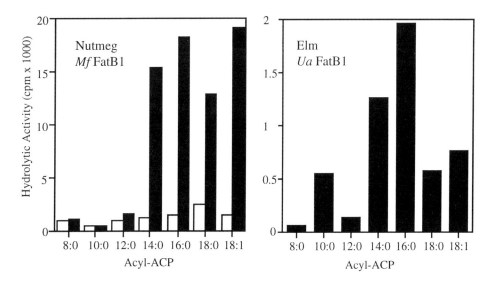

Figure 1 Substrate Specificity of cDNA-Expressed Thioesterases

Nutmeg: *E. coli* was transformed with vector only (Ctrl) or with the *Mf FatB1* expression plasmid. Liquid cultures were grown to equal density and crude extracts prepared. Enzyme extracts representing equal proportions of cultures were assayed with the different acyl-ACP substrates. Elm: *E. coli* was transformed with His-tag *Ua FatB1* expression plasmid. The affinity-purified *Ua* FatB1 enzyme was diluted to a concentration appropriate for the assay conditions and measured. At this dilution the vector-only control did not show any activity above background

In order to study the *in vivo* impact on interception levels of a class II FAS with a binary enzyme, the mature portion of *Ua* FatB1 was expressed in *E. coli* and we manipulated the extent of saturated pathway interception. Under low-interception conditions Ua FatB1 produced mainly 14:0 and 16:0. When the interception rate was increased, the relative production of 8:0 and 10:0 increased, and especially 16:0 was diminished. This result derived in a heterologous model organism represents an *in vivo* demonstration that a wide-specificity enzyme like *Ua FatB1* can cause the production of very different fatty acids depending on the its activity relative to the elongation capacity of the fatty acid synthase. In summary, these results support the *in vitro* characterization of Ua FatB1 as a thioesterase acting on many saturated acyl-ACP substrates. More importantly, they support our model that this broad-specificity enzyme might be the only thioesterase needed for the near-exclusive production of 10:0 in elm seeds.

The absence of specialized, single-specificity medium-chain thioesterases in these two species shows a second evolutionary mechanism by which a normal long-chain fatty acid synthase can be modified to the production of medium chains. It also represents strong indirect evidence for the existence of modified elongation reactions in these seeds.

Acknowledgments

The authors would like to thank Robert T. Hirano, Harold D. Lyon Arboretum, for harvesting and shipping the developing nutmeg seeds. We are grateful to Janet Nelsen, Calgene, prepared most of the acyl-ACP substrates, and to Aaron Dubberley who helped with the screening of the elm library. Our special thanks go to the Calgene fatty acid analysis, plant transformation , growth chamber and greenhouse groups for their outstanding support.

References

(1) Hilditch, T.P. and Williams, P.N. (1964) The chemical constitution of natural fats. (John Wiley, London).
(2) Voelker, T.A., Worrell, A.C., Anderson, L., Bleibaum, J., Fan, C., Hawkins, D., Radke, S. and Davies, H.M. (1992) Fatty acid biosynthesis redirected to medium chains in transgenic oilseed plants. Science 257, 72-74.
(3) Dehesh, K., Jones, A., Knutzon, D.S., Voelker, T.A. (1996) Production of high levels of 8:0 and 10:0 fatty acids in transgenic canola by over-expression of *Ch FatB2*, a thioesterase cDNA from *Cuphea hookeriana*. The Plant Journal 9, 167-172.
(4) Jones, A., Davies, H.M. and Voelker, T.A. (1995) Palmitoyl-acyl carrier protein (ACP) thioesterase and the evolutionary origin of plant acyl-ACP thioesterases. The Plant Cell 7, 359-371.
(5) Dörmann, P., Voelker, T.A. and Ohlrogge, J.B. (1995) Cloning and expression in *Escherichia coli* of a novel thioesterase from *Arabidopsis thaliana* specific for long-chain acyl-acyl carrier proteins. Archives of Biochemistry and Biophysics 316, 612-618.

CARBON PARTITIONING IN PLASTIDS DURING DEVELOPMENT OF *B. NAPUS* EMBRYOS

S. RAWSTHORNE, P.J. EASTMOND, F. KANG, P.M.R. DA SILVA[1], A.M. SMITH[1], D. HUTCHINGS[2] AND M.J. EMES[2]

Brassica and Oilseeds Research and [1]Applied Genetics Departments, John Innes Centre, Norwich, UK and [2]Department of Biological Sciences, University of Manchester, UK.

Summary

Our studies have shown that the metabolism of plastids isolated from developing embryos of *Brassica napus* changes as the pattern of storage product accumulation changes. At the early cotyledonary stage the embryo is accumulating starch and the plastids import glucose 6-phosphate (Glc6P) and utilise it for starch and fatty acid synthesis, and for the oxidative pentose phosphate pathway (OPPP). The metabolism of pyruvate to fatty acids by isolated plastids is negligible at this stage. As the embryo develops to the mid-late cotyledon stage the capacity for import of Glc6P and its subsequent metabolism to fatty acids and starch decreases. In contrast, the capacity of the OPPP increases resulting in a major switch in the partitioning of Glc6P within the plastid during embryo development. Fatty acid synthesis by isolated plastids from embryos at the mid-late stage is highest when pyruvate is supplied as a substrate, again illustrating a major shift in plastidial metabolism. Measurements of the rate of Glc6P metabolism via the OPPP pathway suggests that NADPH generated by the first two steps of the pathway could contribute to the requirement for reducing power from fatty acid synthesis. However, metabolism of pyruvate alone into fatty acids at high rates suggests that there are other endogenous sources of reducing power within the plastid. Measurements of fatty acid synthesis in the light and dark by embryo plastids and chloroplasts from developing leaves have shown that photosynthesis does not provide the ATP and/or reducing power required to saturate *in vitro* starch and fatty acid synthesis by embryo plastids.

Introduction

Embryos of *Brassica napus* (L.) cv. Topas accumulate starch during early cotyledonary filling (up to a fresh weight of 1.5 mg). Subsequently, oil accumulation commences and after a period when both starch and oil accumulate (up to 2.5 mg fresh weight: early-mid cotyledon), the starch content of the embryo declines while the rate of oil accumulation increases (mid-late cotyledon). The synthesis of both starch and fatty acids occurs in the plastids. Both pathways require a carbon source and ATP, and fatty acid synthesis also requires reducing power.

We showed previously (Kang and Rawsthorne, 1994) that glucose 6-phosphate (Glc6P) is the most effective substrate for starch synthesis by plastids isolated from embryos at the early-mid stage and that exogenous ATP is also required. Pyruvate and Glc6P were the best substrates for fatty acid synthesis, both also requiring the supply of exogenous ATP. We also showed that the first two enzymes of the oxidative pentose phosphate pathway (OPPP) were present in plastids. It is also widely known that the developing embryos of brassicas are green and that the plastids in the cotyledon cells contain thylakoid membranes. Taken together these data led us to address several important questions. (1) What leads to the transient accumulation of starch in the early-mid cotyledon embryo? (2) Would incorporation of carbon from pyruvate into fatty acids interfere with that from Glc6P? (3) Does photosynthesis contribute to the carbon and ATP/reducing power requirements of fatty acid or starch synthesis in the embryo? (4) Does plastidial metabolism change during embryo development? (5) Does the plastidial OPPP contribute reducing power to fatty acid synthesis?

To answer these questions we have carried out a range of experiments with embryo extracts and with plastids isolated from embryos of *Brassica napus* (L.) cv. Topas at three distinct developmental stages: early, early-mid, and mid-late cotyledon (corresponding to 1.5, 2.5, and 3.5 mg fresh weight per embryo, respectively). These stages represent starch accumulation only, starch and oil accumulation, and oil and protein accumulation in the developing embryo.

Starch accumulation in *B. napus* embryos

The activities of plastidial phosphoglucomutase, ADP-glucose pyrophosphorylase and soluble and granule bound starch synthases per embryo all increase to the early-mid cotyledon stage and then decline progressively as the embryo enters the main phase of oil accumulation (da Silva and Smith, unpublished). All of these enzymes are required for the conversion of imported Glc6P into starch. These declines in activity clearly reflect a decrease in the capacity of the plastids to make starch. Furthermore, short-term measurements of Glc6P uptake by the plastids reveal that the capacity of the Glc6P translocator on the plastid membrane decreases from the early-mid to mid-late cotyledon stage representing an additional change in capacity (Eastmond and Rawsthorne, unpublished). These changes in capacity are reflected in a significant drop (more than 50%) in the rate of carbon incorporation from Glc6P into starch by plastids from mid-late compared to early-mid embryos. The metabolism behind the degradation of the accumulated starch needs further investigation.

Fatty acid synthesis from multiple metabolites

When pyruvate and Glc6P were provided together (in the presence of ATP) to plastids from early-mid stage embryos the incorporation of carbon from pyruvate into fatty acids had only a slight effect on the incorporation of carbon from Glc6P (Kang and Rawsthorne, 1996). This suggests that (i) pyruvate derived from plastidial glycolysis does not compete with that imported directly from the cytosol, and (ii) that the plastids have a greater capacity for fatty acid synthesis from pyruvate than the capacities to provide it from either glycolysis or the cytosol. Simultaneous supply of Glc6P, pyruvate and acetate led to an *in vitro* rate of fatty acid synthesis that exceeded the *in vivo* rate required for oil accumulation (Kang *et al.*, 1994).

Photosynthesis in developing rapeseed embryos

Plastids isolated from early-mid stage embryos carry out HCO_3^--dependent O_2 evolution/mg chlorophyll at rates comparable to those of chloroplasts isolated from expanding seedling leaves (Eastmond and Rawsthorne, unpublished). However, the light saturation points and chlorophyll contents of the embryo plastids are lower than those of the leaf chloroplasts. We have carried out studies of whole embryo gas exchange and of silique transmittance to light during embryo development. These have revealed that the plastids in cells at the outer face of the embryo might be light-saturated but those within the tissue will not. Embryo plastids were supplied with exogenous substrates or HCO_3^- and the rates of fatty acid and starch synthesis compared in saturating light or in the dark, with or without ATP. Bicarbonate was incorporated into both products but only at very low rates compared to the exogenous substrates. Starch and fatty acid synthesis from the exogenous substrates was greatest in the presence of ATP and light alone gave rates that were less than 30% of the ATP dependent ones. This suggests that photosynthesis can contribute to storage product synthesis but the plastid must be strongly dependent on the cytosol for provision of carbon and ATP to sustain maximal rates.

Developmental changes in plastid metabolism

As described above the enzymes of starch biosynthesis show a strong developmental change. In early cotyledon stage embryos Glc6P is used not only for starch synthesis but is also the most effective substrate for fatty acid synthesis. At this stage carbon from pyruvate was barely incorporated into fatty acids. By the early-mid stage the rate of fatty acid synthesis from pyruvate was comparable with that from Glc6P. As the embryo enters the main phase of oil accumulation (mid-late stage) the plastids have very high rates of fatty acid synthesis from pyruvate and the incorporation of carbon from Glc6P significantly decreases (see Eastmond and Rawsthorne, this volume for details). The change in Glc6P utilization is correlated with the decrease in Glc6P import (see above) and also a decrease in the capacity of plastidial glycolysis. The increase in pyruvate utilization during development correlates with an increase in the capacity of a pyruvate transporter. These studies clearly show that plastidial metabolism changes during development. How these developmental changes are induced requires further investigation.

Oxidative pentose phosphate pathway in embryos

Having shown previously that the first two enzymes of the OPPP were present in the plastids we have investigated the activity of the pathway in intact plastids by feeding [1-^{14}C] Glc6P and measuring $^{14}CO_2$ release. The rate of metabolism of Glc6P via the OPPP is comparable to that into fatty acids up to the early-mid stage. By the mid-late stage the OPPP activity has increased, again revealing a developmental change. The relationship between the OPPP and fatty acid synthesis has been studies in two ways. First, in feeding experiments where more than one substrate is supplied to the plastids (Kang and Rawsthorne, 1996). When Glc6P is supplied alone the measured activity of the OPPP would lead to sufficient NADPH production to drive fatty acid synthesis at the concurrently measured rate. However, when pyruvate was supplied in addition to Glc6P, where the demand for NADPH would be expected to increase, there was only a very small change ($\sim 8\%$) in the OPPP activity while the total rate of fatty acid synthesis increased by more than 60%. To address the linkage more directly plastids were supplied with Glc6P in the presence or absence of ATP which respectively enabled or prevented fatty acid synthesis from this substrate. In the presence of ATP the rate of $^{14}CO_2$ evolution increased significantly (Hutchings and Emes, unpublished). Together, these data suggest that the OPPP can certainly contribute towards the NADPH requirement for fatty acid synthesis and that under *in vitro* conditions the OPPP activity responds to changes in the rate of fatty acid synthesis. However, we still need to establish more precisely the extent of this linkage and the degree of recycling of carbon that occurs via the OPPP.

Conclusion

Metabolism in the plastids of *B. napus* embryos changes during development. In the early stages of cotyledonary filling they are like those from the starch-accumulating pea embryo and utilise Glc6P. As the embryos begin to accumulate oil at high rates the plastids utilise pyruvate like those from castor endosperm. What causes these changes to occur remains an important questions for future research.

References

Kang, F. and Rawsthorne, S. (1994) Starch and fatty acid synthesis in plastids from developing embryos of oilseed rape (*Brassica napus* L.). Plant Journal 6: 795-805.

Kang, F., C.J. Ridout, C.L. Morgan, and S. Rawsthorne (1994) The activity of acetyl-CoA carboxylase is not correlated with the rate of lipid synthesis during development of oilseed rape (*Brassica napus* L.) embryos. Planta **193**, 320-325.

Kang, F. and S. Rawsthorne (1996) Metabolism of glucose-6-phosphate and utilization of multiple metabolites for fatty acid synthesis by plastids from developing oilseed rape embryos. Planta **199**, 321-327.

ISOLATION AND PURIFICATION OF MEMBRANE FRACTIONS INVOLVED IN TRIACYLGLYCEROL METABOLISM IN DEVELOPING SEEDS OF *BRASSICA NAPUS* L.

Dominic J. LACEY*, Paweena PONGDONTRI and Matthew J. HILLS
Department of Brassica and Oilseeds Research
John Innes Centre, Colney, Norwich U.K. NR4 7UH
*present address IARC, Long Ashton, Bristol, BS18 9AF U.K.

Introduction

Storage triacylglycerols (TAG) are synthesized by the action of the acyltransferases of the Kennedy pathway and have been shown to be located in the endoplasmic reticulum (Lacey and Hills 1996). At the previous Plant Lipid Congress (Lacey and Hills 1995) we reported the identification of a low density membrane (LDM) fraction which very efficiently incorporated radiolabeled glycerol-3-phosphate into TAG but contained relatively low activities of the enzymes of phospholipid synthesis. The LDMs have a density of about 1.05 g/ml and thus would float with the oil pad in homogenization media generally used. The LDMs contain only about 5% of the total enzymes of TAG synthesis in the extract but the differences in their metabolic activities leads us to propose that they are contained in a domain of the ER which is involved in TAG accumulation or deposition. The ER has been shown to contain domains which are involved in different aspects of metabolism such as secretory processes of protein synthesis. It is not unreasonable to suggest that upon tissue homogenisation the ER vesiculates and that regions of the ER which are involved in the deposition of storage oil would reach equilibrium at a lower density on a sucrose density gradient than the bulk ER due to the content of neutral lipid. One problem with studies in which cells are homogenised and the sub-cellular compartments separated is the possibility that some parts of the ER become contaminated with oil bodies which causes them to have a lower density. It is clear that the converse is true since the oil body fraction is often heavily contaminated with ER (up to 20% of total marker enzyme activity) and usually contains similar proportions of other organelles such as mitochondria. Most of the contaminants can be removed by washing the oil body pad though a small proportion of ER is difficult to remove. We previously described experiments which show that the formation of the LDMs by contamination of ER by oil bodies is not likely since enzyme composition is different and we could not detect oleosins as an oil body marker. In this paper we describe further work in which we have substantially purified the LDM from contaminating soluble proteins.

Materials and Methods

Embryos were obtained from rape plants as previously described. The embryos were homogenized and membranes and organelles separated and fractionated using sucrose density gradient centrifugation procedures. Analysis of enzyme activities was performed using previously described procedures (Lacey and Hills 1996). N terminal amino acid sequencing was kindly performed by Dr P. Jackson, Applied Biosystems, Cheshire, UK. SDS-PAGE and Western blotting were carried out using standard procedures (Lacey and Hills 1996).

Purification of low density membrane vesicles

SDS-PAGE showed that the low density membrane fraction was substantially contaminated with proteins from the soluble fraction and the polypeptide composition of the soluble and LDM fractions appeared identical (data not shown). A number of methods for purifying the low density membranes from contaminating soluble proteins were assessed. Ultrafiltration and gel filtration using Sephacryl S-1000 proved to be ineffective since large amounts of enzyme activity were lost through the binding of the LDMs to the filtration membranes or the column matrix. Dilution of the LDM gradient fractions with ice cold 100 mM Na_2CO_3 followed by centrifugation at 100,000g x 60 min, initially gave encouraging results and yielded a pellet which contained a number of proteins which were substantially enriched compared to the cytosolic fractions (data not shown). Gels of the purified LDM protein were electroblotted onto PVDF membrane and the N terminal amino acid sequence of three prominent bands were determined by Edman degradation.

N-terminal amino acid sequence of three prominent polypeptides from LDM fraction:

16 kDa   PTGRFETMREWVHDAISARR

20 kDa   GLEET(I+L)C*(T+S)MR(T+C*)(T+H)EN(L+I)DDP(A+P)RADVYKPNLGRRTTLN

25kDa   RQRELLIFKAAV

Assignments eg (I+L) indicate where 50% of the signal was contributed from both residues. The C* represents a peak at pam-CV, therefore this residue is tentatively a C.

BLAST search of these sequences showed the 20 kDa to be almost identical to cruciferin, a vacuolar storage protein in rape seeds. The 16 and 25 kDa polypeptides had little similarity to other proteins. The presence of cruciferin indicated the continued presence of contaminating proteins. Further investigation by Western blotting using antibodies raised against cruciferin and the large sub-unit of ribulose bisphosphate carboxylase-oxygenase (Rubisco) confirmed this suspicion. Such contamination may not be unexpected since Rubisco and the storage proteins occur as very large structures in the plastid and vacuole and would pellet under these conditions. It is also very likely that a prominent polypeptide at 14 kDa (data not shown) was the small subunit of Rubisco.

Purification of LDM by floatation

Since the carbonate washing of the LDMs followed by pelleting by centrifugation yielded a pellet which was still clearly contaminated with some large soluble proteins the LDMs were purified by floatation through a sucrose layer. The sucrose density of the LDM fraction was adjusted to 30% sucrose and this was overlaid by 20% and 5% sucrose layers, each containing 5 mM EDTA (pH 7.5). The yield of activity of the enzymes of TAG synthesis was about 20-25% from the untreated LDM fraction. Protein estimation by Bradford assay showed that the LDMs had been substantially purified with more than 99% of the protein being removed giving purification factor of about 200 fold. In addition the specific activity of the enzymes of TAG synthesis in terms of protein were about 100 fold higher than those in the ER. Thus not only is the LDM fraction very much more efficient in incorporating glycerol-3-phosphate into TAG than the bulk ER (Lacey and Hills 1996) but the much higher specific activity leads us to speculate that the LDM are vesicles derived from regions of the ER which are specifically involved in the synthesis and deposition of TAG. The polypeptide composition of the LDM fraction from the gradient, the floatation purified LDMs and the ER are shown in Fig. 1.

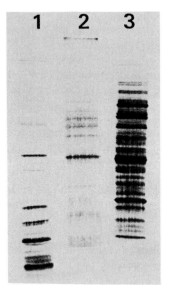

Fig. 1. Silver stained 8-15% SDS-PAGE protein from the ER fraction (1;5µg), the purified LDM fraction (2;2µg) and the unpurified LDM fraction (3;5µg) from the sucrose density gradient.

It can be seen from Fig. 1 that the polypeptide composition of the washed LDM is very different to that of the unwashed LDM (which appears to be identical to the soluble protein fraction, not shown). There are a number of polypeptides which have the same molecular masses as those in the ER but a number of polypeptides present in the ER fraction are missing from the LDM fraction. Identification of the major proteins in the LDMs will allow the production of antibody probes which can be used in ultrastructural studies to determine the location of these region within the ER.

References

Lacey DJ and Hills MJ (1995) Identification of a low density membrane fraction involved in storage triacylglycerol synthesis. In: Plant Lipid Metabolism, Kader J-C and Mazliak P eds pp 482-485
Lacey, D.J. and Hills, M.J. (1996) Heterogeneity of the endoplasmic reticulum with respect to lipid synthesis in developing seeds of Brassica napus L. *Planta*, In Press

ALTERED FATTY ACID COMPOSITION OF MEMBRANE LIPIDS IN SEEDS AND SEEDLING TISSUES OF HIGH-SATURATE CANOLAS

GREGORY A. THOMPSON AND CHINGYING LI
Calgene, Inc.
1920 Fifth Street, Davis, California, USA 95616

Introduction

Genetic modification of rapeseed to produce a high-stearate oil using antisense technology to reduce expression of stearoyl-ACP desaturase was reported in 1992 by Knutzon et al. When tested in the field, lines from the first modified plants showed stearate contents between 20 and 30% with total saturates slightly above 30%. Seed yield and germination under standard conditions were normal.

Selection and breeding with descendants from the best early transformant has led to materials with greater than 30% stearate and total saturate contents in excess of 40%. At this higher level of stearate, germination rates are lower, and seedlings are less vigorous. A suggested cause is the potential accumulation of stearate in membrane lipids, leading to membranes less able to adapt to changes in temperature and moisture content. This study is a first examination of stearate levels in membrane lipids of high-stearate seeds, looking for composition differences that might explain the observed effects on germination.

Materials and Methods

Seeds used in these studies were from material known as 77825-83, an F7 generation breeding line selected from a cross between an original high-stearate transformant and a low-linolenic canola variety.

Seeds were germinated on filter paper soaked in 1/10 MS salts, and seedlings were allowed to grow in sterile closed containers at 25°C under light (120 $\mu E/m^2$, 16 h photoperiod). Tissue was harvested and analyzed immediately or stored frozen at -70°C until analyzed. For analysis of lipid classes, tissues first were boiled for 10 min in isopropanol to kill phospholipases, then extracted in chloroform-methanol (Bligh and Dyer, 1959), and then the combined isopropanol and chloroform-methanol extracts were dried and redissolved in chloroform:acetic acid (100:1 v/v) for separation into neutral lipids, glycolipids (+MAG), and phospholipids by chromatography on silica columns (Lynch and Steponkus, 1987). Fatty acid methyl esters for gas chromatographic analysis were prepared by acidic methanolysis in sulfuric acid-methanol (5% w/v).

Results and Discussion

By 6 days post-imbition, seeds from 77825-83 sorted into three categories: (a) 18% of the seeds germinated and grew into normal seedlings indistinguishable from standard canola. (b) 69% of the seeds germinated but their cotyledons had yellow or brown spots in the center of each cotyledon lobe. Severity of the spots ranged from a minimum of damage to the inner cotyledon, to larger spots on both inner and outer cotyledons, and in the worst cases, to barely germinated seedlings in which the majority of the tissue was abnormal. (c) 13% of the seeds failed to germinate and when dissected were observed to have dark brown tissue in the center of each cotyledon lobe, most severe in the inner cotyledon and the inner surface of the outer cotyledon. Axis tissue was rarely damaged.

The average saturate content of 77825-83 was 43% by weight, with stearate comprising 33%. Individual seeds ranged in total saturates from 32% up to 50%. Seeds with spotted cotyledons were found with total saturates as low as 38%, and their prevalence increased with increasing total saturate content.

Membrane lipids in 77825-83 contained significantly higher levels of saturates than the host canola variety, A112. Phospholipids, which comprise the majority of membrane lipid in mature seed, contained 23 mol% saturates, with 15 mol% stearate. Glycolipids in mature seed, plastid lipids probably remaining from when the embryo was green, contained 33 mol% saturates with 18 mol% stearate. These lipids may be synthesized via the prokaryotic pathway which incorporates 16:0 at sn-2, and potentially could have stearate at sn-2 as well as sn-1.

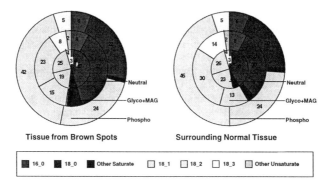

Figure 1. Fatty acid composition of lipid classes from tissues of mature seeds with spotted cotyledons.

Brown tissues from mature seeds with spotted cotyledons were more highly enriched in saturates than the surrounding normal-appearing tissue (Fig. 1). Neutral storage lipids (inner circle) in the spots have 52 mol% total saturates compared with 47 mol% in the surrounding tissue. Glycolipids (middle ring) are similarly affected.

During germination and growth, additional differences were observed. Firstly, seeds which germinated and developed into normal seedlings utilized storage lipids very rapidly, as in normal canola. Secondly, spotted seedlings that showed nearly normal vigor also utilized lipid reserves, but at a slightly slower rate which was particularly noticeable between 4 and 6 days post-imbibition (dpi). Levels of stearate remained somewhat high even after six days. Not unexpectedly, the lipid content of seeds which wouldn't germinate did not change during the 6-day period. Similarly, seeds which germinated but never grew did not utilize storage lipids.

A more detailed examination was made of cotyledon tissues from spotted, but healthy, seedlings at 6 dpi and of imbibed seed that did not germinate by 6 days. Comparison with dry mature seeds showed that by day 6 the healthy spotted seedlings had utilized about 75% of the stored neutral lipid. Furthermore, these healthy seedlings had synthesized sufficient glycolipid during greening to be enriched five-fold over mature dry seed. The amount of phospholipid in healthy seedlings had not increased very much during the 6-day period. By comparison, amounts of lipid in each class did not change much in the ungerminating seed.

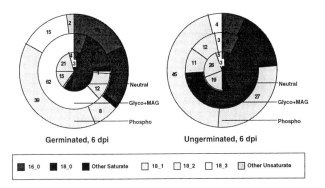

Figure 2: Fatty acid composition of lipid classes in seedlings and ungerminated seeds at 6 dpi.

Several important observations were made from analysis of lipid classes in these 6-day old tissues (Fig. 2). Firstly, neutral lipid (inner circle) remaining in healthy seedlings were >60 mol% total saturates. Since stearate is not incorporated at sn-2, >80% of the remaining triacylglycerols (TAG) must have the structure Sat-Unsat-Sat (SUS). This suggests that SUS may be a less preferred substrate for rapeseed lipases, or that it may have crystallized into a physical state less accessible to lipases. Secondly, newly synthesized glycolipids (middle ring) in healthy seedlings were comprised of 62 mol% 18:3 and were relatively low in saturate content. Thus, stearate probably is not recycled into plastid membranes during seedling growth and greening. Thirdly, glycolipids from the ungerminated seeds have extremely high saturate content (74 mol%), so that nearly half of the molecules have saturates at both sn-1 and sn-2. Palmitate, at only 8 mol%, cannot account for all of the saturated acyl groups at sn-2, which suggests that the prokaryotic pathway does incorporate 18:0 at sn-2. Finally, phospholipids (outer ring) maintain significant stearate in healthy seedlings, but there is a shift to polyusaturates and an increase in 16:0. Unfortunately, without a major net synthesis of phospholipids, the data do not permit a conclusion as to whether stearate is being re-incorporated into newly synthesized phospholipids.

Summary

This study uncovered several key points: (1) Membrane lipids in mature seeds of high-stearate canola contain much higher than normal saturated fatty acids, indicating that rapeseed embryos do not have an effective mechanism to protect membrane lipids from accumulation of stearate. (2) Average saturate contents above 40% in high-stearate canola leads to the appearance of damaged cotyledon tissue (spots) and decreased germination. The damaged tissue itself has nearly 50% total saturates. (3) Plastid glycolipids that are potentially derived from the prokaryotic-pathway may have extremely high saturate contents, especially in seeds that will not germinate. (4) SUS triacylglycerols are degraded more slowly than less saturated TAG. Engineering very high levels of SUS in seeds could have an impact on seedling vigor.

References

Bligh EG, Dyer WJ (1959) *Can. J. Biochem. Physiol.* **37**:911-917.
Knutzon DS et al (1992) *Proc. Natl. Acad. Sci., USA* **89**:2624-2628.
Lynch DV, Steponkus PL (1987) *Plant Physiol.* **83**:761-767.

MUTANTS OF BRASSICA NAPUS WITH ALTERED SEED LIPID FATTY ACID COMPOSITION

B. RÜCKER, G. RÖBBELEN
University of Göttingen, Institute of Agronomy and Plant Breeding
Von Siebold Str. 8, D-37075 Göttingen, Germany

Mutagenesis is an efficient approach to modify seed oil composition in rapeseed. A series of mutant lines with variation in polyenoic fatty acid composition of the seed oil was isolated after ethylmethane sulfonate treatment of low erucic acid *Brassica napus* seeds. Selection was carried out for altered fatty acid composition in the seed oil. C18 fatty acid composition of the mutant lines ranged from 33% to 80% for oleic acid (C18:1), 7% to 40% for linoleic acid (C18:2) and 2% to 13% for linolenic acid (C18:3). Variation was also created for palmitic acid content (C16:0, 3% - 9%). This material can be used to get more information about the biochemistry and regulation of lipid metabolism. The mutant lines are also most valuable for rapeseed breeding programms derected on the development of cultivars with specific seed oil qualities. In this paper we characterize a collection of mutations for their effects on different lipids and for their expression in different tissues. The results are discussed in relation to general field performance and productivity.

Materials and Methods

Plant material: Mutants with low C18:3 and high C18:2 were isolated after ethylmethane sulfonate (EMS) seed treatment of the canadian spring rapeseed cv. Oro in 1975 (Röbbelen, Nitsch 1975). By repeated backcrossing the mutations were introduced into canola winter oilseed rape germplasms (Rücker, Röbbelen 1996). Lines Goe 9, 1420/95 and 1427/95 were selected in this programm. The same low C18:3 mutation was introduced into high yielding spring rapeseed germplasms at the University of Manitoba, Canada, resulting in the cv. Apollo (Scarth et al. 1995). Cultivar Apollo was kindly provided by R. Scarth. The high C18:1 mutant lines M453 and M457 were selected after EMS mutagenesis of a low C18:3 line closely related to line Goe 9. M19661, M19782, M7505, M19517 and M4936/95 were isolated after EMS mutagenesis of the canola winter rapeseed cv. Wotan (Rücker, Röbbelen 1995).
Fatty acid analysis: Fatty acid composition of the seed oil was analysed as described earlier (Rücker, Röbbelen 1996). For the determination of fatty acid composition in leaves and roots plants were grown in hydroponic culture pots in the greenhouse for three weeks. Leaves and roots were harvested (for each genotype 2 mixed samples of 5 plants), dried (40°C, 1 day) and crushed. Sample preparation: extraction of total lipids with 0.5 ml dichlormethane/methanol (2:1) at 20°C over night, transmethylation with 0.5 ml Na-methylate/methanol (0.5 mol/l) for 20 min at 20°C; addition of 50 µl

isooctane and 200 µl 5% NaHSO₄ in water; centrifugation (3 min); injection of 2-3 µl of the upper phase into the gaschromatograph (Perkin Elmer 8600, 0.25 mm x 25 m FFAP Fa. Machery & Nagel, Düren, Germany, 210°C isothermal, 120 kPa H_2).

Results and Discussion

C18:2 desaturation: Mutagenesis created variation of C18:1, C18:2, C18:3 and C16:0 levels in the seed oil (Table 1). Modification of C18:3 or C18:2 resulted in concomitant changes of C18:2 and C18:1, respectively. Reduction of C18:2 desaturation was found in line Goe 9 and cv. Apollo. Line 1427/95 exhibits increase in C18:1 desaturation. From crosses of both genotypes a line with decreased C18:1 and C18:3 and about 40% C18:2 was selected (1420/95).

C18:1 desaturation: In M457 C18:1 is increased at the expense of C18:3. In all the other high C18:1 mutants C18:2 is reduced mainly. A slight decrease of C18:3 seems to be a result of reduced substrate for C18:2 desaturation. In the genotypes with the highest C18:1 levels 1% reduction of C16:0 was observed. Modification of fatty acid composition is more complex in M4963. At the expense of C18:1 the polyunsaturated acids as well as C16:0 are increased.

TABLE 1. Variation of fatty acid composition in mutant lines isolated from EMS treated low erucic acid oilseed rape in comparison to the canola cv. Wotan (% of total fatty acids ±SD)

Genotype	n	C16:0	C18:1	C18:2	C18:3
Goe 9	10	4.3 (±0.5)	58.6 (±2.1)	30.8 (±1.4)	**2.5** (±0.2)
cv. Apollo	1	4.0 (-)	61.2 (-)	28.3 (-)	**1.8** (-)
1420/95	4	5.5 (±0.4)	49.6 (±2.5)	**34.9** (±0.9)	2.8 (±0.3)
1427/95	3	6.0 (±0.6)	33.3 (±1.0)	**39.6** (±1.6)	11.6 (±1.0)
M457	12	4.0 (±0.1)	**71.0** (±0.9)	17.5 (±0.8)	**3.1** (±0.2)
M453	3	4.3 (±0.4)	**72.5** (±1.6)	10.0 (±1.1)	8.7 (±0.2)
M19661	9	3.4 (±0.1)	**78.0** (±1.2)	7.3 (±0.8)	7.0 (±0.7)
M19782	8	3.6 (±0.4)	**80.3** (±3.6)	5.7 (±2.3)	6.2 (±0.9)
M7507	7	3.6 (±0.1)	**78.9** (±0.8)	7.4 (±0.4)	5.8 (±0.5)
M19517	7	3.5 (±0.2)	**76.3** (±1.5)	8.5 (±1.0)	7.5 (±0.4)
M4963	8	**9.0** (±0.5)	46.1 (±1.4)	27.4 (±1.1)	12.8 (±0.9)
cv. Wotan	15	4.7 (±0.4)	60.3 (±1.8)	20.8 (±0.9)	9.9 (±1.2)

To determine whether the same genes control fatty acid composition in the seed oil and in polar lipids of different tissues, we compared the overall fatty acid composition of seeds, roots and leaves (Table 1, Table 2). Deficiencies in C18:2 and C18:3 desaturation were not restricted to the seeds. The mutations were also expressed in roots and leaves. Modifications of leave lipids were low in comparison to roots indicating changes in extrachloroplast lipids. A substantial reduction of C18:3 and concomitant increase of C18:2 was observed in the roots of all low C18:3 mutants. For membrane function this substitution of C18:3 by C18:2 seems to be tolerated: Goe 9 and cv. Apollo have no deficiencies in plant development and they are high yielding.

TABLE 2. Fatty acid composition of totol lipids in roots and leaves of mutants with altered seed oil in comparison to the canola cv. Wotan (% of total fatty acids)

Genotype	Roots				Leaves				
	C160	C181	C182	C183	C160	C163	C181	C182	C183
Goe 9	17,2	8,6	40,3	22,4	13,7	10,9	2,5	16,1	49,6
cv. Apollo	17,1	8,7	41,0	23,0	11,0	14,0	3,1	16,7	50,4
1420/95	16,6	6,8	43,4	21,2	10,8	13,2	2,5	21,7	46,4
1427/95	17,7	8,5	21,3	41,0	14,1	10,6	1,3	13,3	52,6
M457	16,8	11,2	37,7	23,5	14,9	10,4	3,2	14,8	49,7
M453	13,1	34,6	13,8	28,0	13,1	10,3	8,9	7,8	52,9
M19661	15,9	12,3	16,3	45,7	13,6	9,4	6,3	8,9	54,1
M19782	15,8	13,2	16,6	43,6	13,6	9,2	7,6	8,6	53,2
M7507	11,8	34,3	14,8	28,5	12,5	10,5	10,0	8,2	52,1
M19517	15,4	16,0	17,3	40,0	13,6	9,7	9,1	7,9	53,2
M4963	17,3	8,6	13,8	49,6	14,3	10,8	2,1	11,7	53,5
cv. Wotan	16,8	10,2	18,8	44,4	14,3	11,5	1,8	9,5	54,9

In contrast M453 and M7507 exhibit substantial reduction of polyenoic acids in root lipids. Both lines exhibit very poor field performance indicating that a significant increase of C18:1 in membranes affects plant growth.

The high C18:1 mutants M19661, M19782 and M19517 have only slightly changed root and leaf lipids. These mutations are expressed predominantly in the seeds. After crossing the mutants with cv. Wotan and with each other it was found, that the variation is caused by a single mutant allel at the same locus in M19661, M19782 and M19517. The mutant lines are not distinguishable in yield performance from the wild type and they are most valuable for rapeseed breeding programs. M7507 and M19517 were selected from the same M1 plant. In M7507 mutations were induced at two loci. The mutation at the second locus is not expressed seed specific. The phenotype of M7507 is like that of M453 in roots with a higher C18:1 level in seeds.

Increase of C18:2 in 1427/95 is low in roots and leves. Thus, this mutation is expressed in roots and leaves on a very low level only. Fatty acid modifications in M4963 are also expressed predominantly in the seeds.

Acknowledgements: This work was supported by the Gemeinschaft zur Förderung der privaten Deutschen Pflanzenzüchtung e.V., Bonn, Germany (project ÖE 95/95 NR).

References

Röbbelen, G. and Nitsch, A. (1975) Genetical and physiological investigations on mutants for polyenoic fatty acids in rapeseed, *B.napus L. Z. Pflanzenzücht.* **75**, 93-105.

Rücker, B. and Röbbelen, G. (1995) Development of high oleic acid rapeseed *Proc. 9th Int. Rapeseed Congress*, Cambridge, UK, 389-391.

Rücker, B. and Röbbelen, G. (1996) Impact of low linolenic acid content on seed yield of winter oilseed rape (*Brassica napus* L.). *Plant Breeding* **115** (in press).

Scarth, R., S.R. Rimmer and P.B.E. McVetty (1995) Apollo low linolenic summer rape. *Can. J. Plant Sci.* **75**, 203-204.

ERUCIC ACID DISTRIBUTION IN *BRASSICA OLERACEA* SEED OIL TRIGLYCERIDES

S.L. MACKENZIE, E.M. GIBLIN, D.L. BARTON, J.R. McFERSON[1], D. TENASCHUK and D.C. TAYLOR
NRC Canada, Plant Biotechnology Institute, 110 Gymnasium Place, Saskatoon, SK, Canada S7N 0W9. [1]USDA, ARS Plant Genetic Resources Unit, Cornell University, Geneva, NY, USA.

Introduction

Erucic acid is generally excluded from the *sn*-2 position of seed oil triglycerides (TAGs)of Brassicacea species. Stereospecific analyses *of Brassica juncea, B. napus, B. oleracea, B. rapa, Crambe abyssinica, Lunaria annua, Sinapis alba,* and other wild Cruciferae (1-4) have not yet detected significant amounts of erucic acid in the *sn*-2 position.

Brassica genotypes having the ability to insert erucic acid into the sn-2 position of seed oil triglycerides would be of potential value as (a) a source of an erucoyl-specific lysophosphatidic acid acyl transferase (LPAT) gene for transforming high-erucic acid *B. napus* cultivars and (b) for generating high-erucic acid *B. napus* by interspecific crossing. Taylor *et al.* (5) reported the detection of more than 20% *sn-2* erucic acid in three accessions of *B. oleracea.* We now report the results obtained by screening more than 300 accessions in the USDA seed collection.

Experimental Methods

Sample Selection
B. oleracea accessions containing more than 45% seed oil erucic acid, as reported by Mahler and Auld (6), were selected for analysis.

Lipase Screen
Seed samples were ground using a Polytron at high speed for 45 sec. in the presence of 2:1 chloroform:isopropanol + BHT (0.25). The suspensions were filtered and diluted to 10 ml. An aliquot (1 ml) of each total lipid extract was dried under nitrogen. Tris buffer (1M; pH 8.0; 1 ml), calcium chloride (2.2%; 100 µl) and bile salts (0.055; 250 µl) were added, and the mixture sonicated for 10 min at 40°C. Pancreatic lipase (1 mg/100 µl in tris buffer) was added and the mixture was rotated (150 rpm) at 40°C for 10 min. The reaction was terminated by adding 1 ml ethanol followed by 1 ml 6N HCl. The solution was extracted thrice with 2 ml dry diethyl ether followed by centrifugation. The ether layer was washed with 2 X 2 ml water. Chloroform:benzene:methanol (1:1:1) was added and after drying under nitrogen, dissolved in 100µl hexane. The sample was spotted on 250 µm Silica G TLC plates. After pre-running with 100% diethyl ether, the chromatogram was developed using either hexane:ether:acetic acid (70:30:1) or diethyl ether (100%). The *sn*-2 MAG band was removed, and transmethylated directly using 3N HCl/methanol.

GC Analysis
Fatty acid methyl esters were assayed using a DB23 column (30 m X 0.32 mm I.D.; Chromatographic Specialties, Brockville ON). Operating conditions were: Temperature program : 180° hold 1 min; 4°/min to 240°; Injector temperature 200°; split mode, ratio 80:1; Detector temperature 250°; Gases :carrier, He; make up, N.
Triglycerides were assayed using a DB1 column (3 m, wide bore) (Chromatographic Specialties, Brockville, ON). The operating conditions were: Temperature program : 275°C-350°C @ 10°C/min; 5-10 min hold; Injector : 350°C; split mode, ratio 10:1; Detector : 375°C; Gases : carrier, He; make up, N.

Mass Spectrometry
Ammonia chemical ionization mass spectrometry was performed using a VG model 70-250 SEQ hybrid mass spectrometer. Samples were introduced through the solids probe which was heated to 330°C. The source temperature was 250°C. Spectra were acquired by scanning from m/z 1300 to m/z 100 every 10 sec with a 0.5 sec settling time.

Results and Discussion

Seed Selection
The selection of accessions containing more than 45% total seed oil erucic acid was entirely arbitrary because there is no *a priori* reason to believe that high erucic acid accessions contain significant *sn*-2 erucic acid, except when the total erucic acid content is significantly above 67 mol%.

sn-2 Erucic Acid Analyses
The fatty acid compositions of those accessions containing more than 25% erucic acid in the *sn*-2 position of their seed oil triglycerides are shown in Table 1. The majority of the accessions contained less than 10% erucic acid in the *sn*-2 position of TAGs, but seven genotypes contained more than 30% *sn*-2 erucic and two contained more than 50%. There was no significant correlation between the total seed oil erucic acid content and the *sn*-2 proportion for the total population of samples screened. The correlation coefficient was 0.59 for the seven samples containing more than 30% *sn*-2 erucic acid.

TABLE 1. *B. oleracea* accessions : *sn*-2 Fatty acid composition (Wt %)

TAG 22:1	16:0	18:0	18:1	18:2	18:3	20:1	22:1	% Tot 22:1
60.0	9.0	1.8	9.0	13.6	9.0	2.0	**55.6**	30.9
53.6	6.7	1.2	12.0	10.8	7.1	4.6	**50.6**	31.5
58.0	8.3	2.4	22.9	15.2	7.2	-	**43.9**	25.2
53.1	6.1	2.9	13.4	18.1	13.5	4.4	**41.6**	26.1
50.6	5.6	3.4	19.5	18.1	10.4	2.6	**40.4**	26.6
45.0	5.6	2.8	22.3	18.7	7.9	4.6	**35.2**	26.1
55.0	4.4	1.7	23.2	19.3	9.1	3.4	**31.9**	19.3
56.8	2.3	1.1	19.0	27.0	12.9	1.9	**28.1**	16.5
53.4	5.0	1.1	23.4	22.8	17.0	1.6	**27.7**	17.3
50.6	2.9	0.8	22.3	26.6	14.0	4.3	**26.7**	17.6
45.3	13.5	9.3	13.0	13.5	8.3	2.1	**25.5**	18.8
48.8	4.4	1.6	24.3	26.1	14.2	1.5	**25.1**	17.1

When the proportion of erucic acid in the sn-2 position is expressed as a percentage of the total erucic in the seed oil (Table 1, last column), the resulting value is a direct indicator of the ability of the LPAT to insert erucic acid into the sn-2 position. Thus, six of the genotypes are able to insert more than 25% of the total seed oil erucic acid into the sn-2 position. Those genotypes having a large proportion of the total seed oil erucic acid in the sn-2 position are targets for cloning an erucoyl-specific LPAT gene and, subject to their having suitable agronomic characteristics, for use in a breeding program to develop a super-high erucic acid B. napus.

Mass Spectrometry
The seed oils of two genotypes were shown by high temperature GC to contain a triglyceride having a relative retention time identical to a standard trierucin sample. The identification was confirmed by the detection of a molecular adduct ion at m/z 1072 using ammonia CI GC/MS.

Gentoypes
The B. oleracea genotypes having significant sn-2 erucic acid all required extensive vernalization to induce bolting and seed set. They were derived from a variety of locations including Afghanistan, France, Great Britain, Japan and Turkey.

References

1. Appelqvist, L.A. (1976) Lipids in the Cruciferae in J.G. Vaughan, A.J. MacLeod and B.M.G. Jones (eds), *The Biology and Chemistry of the Cruciferae*, Academic Press, London, pp. 221-277.
2. Ackman, R.G. (1983) Chemical composition of rapeseed oil, in J.K.G. Kramer, F.D. Sauer and W.J. Pigden (eds), *High- and Low-Erucic Acid Rapeseed Oils. Production, Usage, Chemistry and Toxicological Evaluation*, Academic Press, Toronto, pp 85-129.
3. Mukherjee K.D., and Kiewitt, I. (1986) Lipids containing very long chain mono unsaturated acyl moieties in seeds of *Lunaria annua*. *Phytochemistry* **25**, 401-404.
4. Yaniv, Z., Elber, Y., Zur, M., and Schafferman, D. (1991) Differences in fatty acid composition of oils of wild Cruciferae, *Phytochemistry* **30**, 841-843.
5. Taylor D.C., MacKenzie, S.L., McCurdy, A.R., McVetty, P.B.E., Pass, E.W., Stone, S.J., Scarth, R., SR Rimmer, S.R., and Pickard, M.D. (1994) Stereospecific Analysis of Seed Triacylglycerols from High Erucic Acid Brassicaceae : Detection of Erucic Acid at the sn-2 Position in *B. oleracea* Genotypes, *J. Am.Oil Chem. Soc.* **71**, 163-167.
6. Mahler, K.A., and Auld, D.L. (1989) *Fatty Acid Compositions of 2100 Accessions of Brassica*, Idaho Agricultural Experiment Station, University of Idaho College of Agriculture, Moscow, pp. 6-38, 51-62.

Acknowledgements: The authors thank L.R. Hogge for determining triglyceride mass spectra.

Addendum
Recently, we conducted a screen of 40 B. rapa accessions. We identified three accessions for which the sn-2 fatty acids contained 51.7, 17.8 and 10.7 % erucic acid (Table 2). These values correspond to, respectively, 34.7, 11.8 and 8.3 % of the seed oil erucic acid. All three accessions originated in India.

TABLE 2. B. rapa accessions: sn-2 Fatty acid composition (Wt %)

TAG 22:1	16:0	18:0	18:1	18:2	18:3	20:1	22:1	% Tot 22:1
49.6	2.7	1.9	15.0	9,6	5.1	8.0	**51.7**	34.7
50.5	1.2	0.7	33.9	25.9	14.5	3.2	**17.8**	11.8
43.2	1.6	0.7	37.9	29.2	15.0	2.8	**10.7**	8.3
48.8*	1.4	0.7	55.9	26.8	10.9		**2.8**	1.9

* B. rapa cv R500

FATTY ACID COMPOSITION OF DIFFERENT TISSUES DURING HIGH STEARIC OR HIGH PALMITIC SUNFLOWER MUTANTS GERMINATION

R. ÁLVAREZ-ORTEGA, S. CANTISÁN, E. MARTÍNEZ-FORCE, M. MANCHA AND R. GARCÉS.
Instituto de la Grasa. CSIC.
Av. Padre García Tejero, 4.
41012-SEVILLA. SPAIN.

Introduction

Mutations in the steps of plastidial fatty acid synthesis related to palmitic and stearic acids formation produce big changes in the seed accumulation pattern of these acids. Mutants showing high palmitic or high stearic contents have been isolated in *Helianthus* (Osorio et al.,1995; Fernández-Martínez et al., 1996), *Glycine* (Graef et al., 1985) and *Arabidopsis* (Wu et al., 1994; Lightner et al., 1994). Those mutations being expressed in the whole plant usually produce no-growth, abnormal growth, no seed formation or seeds having low oil content. On the contrary, those expressed exclusively at the seed level do not show such aberrations. Five sunflower mutants with altered seed fatty acid composition (Very high 18:0, CAS-3; high 18:0, CAS-4 and CAS-8; very high 16:0, CAS-5; and very high 16:0 and 18:1, CAS-12) and two control sunflowers (RHA-274 and the high oleic HAOL-9) have been studied. The expression of the mutations during seed germination was studied by determining the fatty acid composition in TAG, DAG, PI, PC, PE, PG, DGDG and MGDG in different tissues of the plant (cotyledons, roots, leaves and stems).

TABLE 1. Fatty acid composition of total seed lipids from controls (RHA-274 and HAOL-9) and high saturated fatty acid mutants (CAS-3, CAS-4, CAS-8, CAS-5 and CAS-12).

	16:0	16:1	18:0	18:1	18:2	18:3	20:0	22:0
RHA-274	6.4	-	4.9	30.8	56.2	-	0.3	1.0
CAS-3	5.4	-	26.1	14.2	51.3	-	1.4	1.3
CAS-4	5.9	-	11.9	27.8	53.0	-	0.6	0.7
CAS-8	6.3	-	11.7	18.1	61.6	0.2	0.6	1.3
CAS-5	30.7	4.4	4.3	8.1	50.8	0.4	0.2	1.2
HAOL-9	5.2	-	3.7	88.1	1.5	-	0.3	1.1
CAS-12	27.8	7.6	1.7	56.7	4.4	-	0.3	1.3

Results and Discussion

Despite the altered lipid profile in the seed (Table 1), the growth and development of mutant plants resemble that of control plants. Our results from leaves, stems and roots show that the fatty acid composition of mutants and wild types are very similar although they do not share the same genetic background. As expected, the bigger differences were observed in developing cotyledons and also at the beginning of germination.

STEARIC ACID MUTANTS

The levels of stearic acid in the cotyledons of these mutants were higher than that of the control all over the germination period, the high stearic content at 10-15 days after germination came from the not catabolised seed fatty acids. This level start to decreased after 5 days (figure 1a) due to the rapid utilization of reserve fat while linolenic acid content increased (figure 1b) so indicating the switching from reserve to photosynthetic tissue. Initially, the stearic acid content in stems and roots was also higher than in the adult plant while the leaves showed a normal fatty acid composition from the begining (results not shown).

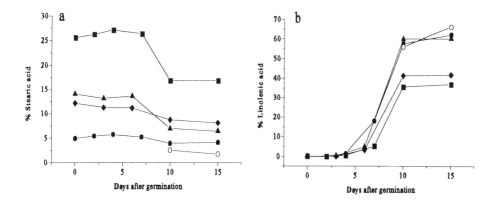

Figure 1. Evolution of stearic acid (a) and linolenic acid (b) levels in cotyledons of the control line (RHA-274, ●), in the very high (CAS-3, ■) and high stearic mutants (CAS-4, ▲; CAS-8, ♦) during germination. Also shown are the levels in control leaves (O).

PALMITIC ACID MUTANTS

Vegetative tissues of all sunflower lines have 15-20% palmitic acid. The level of this acid decreased in the germinating cotyledon of the high palmitic mutants and increased in that of the control line to reach this value (figure 2a). Similar results, than that of the high stearic mutants, were obtained with respect to the linolenic acid content (figure 2b), after 5 days the linolenic acid level increased and the cotyledon became green.

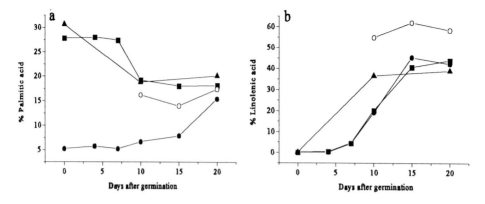

Figure 2. Evolution of palmitic acid (a) and linolenic acid (b) levels in cotyledons of the control line (HAOL-9, ●), high palmitic (CAS-5, ▲) and high palmitic and oleic mutant (CAS-12, ■) during germination. Also shown are the levels in control leaves (O).

These results suggest that the mutations are affecting only genes being expressed during seed development. These mutants were obtained using diferent mutagenic agents but all of them were grown and selected in the field. Only those mutants able to grow and produce enough seeds in the capitulum were chosen, such selection being against all posible mutants afecting the fatty acid composition in the whole plant. The phenotypes observed in these mutants point to at least three enzymes: KASII (ketoacyl-ACP synthetase II), stearoyl-ACP desaturase and thioesterase/s. Enzymatic studies are being carried out to determine the genes responsible for the observed phenotypes.

References

Fernández-Martínez, J.M., Mancha, M., Osorio, J., and Garcés, R. (1996) Isolation of an X-rays induced sunflower mutant containing high levels of palmitic acid in a high oleic background. Submitted.

Graef, G.L., Miller, L.A., Fehr, W.R., and Hammond, E.G. (1985) Fatty acid development in a soybean mutant with high stearic acid, *JAOCS*, **62**, 773-775.

Lightner, J., Wu, J., and Browse, J. (1994) A mutant of *Arabidopsis* with increased levels of stearic acid, *Plant Physiol.*, **106**, 1443-1451.

Osorio, J., Fernández-Martínez, J., Mancha, M., and Garcés, R. (1995) Mutant sunflowers with high concentration of saturated fatty acid in the oil, *Crop Science*, **35**, 739-742.

Wu, J., James, D.W., Dooner, H.K. and Browse, J. (1994) A mutant of *Arabidopsis* deficient in the elongation of palmitic acid, *Plant Physiol.*, **106**, 143-150.

PHOTOSYNTHETIC CARBON METABOLISM OF OLIVES

JUAN SÁNCHEZ and JOAQUÍN J. SALAS
Instituto de la Grasa, C.S.I.C.
Av. P. García Tejero, 4. 41012 Sevilla. Spain

Introduction

Developing olives (*Olea europaea* L.) contain active chloroplasts and are, therefore, able to fix CO_2 in the light (Sánchez, 1994). On the other hand, a high phosphoenolpyruvate carboxylase activity, present in the pulp tissue, enables the fruit to fix bicarbonate in the dark. Thus, tissue slices from from the pulp of developing olives can fix radiolabelled bicarbonate both under illumination and in the dark. It was reported previously that in olives developed under heterotrophic or autotrophic conditions the oil content was reduced by approximately one third as compared with the control (Sánchez, 1995), suggesting that fruit photosynthesis contribute significantly to the final oil content of the fruit. Moreover, it was observed that the rate of lipid synthesis from ^{14}C-bicarbonate in the light was drastically reduced in heterotrophic olives and enhanced in autotrophic ones. This communication deals with the nature of the products formed by tissue slices from the pulp of olives developed under different trophic regimes when incubated with labelled bicarbonate under either photosynthetic or non-photosynthetic conditions.

Material and Methods

Olive trees, cv Picual,, some 25-30 year-old, were used. Some branches were defoliated eight weeks after flowering, when the period of physiological fruit abscission was over and before the lignification of the endocarp and the onset of oil accumulation in the mesocarp. At the same time selected fruits were enclosed in specially designed dark cages to procure non-photosynthetic conditions for the development of the encaged olive (Sánchez, 1995).

Tissue slices prepared from the pulp of olives harvested 20 WAF were incubated in a buffered solution (50 mM potassium phosphate, pH 6.0, 0.5% ascorbate) containing 1.7 mM ^{14}C-bicarbonate (12 Ci/mol). Incubations were carried out at 30°C for three hours. After incubation lipids were thoroughly extracted according to Hara and Radin (1978): the tissue was succesively extrated three times with hexane-isopropanol (3:2) and the extracts combined; the tissue remains were then extracted with a volume of 0.5 M sodium sulfate equal to half the volume of the combined lipid extract, and this aqueous extract was then mixed with the lipid extract to form a two-phases system. The lower phase, containing the non-lipid products, was fractionated by a combination of cationic/anionic gels (Redgwell, 1980) to yield four groups of products: carbohydrates, organic acids, amino acids and phosphoesters. Carbohydrates and organic acids were eventually analyzed by HPLC.

Results and Discussion

As depicted in Table 1 the label of ^{14}C-bicarbonate was incorporated by tissue slices of olive pulp into three main groups of compounds: lipids, carbohydrates and organic acids, altogether accounting for at least 95 % of total radioactivity, with the rest being associated to amino acids and phosphoesters mainly. (Radioactivity associated to insoluble material has not been measured in these experiments). The distribution of label among different groups varied depending on the incubation conditions and the growth regime to which olives had been submitted. Thus, in the control incubated under illumination radioactivity was mostly incorporated into carbohydrates (47%), organic acids (35%) and lipids (13%). These fractions were further analyzed to find that most of the radioactivity incorporated into carbohydrates was associated to sucrose (90%), whereas the label in the organic acids fraction was distributed between malate (60%) and citrate (40%). The lipid fractions consisted mainly of TAG (60%), DAG (20%) and PC (15%).

TABLE 1. Distribution of radioactivity incorporated from ^{14}C-bicarbonate by tissue slices from olive pulp.

Growth Regime	Incubation Conditions	Distribution of label (nmol/h/g)			
		Lipids	Sugars	Org. Acids	Other
Control	Light	16	56	42	5
	Dark	tr	1	16	1
Autotrophic	Light	28	63	43	6
	Dark	tr	1	14	1
Heterotrophic	Light	tr	2	11	2
	Dark	tr	1	14	2

In autotrophic olives (those developed on defoliated branches) the total incorporation of ^{14}C-bicarbonate under illumination was slightly increased, indicating that the photosynthetic activity of the tissue was enhanced to cope with the restriction in the supply of photoassimilates, required for growth and development, brought about by the removal of source leaves. Interestingly, such increase in total incorporation was mainly reflected in the lipid fraction, in agreement with our previous report (Sánchez, 1995), with the percentage distribution among other fractions remaining essentially unchanged. In tissue slices from heterotrophic olives (those developed in the dark) the incorporation of ^{14}C-bicarbonate in the light was drastically reduced as compared with the control, with most the radioactivity incorporated into organic acids and very low incorporation either lipids or carbohydrates.

In the dark, on the other hand, total incorporation and distribution of radioactivity among products was similar for all the three types of tissues, and paralled that measured in tissue slices incubated in the dark, i.e., most of the radioactivity was found to be associated to organic acids.

These results indicate that in olive exist two independent mechanisms to fix CO_2/HCO_3^-. In the light CO_2 is fixed through the Calvin cycle and reduced carbon can be incorporated into carbohydrates and lipids. In the dark bicarbonate is fixed by phosphoenolpyruvate carboxylase to give oxaloacetate, which is then reduced by malate dehydrogenase. Malate can then be decarboxylated in the plastids by malic enzyme to yield pyruvate; in this reaction the same labelled carbon fixed in the carboxylation reaction catalyzed by phosphoenolpyruvate carboxylase from ^{14}C-bicarbonate in the course of the incubation is evolved; so that, although pyruvate as been found to be a good precursor of fatty acid synthesis in olive pulp both in the light and in the dark (del Cuvillo, 1994), the pyruvate produced from malate by malic enzyme is not labelled and can not, therefore, incorporate the radioactivity of ^{14}C-bicarbonate into fatty acids, which accumulates into organic acids, malate and citrate, which are the most abundant acids in the pulp tissue of developing olives (Donaire *et al.*, 1975). Whether or not malate is an *in vivo* source of reduced carbon for the synthesis of fatty acids in olive pulp, as it has been reported to be the case in plastids of developing rape seed (Smith *et al*, 1992), remains to be demonstrated. So far, it has been checked that tissue slices from olive pulp have not been capable of incorporating (U- ^{14}C)malate into lipids under different experimental conditions (results not shown), although it remains to be tested whether isolated plastids are capable of using malate as the precursor of fatty acid synthesis.

References

del Cuvillo, M.T. (1994) Lipid biosynthesis in the pulp tissue of developing olives. Ph D Thesis. University of Seville.
Donaire, J.P., Sánchez, A.J., López-Gorgé, J. and Recalde, L. (1975) Metabolic changes in fruit and leaf during ripening in the olive, *Phytochemistry* **14**, 1167-1169.
Hara, A. and Radin, N.S. (1978) Lipid extraction with a low toxicity solvent, *Anal. Biochem.* **90**, 420-426.
Redgwell, R.J. (1980) Fractionation of plant extracts using ion-exchange Sephadex, *Anal. Biochem.* **107**, 44-50.
Sánchez, J. (1994) Lipid photosynthesis in olive fruit, *Prog. Lipid Res.* **33**, 97-104.
Sánchez, J. (1995) Olive oil biogenesis: Contribution of fruit photosynthesis, in J.C. Kader and P. Mazliak (eds.), *Plant Lipid Metabolism*, Kluwer Academic Publishers, Dordrecht, pp. 564-566.
Smith, R.G., Gauthier, D.A., Dennis, D.T. and Turpin, D.H. (1992) Malate- and pyruvate-dependent fatty acid synthesis in leucoplasts from developing castor endosperm., *Plant Physiol.* **98**, 1233-1238.

BIOGENESIS OF ALCOHOLS PRESENT IN THE AROMA OF VIRGIN OLIVE OIL

JOAQUIN J. SALAS and JUAN SANCHEZ
Instituto de la Grasa, CSIC
Av. Padre Garcia Tejero nº4. 41012 Sevilla. Spain

Introduction

Contrary to most vegetable oils, which are extracted from seeds, virgin olive oil is prepared from the fleshy pulp of a fruit using mild physical procedures, thus resulting in a product highly prized for its nutritional properties and delicate aroma. Six carbon alcohols - hexanol, E-2-hexenol and Z-3-hexenol-, together with their ethyl esters, are important components of the aroma of virgin olive oil (Olias *et al.*, 1980). Investigations carried out in other plants species indicate that such volatile alcohols are formed from polyunsaturated fatty acids through the lipoxygenase pathway, which is triggered upon cutting or crushing the tissue, and involves the participation of lipoxygenase, hydroperoxide lyase and alcohol dehydrogenase (Hatanaka *et al.*, 1973; Vick *et al.*, 1987). The latter catalyzes the reduction of aldehydes produced by the action of hydroperoxide lyase to yield alcohols and is, therefore, responsible for the pattern of volatile alcohols characteristic of each plant species.

Three different alcohol dehydrogenases (ADHs) have been detected in enzyme extracts prepared from the pulp of ripening olives: an alcohol:NAD:oxidoreductase (EC 1.1.1.1) and two alcohol:NADP:oxidoreductases (EC 1.1.1.2), their kinetic parameters have been determined and are reported in this comunication, together with a discussion on the potential implication of these enzymes in the biogenesis of volatile alcohols constituent of the aroma of virgin olive oil.

Material and Methods.

Olive (*Olea europaea*) fruits, cv. Picual, were harvested at the beginning of the ripening period, some 30 weeks after flowering, which is characterized by a change in the colour of the fruit, which turns from green to purple. After removing the stone, 100g of pulp tissue was ground in 750 ml of cold acetone (-20°C) to prepare an acetone powder which was stored at -20°C until use. An enzyme extract was prepared, by using a glass homogenizer, from 3g of acetone powder in 90 ml of 50mM potassium phosphate, pH 7.2, 14 mM 2-mercaptoethanol, 2 mM DTE, 1 mM PMSF and 10% glycerol. After centrifugation the extract was fractionated by ammonium sulphate precipitation; the 30 to 60% cut, containing most of the activity, was further fractionated by ion exchange and affinity chromatography to yield three different enzyme forms: NADP-ADH I, NADP-ADH II and NAD-ADH.

Enzyme activities were determined by measuring the aldehyde dependent oxidation of NAD(P)H.

Results and discussion.

Enzyme extracts prepared from the acetone powders described above displayed a total NADP-ADH close to 1 U/mg prot, whereas NAD-ADH activity was some 20-fold lower. Such crude extract was submitted to ion-exchange chromatography on DEAE-Sepharose yielding two peaks of activity, the first one was NADP-dependent and was named NADP-ADH I, the second one contained both NAD- and NADP- dependent activities. These two activities were further separated by affinity chromatography on Cibacrom blue thus yielding two enzyme forms called NAD-ADH and NADP-ADH II. It was found that NADP-ADH II was the most abundant, accounting for about 90 % of total ADH activity in in the enzyme extract, and was purified some 30-fold by the chromatographic steps described. On the other hand, it was found that the purified enzyme solution in 50 % glycerol maintained full activity after a month at -20ºC.

Kinetic parameters of all the three enzymes were determined using hexanal as the substrate. Results (Table 1) showed that both NAD-ADH and NADP-ADH I displayed rather low affinity for hexanal, with apparent Kms in the order of millimolar, while the third and most abundant enzyme form, NADP-ADH II, showed an apparent Km for hexanal in the micromolar range as well as the highest V_{max}. These results point to the fact that it is NADP-ADH II the enzyme responsible for the biosynthesis of hexanol in ripening olives, so that it was further studied to find the especifity of the enzyme towards other putative substrates.

TABLE 1. Alcohol dehydrogenases, kinetic parameters of hexanal

Enzyme	Km (mM)	Vmax (U/mg prot)
NAD-ADH	2.0	19.0
NADP-ADH I	1.9	17.0
NADP-ADH II	0.04	44.0

As shown in table 2 the enzyme displayed maximal activity with hexanal, but still showing high activity with nonanal, an alternative intermediate in the oxidation of linoleate through the lipoxygenase pathway, as well as with E-2-hexenal and E-2-Z-6-nonadienal, wich are alternative intermediates in the degradation of linoleate through the same pathway. The activity with E-2-nonenal, which is not an intermediate of the lipoxygenase pathway, was clearly reduced in comparison to the others C9 aldehydes, whereas short chained aldehydes were poor substrates of this enzyme. Moreover, kinetic parameters were determined for some of these aldehydes and was found that the enzyme showed low affinity for propionaldehyde whereas it yielded similar Km values for the C6 and C9 aldehydes which are involved in the lipoxygenase pathway. On the other hand, an apparent Km for NADPH of 0.006 mM was determined for this enzyme using hexanal as the second substrate.

TABLE 2. NADP-alcohol dehydrogenase II, especificities and kinetic parameters of several substrates.

Substrate	Relative activity (%)	Km (mM)	Vmax (U/mg prot)
Acetaldehyde	0	nd	nd
Propionaldehyde	12	4.4	11.3
Butyraldehyde	17	nd	nd
Hexanal	100	0.040	44.0
E-2-hexenal	68	0.012	31.0
Nonanal	80	0.030	33.4
E-2-nonenal	40	nd	nd
E-2-Z-6-nonadienal	70	nd	nd

All together the result showed in this comunication suggest that carbonilic compounds generated from linoleate and linolenate in ripening olives through the secuential action of lipoxygenase and hydroperoxide lyase, can be reduced to alcohols by a NADPH dependent alcohol dehydrogenase which displays high activity and affinity for C6 and C9 aldehydes involved in the pathway. This enzyme, therefore, accounts for the formation of the C6 alcohols and, upon subsecuent ethylation catalyzed by specific acyl transferases, C6 ethyl esters in the pulp of ripening olives submitted to milling for oil extraction.

References

Hatanaka, A. and Harada, T. (1973), *Phytochemistry* **12**, 2341-2346.
Olías, J.M., Gutierrez, F., Dobarganes, M.C. and Gutierrez, R. (1980), *Grasas y Aceites* **31**, 391- 402
Vick, B.A. and Zimmerman, D.C. (1987), *The Biochemistry of Plants*, Vol 9. P.K. Stumpf (ed.), Academic Press, New York. 53-90.

DYNAMICS OF TRIACYLGLYCEROL COMPOSITION IN DEVELOPING SEA BUCKTHORN (*HIPPOPHAË RHAMNOIDES* L.) FRUITS

A.G. VERESHCHAGIN, O.V. OZERININA, V.D. TSYDENDAMBAEV

Laboratory of Lipid Metabolism, Institute of Plant Physiology, Russian Academy of Sciences, Botanicheskaya 35, 127276 Moscow, Russia

Previously, maturation of sea buckthorn fruits was shown to be accompanied by considerable changes in the fatty acid (FA) composition of fruit mesocarp triacylglycerols (TAGs) [1]. In the present study, we determined the dynamics of the content of separate TAGs themselves and the kinetics of their accumulation in water-rich pulp of a maturing fruit.

METHODS

Concentrations of FA species and types (palmitic, P; hexadecenoic, H; octadecenoic, O; linoleic, L; total saturated, S; total unsaturated, U) in TAGs at 53-107 days after pollination (DAP), absolute content of separate TAGs (by lipase hydrolysis), and a constant of relative rate of TAG accumulation (k, d^{-1}) were determined as described earlier [1,2].

RESULTS AND DISCUSSION

As shown in Fig. 1, group B TAGs (SUS+SUU+U_3 positional types) predominate in mesocarp, and the amount of group A ones does not exceed 1-2%. The absolute content of each TAG type is increased during maturation, but, as shown by k values, group A TAGs are characterized by a greater accumulation rate as compared to those of group B. This is consistent with the predominance of S over U in the rate of their biosynthesis in developing mesocarp [1]. Interestingly, at the terminal stage of growth, symmetric 1,3-S TAG types (S_3 and SUS) begin to give way to the symmetric 1,3-U ones (U_3 and USU) respectively in the rate of their formation.

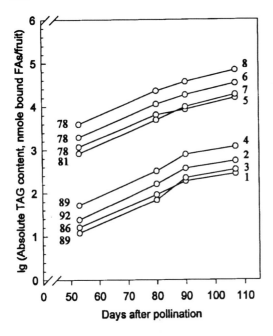

Fig.1. Dynamics of absolute content of separate TAG positional types and their groups in maturing mesocarp.

1, S_3; 2, SSU; 3, USU, 4, group A; 5, SUS, 6, SUU, 7, U_3; 8, group B. 78-92, $k \cdot 10^3$ values (d^{-1}).

To assess the dynamics of accumulation of separate TAGs in more detail, the composition of 17 major (>1%) positional species of group B TAGs in maturing fruits was determined. At every DAP, found TAG composition is close (r=0.998) to that calculated according to the theory of 1,3-Random,2-Random Distribution of FAs between the respective TAG positions [2]. As shown in Fig 2, the absolute content of all TAG species increases during growth, but at a different rate. Moreover, *rac*-1-O TAG species (PHO, HHO, POO, HOO, PLO, and HLO) stop to accumulate long before the end of maturation due to a limited O supply [1], which provides for the formation of only *sn*-2-O TAGs devoid of *rac*-1-O, i.e. POP, POH, and HOH.

Major TAG species, in their qualitative FA composition, can be divided into groups I, II, and III of 5 species each (53-69, 7-17, and 10-12% of total TAGs). The groups include, in the mid-position, H, O, and L, respectively, but are identical in the qualitative FA pattern of their putative *sn*-1,3-diacyl moieties ("*rac*-1-components") and in decreasing order of their quantitative content: *rac*-1-PH-TAG > -PP- > -HH- > -PO- > -HO- (Fig. 2).

In decreasing order of their k value, group I-III TAGs can be arranged as I > III > II. This order is only partially consistent with the respective rate order for the FAs present in the *sn*-2 position of maturing mesocarp TAGs: H > O > L [1]. On the hypothesis that *sn*-2-acylation of *rac*-1-components is a rate-limiting factor of TAG accumulation, the above facts account for the predominance of group I as compared to the II and III ones and for the relationship between the SUS species in their quantitative content (PHP > POP > PLP). At the same time the fact that group III exceeds group II in the rate of its formation can be explained only by a greater affinity of L as opposed to O to the *sn*-2 position of TAGs in spite of a lower velocity of L biosynthesis [2].

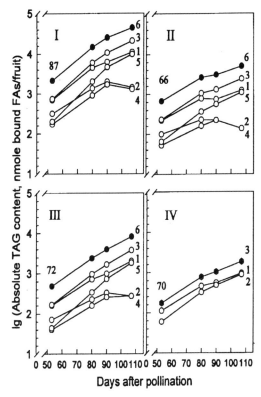

Fig. 2. Dynamics of absolute content of major TAG positional species and their groups in maturing mesocarp.

I: 1, PHP; 2, PHO; 3, PHH; 4, HHO; 5, HHH; 6, group I. II: 1, POP; 2, POO; 3, POH; 4, HOO; 5, HOH; 6, group II. III: 1, PLP; 2, PLO; 3, PLH; 4, HLO; 5, HLH; 6, group III. IV: 1, PHL; 2, HHL, 3, group IV.
66-87, $k \cdot 10^3$ values of TAG groups I-IV (d^{-1}).

Thus, the formation of group I-III TAGs can be suggested to involve, in the first place, esterification of *rac*-1-OH moieties of glycerol yielding 5 different *rac*-1-components as TAG precursors; then, the latter are subjected to a selective *sn*-2-acylation by H, O, and L to bring about the formation of major mesocarp TAG species. A mesocarp enzyme responsible for the acylation can be supposed to be similar to the lysophosphatidic acid acyltransferase, which incorporates erucic acid into the *sn*-2 position of *Brassica oleracea* seed TAGs [3]. Rate-limiting factors of mesocarp TAG accumulation are both FA supply and the affinity of FAs to the *rac*-1 and *sn*-2 positions of TAGs. The possible biosynthetic mechanism of group IV TAGs (Fig. 2) is yet to be established.

This work was supported by the grant 96-04-49320 of the Russian Foundation for Basic Research.

1. Berezhnaya, G.A., Yeliseev, I.P., Ozerinina, O.V., Tsydendambaev, V.D., and Vereshchagin, A.G.: Developmental changes in the absolute content and fatty acid composition of acyl lipids in sea buckthorn fruits, *Plant Physiol. Biochem.* **31** (1993), 323-332.

2. Ozerinina, O.V., Berezhnaya, G.A., Yeliseev, I.P., and Vereshchagin, A.G.: Composition and structure of triacylglycerols from sea buckthorn fruit mesocarp, *Prikl. Biokhim. Mikrobiol.* **24** (1988), 422-429.

3. Taylor, D.C., Barton, D.L., Giblin, E.M., MacKenzie, S.L., van den Berg, C.G.G., and McVetty, P.B.E.: Microsomal lysophosphatidic acid acyltransferase from a *Brassica oleracea* cultivar incorporates erucic acid into the *sn*-2-position of seed triacylglycerols. *Plant Physiol.* **109** (1995), 409-420.

Section 8:

Molecular Biology and Biotechnology

MOLECULAR BIOLOGY OF GENES INVOLVED IN CUTICULAR WAX BIOSYNTHESIS

J.D. HANSEN, C. DIETRICH, Y. XIA, X. XU, T-J. WEN, M. DELLEDONNE, D.S. ROBERTSON, P.S. SCHNABLE, B.J. NIKOLAU, *Departments of Biochemistry and Biophysics, Zoology and Genetics, and Agronomy, Iowa State University, Ames, Iowa.*

The cuticle is the outermost interface between a plant and its environment, and as such plays a crucial role in the plant's survival (Martin and Juniper, 1970; Kolattukudy, 1981). It consists of cutin embedded in a complex mixture of lipids commonly referred to as the cuticular waxes. Cuticular waxes are a complex mixture of lipids including hydrocarbons (n-alkanes, branched alkanes, cyclic alkanes, alkenes), ketones, ketols, alcohols, aldehydes, diols, acids, and esters, which are derived from cellular fatty acids (Kolattukudy et al., 1976; Tulloch, 1976). Early biochemical investigations have identified reactions by which many of the various components of the cuticular wax are synthesized (Kolattukudy et al., 1976). However, the isolation and characterization of enzymes postulated to be involved in these processes has been technically difficult to achieve because plant epidermal tissue, the site of cuticular wax biosynthesis (Kolattukudy et al., 1976), is difficult to obtain in large quantities. To overcome this technical barrier, a molecular genetic approach was taken to isolate genes involved in cuticular wax biosynthesis (Schnable et al., 1995). In maize and *Arabidopsis* mutations that affect the normal accumulation of cuticular waxes are known as *glossy* (*gl*) and *eceriferum* (*cer*), respectively. Using the *Mutator* transposon system, we have cloned the *gl1* and *gl8* loci of maize via transposon tagging. In addition, we have cloned the *CER2* locus of *Arabidopsis* via chromosome walking.

1. The *glossy1* locus of maize

We were able to molecularly clone the *gl1* locus of maize due to the insertional mutagenesis of this locus by the *Mu1* transposon (Hansen et al., 1996). The *gl1* cDNA is predicted to code for a protein of 319 amino acids with an estimated molecular mass of 35.3 kDa and a theoretical isoelectric point of 9.38. The GL1 protein is homologous to the N-terminal portions of the EPI23 protein of *Kleinia odora*, the CER1 protein of *Arabidopsis* and the protein coded by the rice EST RICS2751A. These homologous proteins are predicted to be have 625 amino acids residues.

Analyses of these proteins with the TMPredict algorithm and comparison of these outputs to the Kyte-Doolittle hydrophobicity plots indicate that these proteins are integral membrane proteins. In addition, the PSort algorithm identified the NH_2-terminal 40 residues of the GL1 protein as a plasma membrane-targeting signal peptide.

The predicted secondary structure and membrane topology of the GL1 and homologous proteins are similar to a superfamily of receptors, including the chemokine receptors, that have been characterized from animals (Donnelly and Findlay, 1994). These receptors consist of seven transmembrane helices, which together represent the ligand-binding domain. Due to the fact that the GL1 protein is considerably shorter than the EPI23, CER1 and the RICS2751A proteins, the former may contain only five transmembrane helices. Analogous to the animal receptors, the GL1-homologous proteins are predicted to have the transmembrane helices at the NH_2-terminal portion of the polypeptides, and the COOH-terminal portion of these proteins have a hydrophobicity profile characteristic of a globular, water-soluble protein. In animals, this COOH-terminal domain interacts with G-proteins, transducing the signal that an extracellular ligand has bound to the receptor.

Despite the sequence similarity between the CER1 and GL1 proteins, their biochemical function is unclear since mutations at each gene affect the respective cuticles differently. *CER1* has been predicted to code for an aldehyde decarbonylase (Aarts et al., 1995), an enzyme required for the production of the alkane fraction of the cuticular waxes of this species. Indeed, mutations at the *CER1* locus cause an enrichment of the aldehydes and a depletion of the alkanes and alkane-derived metabolites (ketones and secondary alcohols). In maize however, its unlikely that *gl1* codes for an aldehyde decarbonylase. This conclusion is based upon the fact that alkanes account for a very small portion of the cuticular waxes of wild-type maize seedlings (about 1%), and because mutations at the *gl1* locus qualitatively and/or quantitatively affect the accumulation of fatty aldehydes, alcohols, and the ester components of the maize cuticular waxes. Therefore, even though the GL1 and CER1 proteins share similar structures, they may in fact perform different functions in cuticular wax biosynthesis. Alternatively, both proteins may perform similar, but as yet unidentified, function(s).

2. The *glossy8* locus of maize

The *gl8* locus of maize was cloned by the same strategy used to clone *gl1*, with the exception that the *gl8* locus was "tagged" by a *Mu8* transposon. The *gl8* gene is interrupted by two introns and codes for a protein of 327 amino acids with a predicted molecular mass of 36 kDa. The N-terminal 29 residues of the GL8 protein has the characteristics of a transit peptide, which the PSort algorithm predicts would target this protein to the plasma membrane. The hydrophobicity profile, and the TMPredict algorithm indicate that GL8 is an integral membrane protein. The predicted amino acid sequence of the GL8 protein is homologous to β-ketoacyl reductases, such as the fabG protein of *Escherichia coli*, and the Cuphea β-ketoacyl-ACP reductase. This sequence similarity indicates that the GL8 protein is part of the membrane-bound fatty acid elongase system that generates the very long chain fatty acids required for cuticular waxes. Furthermore, this finding indicates that this fatty acid elongase system is probably heteromeric in structure being composed of a "complex" of individual gene products, analogous to the bacterial and chloroplastic *de novo* fatty acid synthase, and unlike the animal and yeast fatty acid synthase.

3. The *CER2* locus of *Arabidopsis*

We have cloned the *CER2* locus of *Arabidopsis* by chromosome walking (Xia et al., 1996). Despite the intimate knowledge of the molecular structure of the *CER2* gene and encoded protein afforded by this achievement, its biochemical function is not deducible from these data. We have therefore utilized the *CER2* gene as a biochemical marker to investigate the expression pattern of a cuticular wax gene.

The *cer2* phenotype is expressed in stems and siliques of *Arabidopsis*. This restriction of the phenotype is due to the pattern of *CER2* expression. Namely, *in situ* hybridization demonstrate that the *CER2* mRNA accumulates in the epidermal cells of stems and siliques. Furthermore, the *CER2* promoter drives *GUS* expression in transgenic *Arabidopsis* plants in the epidermis of stems and siliques; the most prominent expression occurs in young siliques, but only the upper portions of stems and large siliques stain for GUS activity. Hence, the localization of GUS activity roughly corresponds to those regions that are undergoing elongation. Expression of the *CER2-GUS* gene was not detected in rosette leaves (before or after bolting), cauline leaves, sepals, petals, or roots (organs which do not exhibit the *cer2* phenotype).

4. References

Aarts, M.G.M., Keijzer, C.J., Steikema, W.J., and Pereira, A. (1995) Molecular characterization of the *CER1* gene of *Arabidopsis* involved in epicuticular wax biosynthesis and pollen fertility. *Plant Cell* **7**, 2115-2127.

Donnelly, D. and Findlay, J.B.C. (1994) Seven-helix receptors: structure and modelling, *Curr. Opin. Struct. Biol.* **4**, 582-589.

Hansen, J.D., Pyee, J., Xia, Y., Wen, T-J., Robertson, D.S., Kolattukudy, P.E., Nikolau, B.J., and Schnable, P.S. (1996) The *glossy1* locus of *Zea mays* L. and an epidermis-specific cDNA from *Kleinia odora* define a novel class of plant receptor-like proteins required for the normal accumulation of cuticular waxes, submitted for publication.

Kolattukudy, P.E. (1981) Structure, biosynthesis, and biodegradation of cutin and suberin, *Ann. Rev. Plant Physiol.* **32**, 539-567.

Kolattukudy, P.E., Croteau, R., and Buckner, J.S. (1976) Biochemistry of plant waxes, in P.E. Kolattukudy (ed.), *Chemistry and Biochemistry of Natural Waxes*, Elsevier Press, New York, pp. 289-347.

Martin, J.T. and Juniper, B.E. (1970) *The Cuticle of Plants*, Edward Arnold Ltd., Edinburgh

Schnable, P.S., Stinard, P.S., Wen, T-J., Heinen, S., Weber, D., Schneerman, M., Zhang, L., Hansen, J.D., and Nikolau, B.J. (1995) The genetics of cuticular wax biosynthesis, *Maydica* **39**, 279-287.

Tulloch, A.P. (1976) Biochemistry of plant waxes, in P.E. Kolattukudy (ed.), *Chemistry and Biochemistry of Natural Waxes*, Elsevier Press, New York, pp. 235-287.

Xia, Y., Nikolau, B.J., and Schnable, P.S. (1996) Cloning and characterization of *CER2*, an *Arabidopsis* gene that affects cuticular wax accumulation, *Plant Cell*, in press.

EXPRESSION OF GENES INVOLVED IN WAX BIOSYNTHESIS IN LEEK

Y. RHEE, A. HLOUSEK-RADOJICIC, P.S. JAYAKUMAR, D. LIU,
AND D. POST-BEITTENMILLER
The Samuel Roberts Noble Foundation,
P.O. Box 2180, Ardmore, OK 73402

Introduction

Accumulation of waxes on the outermost surface of plant cells is a complex process including perception of the necessary need for, biosynthesis of, and transportation and deposition of the waxes. A fair amount of biochemical data is available suggesting that epicuticular plant waxes are derived from stearate and that elongases are involved in lengthening the chain length followed by a several enzyme reactions responsible for making many other derivatives such as aldehydes, alkanes, ketones, and wax esters (3, 4). Very little attention, however, has been given to regulation of overall wax biosynthesis. The objective for this study is to provide insights regarding regulation of wax biosynthesis and accumulation in a model system by comprehensively examining wax accumulation and expression of genes involved in wax biosynthesis.

Materials and Methods

The third leaf from leeks regrown for 14 days as described in previous work (1) was cut sequentially from above the intercalary meristem 3cm, 2cm, 2cm, 2cm, 5cm, 5cm, 5cm (segment I, II, III, IV, V, VI, VII, respectively). Each segment was analyzed for epicuticular wax contents using a HP5 30m column and GC/MS. Wax was extracted from the surface of the segments by dipping them in $CHCl_3$ for 60 secs. C30 alkane was added as an internal standard. For microsomal preparation, 50mg of fresh leek epidermal tissues were homogenized in buffer (80mM HEPES-KOH pH7.2, 2mM EDTA, 320mM sucrose, 2mM DTT, and 3mM PMSF) followed by a centrifugation at 5K rpm for 15 minutes at 40C. The supernatant then was spun again in airfuge (15 minutes at 22 psi). The resulting pellet was resuspended gently in buffer (80mM HEPES-KOH pH7.2, 15% glycerol, and 2mM DTT) and used for elongases assay (1). The supernatant from the airfuge spinning was combined with 1/10 volume of 50% glycerol and stored frozen until later usage for stearoyl-ACP thioesterase assay(2).

Results and Discussion

Wax analysis on the surface of the leek leaf revealed that a C31 ketone was a major component in the wax demonstrating that the decarbonylation pathway was dominant in the leek. The C31 ketone, therefore, was used as a marker to monitor wax content in subsequent experiments. Other detectable components included aldehydes and alkanes with chain length ranging from C26 to C31 as well as free fatty acids from C16 to C22. The analyses also showed that there was a differential accumulation of the total wax along the length of the leaf (Fig. 1). The lower section of the leaf has little wax compared to the upper portions of the leaf. Increased wax accumulation began in segment III (Fig. 1) and continued until it plateaued near the top of the leaf. Due to unsynchronized development of the plants, it was essential to repeat the experiment several times. Eventhough the area where the increase of the accumulation startedvaried depending upon the leaves, in general it was shown in segment III or IV. Therefore, the region showing the initiation of wax accumulation was between 5 cm and 9cm from the intercalary meristem region. Light microscopic examination indicated that the initiation of wax accumulation occurred after cell elongation was completed (data not shown). Most importantly, this clear differential accumulation may permit differential screening or differential display to clone additional genes involved in wax biosynthesis. The factors responsible for the induction were not determined.

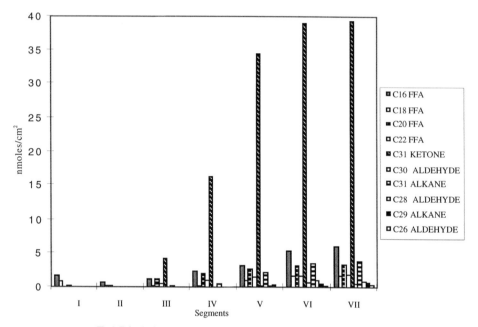

Fig.1 Epicuticular wax analysis along the length of the leek leaf.

Since elongases are an essential part of wax biosynthesis (3,4) their activities were examined along the length of the leaf. As shown in Fig. 2, the highest activity was observed in segment III where initiation of the wax accumulation was also shown. This correlation between the highest elongation activity and the initiation of the wax accumulation was observed in several leaves. These results strongly indicate that expression of elongase activity was closely correlated with wax biosynthesis. The reason for decrease of elongase activity after the peak while wax accumulation is maintained at high level is unknown at this moment. It may indicate that the elongase activities at an early stage were enough to maintain the pool of very ling chain fatty acids for wax production. In addition, there may be little turn over of the surface wax. The highest activity of stearoyl-ACP thioesterase was also observed in segment III (data not shown) which is consistent with an observation that C18 saturated fatty acid is the precursor for wax biosynthesis (1,3).

Northern analysis of RNA isolated from different segments (I+II, III+IV, and VI+VII) of the leaf showed no significant change in the level of KAS III transcript (data not shown). This indicated that the steady state level of KASIII transcript was not increased eventhough there may be an increased demand for fatty acid biosynthesis due to wax production in the epidermal cells.

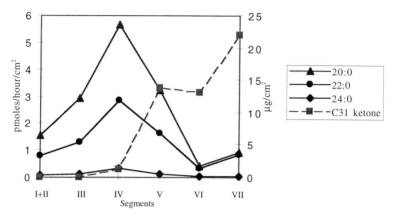

Fig.2. Elongases assay along the length of the leek leaf.

Acknowledgments

This project was supported by the Noble Foundation. The authors sincerely acknowledge David Huhman for helping GC/MS analysis of waxes and Yuling Liu for peeling epidermal tissues.

References

[1] Evenson, K.J. and Post-Beittenmiller, D. (1995) Fatty acid elongating activity in rapidly-expanding leek epidermis, *Plant Physiol.* **109**: 707-716.
[2] Liu, D. and Post-Beittenmiller, D. (1995) Discovery of an epidermal stearoyl-acyl acrrier protein thioesterase: its potential role in wax biosynthesis, *J. Biol. Chem.* **270**: 16962-16969.
[3] Post-Beittenmiller, D. (1996) Biochemistry and Molecular biology of Wax production in Plants, *Annu. rev. Plant physiol. plant Mol. Biol.* **47**: 405-430.
[4] von Wettstein-Knowles P.M. (1995) *Biosynthesis and Genetics* pp91-129 in *Waxes: Chemistry, molecular bilogy and Functions*, Hamilton, R.J. ed., Dundee, Scotland: The Oily Press

EXPRESSION OF CASTOR AND *L. FENDLERI* OLEATE 12-HYDROXYLASES IN TRANSGENIC PLANTS

Effects on lipid metabolism and inferences on structure-function relationships in fatty acid hydroxylases

P. BROUN, N. HAWKER and C.R. SOMERVILLE
Carnegie Institution of Washington
Department of Plant Biology
290, Panama street
Stanford CA 94305

Introduction

Ricinoleic acid (D-12-hydroxyoctadec-cis-9-enoic acid), is an hydroxylated fatty acid which constitutes 85-90% of the seed fatty acids in castor bean plants (*Ricinus communis L.*). This unusual fatty acid is also one of a series of related Hydroxy Fatty Acids (HFAs) produced in the seeds of *Lesquerella* species. In these species, which, like *A. thaliana* and rapeseed belong to the *Brassicacae* family, ricinoleic acid is generally a minor constituent. Major HFAs include densipolic (12-OH, 18:2 (3,9)), lesquerolic (14-OH, 20:1 (9)) and auricolic (14-OH, 20:2 (3,9)) acids.

In castor, where metabolism of HFAs has been studied in most detail, ricinoleic acid is synthesized in seeds on phosphatidyl choline, then very efficiently removed from membranes and transferred to the triacylglycerol pool (Bafor et al., 1991).

Recently, we have reported the isolation of a cDNA clone encoding the oleate 12- hydroxylase from castor (van de Loo et al., 1995). Constitutive expression of the hydroxylase cDNA in transgenic tobacco resulted in accumulation of low levels of ricinoleate in seed lipids, but not in leaves and roots. In order to further characterize metabolism of HFAs in transgenic plants, we have introduced this cDNA into *A. thaliana*. We report here how lipid metabolism is affected in transgenic plants.

Extraplastidial ω-6 desaturases and castor oleate 12- hydroxylase share a number of biochemical characteristics. Cloning of a cDNA encoding the castor hydroxylase has also confirmed that the two enzymes are closely related (van de Loo et al., 1995). Although reaction mechanisms are expected to be similar, they lead to a different outcome. In order to investigate what structural components in these enzymes are responsible for this difference, we proceeded to isolate the gene encoding another oleate 12-hydroxylase, from *Lesquerella fendleri*. Multiple comparisons of desaturase and hydroxylase sequences revealed key differences between the two categories of enzymes. We report here investigation of their functional significance.

1. Expression of Castor and *L. fendleri* Hydroxylase Genes in transgenic *A. thaliana*

We designed degenerate primers based on the sequence of castor fatty acid hydroxylase CFah12 and used them to PCR-amplify cDNAs from *L. fendleri*. One such cDNA detected an abundant seed specific transcript on Northern blots of *L. fendleri* RNA. Its sequence had extensive similarity with the CFah12 gene. This cDNA was used to isolate a genomic clone which was introduced into *A. thaliana*.

Expression of the *L. fendleri* gene resulted in accumulation of HFAs in transgenic plants, up to 15% of seed fatty acids, thus establishing the gene encodes *L. fendleri* hydroxylase LFah12.

Transgenic plants expressing CFah12 under the control of a strong seed specific promoter were also obtained. In these plants, HFAs constitute up to 20% of the seed fatty acids. Seed fatty acid composition of *A. thaliana* plants expressing CFah12 or LFah12 is very similar, suggesting the two enzymes have comparable activities in transgenic plants.

Ricinoleic acid is only one of four HFAs produced in transgenic seeds, which also accumulate densipolic, lesquerolic and a small amount of auricolic acid. This suggests that *Arabidopsis* and related *Lesquerella* species metabolize ricinoleic acid in a similar way.

Expression of LFah12 under the control of the CaMV 35S promoter did not affect fatty acid composition of vegetative organs, even though hydroxylase activity could be detected. This implies either poor enzyme activity or efficient turnover of HFAs in non-seed tissues.

Accumulation of HFAs was accompanied with an increase in oleate levels and a concurrent decrease in 18:2 and 18:3, suggesting the oleate 12- desaturase is inhibited in transgenic plants expressing either gene.

2. Investigations of Structure-Function Relationships in Hydroxylases and related Desaturases

We performed multiple comparisons between oleate 12- desaturases, CFah12 and LFah12, and identified six residues conserved among desaturases which differ in fatty acid hydroxylases. Using appropriate growth conditions (Covello and Reed, 1996), we were able to express LFah12 in yeast, under the control of the *GAL1* promoter. Yeast strains over-expressing the *Lesquerella* gene accumulated a small amount of ricinoleic acid. We could also detect small levels of 18:2, indicative of some desaturase activity of the enzyme in this context.

In order to establish the functional significance of observed residue differences between desaturases and hydroxylases, we used site-directed mutagenesis to substitute desaturase residues for the corresponding residues in LFah12 at all six positions. In yeast strains expressing the mutant hydroxylase, ratios of 18:2 to ricinoleic acid levels were more than 30-fold higher than in control strains expressing the wild-type gene (fig. 1). This result indicates that these residues are essential in LFah12 in determining the outcome of fatty acid oxidation.

Conclusion

We described here the isolation of a novel fatty acid hydroxylase from *L. fendleri*. We also presented some results from the analysis of transgenic *A. thaliana* plants expressing the castor and *L. fendleri* genes. We plan to use these transgenic plants to dissect mechanisms involved in removing HFAs from membranes, channeling them to storage lipids or breaking them down. We also hope to gain understanding of what controls such mechanisms.

We also reported here the critical role played by a small number of residues in controlling the outcome of fatty acid oxidation. Narrowing down on fewer residues will make it easier to rationalize structural differences between hydroxylases and desaturases, and understand how these differences affect reaction mechanisms.

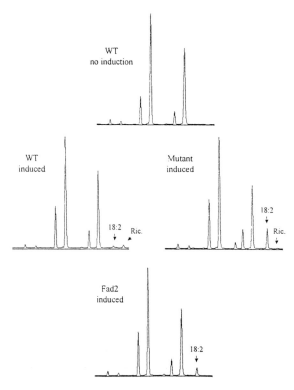

Figure 1: Gas chromatogram of fatty acid methyl esters from yeast strains expressing a wild-type or a mutant fatty acid hydroxylase from *L. fendleri*

Bafor, M., Smith, M.A., Jonsson, L., Stobart, K. and Stymne, S. (1991) Ricinoleic acid biosynthesis and triacylglycerol assembly in microsomal preparations from developing castor-bean (Ricinus communis) endosperm, *Biochem. J.* **280**, 507-514.

Covello, P.S. and Reed, D.W. (1996) Functional expression of the extraplastidial Arabidopsis thaliana oleate desaturase gene (FAD2) in Saccharomyces cerivisiae, *Plant Physiol.*. **111**, 223-226.

van de Loo, F.N., Broun, P., Turner, S., Somerville, C.R. (1995) An oleate 12- hydroxylase from castor (Ricinus communis L) is a fatty acyl desaturase homologue, *Proc. Natl. Acad. Sci. USA* **92**, 6743-6747.

CHARACTERIZATION OF PHOSPHOLIPASE D-OVEREXPRESSED AND SUPPRESSED TRANSGENIC TOBACCO AND ARABIDOPSIS

Xuemin Wang, Suqin Zheng, Kirk Pappan, and Ling Zheng
Department of Biochemistry, Kansas State University, Manhattan, Kansas 66506, USA.

Introduction

PLD hydrolyzes membrane phospholipids to phosphatidic acid, which, along with its metabolites, is an important mediator in cell signalling and regulation. PLD in mammals has been proposed to be involved in a broad range of cellular processes including cell proliferation, protein secretion, metabolic regulation, immune and inflammatory responses (2). PLD in yeast is required for sporulation and thought to play a role in meiotic signaling (2). PLD activity in plants has been implicated in different physiological processes that include membrane breakdown during stress injuries, pathogenesis, senescence, and ageing; messenger production in response to environmental cues; lipid mobilization during seed development and germination (3, 4). Recent cloning of phospholipase D cDNA from plants (1, 5), human, and yeast (2) reveals that PLD is a novel but highly conserved gene family. The availability of PLD cDNAs provides molecular tools needed to improve the understanding of PLD function. This report describes transgenic manipulations of PLD gene expression in tobacco and *Arabidopsis*.

Materials and Methods

A 2.8 kb cDNA fragment encoding full length amino acid sequence of castor bean PLD (5) was inserted into the *Agrobacterium tumefaciens* transfer vector pKYLX7 in the sense and antisense orientation under the control of the cauliflower mosaic virus 35S promoter. A 780-bp fragment of *Arabidopsis* PLD cDNA (1) was inserted into pKYLX7 in the antisense orientation. The T-DNA regions of pKYLX7 tobacco were introduced into plants through *Agrobacterium*-mediated gene transfer. Transformation of tobacco and *Arabidopsis* was achieved through leaf discs inoculation and vacuum infiltration, respectively.

Results and Discussion

Among 34 sense transgenic lines of tobacco generated, 16 overexpressed PLD activity at least 20-fold higher than control plants (transformed with vector alone). The overexpressed castor PLD in tobacco was readily detectable by measurements of PLD activity and protein (Fig. 1 A and B) since PLD activity in control leaves was relatively low and the antibody raised against castor bean PLD weakly crossreacted with tobacco PLD.

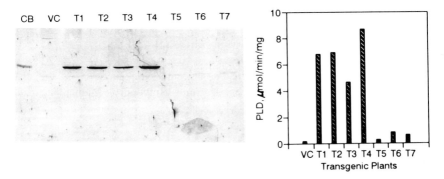

Figure 1. Overexpression of PLD in tobacco. A. immunoblot of PLD protein from different transgenic plants. Each lane was loaded with 20 µg protein. B. PLD activity in the corresponding transgenic plants.

Thirty-seven antisense transgenic lines of tobacco were produced. The antisense mRNA was made in those plants, but none of them showed substantial decrease in PLD activity. Sequence differences between the castor and tobacco PLD genes might be one of the possible explanations for the inefficient antisense suppression.

Subsequent cloning of a PLD cDNA from *Arabidopsis* enabled us to suppress PLD in *Arabidopsis* using its homologous antisense DNA. Among 15 antisense-transgenic lines, one exhibited less that 3 % PLD activity in leaves when compared to wild type (Table 1). Consistent with the results of activity assay, PLD protein was undetectable with immunoblotting using PLD antibodies raised against a 12-amino acid peptide corresponding to the C-terminus of the reported PLD sequence (1). Small variations occurred in the degree of PLD suppression among different tissues, with the percentage of suppression in roots being the smallest. The low PLD activity co-segregated with kanamycin resistance in a 3:1 ratio, suggesting a single T-DNA insertion in genome.

Table I. Antisense Suppression of PLD in *Arabidopsis*

Tissue	PLD Activity, *nmol/min/mg*		Suppression %
	Antisense	Wild Type	
Leaf	25	850	97
Flower	159	1584	90
Silique	29	443	94
Root	293	1830	84

Fig. 2. Immunoblot of PLD in soluble (S) and membrane (M) fractions

The production of PLD-overexpressed and suppressed transgenic plants provides an excellent system to investigate the control and cellular function of the major phospholipid-hydrolyzing enzyme. An initial concern using the 35S promoter was that constitutive and high levels of PLD expression in all the tissues might be detrimental to plants due to possible excessive degradation of membrane lipids. In the present study, the transgenic tobacco expressed PLD protein more than 50-fold higher than in wild type, and yet the transgenic plants grew and developed to maturity. No significant differences in phospholipid content and composition were observed between the PLD-overexpressed and wild type plants. The introduced PLD was found in plasma membrane, microsomal membrane, and soluble fractions. These results indicate that PLD activity *in vivo* is tightly regulated after transcription and translation.

The PLD antisense-suppressed *Arabidopsis* also exhibited no dramatic phenotypical alterations. This suggests that either the residual PLD is sufficient for sustaining plant growth or other types of PLD compensate for the loss of this PLD. During the process of investigating this latter hypothesis, a novel type of PLD activity was identified that requires polyphosphoinositides (PPI), phosphatidylinositol 4,5-bisphosphate (PIP_2) or phosphatidylinositol 4-phosphate, for activity. The PLD that has been suppressed (Table 1 and Fig. 2) is the "conventional" plant PLD that requires the use of millimolar concentrations of Ca^{2+} and SDS for maximal activity (4, 5). However, the antisense PLD transgenics showed the PPI-requiring activity comparable to that in wild type. This PLD activity is functional at nanomolar concentrations of Ca^{2+} and reached plateau at 5 μM Ca^{2+}. These results indicate the presence of other types of PLD and also point to the significance and challenge of understanding the complexity of PLD regulation and function.

Acknowledgements

This work was supported by grants to X.W. from the U.S. Department of Agriculture (92-37304-7895 and 95-37304-2220) and the National Science Foundation (IBN-9511623). This is contribution 96-131-B of the Kansas Agricultural Experiment Station.

References

1. Dyer JH, Zheng L, Wang X: Cloning and nucleotide sequence of a cDNA encoding phospholipase D from Arabidopsis (Accession No. U36381) (PGR 95-096). Plant Physiol 109:1497 (1995).
2. Morris AJ, Engebrecht J, Frohman MA: Structure and regulation of phospholipase D. TIBS. 1996 (in press)
3. Wang X: Phospholipases. In Moore TS (ed), Lipid Metabolism In Plants, pp. 499-520. CRC press, Boca Raton (1993).
4. Wang X, Dyer JH, Zheng L: Purification and immunological analysis of phospholipase D from castor bean endosperm. Arch Biochem Biophys 306: 486-494 (1993).
5. Wang X, Xu L, Zheng L: Cloning and expression of phosphatidylcholine-hydrolyzing phospholipase D from *Ricinus communis* L. J Biol Chem 269: 20312-20317 (1994).

PURIFICATION AND MOLECULAR CLONING OF BELL PEPPER FRUIT FATTY ACID HYDROPEROXIDE LYASE

K. MATSUI, M. SHIBUTANI, Y. SHIBATA, AND T. KAJIWARA
Department of Biological Chemistry, Faculty of Agriculture, Yamaguchi University, Yamaguchi 753, Japan

1. Introduction

Fatty acid hydroperoxide lyase (HPO lyase) is an enzyme that cleaves a C-C bond of the hydroperoxides (HPOs) of polyunsaturated fatty acids to generate aldehydes and ω-oxo-acids (Matsui et al., 1991). The former, short-chain volatile aldehydes, are important constituents of the characteristic flavors of crops. Although the physiological role of HPO lyase has not been fully elucidated, rapid formation of C6-aldehydes during a hypersensitive-resistance response has been recently reported (Croft *et al.* 1993). Thus, involvement of the enzyme in anti-bacterial responses has been suggested. Elucidation of the properties and structures of the enzyme would shed light on its role and develop practical uses of the enzyme, *e.g.*, enhancement of resistance of a plant to pathogens and improvement of flavor character of a crop by genetic engineering. Thus, in this study, we purified and characterized HPO lyase. Furthermore, a cDNA coding the enzyme was isolated, and its primary structure has been determined.

13(*S*)-hydroperoxy-(9*Z*,11*E*,15*Z*)-octadecatrienoic acid

↓ **Fatty acid hydroperoxide lyase**

(3*Z*)-hexenal   12-oxo-(9*Z*)-dodecenoic acid

2. Results and Discussion

HPO lyase was purified to a homogeneous state from immature fruits of green bell pepper (Shibata *et al.* 1995a). It was deduced to be a homo-trimer of 55-kDa subunits. The activity was considerably inhibited by lipophilic antioxidants such as nordihydroguaiaretic acid and α-tocopherol. Furthermore, HPO lyase was rapidly and irreversibly inactivated by fatty acid hydroperoxide (Matsui et al., 1992). These imply the involvement of radical species in its catalysis. The enzyme showed strict substrate specificity, and 13-hydroperoxide of α-linolenic acid is thought to be a sole substrate *in vivo*. Spectrophotometric analyses of HPO lyase indicated that it was a heme protein (Shibata *et al.* 1995b). Diagnostic features include the Soret band at 391 nm and a broad α band at around 500 to 540 nm. Reduction with dithionite led to the shift in the Soret band to 408 nm and the α band to around 550 nm. Precise spectrophotometric analysis on the pyridine hemoferrochrome prepared from bell pepper HPO lyase exhibited a sharp γ band at 419 nm, an α band at 556 nm and a β band at 523 nm. Difference spectrum of dithionite reduced minus non-treated also showed a distinct α band at 556 nm and a β band at 523 nm. These are typical diagnostic features of heme *b* (protoheme IX). This spectroscopic feature highly resembles a cytochrome P450, allene oxide synthase (Song and Brash, 1991). Extensive CO treatment of the enzyme caused essentially no appearance of a peak at 450 nm, which is an important diagnostic feature of P450 protein.

The finding that the bell pepper HPO lyase has the spectral characteristics of a cytochrome P450 raises the question of the relationship of this enzyme to other P450s. To this end, molecular cloning of the enzyme was attempted. A cDNA library was constructed from immature bell pepper fruit mRNA and was screened by the antiserum against HPO lyase. Of 3×10^5 clones, six were positive and shared essentially identical sequences. The same sequences as those determined from the purified enzyme could be seen in the sequence of the longest clone, λPF23m. Twenty positives were isolated by further screening with λPF23m as a probe. Because partial sequences determined were identical, the longest one, PL22, was analyzed further (Matsui *et al.*, 1996). PL22 had an insert of 1647 bp and coded 480 amino acids. The deduced molecular weight was 54055, which suggested that this clone encodes near full-length sequence. Expression of the coding region of PL22 in *E. coli* under the control of *trc* promoter resulted in appearance of HPO lyase activity in the cells. Immunoblot analysis of the cell lysate showed that the heterologously expressed HPO lyase had the same subunit size as that purified from bell pepper fruits. Sequence data base searches using the BLAST algorithm program (Altschul *et al.*, 1990) showed that PL22 shared homology with cytochrome P450s. In the P450 family the enzyme is most closely related to allene oxide synthase (CYP74A, with 40% sequence similarity). Although incomplete, the segments of the sequence highly conserved within P450s could

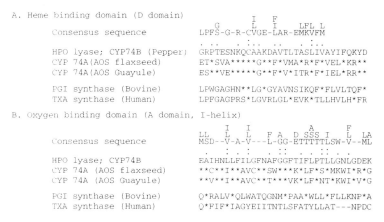

Figure 1. Comparison of heme binding domain and oxygen binding domain of the bell pepper HPO lyase and P450 consensus sequence, or the other P450s specialized for lipid peroxides.

could be found in the molecule, characterizing the bell pepper HPO lyase as a member of the cytochrome P450 superfamily of hemoproteins. The enzyme could be classified in the same family with allene oxide synthase (CYP74). However, each enzyme has discriminative catalytic property, from which a novel subfamily, CYP74B has been created for the bell pepper HPO lyase. As a result, allene oxide synthase was renamed as CYP74A (D. R. Nelson, personal communication). There exist several unique but important modifications to the typical P450 consensus sequence. As the heme binding domain P450s retain the conserved residues FXXGXXXCXG (Song, et al., 1993). Three of the four conserved residues are changed as shown in Fig. 1. Interestingly, five residues surrounding heme binding cysteine (NKQCA) are conserved between bell pepper HPO lyase and two allene oxide synthases. Within the I-helix domain, which is thought to help form the oxygen-binding pocket above the distal surface of the heme, the bell pepper HPO lyase showed little homology to other P450s. As noted with flaxseed allene oxide synthase (Song, et al., 1993), one of the key elements of this oxygen-binding pocket, i.e., G248 in P450CAM, is conserved, but the other, T252 is absent and replaced with isoleucine in the expected position. Purified bell pepper HPO lyase needs neither reducing equivalent of NADPH nor molecular oxygen to accomplish its catalysis. Furthermore, bell pepper HPO lyase showed essentially no affinity against CO. These differences in the structure may contribute to the unique enzymatic properties of HPO lyase. In addition, there are another examples of enzymes that are completely specialized for the metabolism of endogenous peroxide substrates, such as thromboxane synthase and prostacyclin synthase. It should be noted that in the reaction of thromboxane synthase, the C-C bond attaching the endoperoxide is cleaved (Hecker and Ulrich, 1989). The mechanism of HPO lyase can then be speculated in close analogy to thromboxane synthase. The heterologous expression system constructed in this study must be a powerful tool to confirm this mechanism. Furthermore, genetic engineering of a crop with this gene may reveal the unknown physiological role of HPO lyase.

3. References

Altschul, S. F., Gish, W., Miller, W., Myers, E. W. and Lipman, D. J. (1990) Basic local alignment search tool, *J. Mol. Biol.* **215**, 403-410.

Croft, P. K. C., Jüttner, F. and Slusarenko, A. J. (1993) Volatile products of the lipoxygenase pathway evolved from *Phaseolus vulgaris* (L.) leaves inoculated with *Pseudomonas syringae* pv. *phaseolicola*, *Plant Physiol.* **101**, 13-24.

Hecker, M. and Ulrich, V. (1989) On the mechanism of prostacyclin and thromboxane A2 biosynthesis, *J. Biol. Chem.* **264**, 141-150.

Matsui, K., Toyota, H., Kajiwar, T., Kakuno, T. and Hatanaka, A. (1991) Fatty acid hydroperoxide cleaving enzyme, hydroperoxide lyase, from tea leaves, *Phytochemistry* **30**, 2109-2113.

Matsui, K. Kajiwara, T. and Hatanaka, A. (1992) Incativation of tea leaf hydroperoxide lyase by fatty acid hydroperoxide, *J. Agric. Food Chem.* **40**, 175-178.

Matsui, K., Shibutani, M., Hase, T. and Kajiwara, T. (1996) Bell pepper fruit fatty acid hydroperoxide lyase is a cytochrome P450 (CYP74B), *FEBS Lett.* submitted.

Shibata, Y., Matsui, K., Kajiwara, T. and Hatanaka, A. (1995a) Purification and properties of fatty acid hydroperoxide lyase from green bell pepper fruits. *Plant Cell Physiol.* **36**: 147-156.

Shibata, Y., Matsui, K., Kajiwara, T. and Hatanaka, A. (1995b) Fatty acid hydroperoxide lyase is a heme protein. *Biochem. Biophys. Res. Commu.* **207**: 438-443.

Song, W.-C. and Brash, A. R. (1991) Purification of an allene oxide synthase and identification of the enzyme as a cytochrome P-450, *Science* **253**, 781-784.

Song, W.-C., Funk, C. D. and Brash, A. R. (1993) Molecular cloning of an allene oxide synthase: A cytochrome P450 specialized for the metabolism of fatty acid hydroperoxides, *Proc. Natl. Acad. Sci. USA* **90**, 8519-8523.

PRODUCTION OF γ-LINOLENIC ACID BY TRANSGENIC PLANTS EXPRESSING CYANOBACTERIAL OR PLANT Δ^6-DESATURASE GENES

PHILLIP D. BEREMAND, ANDREW N. NUNBERG, AVUTU S. REDDY, AND TERRY L. THOMAS
Department of Biology, Texas A&M University
College Station, TX 77843

Δ^6-desaturase genes were isolated from cyanobacteria and *Borago officinalis*. Introduction of these genes into tobacco under the control of the cauliflower mosaic virus 35S promoter resulted in γ-linolenic acid (GLA) production in leaf lipids. The use of seed specific promoters also resulted in significant levels of GLA in seed lipids.

1. Introduction

The manipulation of plant lipid composition for the production of specialty oils is a rapidly growing field of biotechnology. Notable success has been reported using classical plant breeding techniques [1] and molecular genetic approaches [2]. We have focused on genetically engineering the production of plant oils that contain GLA and octadecatetraenoic acid (OTA). Oils containing GLA and OTA are valuable for nutritional, medical, and industrial applications [3,4]. Few plant oils (evening primrose, borage, and black currant being exceptions) contain GLA due to the lack of a functioning Δ^6-desaturase. This enzyme catalyzes both the desaturation of linoleic acid to form GLA and of α-linolenic acid to form OTA. While rare in higher plants, this activity does occur in several other groups of organisms including certain algae, fungi, and cyanobacteria [5,6,7]. We have cloned genes for Δ^6-desaturase from both cyanobacteria [8,9] and borage [Nunberg, Beremand, and Thomas, manuscript in preparation]. Both genes have been introduced into tobacco resulting in an acquired ability to produce GLA.

2. Cloning and expression of cyanobacterial Δ^6 desaturase

The gene for Δ^6- desaturase was first cloned in our laboratory from the cyanobacterium *Synechocystis* sp strain PCC6803 using an expression based cloning method. A cosmid library was created and clones were mobilized into *Anabaena* sp strain PCC

7102, a cyanobacterium that lacks Δ^6-desaturase activity. Recipient cells were screened for GLA production. Positive clones showed sequence homology to other known desaturase genes [8]. The cyanobacterial gene was placed under the control of a 35S promoter and introduced into tobacco plants. Significant, but low, levels of GLA were produced in leaves of transgenic tobacco [9]. We then directed our efforts at obtaining a Δ^6-desaturase clone from a higher plant source. We focused primarily on cloning and expressing the gene for Δ^6-desaturase from borage, the plant that produces the highest levels of GLA.

TABLE 1. GLA and OTA content of transgenic tobacco leaves expressing cyanobacterial or borage Δ^6-desaturase genes.

Source of gene	Percent of C18 fatty acids ± SD	
	GLA	OTA
Synechocystis	1.2 ± 0.14	2.9 ± 0.14
Borago officinalis	20.4 ± 1.36	9.7 ± 0.97
None	0	0

3. Isolation and expression of a borage Δ^6-desaturase gene

We observed that maximum levels of Δ^6-desaturase activity in borage seeds occurred about 15 days post pollination (DPP). Seeds at this stage of development were therefore collected from hand pollinated flowers for RNA isolation. Total RNA and membrane bound polysomal RNA were isolated. Poly A+ RNA from both RNA populations was used to direct the synthesis of separate cDNA libraries (the vector in both cases was 1 ZAPII from Stratagene). Screening the borage cDNA library with probes derived from the cyanobacterial Δ^6-desaturase gene proved to be ineffective. This may be due to the considerable divergence between cyanobacteria and higher plants. We, therefore, employed several alternative approaches to identify Δ^6-desaturase genes from borage. The most successful of these was the generation of an EST database of 5' sequence tags for messages expressed in borage embryos. Random sequencing of a subtracted, seed library identified a truncated cDNA that had significant similarity to the *Synechocystis* Δ^6-desaturase gene in the GenBank database using the BLASTX algorithm. This cDNA was used to screen the membrane bound polysomal library for a full length clone. A cDNA was identified that had an open reading frame encoding a 448 amino acid protein that has significant similarity to the protein encoded by the *Synechocystis* Δ^6-desaturase gene. This clone was chosen for further examination.

Northern blot analysis indicated that RNA homologous to the putative Δ^6-desaturase began to accumulate at 10 days post pollination. This is consistent with the accumulation of GLA in borage embryos. The putative Δ^6-desaturase was compared to other known desaturases in both higher plants and in cyanobacteria. These alignments also suggested that this isolate was the borage Δ^6 desaturase homologue. A hydropathy plot of the predicted borage protein shares many similar features with plots derived from desaturases that are thought to be localized in the endoplasmic reticulum. Ultimate confirmation of the identity of the borage Δ^6-desaturase gene came from the ability of the clone to confer GLA production in plant systems otherwise lacking that ability. We were able to do this in both transient and stable transformation systems.

A transient assay system was developed using carrot suspension cell protoplasts to detect the activity of the putative gene. Carrot suspension cells were chosen as the source for the protoplasts because of an abundance of the 18:2 fatty acid. Introduction of the borage gene resulted in the transient production of GLA in the transformed carrot protoplasts. These results confirmed the identity of the borage Δ^6-desaturase gene.

The borage Δ^6-desaturase gene was cloned into Bin 19 based vectors under the control of the 35S promoter and introduced into tobacco. Stable transformants were obtained. Analysis of the fatty acid composition of leaf lipids is shown in Table 1. Comparison to the expression of the cyanobacterial enzyme indicates that expressing the borage enzyme resulted in significantly higher levels of GLA. We have also transformed plants with the borage Δ^6-desaturase gene under the control of seed specific promoters. This has resulted in the production of seed oils with significant levels of GLA.

Our results indicate that the engineering of oil seed crops to produce specialty oils containing GLA and/or OTA can be rapidly achieved. We are currently optimizing the expression of the borage gene in seeds of transgenic plants.

4. Acknowledgments

This work was supported in part by grants from the Bio Avenir Program sponsored by Rhône-Poulenc, the Ministry in Charge of Research and the Ministry in Charge of Industry (France).

5. References

1. Ohlrogge,J., J. Browse, and C.R. Somerville.1991. The genetics of plant lipids. Biochim Biophys Acta **1082**:1-26.
2. Töpfer, R., N. Martini, and J. Schell.1995. Modification of plant lipid synthesis. Science **268**:681-686.
3. Craig, M. and M.K. Bhatty.1964. Naturally occurring all cis 6,9,12,15-octadecatetraenoic acid in plant oils. J Amer Oil Chem Soc **41**:209-211.
4. Fan, Y-Y. and R.S. Chapkin.1992.Mouse peritoneal macrophage prostaglandin E_1 synthesis is altered by dietary gamma-linolenic acid. J Nutr **122**:1600-1606.
5. Kyle, D.J.1991. Specialty oils from microalgae: new perspectives, in Rattray, J. (ed), *Biotechnology of Plant Fats and Oils,* American Oil Chemists Society, Champaign, pp130-143.
6. Kennedy, M.J., S.L. Reader, and R.J. Davies.1993. Fatty acid production characteristics of fungi with particular emphasis on gamma linolenic acid production. Biotech and Bioeng **42**:625-634.
7. Murata, N.1989. Low temperature effects on cyanobacterial membranes. J Bioenerget and Biomembranes **21**:61-75.
8. Reddy, A.S., M.N. Nuccio, L.M. Gross, and T.L. Thomas.1993. Isolation of a Δ^6-desaturase gene from the cyanobacterium *Synechocystis* sp. strain PCC 6803 by gain-of-function expression in *Anabaena* sp strain PCC 7120. Plant Mol Biol **27**:293-300.
9. Reddy, A.S. and T.L. Thomas.1996. Expression of a cyanobacterial Δ^6-desaturase gene results in γ-linolenic acid production in transgenic plants. Nature Biotech **14**:639-642.

cDNA CLONING OF CUCUMBER MONOGALACTOSYL DIACYLGLYCEROL SYNTHASE AND THE EXPRESSION OF THE ACTIVE ENZYME IN *ESCHERICHIA COLI*

Hiroyuki Ohta[1], Mie Shimojima[1], Akihiro Iwamatsu, Tatsuru Masuda[1], Fumijiro Kitagawa[1], Yuzo Shioi[1,3] and Ken-ichiro Takamiya[1], [1]Fac. of Biosci. & Biotech., Tokyo Insti. of Tech., Nagatsuta, Yokohama 226, Japan. [2]Kirin Brewery Co. Ltd. Central Labo. for Key Tech., 1-13-5 Fukuura, Kanagawa-ku, Yokohama 236, Japan. [3]Present adress; Dep. of Biol. and Geosci., Fac. of Sci., Shizuoka Univ, 836 Ohya, Shizuoka 422, Japan.

Introduction

Monogalactosyldiacylglycerol, MGDG, is the most abundant lipids in photosynthetic membrane in higher plants and eukaryotic argae. MGDG reached to almost 40 to 50% in thylakoid lipids of the chloroplasts, and therefore, this lipid is considered as the most abundant lipid in the earth. The last step of the MGDG synthesis was catalyzed by MGDG synthase (UDP-galactose:diacylglycerol galactosyltransferase EC2.4.1.46). Digalactosyldiacylglycerol (DGDG), another major lipid in thylakoid membranes, is synthesized by dismutation between two molecules of MGDG. And thus, UDGT is a key enzyme in the formation of thylakoid membranes.

In spite of such an importance of this enzyme, the molecular aspect of the protein still remained unclear. Purification of the enzyme have been tried by several groups for more than ten years. In spinach, the isolation of the UDGT as a single protein was reported by Teucher and Heinz [1] and Maréchal *et al* [2], independently. According to their results, molecular mass of the purified proteins and the properties of the solubilized enzymes were different from each other. Teucher and Heinz [1] proposed that a polypeptide of 22 kDa was associated with the activity. In contrast, Maréchal *et al* [2] reported that they led to 90% enrichment of a 19 kDa polypeptide as the enzyme, and furthermore, no 22 kDa polypeptide could be visualized in the purest fraction. Since these two groups used chloroplast envelopes as the enzyme source, the final amount of the enzyme fraction obtained was quite low, and therefore, further analyses of the purified protein have been almost impossible work.

We have purified UDGT using microsome fraction of cucumber cotyledons [3]. Although the specific activity of the enzyme in microsome fraction was 100 fold lower than that in envelope fraction which has been used in spinach as an enzyme source, total activity obtained was much higher than envelope fraction. Finally, we have found that the purified protein had 47 kDa of the molecular weight which was completely different from those in spinach.

Recently, we cloned this MGDG synthase cDNA for the first time. Subsequently, by the functional expression in *Escherichia coli*, we determined the 47 kDa protein was cucumber MGDG synthase. Here we describe our results of the cloning and functional expression of this enzyme.

Materials and Methods

CLONING OF MGDG SYNTHASE cDNA

A mixture of ^{32}P-labeled degenerate oligonucleotide probes from a 10-amino acid sequence of a peptide (CYCPSTEVAK in Figure) was used to screen a cucumber λgt11 cDNA library (constructed from poly A$^+$ RNA of 5-day-old seedlings illuminated for 6 h).

EXPRESSION OF MGDG SYNTHASE IN E. COLI.

The region for mature protein of MGDG synthase cDNA was amplified by PCR. The amplified cDNA was inserted into the expression vector pGEX-3X to express the MGDG synthase as a fusion to glutathione S-transferase (GST). XL1-Blue was transformed with the constructs for the expression.

ASSAY OF MGDG SYNTHASE ACTIVITY

MGDG synthase activity of cell-free extract of *E. coli* was assayed using 1,2-dioleoyl-*sn*-glycerol (200 μg/ml) and UDP-[^{3}H]-galactose (400 μM) as substrates according to Teucher and Heinz [1]. The analysis of lipids as the reaction products was performed by thin layer chromatography. UDP-[^{14}C]-galactose was used as the substrate to detect the products by image analyzer (fujix BAS2000; Fuji Photo Film).

Results and Discussion

After the purification of the 47 kDa protein as described previously [3], the protein was blotted on PVDF membrane and digested the blotted protein directly by *Achromobacter* protease I which specifically cleaved carboxyl end of Lys residue of the protein. Subsequently, the peptides were separated by HPLC, and these peptides were sequenced by gas-phase protein sequencer [4]. According to the sequences determined, mixed oligonucleotides were synthesized and, a cucumber lgt11 cDNA library was screened. By the screening of 320,000 plaques, an positive clone was isolated.

The *Not*I fragment of the positive clone was subcloned into *Not*I site of pBluescript II SK$^+$ vector, and sequenced. The DNA sequencing revealed that the clone had 1,647 bp and the deduced amino acid sequence had an open reading frame which encodes 422 amino acids (Figure). Most of the peptides had a Lys residue just before the each peptide because of the strict specificity of the *Achromobacter* protease I. However, an peptide which exists in the most upstream region had no lysine residue. Therefore, we determined that here was the N-terminal end of the mature protein. The mature protein has 46,552 Da of the molecular weight. The molecular weight is almost identical to that estimated by SDS-PAGE [3]. The deduced amino acid sequence also indicated that the open reading frame encoded extra N-terminal region corresponding to the transit peptide. However, the clone lacked the initiation codon and 5'-noncoding region.

From these results, it was confirmed that we cloned cDNA which encodes the 47 kDa protein. However, it was still obscure whether this 47 kDa protein was MGDG synthase or not. To clarify it, we expressed the mature region of the cloned cDNA as a fusion protein with GST.

We amplified mature region by PCR using two primers which contained *Bam*HI and *Eco*RI restriction enzyme sites, respectively. After the digestion of the amplified cDNA by *Bam*HI and *Eco*RI, the fragment was ligated with the expression vector pGEX-3X which was also digested with two enzymes. *E. coli*, XL1-Blue was transformed with the vector. Expression of the fusion protein was confirmed by SDS-PAGE.

The enzyme activity was measured as the incorporation of [^{3}H]-galactose into ethyl acetate soluble fraction. The galactose incorporation activity was clearly observed in 1b-2 which expressed the fusion protein. Additionally, after addition of IPTG (1 mM),

```
PRSGASLSLSSRGSSSLRRFVNEFNNVIKFHCHKPPLGFASLGGVSDETN
GIRDDGFGVSQDGALPLNKIEAENPKRVLILMSDTGGGHRASAEAIKAAF
NEEFGNNYQVFITDLWTDHTPWPFNQLPRSYNFLVKHGTLWKMTYYVTAP
KVIHQSNFAATSTFIAREVAKGLMKYRPDIIISVHPLMQHVPIRILRSKG
LLNKIVFTTVVTDLSTCHPTWFHKLVTRCYCPSTEVAKRALTAGLQPSKL
KVFGLPVRPSFVKPIRPKIELRKELGMDENLPAVLLMGGGEGMGPIEATA
KALSKALYDENHGEPIGQVLVICGHNKKLAGRLRSIDWKVPVQVKGFVTK
MEECMGACDCIITKAGPGTIAEAMIRGLPIILNDYIAGQEAGNVPYVVEN
GCGKFSKSPKEIANIVAKWFGPKADELLIMSQNALRLARPDAVFKIVHDL
HELVKQRSFVPQYSG
```

Fig. 1 Deduced amino acid sequence of cucumber MGDG synthase. Amino acid sequences of peptides obtained from the purified protein are underlined. N-terminus of the mature region determined from the N-terminal peptide was indicated by arrow. The highlighted peptide was used to screen cDNA library.

approximate 8-fold increase in MGDG synthetase activity was observed. By contrast, we could not detect any activity of incorporation of [^{3}H]-galactose into lipid fraction in the transformant with vector only. By the TLC analysis of the reaction product, we confirmed that the transformant 1b-2 certainly produced MGDG, dramatically, and that the production was increased by the addition of IPTG. These results described here provide direct evidence that the 47 kDa protein is cucumber MGDG synthase.

References

[1] Teucher, T and Heinz, E. Purification of UDP-galactose:diacylglycerol galactosyl-transferase from chloroplast envelopes of spinach (*Spinacia oleracea* L.) *Planta* 1991;184: 319-326.
[2] Maréchal, É, Block, M.A., Joyard, J, Douce, R. Purification of UDP:galactose 1,2-diacylglycerol galactosyltransferase from spinach chloroplast envelope membranes. *C. R. Acad. Sci. Paris* 1991;313:521-528.
[3] Ohta, H, Shimojima, M, Arai. T., Masuda, T., Shioi, Y., Takamiya, K. UDP-galactose:diacylglycerol galactosyltransferase in cucumber seedlings:Purification of the enzyme and the activation by phosphatidic acid. *Plant Lipid Metabolism* (eds by Kader, J.-C and Mazliak, P.) 1995;152-155.
[4] Iwamatsu, A. S-Carboxymethylation of proteins transferred onto polyvinylidene difluoride membranes followed by in situ protease digestion and amino acid microsequencing.*Electrophoresis* 1992;13:142-147.

DIFFERENTIAL DISPLAY OF mRNA FROM OIL-FORMING CELL SUSPENSION CULTURES OF *BRASSICA NAPUS*

R. J. WESELAKE[1], J.M. DAVOREN[1], S.D. BYERS[1], A. LAROCHE[2], D.M. HODGES[1], M.K. POMEROY[3] AND T.L. FURUKAWA-STOFFER[1]
[1]*Chemistry Department, University of Lethbridge, Lethbridge, AB, T1K 3M4, Canada;* [2]*Agriculture and Agri-Food Canada, Lethbridge, AB, T1J 4B1;* [3]*Plant Research Centre, Agriculture and Agri-Food Canada, Ottawa, ON, K1A 0C6*

1. Introduction

Diacylglycerol acyltransferase (DGAT, EC 2.3.1.20) catalyzes the acylation of *sn*-1,2-diacylglycerol to generate triacylglycerol (TG) [1]. The initial step in the biosynthesis of the fatty acid precursors is the formation of malonyl-CoA from acetyl-CoA and bicarbonate, which is catalyzed by acetyl-CoA carboxylase (ACCase, EC 6.4.1.2.) [1]. In developing seeds of *Brassica napus* L., DGAT [2,3] and ACCase [4] may both represent important control points in the biosynthesis of oil. The current study has examined TG accumulation, the activities of DGAT and ACCase, and gene expression in microspore-derived (MD) cell suspension cultures of *B. napus* L. cv Jet Neuf grown in different sucrose concentrations.

2. Materials and Methods

MD cell suspension cultures of *B. napus* L. cv Jet Neuf were maintained according to Orr *et al.* [5] except the concentrations of sucrose in the culture media were 2, 6 and 14% (w/v). Total lipid (TL) was extracted [6] and TL and TG were quantified by GLC [3]. Microsomal DGAT activity was assayed according to Weselake *et al.* [3] but 5 mg BSA/mL were incorporated in to the reaction mixture and ATP and CoA were omitted. ACCase was assayed in the homogenate according to Kang *et al.* [7] except BSA was omitted from the extraction buffer. Protein content was determined using the Bradford [8] microassay (Bio-Rad). RNA was purified using TRIZOL™ Reagent (Life Technologies). Differential display of mRNA via PCR [9] was conducted using the primer $T_{12}AA$ in combination with arbitrary decamers. cDNA fragments, representing apparently upregulated mRNAs, were excised from the gel, cloned, sequenced, and compared to sequences in GenBank.

3. Results and Discussion

The TG content of the cells increased about 3-fold and 2-fold on a fresh weight (FW) and dry weight basis, respectively, when sucrose concentration was increased from 2 to 6%. The fraction of TG in TL increased from 35 to about 60% when sucrose concentration was increased from 2 to 6% suggesting that more fatty acyl groups were being directed to TG at the expense of other glycerolipids and fatty acids. The activities per unit FW and specific activities of DGAT and ACCase for cells grown at sucrose concentrations of 2, 6 and 14% are shown in TABLE 1. Enhanced TG accumulation was accompanied by a 6-fold increase in microsomal DGAT activity per unit FW when sucrose concentrations were increased from 2 to 14%, whereas ACCase activity per unit FW decreased by 35%.

TABLE 1. Effect of sucrose concentration on the activities of DGAT and ACCase in MD cell suspension cultures. DGAT and ACCase were assayed in the microsomal fraction (10,000 -100,000 g) and homogenate, respectively. Standard errors are based on 4 samples.

Sucrose (%, w/v)	DGAT (pmol/min/ g FW)	(pmol/min/ mg protein)	ACCase (nmol/min/ g FW)	(nmol/min/ mg protein)
2	25.4 ± 11.5	12.2 ± 7.0	3.2 ± 0.2	1.3 ± 0.3
6	99.0 ± 25.1	35.7 ± 4.6	3.4 ± 0.4	1.6 ± 0.5
14	151.7 ± 35.6	31.5 ± 3.9	2.0 ± 0.3	0.7 ± 0.3

Gene expression was analyzed using the differential display method [9] in an attempt to identify mRNA responsible for the increase in the level of DGAT when sucrose concentration was increased from 2 to 14%. cDNA fragments representing mRNAs were selected on the basis of their apparent increasing abundance. Results with the anchored primer $T_{12}AA$ in combination with an arbitrary decamer are shown in *Figure 1*. The deduced amino acid sequence of the open reading frame of the cDNA fragment indicated with an arrow exhibited 96% identity to the last 25 amino acid residues of a mitochondrial acyl carrier protein isoform from *Arabidopsis thaliana* [10]. Thus, the identification of a transcript encoding a protein implicated in lipid biosynthesis suggests that the cell suspension system might be useful for identifying other genetic components representing lipid synthetic machinery, such as an mRNA encoding DGAT.

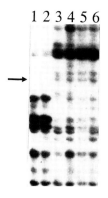

Figure 1. Partial autoradiogram representing differential display of mRNA from MD cell suspension cultures. Lanes 1 and 2 are from cells cultured in 2% (w/v) sucrose, lanes 3 and 4 from cells cultured in 6% sucrose, and lanes 5 and 6 from cells cultured in 14% sucrose. All bands were obtained using the specific anchored primer $T_{12}AA$ in combination with the primer 5'-CAGGCCCTTC-3'. α-[^{32}P]dCTP was used for radiolabeling of DNA. Autoradiography was performed using KODAK X-OMAT AR film for 2 h at -80°C using one intensifying screen.

4. Acknowledgements

This research was supported by the Natural Sciences and Engineering Research Council of Canada and the Alberta Agricultural Research Institute.

5. References

1. Ohlrogge, J.B. and Browse, J.: Lipid biosynthesis, *The Plant Cell* **7** (1995), 957-970.
2. Perry, H.J. and Harwood,J.L.: Radiolabelling studies of acyl lipids in developing seeds of *Brassica napus*: use of [1-^{14}C]acetate precursor, *Phytochemistry* **33** (1993), 329-333.
3. Weselake, R.J., Pomeroy, M.K., Furukawa, T.L., Golden, J.L., Little, D.B., and Laroche, A.: Developmental profile of diacylglycerol acyltransferase in maturing seeds of oilseed rape and safflower and microspore-derived cultures of oilseed rape, *Plant Physiol* **102** (1993), 565-571
4. Turnham, E. and Northcote, D.H.: Changes in the activity of acetyl-CoA carboxylase during rape-seed formation, *Biochem. J.* **212** (1983), 223-229.
5. Orr, W., Keller, W.A., and Singh, J.: Induction of freezing tolerance in an embryogenic cell suspension culture of *Brassica napus* by abscisic acid at room temperature, *J. Plant Physiol.*, **126** (1986), 23-32.
6. Hara, A. and, Radin, S.: Lipid extraction of tissues with a low-toxicity solvent, *Anal. Biochem.*, **90** (1979), 420-426.
7. Kang, F., Ridout, C.J., Morgan, C.L., Rawsthorne, S. The activity of acetyl-CoA carboxylase is not correlated with the rate of lipid synthesis during development of oilseed rape (*Brassica napus* L.) embryos, *Planta* **193** (1994), 320-325.
8. Bradford, M.M.: A rapid and sensitive method for the quantitation of microgram quantities of protein utilizing the principle of protein-dye binding, *Anal. Biochem.* **72** (1976), 248-254.
9. Liang, P. and Pardee, A.B.: Differential display of eukaryotic messenger RNA by means of the polymerase chain reaction, *Science* **257** (1992), 967-971.
10. Shintani, D.K. and Ohlrogge, J.B.: The characterization of a mitochondrial acyl carrier protein isoform isolated from *Arabidopsis thaliana*, *Plant Physiol.* **104**, (1994) 1221-1229.

ANALYSIS OF THE ARABIDOPSIS ENOYL-ACP REDUCTASE PROMOTER IN TRANSGENIC TOBACCO

Gert-Jan de Boer[1], Tony Fawcett[2], Antoni R. Slabas[2], H. John J. Nijkamp[1] and Antoine R. Stuitje[1]
[1]Department of Genetics, Vrije Universiteit, Institute for Molecular Biological Sciences, BioCentrum Amsterdam, de Boelelaan 1087, 1081 HV Amsterdam, The Netherlands.
[2]Department of Biological Sciences, The University of Durham, Durham DH1 3LE, United Kingdom.

Introduction

During the life cycle of a plant, the demand for the de novo synthesis of lipids is variable and largely dependant on the type and developmental state of the tissue. For example, during seed development the expression of genes encoding the different Fatty Acid Synthetase (FAS) components is increased dramatically (1). Whilst the synthesis of both storage lipids and membrane lipids relies on the action of the FAS complex, lipid synthesis is most likely controlled by differential expression of housekeeping FAS genes rather than switching on (a) distinct seed specific gene(s). However, some FAS components like the acyl carrier protein (ACP) are encoded by a small gene family whose members are tissue specifically expressed.(2,3), whereas for other components of FAS encoding components there is no apparent distinction between housekeeping genes and tissue specific or seed expressed genes. For example, the NADH-dependent enoyl-ACP reductase (ENR), which catalyses the last of two reducing steps in each cycle of FAS to generate a saturated acyl-ACP, was shown to be encoded by a single gene in *Petunia hybrida* and *Arabidopsis thaliana*. (1,5). cis-acting elements are therefore expected to be present in the promoter of the *enr* gene, which allow trans-acting transcriptional control elements to coordinate the expression of the *enr* gene with that of other FAS components during life cycle of the plant. In the present study we have attempted to identify these cis-acting elements, by fusing specific regions of the Arabidopsis *enr* promoter to the *uidA* (GUS) reporter gene (4) and analysing the GUS activity in transgenic tobacco.

Results
Construction of a 5' promoter deletion series of the Arabidopsis enoyl-ACP reductase promoter.
A promoter deletion series was constructed, either by PCR or by making use of appropriate restriction sites present in the Arabidopsis *enr* promoter (fig 1). In general, the position of the 5' end-points of the deletion series was based on short regions of homology, which had been identified by comparison of the promoter sequences of the Arabidopsis *enr* with that of the *enr* genes of *Nicotiana tabacum* (5). In addition, a gene

construct was made that only contained 45 base pairs upstream of the transcription start site and also lacks the intron which is normally present in the untranslated leader of the gene.

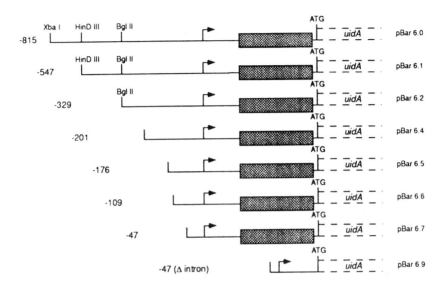

Fig. 1) Schematic representation of a 5' deletion series of the Enoyl-ACP reductase promoter from *Arabidopsis thaliana* fused to the *uidA* (GUS) reporter gene. Numbers indicate the relative distances to the transcription start site.

Expression patterns of the different promoter deletion constructs.

Transgenic Tobacco plants were analysed for GUS activity, using both fluorimetric assays and histochemical staining using X-Gluc. In general, the constructs showed no detectable expression in leaf tissue but high expression in developing seeds. The expression pattern of an 815bp ENR promoter-*uidA* construct was analysed during seed development. Four different stages were defined essentially as described by deSilva et al. (1992). Results indicate that the highest uidA activity was found in stage 3 of seed development. These findings are in agreement with the results found for the ACP05 promoter (2) (data not shown). Analysis of the GUS activity at this stage of seed development, of the various promoter deletion constructs, showed no significant difference in activity (data not shown). This indicates that the element responsible for the marked increase in expression during seed maturation, may either be located within 47 bp upstream of the transcription start or in the untranslated leader of the gene.

Expression of ENR *uidA* constructs during seedling development.

Since Enoyl-ACP reductase is encoded by a single gene, we examined the expression pattern of the reporter gene driven by the different promoter deletions during seedling development, in order to follow the developmental expression of this housekeeping gene. In very young seedlings GUS activity can be observed in the roots, stem and the cotyledons. But staining in the roots decreases with the ageing of the seedlings and only remains in the vascular tissue and the root apical meristem. A similar effect is observed in the other parts of the plant, where GUS activity is present in the vascular tissue which fades away when the tissue matures. Upon the development of the shoot apical meristem and the onset of primary leaf development, intense staining is observed which fades away during maturation of the primary leaves. This expression pattern is repeated upon the formation of the next set of leaves and so on. Subsequent deletion of a 128 bp fragment (-329 to -201 relative to the transcription start site) leads to the loss of expression in the shoot apical meristem but does not appear to affect the expression in the root meristem. Therefore, a cis-acting element that specifically regulates *enr* expression in the shoot apical meristem may be confined within this region. Another striking change in the expression pattern was observed after deletion of the intron from the untranslated leader in the promoter fragment. This construct clearly shows a more intense staining of the tissue in the mature parts of the root when compared to the identical construct in which the intron is still present.

Conclusions

In arabidopsis and other plant species like petunia and tobacco, the NADH-dependent enoyl-ACP reductase, which is a key component in each cycle of FAS, is encoded by a single gene (*enr*). As a consequence, the expression of this housekeeping gene must be tightly regulated to satisfy the demands for the enzyme during plant development in tissues with a high level of FAS activity, e.g. meristem and seed. In this study we have shown the presence of at least two distinct elements, within the arabidopsis *enr* promoter, which can positively influence the expression of the gene in the shoot apical meristem or in developing seed. Another aberrant pattern of expression was observed after deletion of an intron, which is normally present in the untranslated leader of the gene. Although it cannot be excluded that this altered expression pattern is due to a difference at the post-transcriptional level, this deletion does not seem to affect the expression of the construct in parts of the plant other than the roots. Therefore the intron might contain a cis-acting element which is involved in controlling transcriptional activity to a level sufficient for housekeeping functions. Another striking observation is the low or non-detectable expression of the transgene in leaves of transgenic Tobacco using fluorimetric assays. Preliminary studies with similar constructs in rapeseed, however, do seem to indicate that the transgene is expressed in the leaves of these plants (Wyatt Paul, personal communication). These results are a starting point for future experiments aimed at the indentification of genes involved in the transcriptional control of enoyl-ACP reductase.

References

1 Kater, M.M., Koningstein, G.M., Nijkamp, H.J.J. and Stuitje, A.R. (1991) cDNA cloning and expression of *Brassica napus* enoyl-acyl carrier protein reductase in *Escherichia coli*, Plant Mol. Biol. **17**, 895-909

2 deSilva, J., Robinson, S.J. and Safford, R. (1992) The isolation and functional characterisation of a B. napus acyl carrier protein 5' flanking region involved in the regulation of seed storage lipid synthesis, Plant Mol. Biol. **18**, 1163-1172

3 Baerson, S.R. and Lamppa, G.K. (1993) Developmental regulation of an acyl carrier protein gene promoter in vegetative and reproductive tissues, Plant Mol Biol. **22**, 255-267

4 Jefferson, R.A., Kavanaugh, T.A. and Bevan, M.W. (1987) GUS fusions: ß-glucuronidase asa sensitive and versatile gene fusion marker in higher plants, EMBO J. **6**, 3901-3907

5 de Boer, G.J., Kater, M.M. Wagenaar, A.R. Fawcett, T. Slabas, T.R. Nijkamp, H.J.J. and Stuitje , A.R. (1995) Structure of plant enoyl-ACP reductase genes. Plant Lipid Metabolism, Kluwer Academic Publishers, Dordrecht, pp 467-469

MOLECULAR BIOLOGY OF BIOTIN-CONTAINING ENZYMES REQUIRED IN LIPID METABOLISM

JOONG-KOOK CHOI, JINSHAN KE, ANGELA L. MCKEAN, LISA M. WEAVER, TUAN-NAN WEN, JINGDONG SUN, TOMAS DIEZ, FEI YU, XUENI GUAN, EVE S. WURTELE AND BASIL J. NIKOLAU,
Departments of Biochemistry and Biophysics, and Botany, Iowa State University, Ames, Iowa.

The water soluble vitamin, biotin, acts as a cofactor for a set of enzymes that catalyze carboxylation, decarboxylation or transcarboxylation reactions (Moss and Lane, 1971). In plants we have characterized four biotin-containing enzymes each of which catalyze reactions required in lipid metabolic processes. These enzymes are the homomeric and heteromeric isozymes of acetyl-CoA carboxylase (ACCase), methylcrotonyl-CoA carboxylase (MCCase) and geranoyl-CoA carboxylase (GCCase). The studies of these biotin-containing enzymes has led to an interest in biotin biosynthesis, and we have cloned the gene coding for biotin synthase.

1. Acetyl-CoA carboxylase

ACCase catalyzes the ATP-dependent carboxylation of acetyl-CoA, to form malonyl-CoA. Malonyl-CoA is the activated two-carbon unit which is used as the substrate in the biosynthesis of a variety of polyketide derivatives, including fatty acids, flavonoids, and stilbenoids. In addition, malonyl-CoA is used as the substrate for the malonation of a variety of plant phytochemicals. These metabolic fates of malonyl-CoA are separated into spatially and temporaly distinct compartments of the plant. Since malonyl-CoA cannot readily cross membranes, isozymes of ACCase that accumulate in different compartments of the plant provide an independent means of generating malonyl-CoA in each compartment. Depending upon the plant species that is examined, these isozymes of ACCase have homomeric or heteromeric quaternary structures.

1.1 HOMOMERIC ACCase

All plants contain the homomeric ACCase, which is a dimeric enzyme composed of 250-kDa biotin-containing subunits. With the exception of Gramineae, the homomeric ACCase appears to be located predominantly in the cytosol of plant cells. Furthermore, in pea (Alban et al., 1994) and leek (Caffrey, 1995) the homomeric ACCase is concentrated in the epidermal cells of the leaf.

In *Arabidopsis thaliana* two genes code for the homomeric ACCase, *ACC1* and *ACC2* (Yanai et al., 1995). These two genes are tandemly arranged in a 25-kb interval in the middle of the chromosome 1. The two genes code for proteins that are very similar in primary sequence (over 90% identical sequence). The major difference between these two genes is in the structure of the fourth intron; in *ACC2* this intron contains a 2-kb insertion relative to the homologous intron of *ACC1*. The origin of this insertion is unclear.

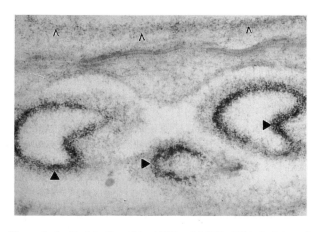

Figure 1. *In situ* detection of the *ACC1* and *ACC2* mRNAs in tissues of developing siliques. These mRNAs, which code for the homomeric ACCase accumulate to highest level in a subset of the cells of the inner integument of the ovules (▲) and the epidermis of siliques (^).

We have used *in situ* hybridization to investigate the spatial and temporal distribution of the *ACC1* and *ACC2* mRNAs in flowers, leaves, and developing siliques of *A. thaliana*. The probe that was used in these studies could not distinguish between the two mRNAs, thus both gene products were detected. The *ACC1* and *ACC2* mRNAs accumulate to high levels in developing embryos, developing carpels, and epidermal cells of the petals and leaves. The highest level of accumulation is in cells of the inner integument of the ovules, just prior to deposition of an as yet chemically-undefined brown testal wall component by these cells.

1.2 HETEROMERIC ACCase

The plastid-localized ACCase generates malonyl-CoA for *de novo* fatty acid biosynthesis. In most non-Gramineae species this ACCase has a heteromeric structure being composed of four different types of subunits. These are the biotin carboxyl carrier (BCC) subunit, the biotin carboxylase (BCase) subunit, and the α (α-CTase) and β (β-CTase) subunits of carboxyltransferase. The first three of these subunits are coded by nuclear genes, which in *A. thaliana* are *CAC1*, *CAC2*, and *CAC3*, respectively. The β-CTase subunit is coded by the chloroplastic *accD* gene. We have isolated *CAC1*, *CAC2* and *CAC3* and are determining their structures, and characterizing their expression patterns during seed development.

The *CAC1* gene is encompassed on a 3.7-kb genomic fragment, and the gene is interrupted by six introns (Ke et al., 1996). The *CAC2* gene is encompassed on a 5-kb genomic fragment, and is interrupted by at least 14 introns. The *CAC3* gene is encompassed on a 8-kb genomic fragment, and is interrupted by at least 10 introns. *In situ* investigations of the expression of these three genes indicate that they are coordinately expressed during silique development. Immunohistochemical examinations of the developing siliques of *A. thaliana* indicate that the BCC subunit accumulates to highest levels within developing embryos in which seed oils are being deposited. *In situ* hybridization analyses of the *CAC1*, *CAC2* and *CAC3* mRNAs indicate that they accumulate coordinately during silique development, and coincidentally increase as seed oil is deposited. Namely, early in silique development, between 1 and 3 days after flowering (DAF) all three mRNAs accumulate relatively evenly throughout the tissues of the silique. Starting at 5 DAF, until 9 DAF, as the embryo develops from the heart stage to the torpedo and elongated torpedo stages (the period during which seed oil is deposited) the accumulation of these three mRNAs dramatically increase within the embryos. By 11 DAF, when seed oil deposition is, the accumulation of the *CAC1*, *CAC2* and *CAC3* mRNAs decreases, and by 12 and 13 DAF they are barely detectable. Transgene experiments are underway to determine the molecular genetic basis for this developmental pattern of expression of the genes coding for the heteromeric ACCase.

2. Methylcrotonyl-CoA carboxylase

The carboxylation of methylcrotonyl-CoA to form methylglutaconyl-CoA is catalyzed by MCCase. The role of this reaction in plant metabolism is as yet unclear. However, extrapolation from studies with animals and bacteria would suggest that MCCase may have a role in a set of metabolic processes that link amino acid and lipid metabolism. Namely, MCCase may be involved in the mitochondrial metabolism of leucine, mevalonate and noncyclic isoprenoids.

MCCase is a dodecomeric enzyme with an $\alpha_6\beta_6$ quaternary structure (Chen et al., 1993; Diez et al., 1994). The larger of the two subunits (75-85 kDa) contains biotin, whereas the smaller subunit (55-65 kDa) does not. We had previously cloned cDNAs and genes coding for the biotin-containing subunit of MCCase (Wang et al., 1994; Song et al., 1994; Weaver et al., 1995). From the deduced amino acid sequence we surmised that the larger subunit contained the biotin carboxylase and biotin-carrier domains. We have now cloned and sequenced the full-length cDNA and gene coding for the smaller, non-biotinylated subunit of MCCase of *A. thaliana*. The gene coding for this subunit is encompassed on a 4.5-kb genomic fragment, and is interrupted by 9 introns. The deduced amino acid sequence shows a high degree of homology with carboxyltransferase domains of other biotin-containing enzymes. The highest homology is with enzymes that either utilize as a substrate, or make as a product, branched-chain acyl-CoAs, as does MCCase. Namely, the highest homology is with methylmalonyl-CoA decarboxylase, propionyl-CoA carboxylase, and methylmalonyl-CoA-pyruvate transcarboxylase.

3. Geranoyl-CoA carboxylase

GCCase is a biotin-containing enzyme that carboxylates the γ-methyl group of geranoyl-CoA to form carboxygeranoyl-CoA. This enzyme has been characterized only from *Pseudomonas* species, where it is involved in the degradation of non-cyclic isoprenoids such as farnesol, geraniol and citronnellol. In plants, non-cyclic isoprenoids, such as the phytol chain of chlorophyll, are significant cellular constituents, which undergo turnover during the normal growth cycle of the organism. However, little is known about the biochemistry of isoprenoid degradation in plants.

Our recent finding that plants contain GCCase indicates that in these organisms the catabolism of non-cyclic isoprenoids may be analogous to that described for bacterial species. Namely, following the carboxylation of the γ-methyl group, the carboxymethyl branch-group is eliminated. The resulting β-keto thioester would then be amenable to β-oxidation. Ultimately such a process would generate methylcrotonyl-CoA, whose catabolism would require MCCase.

GCCase partially purified from maize seedlings has a biotinylated subunit of about 120 kDa. GCCase activity has been detected in extracts of all plant species examined, including soybeans and carrot. Furthermore, a 120-kDa biotin-containing polypeptide has been detected widely in plant extracts, which would indicate that GCCase is widely distributed in the plant kingdom.

4. Biotin synthase

The post-translational biotinylation of biotin-containing enzymes is absolutely required for the activation of these enzymes. Indeed, biotinylation of biotin-containing enzymes is a potential locus for the regulation of their activities. We have previously shown that biotinylation of MCCase is an important regulatory mechanism by which the activity of this enzyme is determined in leaves and roots of tomato. Specifically, we found that even though the biotinylated subunit of MCCase accumulates to equal levels in roots and leaves, in the latter a significant proportion of the protein accumulated in its apo-form, which explained the lower MCCase activity that was found in extracts of leaves in comparison to roots.

These observations stimulated our interest to investigate the supply of biotin in plants. Biotin synthase catalyzes the terminal reaction of biotin biosynthesis, the insertion of a sulfur atom into dethiobiotin. We identified an *A. thaliana* EST cDNA clone that showed a high degree of sequence identity with the bacterial biotin synthase. We isolated a full-length cDNA, and the single *A. thaliana* gene (*BIO2*) that codes for this EST (Weaver et al., 1996). Proof that this cDNA did indeed code for biotin synthase was obtained by expressing it in *Escherichia coli* and genetically complementing a mutation in the *bioB* gene of this bacterium, which codes for biotin synthase.

The *BIO2* gene of *A. thaliana* is interrupted by five introns. The gene codes for a protein of 379 amino acids, with a predicted molecular mass of 41 kDa. The *A. thaliana* biotin synthase has an N-terminal extension (compared to the bacterial enzyme), which has features indicating it may act as a mitochondrial transit peptide. This would indicate that biotin biosynthesis occurs within mitochondria of plant cells. Indeed, biotin biosynthesis may be derived from pimelic acid, which may be a product of a mitochondrial fatty acid synthase. *In situ* analyses of the accumulation of the biotin synthase mRNA indicates that the spatial pattern of accumulation changes during the development of the siliques. Namely, whereas the mRNA is present throughout the tissues of the silique during the early stages of development (1-10 DAF), later as the silique matures and dessicates the biotin synthase mRNA concentrates in the outer epidermal cells of the silique.

5. References

Alban, C., Baldet, P., and Douce, R. (1994) Localization and characterization of two structurally different forms of acetyl-CoA carboxylase in young pea leaves, of which one is sensitive to aryloxyphenoxyproprionate herbicides, *Biochem. J.* **300**, 557-565.

Caffrey, J.J. (1995) Biochemical and Molecular Biological Characterization of Plant Acetyl-CoA Carboxylases. Ph.D. Thesis, Iowa State University.

Chen, Y., Wang, X, Nikolau, B.J., and Wurtele, E.S. 1993. Purification and characterization of 3-methylcrotonyl-CoA carboxylase from carrot. *Arch. Biochem. Biophys.* **305**, 103-109.

Diez, T.A., Wurtele, E.S., and Nikolau, B.J. (1994) Purification and characterization of 3-methylcrotonyl-Coenzyme A carboxylase from leaves of *Zea mays*. *Arch. Biochem. Biophys.* **310**, 64-75.

Ke, J., Choi, J-K., Smith, M., Horner, H.T., Nikolau, B.J., and Wurtele, E.S. (1996) Structure and *in situ* characterization of the expression of *CAC1*: the *Arabidopsis thaliana* gene coding for the biotin-containing subunit of the plastidic acetyl-CoA carboxylase, submitted for publication.

Moss, J. and Lane, M.D. (1971) The biotin-dependent enzymes, *Adv. Enzymol.* **35**, 321-442.

Song, J., Wurtele, E.S., and Nikolau B.J. (1994) Molecular cloning and characterization of the cDNA coding for the biotin-containing subunit of 3-methylcrotonyl-CoA carboxylase: Identification of the biotin carboxylase and ciotin-carrier domains. *Proc. Nat. Acad. Sci. U.S.A.* **91**, 5779-5783.

Wang, X., Wurtele, E.S., Keller, G., McKean, A.L., and Nikolau B.J. (1994) Molecular cloning of cDNAs and genes coding for β-methylcrotonyl-CoA carboxylase of tomato. *J. Biol. Chem.* **269**, 11760-11769.

Weaver L.M., Yu, F., Wurtele, E.S., and Nikolau B.J. (1996) Characterization of the cDNA and gene coding for the biotin synthase of *Arabidopsis thaliana*. *Plant Physiol.* **110**, 1021-1028.

Weaver, L.M., Lebrun, L., Wurtele, E.S., and Nikolau B.J. (1995) 3-Methylcrotonyl-CoA carboxylase of *Arabidopsis thaliana*: Isolation and characterization of cDNA coding for the biotinylated subunit. *Plant Physiol.* **107**, 1013-1014.

Yanai, Y., Kawasaki, T., Shimada, H., Wurtele, E.S., Nikolau, B.J., and Ichikawa, N. (1995) Genetic organization of the 251 kDa acetyl-CoA carboxylase genes in *Arabidopsis*: Tandem gene duplication has made two differentially expressed isozymes. *Plant Cell Physiol.* **36**, 779-787.

OVER-EXPRESSION, PURIFICATION AND CHARACTERIZATION OF AN ACYL-CoA BINDING PROTEIN FROM *BRASSICA NAPUS* L.

ADRIAN P. BROWN, PHILIP E. JOHNSON and MATTHEW J. HILLS
Department of Brassica and Oilseeds Research, John Innes Centre, Colney, Norwich NR4 7UH, U.K.

Introduction

Acyl-CoA binding proteins (ACBPs) are small proteins of about 10kDa which bind acyl-CoAs but not free fatty acids. ACBPs have been purified from a number of animal sources and more recently gene sequences from yeast and *B. napus* reported. The binding of acyl-CoAs by bovine ACBP has been extensively studied and the results show that long chain acyl-CoAs (14-22 carbon atoms) are bound with high affinity [1]. ACBP can facilitate the transfer of acyl-CoA moieties from phosphatidylcholine membranes immobilised on nitrocellulose to mitochondria or microsomes and the transported acyl-CoAs can be utilised for β-oxidation or glycerolipid biosynthesis [2]. In order to aid investigation of the possible role of ACBPs in plant lipid biosynthesis we wanted to obtain large amounts of pure ACBP encoded by a cDNA from *B. napus* [3].

The vector pET15b was used to over-express the protein in *E. coli* and ammonium sulphate precipitation followed by gel filtration enabled isolation of apparently pure ACBP. It was subsequently shown however that this consisted of a mixture of two isoforms which could be separated by chromatofocusing. An antibody was raised against the mixed protein isolated after gel filtration and western blot analysis used to investigate the expression of ACBP in different tissues of *B. napus*. The effect of adding ACBP to glycerol-3-phosphate-acyltransferase (G-3-P-AT) assays using microsomes isolated from developing *B. napus* embryos was also studied. At a fixed concentration of oleoyl-CoA, addition of ACBP stimulates the incorporation of labelled glycerol-3-phosphate into lipids initially but the protein then appears to become inhibitory as the ratio of ACBP to acyl-CoA increases.

Materials and Methods

Over-expression of ACBP using the vector pET15b was carried out using protocols in the pET system handbook (Novagen). Gel filtration was carried out with a HiLoad™ 16/60 column containing Superdex™ 75 (Pharmacia) with elution in 100mM NaCl, 20mM Tris.HCl pH8.0. Chromatofocusing used a Mono P HR5/20 column with a pH gradient of 6.0 to 4.0 was as described in the manufacturers instructions (Pharmacia).

For G-3-P-AT assays, microsomes were prepared from developing embryos of *B. napus* (cv. Topas) as follows. Embryos were ground in extraction buffer containing 0.24M sucrose, 10mM KCl, 5mM EDTA, 2mM DTT and 100mM HEPES-NaOH (pH7.5) and the extract centrifuged at 10 000g for 10 minutes. The supernatant was then centrifuged at 20 000g for 20 minutes and then microsomes pelleted from the supernatant at 200 000g for 30 minutes. The microsomal fraction was resuspended in a small volume of extraction buffer for use in assays. G-3-P-AT activity was assayed in 200µl reactions containing 500µM G3P, 8mM $MgCl_2$, 60mM KH_2PO_4-NaOH (pH7.5), 1.85 kBq [^{14}C]-G-3-P and varying concentrations of oleoyl-CoA and ACBP. Reactions were started by addition of 100µg microsomal protein and incubated at

30°C for 10 minutes before extraction of lipid following addition of 0.7ml chloroform:methanol (1:1) and 0.6ml 1M KCl, 0.2M phosphoric acid. Assays were performed in triplicate and lipid extracts dried and analysed by scintillation counting.

Results

OVER-EXPRESSION AND PURIFICATION OF ACBP

The 92 amino acid open reading frame from Bn411 [3] was inserted into pET15b to express a protein without an additional histidine tag. Transfer of the plasmid into *E. coli* strain BL21(DE3) allowed efficient expression of *B. napus* ACBP, which was in the supernatant fraction after cell lysis and centrifugation at 15000g. Initial purification was by ammonium sulphate precipitation and ACBP was predominantly in the 80-100% saturation pellet. This contained more than 90% ACBP and contaminating proteins were of higher molecular mass. Gel filtration resulted in a pure protein of about 10kDa as analysed by SDS-PAGE and typical yields were approximately 20 mg ACBP per litre of *E. coli* culture. This protein sample was used to raise an antibody in a rabbit.
Western analysis of native isoelectric-focusing (IEF) gels showed that the apparently pure ACBP contained two proteins with differing pI's. Chromatofocusing allowed resolution of the two proteins and these, together with the mixture after gel filtration were analysed by electrospray mass spectrometry. Two proteins, of relative molecular mass 10170 and 10039 were present in the sample after gel filtration and these corresponded to the two peaks resolved after chromatofocusing. The difference in mass between the two proteins is equivalent to one methionine residue and protein sequencing showed that the proteins eluted from the chromatofocusing column had the N-terminal sequences MGLK and GLKD. This data strongly suggests that the ACBP eluting from the gel filtration column contains two isoforms which are the result of incomplete cleavage of methionine from ACBP by the methionine amino peptidase of *E. coli*. Incomplete cleavage of the N-terminal methionine has been noted before when yeast ACBP was over-expressed in *E. coli* [4]. Analysis of binding of acyl-CoAs by the two isoforms of ACBP using the lipidex 1000 binding assay [5] demonstrated that both isoforms can bind palmitoyl- or oleoyl-CoAs at a binding ratio of 1:1. Amino acid analysis was carried out on the mixed ACBP protein to enable calibration of the amount of ACBP added to assays and used as standards on western blots. The results showed that Bradford assays using gamma-globulin as a standard over-estimated the amount of ACBP present by a factor of approximately 1.9.

WESTERN ANALYSIS

The *B. napus* ACBP antibody was used in analysis of both native and denaturing IEF gels (pH range 4.0-6.0) of the ACBP eluting from the gel filtration column. Native gels demonstrated the presence of two bands whereas denaturing gels resulted in only one band. This result suggests that there are two protein isofoms present which have different pI's due to different conformational states. Denaturing IEF westerns of protein samples from a number of different plant organs indicated the presence of only one major ACBP isoform throughout *B. napus*, even though there are six ACBP genes in the genome. ACBP is present in all the organs of *B. napus* tested and in developing embryos is apparent before oil biosynthesis starts. Similarly in cotyledons of germinating seedlings, ACBP appears three days after imbibition, at which stage approximately half of the lipid present initially has been catabolised. These results suggest that ACBP expression is not closely correlated with oil deposition or breakdown in *B. napus* and the protein has a more fundamental housekeeping role.

EFFECT OF ACBP ON G-3-P-AT ACTIVITY

The mixed ACBP protein was added to assays in increasing amounts at fixed oleoyl-CoA concentrations. The results are shown in Fig. 1. It can be seen that addition of ACBP increases the activity of G-3-P-AT initially but as the ratio of ACBP to acyl-CoA increases the protein appears to become inhibitory. It is not clear if the stimulatory effect of ACBP is due to the fact that acyl-CoA/ACBP complexes are a preferred substrate for the enzyme compared to free CoA or whether ACBP prevents the formation of micelles, which cannot be used by the acyltransferase enzymes. Inhibition of acyltransferase activity by ACBP was unexpected and a possible explanation for this is that free ACBP and acyl-CoA/ACBP complexes compete for binding to receptors or acyltransferase enzymes in the microsomes. Experiments to further investigate the effects of added ACBP on lipid synthesis by *B. napus* microsomes are currently underway.

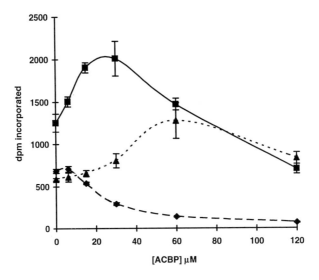

Figure 1. Effect of [ACBP] on microsomal G-3-P-AT activity from rape embryos. The graph shows the amount of [^{14}C]-G-3-P converted into lipid with varying amounts of ACBP added at fixed oleoyl-CoA concentrations. ♦— — —♦ corresponds to an oleoyl-CoA concentration of 10μM, ■———■ to 50μM and ▲- - - -▲ to 100μM.

Acknowledgements

We thank Dr Mike Naldrett at the Protein Sequencing Facility, Nitrogen Fixation Laboratory, John Innes Centre for N-terminal sequencing and Ian Moss at the Advanced Centre for Biotechnology at the Charing Cross and Westminster Medical school for amino acid analysis.

References

1. Rosendal J, Ertbjerg P and Knudsen J (1993) Characterization of ligand binding to acyl-CoA-binding protein, *Biochem J.* **290**: 321-326
2. Rasmussen JT, Faergeman NJ, Kristiansen K and Knudsen J (1994) Acyl-CoA-binding protein (ACBP) can mediate intermembrane acyl-CoA transport and donate acyl-CoA for β-oxidation and glycerolipid synthesis, *Biochem J.* **299**: 165-170
3. Hills MJ, Dann R, Lydiate D and Sharpe A (1994) Molecular cloning of a cDNA from *Brassica napus* L. for a homologue of acyl-CoA-binding protein, *Plant Mol. Biol.* **25**: 917-920
4. Knudsen J, Faergeman NJ, Skøtt H, Hummel R, Børsting C, Rose TM, Andersen JS, Højrup P, Roepstorff P, Kristiansen K (1994) Yeast acyl-CoA-binding protein: acyl-CoA binding affinity and effect on intracellular acyl-CoA pool size, *Biochem J.* **302**: 479-485
5. Rasmussen JT, Börchers T and Knudsen J (1990) Comparison of the binding affinities of acyl-CoA-binding protein and fatty-acid-binding protein for long-chain acyl-CoA esters, *Biochem J.* **265**: 849-855

SEQUENCE ANALYSIS OF CTP:PHOSPHOCHOLINE CYTIDYLYLTRANSFERASE cDNA FROM *ARABIDOPSIS THALIANA*

SUNG HO CHO[1], SANG-BONG CHOI[1,2], JOON CHUL KIM[3], AND KWANG-WOONG LEE[2]

[1]Department of Biology, Inha University, Inchon, Korea; [2]Department of Biology, Seoul National University, Seoul, Korea; and [3]Department of Biology, Kangwon National University, Choonchon, Korea

Introduction

Phosphatidylcholine (PC) is the major lipid component of eukaryotic membranes. The pathway of *de novo* synthesis of PC is composed of consecutive three reactions, and CTP:phosphocholine cytidylyltransferase (CT; EC 2.7.7.15) catalyzes the second reaction, which is generally believed to be the key regulatory step of the pathway. In mammals CT exists both as an active membrane-bound form and as an inactive cytosolic form, and the interconversion between them is believed to regulate the overall formation of PC. As one of the steps to elucidate the regulatory mechanism for PC synthesis in plants, the sequence of *A. thaliana* CT cDNA was determined.

Materials and Methods

cDNA library was obtained from ABRC, Ohio State University, and screened with a DIG-labelled probe made from P11T7 containing partial sequence of CT.

Results and Discussion

The *A. thaliana* CT cDNA is 1447 bp long and contains an open reading frame of 993 bp coding for a protein of 331 amino acids (Fig. 1). The deduced structure of *A. thaliana* CT is composed of three main regions, the catalytic domain in the N-terminal half, the hydrophilic C-terminal region, and the amphipathic domain in the middle.

The catalytic domain region (32~195) was highly conserved among different organisms, showing 76 and 72% homology with the rat and yeast protein sequences, respectively. The hydropathy profile revealed that the C-terminal non-catalytic portion of the protein was very hydrophilic, highly enriched in negatively charged aspartic acid and glutamic acid residues. In the region between the catalytic domain and the C-terminal region, there was an amphipathic α-helical domain (225~260), which was believed to bind the membrane surface in the active formation (Kalmar *et al.*, 1990) (Fig. 2). This may suggest that the regulation of the enzyme activity in plants is also controlled by reversible translocation between the cytosol and the microsomal membranes.

```
AAAATTAAAAAAAAAAAGAGTAGGAGAAGAGGGAAGCGACTAGCACCTTTTGTAGTTTTC    60
CGTTTATTTTCTGTATAAGGCGGGTGATCTCGGCTCCTTCATCGGAAATTATGAGCAACG   120
                                                  M S N V       4
TTATCGGCGATCGAACTGAAGACGGCCTTTCCACCGCCGCTGCCTCTGGCTCTACGGCTG   180
 I G D R T E D G L S T A A A S G S T A V                         24
TCCAGAGTTCTCCTCCCACTGATCGTCCTGTCCGCGTCTACGCCGATGGGATCTACGATC   240
 Q S S P P T D R P V R V Y A D G I Y D L                         44
TTTTCCACTTTGGTCATGCTCGATCTCTCGAACAAGCCAAATTAGCGTTTCCAAACAACA   300
 F H F G H A R S L E Q A K L A F P N N T                         64
CTTACCTTCTTGTTGGATGTTGCAATGATGAGACTACCCATAAGTACAAGGGAAGGACTG   360
 Y L L V G C C N D E T T H K Y K G R T V                         84
TAATGACTGCAGAAGAGCGATATGAATCACTTCGACATTGCAAGTGGGTGGATGAAGTCA   420
 M T A E E R Y E S L R H C K W V D E V I                        104
TCCCTGATGCACCATGGGTGGTCAACCAGGAGTTTCTTGACAAGCACCAGATTGACTATG   480
 P D A P W V V N Q E F L D K H Q I D Y V                        124
TTGCCCACGATTCTCTTCCCTATGCTCATTCAAGCGGACGTGGAAAGGATGTCTATGAAT   540
 A H D S L P Y A H S S G R G K D V Y E F                        144
TTGTTAAGAAAGTTGGGAGGTTTAAGGAAACACAGCGAACTGAAGGAATATCGACCTCGG   600
 V K K V G R F K E T Q R T E G I S T S D                        164
ATATAATAATGAGAATAGTGAAAGATTACAATCAGTATGTCATGCGTAACTTGGATAGAG   660
 I I M R I V K D Y N Q Y V M R N L D R G                        184
GATACTCAAGGGAAGATCTTGGAGTTAGCTTTGTCAAGGAAAAGAGACTTAGAGTTAATA   720
 Y S R E D L G V S F V K E K R L R V N M                        204
TGAGGCTAAAGAAACTCCAGGAGAGGGTCAAAGAACAACAAGAAAGAGTGGGAGAAAAGA   780
 R L K K L Q E R V K E Q Q E R V G E K I                        224
TCCAAACTGTAAAAATGCTGCGCAACGAGTGGGTAGAGAATGCAGATCGATGGGTCCCTG   840
 Q T V K M L R N E W V E N A D R W V P G                        244
GATTTCTTGAAATATTTGAAGAAGGTTCCCATAAGATGGGAACTGCAATCGTAGATCAGT   900
 F L E I F E E G S H K M G T A I V D Q Y                        264
ATCCAAGAAAGGTTAATGAGGCAAAGTCGGCAGAGAGGCTGGAGAACGGTCAGGATGATG   960
 P R K V N E A K S A E R L E N G Q D D D                        284
ACACAGACGACCAGTTCTATGAAGAATACTTCGATCATGACATGGGTAGTGACGATGATG  1020
 T D D Q F Y E E Y F D H D M G S D D D E                        304
AAGGGGAAAAATTCTACGACGAGGAAGAAGTAAAGGAAGAAGAGACAGAGAAAACCGTTA  1080
 G E K F Y D E E E V K E E E T E K T V M                        324
TGACGGATGCTAAAGACAACAAGTAAGAACAAATTTGGCTTGCAGAAACCTCAGATTAGC  1140
 T D A K D N K *                                                331
TCTACTTATGGCCACTTCTACTAAACTCCCTTAAGCCTCGCACTCTCTCTCGAAATTCAT  1200
CTACTTAACATATAATACCAATGTTTAGAAAGAGAGAGTGTGTGATGTGTTTGTTTGTGT  1260
GTGTTGAACAAACGAACGTGTGTGGTTGTCTTTGGTGAGTTGGTCTCATCTTTGTTGATT  1320
TTTGAATGCGCATGTATTTTTTTCTTCTTTTTCTAGACGGGCAAAGTGTTATACAAGAAC  1380
AATGCAATTGTCTAAAACAGGATAAGTCAATGGTTCGTGTGTGCCATAAAGTAAAAAAAA  1440
AAAAAAA                                                        1447
```

Fig. 1. Nucleotide and amino acid sequences of *A. thaliana* CTP:phosphocholine cytidylyltransferase cDNA.

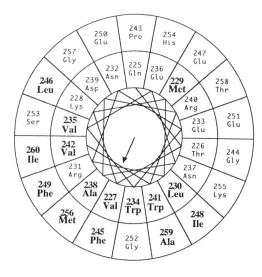

Fig. 2. Helical wheel projection of *A. thaliana* CTP:phosphocholine cytidylyltransferase in the amphipathic helix region. Hydrophobic residues are in bold characters.

Although there were 5 potential sites of phosphorylation by protein kinase C and 2 potential sites of phosphorylation by Ca^{2+}/calmodulin protein kinase II over the entire sequence, only one of those sites lied in the C-terminal region. The Ser-Pro motifs, where phosphorylation-dephosphorylation was suggested to play a role in the translocation of the enzyme in animal tissues (Wang et al., 1993; Kalmar et al., 1994), were not found.

Acknowledgements

This research was supported by a research grant from Inha University, 1995, to S.H. Cho.

References

Kalmar, G.B., Kay, R.J., LaChance, A., Aebersold, R., and Cornell, R.B. (1990) Cloning and expression of rat liver CTP:phosphocholine cytidylyltransferase: an amphipathic protein that controls phosphatidylcholine synthesis, *Proc. Natl. Acad. Sci.* USA. **87**, 6029-6033.

Kalmar, G.B., Kay, R.J., LaChance, A.C., and Cornell, R.B. (1994) Primary structure and expression of a human CTP:phosphocholine cytidylyltransferase, *Biochim. Biophys. Acta* **1219**, 328-334.

Wang, Y., MacDonald, J.I.S., and Kent, C. (1993) Regulation of CTP:phosphocholine cytidylyltransferase in Hela Cells, *J. Biol. Chem.* **268**, 5512-5518.

APPROACHES TO THE DESIGN OF ACYL-ACP DESATURASES WITH ALTERED FATTY ACID CHAIN-LENGTH AND DOUBLE BOND POSITIONAL SPECIFICITIES

Edgar B. Cahoon and John Shanklin
Biology Department, Brookhaven National Laboratory
Upton, New York 11973 USA

Introduction

The chain-lengths and double bond positions of monounsaturated fatty acids strongly influence their physical properties as well as their industrial and nutritional uses. Acyl-acyl carrier protein (ACP) desaturases are the primary determinants of the monounsaturated fatty acid composition of seed oils. The goal of our research is to understand how these enzymes recognize the chain-lengths of substrates and position the placement of double bonds in fatty acids. This information will provide the basis for the rational design of acyl-ACP desaturases that are capable of producing new types of monounsaturated oils in transgenic crops. Outlined below are methods that we are currently using to characterize the substrate and regio-specificities of acyl-ACP desaturases.

Identification of Variant Acyl-ACP Desaturases

The Δ^9-stearoyl (18:0)-ACP desaturase is found in nearly all plant tissues. In addition, several other acyl-ACP desaturases have been identified in plants, including the Δ^4-palmitoyl (16:0)-ACP desaturase of coriander seed (1), the Δ^6-16:0-ACP desaturase of *Thunbergia alata* seed (2), and the Δ^9-myristoyl (14:0)-ACP desaturase of geranium trichomes (3). These enzymes share $\geq 70\%$ amino acid sequence similarity. Given this high degree of structural relatedness, comparisons of the amino acid sequences of these enzymes may provide clues as to which residues dictate the fatty acid chain-length and double bond positional specificities of acyl-ACP desaturases, especially when used in conjunction with the recently determined crystal structure of the Δ^9-18:0-ACP desaturase (4).

To enhance the value of amino acid sequence alignments, we have continued to pursue the isolation of cDNAs for acyl-ACP desaturases with novel activities. In this regard, we have isolated a cDNA for a diverged acyl-ACP desaturase from milkweed (*Asclepias syriaca*) seed, a tissue that contains approximately 25% palmitoleic ($16:1\Delta^9$) and *cis*-vaccenic ($18:1\Delta^{11}$) acids. This cDNA encodes a Δ^9

desaturase that displays seven- to ten-fold greater relative activity with 16:0-ACP and 30-fold greater relative activity with 14:0-ACP than do known Δ^9-18:0-ACP desaturases. The mature milkweed acyl-ACP desaturase shares 61% amino acid identity with the castor Δ^9-18:0-ACP desaturase and, interestingly, is the most structurally diverged acyl-ACP desaturase identified to date.

Alteration of Acyl-ACP Desaturase Activity by Site-Directed Mutagenesis: Conversion of a Δ^6-16:0-ACP Desaturase to a Δ^9-18:0-ACP Desaturase

Mutagenesis studies have been conducted to identify amino acids associated with the substrate and regio-specificities of acyl-ACP desaturases. These studies were performed using cDNAs for a Δ^6-16:0-ACP desaturase and a Δ^9-18:0-ACP desaturase from *Thunbergia alata*. Based on the activities of chimeric mutants of these enzymes, a 30-amino acid domain was identified that contains determinants for fatty acid chain-length and double bond positional specificities. This domain corresponds to amino acids 178-207 of the Δ^9-18:0-ACP desaturase. When this domain was replaced in the Δ^6-16:0-ACP desaturase with the analogous portion of the Δ^9-18:0-ACP desaturase, the resulting enzyme catalyzed both the Δ^6 and Δ^9 desaturation of 16:0- and 18:0-ACP.

The Δ^6-16:0- and Δ^9-18:0-ACP desaturases contain nine non-conserved residues in the 30-amino acid domain described above. Using site-directed mutagenesis, these amino acids were replaced, individually or in combination, in the Δ^6-16:0-ACP desaturase with those present in the Δ^9-18:0-ACP desaturase. Several of the resulting mutants of the Δ^6-16:0-ACP desaturase displayed new activities. These included an enzyme generated by replacement of two amino acids that functioned as a Δ^6 desaturase with nearly equal specificity for 16:0- and 18:0-ACP. Another mutant, which was formed by the replacement of five amino acids of the Δ^6-16:0-ACP desaturase, displayed primarily Δ^9-18:0-ACP desaturase activity. We are currently using the crystal structure of the Δ^9-18:0-ACP desaturase (4) to interpret the structural basis for the activities of these mutants.

Substrate-Dependent Selection of Acyl-ACP Desaturases by Complementation of an *E. coli* Unsaturated Fatty Acid Auxotroph

We have previously shown that plant acyl-ACP desaturases can function in *E. coli* to produce novel monounsaturated fatty acids (3, 5). In addition, we have found that an *E. coli* unsaturated fatty acid auxotroph can be complemented by the expression of acyl-ACP desaturases, and the effectiveness of this complementation is dependent on the fatty acid chain-length specificity of the acyl-ACP desaturase. In these experiments, an *E.coli fabA*⁻ (Ts), *fadR*⁻ strain was used, which requires exogenous unsaturated fatty acids for growth at all temperatures. Acyl-ACP desaturases were co-expressed with a plant-type ferredoxin in the *E. coli* unsaturated fatty acid auxotroph. The expression of plant-type ferredoxin has been demonstrated to stimulate the *in vivo* activity of an acyl-ACP desaturase in wild-type *E. coli* (5).

The long-/medium-chain acyl-ACP pools of *E. coli* are enriched in 14:0- and 16:0-ACP but contain little or no 18:0-ACP (6). Consistent with this, we have found that 14:0- and 16:0-ACP desaturases (*e.g.*, Δ^9-14:0-ACP desaturase, Δ^6-16:0-ACP desaturase, and the milkweed acyl-ACP desaturase) complement the *E. coli* unsaturated fatty acid auxotroph when grown at 30°C, but 18:0-ACP desaturases (*e.g.*, Δ^9-18:0-ACP desaturase) do not complement these cells at this temperature (Fig. 1). As an extension of this observation, we are now in the process of introducing random and site-specific mutations into the Δ^9-18:0-ACP desaturase and then selecting for mutant enzymes with increased specificity for 14:0- and 16:0-ACP by complementation of the *E. coli* unsaturated fatty acid auxotroph.

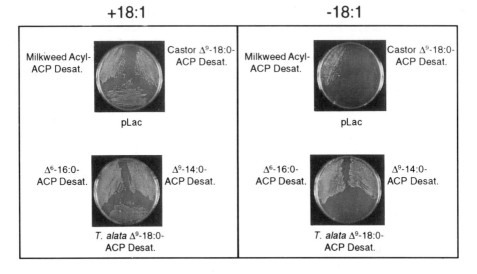

Figure 1. Complementation of an *E. coli* unsaturated fatty acid auxotroph with 14:0- or 16:0-ACP desaturases. Acyl-ACP desaturases were expressed using the vector pLac3d. Cells were grown in the presence of IPTG with or without exogenous oleic acid (+/-18:1).

It is anticipated that information gained from these experiments as well as from crystallographic studies of the Δ^9-18:0-ACP desaturase, will advance us towards the ultimate goal of rationally designing the activities of acyl-ACP desaturases with respect to substrate chain-length and position of double bond insertion.

References

1. Cahoon, E.B., Shanklin, J., and Ohlrogge, J.B. (1992) *Proc. Natl. Acad. Sci. U.S.A.* **89**, 11184-11188.
2. Cahoon, E.B., *et al.* (1994) *J. Biol. Chem.* **269**, 27519-27526.
3. Schultz, D.J., *et al.* (1996) *Proc. Natl. Acad. Sci. U.S.A.* (in press).
4. Lindqvist Y., *et al.* (1996) *EMBO J.* (in press).
5. Cahoon, E.B., Mills, L.A., and Shanklin, J. (1996) *J. Bacteriol.* **178**, 936-939.
6. Ohlrogge, J., *et al.* (1995) *Arch. Biochem. Biophys.* **317**, 185-190.

EFFECT OF A MAMMALIAN Δ-9 DESATURASE ON SPECIFIC LIPIDS IN TRANSGENIC PLANT TISSUES

HANGSIK MOON, MICHAEL SCOWBY, SERGEI AVDIUSHKO, AND DAVID F. HILDEBRAND
Dept. of Agronomy, University of Kentucky, Lexington, KY 40546

INTRODUCTION

The properties of fats and oils are determined by the fatty acid composition, which in turn affects nutritional quality and oxidative stability. Because of their influence on food quality and significance in biological processes, there is interest in the alteration of fatty acid desaturation in plants.

In plants, de novo fatty acid biosynthesis occurs exclusively in the stroma of plastids, whereas, with the exception of plastidial desaturation, modification of fatty acid residues including further desaturation and triacylglrycerol (TAG) assembly are localized in the cytosol/endoplasmic reticulum (ER). The primary fatty acids formed in the plastid (palmitic, stearic, and oleic acid) are used in the plastidic "prokaryotic" pathway for membrane lipid synthesis or diverted to the cytoplasmic "eukaryotic" pathway for the synthesis of membrane lipids or storage TAGs (1). Movement of glycerolipids is believed to occur in the reverse direction between the cytosol/ER and the plastids in the highly regulated manner (2).

In order to study the processes of plant fatty acid desaturation and glycerolipid biosynthesis and to develop a crop seed oil with reduced level of saturated fatty acids, a rat liver stearyl-CoA Δ-9 desaturase (D9DS) gene was introduced into *Nicotiana tabacum* under the control of the 35S promoter via *Agrobacterium* transformation (3) and soybean somatic embryos with the seed-specific phaseolin promoter by particle bombardment. In this article, we decribe the effects of the mammalian Δ-9 desaturase on fatty acid composition of membrane and storage lipids.

MATERIALS AND METHODS

Transgenic Tobacco and Soybean.

The transgenic tobaccos made by Grayburn et al. (3) were used to analyze fatty acid composition. For transformation of soybean, soybean (*Glycine max* Merrill. cv. "J103") somatic embryos were perpared as descrived by Liu et al. (4), and bombarded with pDGN (35S promoter-driven GUS::NPTII and phaseolin promoter-driven D9DS) for the Δ-9 desaturase expressor and pBI426 (35S promoter-driven chimeric GUS::NPTII) for control. Intracellular distribution of D9DS protein in the transgenic tissues was examined by standard western immunoblotting with subcellular fractions prepared by stepwise sucrose density gradients (5).

Lipid and Fatty Acid Analysis.

Total lipids were extracted and fatty acids were analyzed as described by Grayburn et al. (3). Individual lipids were separated on TLC plates (Silica Gel 60) in a solvent system containing acetone:toluene:H_2O (120:30:16). The chloroplast fraction from tobacco leaves were prepared as described by Palmer et al. (6). Positional analyses for the fatty acids of major chloroplast lipids were carried out using lipase from *Rhizopus arrhizus* (Boehringer Mannheim) as described by Ramesha et al. (7) with minor modifications.

RESULTS AND DISCUSSION

Transgenic tobacco leaves and soybean somatic embryos expressing this mammalian Δ-9 desaturase showed siginificant conversions of saturated to monounsaturated fatty acids. The use of seed-specific phaseolin promoter resulted in higher accumulation of monosaturated fatty acids in cotyledons than hypocotyl/radicles of the D9DS-transgenic soybean somatic embryos. Western blot analysis of subcellular fractions from both tissues indicated that the foreign desaturase was associated mainly with the ER. Subcelluar fractionation of mature transgenic soybean somatic embryos showed that the highest ratio of monounsaturated/saturated fatty acids was obtained with lipid bodies, indicating lesser effects on membrane lipids and lower possible negative physiological consequences.

Analysis of individual lipid classes from transgenic tobacco leaf tissue demonstrated that palmitoleic acid (16:1) produced by D9DS was incorporated into most of the major membrane lipid classes (Table I). The highest levels were found in the galactolipids and phosphatidylcholine (PC). In these lipids, increases in 16:1 were accompanied by decreases in linolenic acid (18:3) levels. However, the level of 16:1 was not altered significantly in phosphatidylglycerol (PG), which is considered to be a product of the prokaryotic pathway (8). Since the rat liver stearyl-CoA Δ-9 desaturase is a cytochrome b_5-dependent ER-localized enzyme, the accumulation of 16:1 in a significant amount in monogalactosyldiacylglycerol (MGD) and digalactosyldiacylglycerol (DGD), which are the major chloroplast lipids, indicated metabolic flux of fatty acids between the cytosol/ER "eukaryotic" pathway and the plastid-localized "prokaryotic". In addition, the fact that PC as well as the galactolipids such as MGD and DGD contained high 16:1 levels, supported the theory that PC can be a source of diacylglycerol (DAG) pool, which in turn contributes to the synthesis of plastid lipids (8). Although both palmityl (16:0)-CoA and stearyl (18:0)-CoA are the substrates for D9DS (3), increases in oleic acid (18:1) was not as appreciable as increases in 16:1. This is due to the limited amount of 18:0-CoA available in the cytosol of tobacco cell compared to the amount of 16:0-CoA and rapid further desaturation of 18:1 into polyunsaturated fatty acids.

Table I. Fatty Acid Composition[a] of Chloroplast Lipids from Control- and D9DS-Transgenic Tobacco Leaf

Fatty Acid[c]	MGD		DGD		PG		PC	
	Control	D9DS	Control	D9DS	Control	D9DS	Control	D9DS
16:0	2.7	1.5	14.0	10.0	33.7	24.9	26.8	12.1
16:1c	0.4	9.3	0.4	8.8	0.2	1.3	nd	17.3
16:1t	nd[b]	nd	nd	nd	31.8	34.2	nd	nd
16:3	16.2	15.6	2.7	3.7	nd	nd	nd	nd
18:0	0.4	0.2	2.6	1.0	3.2	2.8	9.6	4.4
18:1	0.3	0.3	0.4	0.5	1.9	3.2	4.6	5.6
18:2	3.1	2.7	2.8	2.6	8.3	9.6	18.8	25.9
18:3	72.7	63.6	72.3	67.0	16.6	19.2	40.2	34.6

[a] Values (% of total fatty acid) are means obtained from three replications.
[b] nd, Not detected.   [c] 16:1c, 16:1 (Δ-9, *cis*); 16:1t, 16:1 (Δ-3, *trans*)

REFERENCES

1. Töpfer, R., Martini, N., and Schell, J. (1995) Modification of plant lipid synthesis, *Science* **268**, 681-686
2. Polashock, J.J., Chin, C.K., and Martin, C.E. (1992) Expression of the yeast Δ-9 fatty acid desaturase in Nicotiana tabacum, *Plant Physiology* **100**, 894-901
3. Grayburn, W.S., Collins, G.B., and Hildebrand, D.F. (1992) Fatty acid alteration by Δ-9 desaturase in transgenic tobacco tissue, *Biotechnology* **10**, 675-678
4. Liu, W., Moore, P.J., and Collins, G.B. (1992) Somatic embryogenesis in soybean via somatic embryo cycling, *In Vitro Cell. Dev. Biol.* **28P**, 153-160
5. Lord, J.M., Kagawa, T., and Beevers, H. (1972) Intracellular distribution of enzymes of the cytidine diphosphate choline pathway in castor bean endosperm, *Proc. Nat. Acad. Sci. USA* **69**, 2429-2432
6. Palmer, J.D. (1986) Isolation and structural analysis of chloroplast DNA, *Methods in Enzymology* **118**, 167-186
7. Ramesha, C.S. and Thompson, G.A. Jr. (1983) Cold stress induces in situ phospholipid molecular species changes in cell surface membranes, *Biochimica et Biophysica Acta* **731**, 251-260
8. Ohlrogge, J. and Browse, J. (1995) Lipid biosynthesis, *Plant Cell* **7**, 957-970

TWO ACYL-LIPID Δ9 DESATURASE GENES OF THE CYANOBACTERIUM, *SYNECHOCOCCUS* SP. STRAIN PCC7002

T. SAKAMOTO, V.L. STIREWALT and D.A. BRYANT
Department of Biochemistry and Molecular Biology
Pennsylvania State University, University Park, PA 16802, USA

Abstract. In cyanobacteria, lipid-bound saturated fatty acids are converted to monounsaturated fatty acids by membrane-bound Δ9 desaturases. We have cloned and characterized two Δ9 desaturase genes (*desC* and *desE*) from the cyanobacterium *Synechococcus* sp. strain PCC7002. The structural similarity of these two cyanobacterial Δ9 desaturases, and their gene expression patterns in response to temperature will be described.

1. Introduction

The unicellular, marine cyanobacterium *Synechococcus* sp. strain PCC 7002 is classified as member of Group 2 based upon the fatty acid composition of its lipids and their pattern of desaturation (Murata *et al.*, 1992). This cyanobacterium synthesizes lipids containing C_{18} fatty acids with none, one, two or three double bonds at the Δ9, Δ12, and ω3 (or Δ15) positions at the *sn*-1 position, and C_{16} fatty acids containing none or one double bond at the Δ9 position of the *sn*-2 fatty acid (Murata and Wada, 1995). Double bonds in the *sn*-1 fatty acid are added sequentially starting with desaturation at the Δ9 position and proceeding to the ω3 position. These desaturations are catalyzed by membrane-bound desaturases. Three desaturase genes, *desC* (Δ9 desaturase), *desA* (Δ12 desaturase) and *desB* (ω3 desaturase) are responsible for the conversion of stearate to α-linolenate (Sakamoto *et al.*, 1994a; 1994b; 1994c). We have isolated and characterized the *desA*, *desB* and *desC* genes from *Synechococcus* sp. strain PCC7002 and studied the patterns of expression of these three desaturase genes in response to changes in the ambient temperature (Sakamoto and Bryant, 1996). In the work presented here, we isolated a novel Δ9 desaturase gene (*desE*) and characterized its expression in response to temperature.

2. Isolation of Two Δ9 Desaturase Genes from *Synechococcus* sp. strain PCC7002

A 0.35-kbp DNA fragment was amplified by polymerase chain reaction (PCR) from the genomic DNA of *Synechococcus* sp. strain PCC 7002 using the degenerate primers designed from the conserved regions among *desC* genes (Sakamoto *et al.*, 1994c). The degenerate primers (forward primer: 5' AA(T/C) AA(A/G) GG(A/C/T/G) TT(T/C) TGG TGG, 32-fold degenerate 18-mer; the reverse: 5' TG (A/G)TG (A/G)TT (A/G)TT (A/G)TG CCA, 16-fold degenerate, 17-mer) correspond to the conserved amino acid sequences of "NKGFWW" (positions 155 to 160) and "WHNNHH" (positions 267 to 272 in the deduced amino acid sequence of the DesC protein of *Synechocystis* sp. strain PCC 6803). Using the 0.35-kbp PCR product as the hybridization probe, five positive clones containing the 2.8-kbp *Eco*RI insert were isolated from 600 independent recombinant clones of the partial library that composed of *Eco*RI fragments of 2 to 4 kbp in a low-copy number vector, pHSG576 (GenBank accession number, D78513). The nucleotide sequence of 1.7 kbp of one positive clone was determined, and an open reading frame was found in the region sequenced. This open-reading frame is predicted to encode a polypeptide of 277 amino acids that show a high degree of sequence similarity to the deduced amino acid sequences of the DesC proteins of cyanobacteria (65-71% identity).

Two DNA fragments were detected in a total genomic DNA blot of *Synechococcus* sp. strain PCC7002 during Southern hybridization analyses at the low stringent conditions using a 1.1 kbp probe derived from the *desC* gene of *Anabaena variabilis* M-3 (Sakamoto *et al.*, 1994c). One DNA fragment strongly hybridized with the probe corresponded to the *desC* gene. To isolate the other DNA fragment, an *Eco*RI-*Xba*I 1.5 kbp that hybridized weakly, a partial genomic library containing *Eco*RI-*Xba*I 1.5 kbp in pUC19 was screened with the *desC* gene probe of *A. variabilis*. Nine putative positive clones were selected from 600 independent colonies, and nucleotide sequencing of the clones was performed. Two clones contained an identical sequence in the insert DNA showing a high degree of sequence similarity to the known Δ9 desaturases of cyanobacteria, mammal and yeast. A DNA fragment (2.7-kbp *Bgl*II/*Xba*I) containing the full length of the gene was isolated, and the nucleotide sequence of a 2.0 kbp region was determined. An open reading frame was found in the region sequenced. This open-reading frame is predicted to encode a polypeptide of 300 amino acids that shows 37% identity to the deduced amino acid sequence of the DesC protein of *Synechococcus* sp. strain PCC7002. Thus, this gene is assumed to encode the second Δ9 desaturase of the cyanobacterium and was designated the *desE* gene.

3. Comparison of the Primary Structures of Δ9 Desaturases

The deduced amino acid sequences of the DesC and DesE proteins of *Synechococcus* sp. strain PCC 7002 showed a high degree of similarity to the Δ9 desaturases of *Rattus norvegicus* (rat) and *Saccharomyces cerevisiae* (34-38%). Hydropathy profiles of these DesC and DesE proteins have two highly hydrophobic regions, suggesting that they are membrane-integral proteins. Three domains are conserved among all of the deduced amino acid sequences of the known Δ9 desaturases. Each domain contains the highly conserved histidine motif found in all enzymes containing the Fe-O-Fe cluster. The second domain contains the highly conserved tryptophan and aspartate residues (position 112 and position 128 in the deduced amino acid sequence of the DesE protein of *Synechococcus* sp. PCC7002). The conserved motif, i.e., $W-X_3-H-X_2-HH-X_{(7 \text{ or } 8)}-D$, found in all fatty acid desaturases. These conserved amino-acid residues might be involved in the common reactions in the reduction of hydrocarbon chain.

4. Expression Patterns of *desC* and *desE* Genes

The growth-temperature dependence of expression of the *desC* and *desE* genes was investigated by RNA blot (northern) hybridization analyses. Two hybridizing *desC* transcripts of 0.95 and 1.2 knt were detected in total RNAs isolated from cells grown at 38°C and 22°C. The *desC* gene may have two transcription initiation sites, or the mRNA might be cleaved after transcription. The relative intensities of these two hybridization signals were similar for RNAs from both growth temperatures, indicating that the steady-state mRNA levels for the *desC* gene are similar at both growth temperatures. The accumulation of *desC* transcripts was increased after a temperature shift-down to 22°C (1.5 to 2.5-fold) and the increased abundance of transcripts at 22°C recovered by a temperature upshift to 38°C. The calculated half-life for *desC* transcripts was about 5 min at either temperature. Since the stability of the *desC* mRNA is not altered by temperature, the rate of mRNA synthesis is transiently increased by temperature shift-down.

A single *desE* transcript of 1.1 knt was detected in the total RNA of cells grown at 22°C but was barely detectable in the RNA of cells grown at 38°C. The relative abundance of the *desE* transcripts increased after temperature shift-down from 38°C to 22°C, then after 15-min treatment at 22°C reached the same steady-state level as in cells grown at 22°C. The *desE* mRNA abundance decreased rapidly after temperature shiftup from 22°C to 38°C, and reached an undetectable level after only 5 min at 38°C. The estimated half-life of the *desE* mRNA was about 1 min at 38°C. Transcripts of the *desE* gene were stabilized at 22°C, and the calculated half-life time was 21 min. Thus, an important component determining *desE* transcript abundance at 22°C is the increased stability of these transcripts at lower temperature.

To identify the function of the *desE* gene, a *desE* mutant strain has been constructed by interposon mutagenesis; physiological and biochemical analyses of this mutant are ongoing.

5. References

Murata, N. and Wada, H. (1995) Acyl-lipid desaturases and their importance in the tolerance and acclimation to cold of cyanobacteria. *Biochem. J.* **308**: 1-8.

Murata, N., Wada, H. and Gombos, Z. (1992) Modes of fatty-acid desaturation in cyanobacteria. *Plant Cell Physiol.* **33**: 933-941.

Sakamoto, T., Wada, H., Nishida, I., Ohmori, M. and Murata, N. (1994a) Identification of conserved domains in the $\Delta 12$ desaturases of cyanobacteria. *Plant Mol. Biol.* **24**: 643-650.

Sakamoto, T., Los, D. A., Higashi, S., Wada, H., Nishida, I., Ohmori, M. and Murata, N. (1994b) Cloning of $\omega 3$ desaturase from cyanobacteria and its use in altering the degree of membrane-lipid unsaturation. *Plant Mol. Biol.* **26**: 249-263.

Sakamoto, T., Wada, H., Nishida, I., Ohmori, M. and Murata, N. (1994c) $\Delta 9$ acyl-lipid desaturases of cyanobacteria: Molecular cloning and substrate specificities in terms of fatty acids, *sn*-positions, and polar head groups. *J. Biol. Chem.* **269**: 25576-25580.

Sakamoto, T. and Bryant, D. A. (1996) Temperature-regulated transcription accumulation and mRNA stabilization for fatty acid desaturases genes in the cyanobacterium *Synechococcus* sp. strain PCC7002. submitted.

ISOLATION AND CHARACTERIZATION OF TWO DIFFERENT MICROSOMAL ω-6 DESATURASE GENES IN COTTON (*GOSSYPIUM HIRSUTUM* L.)

Q. LIU[1,2], S.P. SINGH[1], C.L. BRUBAKER[1], P.J. SHARP[2], A.G. GREEN[1] AND D.R. MARSHALL[2]
[1]*Commonwealth Scientific and Industrial Research Organisation, Division of Plant Industry, GPO Box 1600, ACT 2601, Australia*
[2]*University of Sydney, Plant Breeding Institute, Cobbitty, PMB11, Camden, NSW 2570, Australia*

1. Introduction

Microsomal ω-6 desaturases use cytochrome b_5 as electron donor to introduce a double bond into the ω-6 position of monounsaturated oleic acid to produce polyunsaturated linoleic acid. Thus microsomal ω-6 desaturases play a vital role in the polyunsaturated fatty acid synthesis in angiosperms. It has been estimated that these enzymes are responsible for more than 90% of the polyunsaturated fatty acid synthesis in non-photosynthetic tissues and developing seeds of oil crops (1).
We describe here the cloning and characterisation of two different microsomal ω-6 desaturases from tetraploid cotton (*Gossypium hirsutum* L.). This was done as a prelude to applying antisense techniques to modify the relative proportion of unsaturated fatty acids with the ultimate goal of producing POS type of fatty acid profile in cottonseed oil.

2. Materials and Methods

2.1. CONSTRUCTION OF λ cDNA LIBRARY FROM COTTONSEED

To isolate the cDNAs encoding fatty acid desaturases, a cDNA library was constructed using developing embryos of cotton (*Gossypium hirsutum* L. cv. Deltapine-16). Total RNA was isolated from 21-30 days old (days after anthesis) embryos. The cDNAs were ligated into EcoRI-predigested Lambda ZAP II vector (Stratagene), and packaged according to manufacturer's instructions. One million plaques were screened with an [α-^{32}P]dCTP-labelled *Brassica juncea* microsomal ω-6 desaturase cDNA (2). Several positive plaques were identified and their cDNA inserts were characterised by nucleotide sequence determination in pBluescript following excision from the phagemids.

2.2. GENOMIC DNA SOUTHERN BLOT ANALYSIS

Genomic DNA was prepared from young leaves of *G. hirsutum* and *G. barbadense* following Paterson et al. (3). 20μg DNA were digested with EcoRI or HindIII and electrophoresed through a 0.7% agarose gel. The gel was blotted onto a Hybond-N$^+$ nylon membrane (Amersham). The filters were probed with [α-^{32}P]dCTP-labelled DNA fragments corresponding to the 3' non-coding sequences, unique to the two genes. The hybridisation was performed overnight at 42°C, and the filters were washed in 2x SSC and 0.5x SSC, 0.1% SDS at 55°C prior to autoradiography.

2.3. RNA NORTHERN ANALYSIS

Total RNA was extracted from young leaves and immature embryos (25, 30, 36, 45 days after fertilisation) of *G. hirsutum*. The RNAs were denatured with formaldehyde, resolved by electrophoresis in 1% agarose gel containing formaldehyde, transferred to Hybond-N$^+$ nylon membranes, and probed with the [α-^{32}P] dCTP -labelled gene specific DNA fragments. Hybridisation and washing were carried out in the same conditions as described above for genomic Southern blot hybridisation.

2.4. EXAMINATION OF SEED-SPECIFIC PROMOTERS AND MAKING OF ANTISENSE CONSTRUCT

Expression of GUS constructs with various seed specific promoters, including lectin (4) were examined by means of bombardment with DNA-coated-gold particles. 1 μg Plasmid DNA was coated onto 1 mg gold particles and delivered onto the surface of embryo tissues by a He-driven Biolistic Particle Delivery System (Bio-rad, Model PDS-1000). Transient expression of GUS was assessed by histochemical staining according to Jefferson et al. (5).

3. Results and Discussion

The microsomal ω-6 desaturase gene in *G. hirsutum* is encoded by a gene family comprising of at least two distinct members, pghD12-1 and pghD12-2. As indicated in Table1, both these genes have homologous coding sequences (77% identical at the amino acid level) but unique 5' and 3' untranslated regions. The open reading frame of both genes showed a high degree of conservation (75% at the amino acid level) to other microsomal ω-6 desaturases. This conservation fell away when compared with a plastidial ω-6 desaturase and microsomal ω-3 desaturase (<40%) indicating the identity of the genes encoded by the two cotton sequences to be microsomal ω-6 desaturase.

Table 1. Homology between the deduced amino acid sequences of pghD12-1 (cotton microsomal ω-6 desaturase) and other membrane-bound desaturases

sequence	similarity[a] (%)	identity[a] (%)
pghD12-2 (cotton microsomal ω-6)	89	77
FAD2-1 (soybean microsomal ω-6)[6]	87	75
FAD2-2 (soybean microsomal ω-6)[6]	88	74
pTCCC1 (rapeseed plastidial ω-6)[7]	50	22
fad3 (*Arabidopsis* microsomal ω-3)[8]	60	37

[a]Comparisons are with the pghD12-1 sequence employing the Gap program of the GCG package using a gap weight of 3.0 and length of 0.1.

Genomic Southern blot analysis using pghD12-1- and pghD12-2-specific probes confirmed that these genes are nonallelic (Fig.1). EcoRI and HindIII digested genomic DNA from *G. hirsutum* revealed two bands when probed with pghD12-1 and pghD12-2 probes. The 6kb band observed in the EcoRI lane of pghD12-2 probed blot is probably the result of incomplete cutting. These results suggest the existence of at least two copies each of the genes in *G. hirsutum* genome. This is consistent with the tetraploid nature of *G. hirsutum* genome. Interestingly, similar genomic blots from another tetraploid species, *G. barbadense*, revealed pghD12-1 to be a single copy gene while pghD12-2 had two copies.

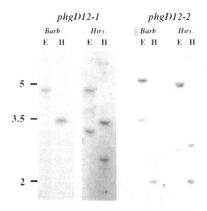

Fig.1. Southern blots of genomic DNA probed with pghD12-1- and pghD12-2-specific sequence fragments. *Barb=G. barbadense*; *Hir=G. hirsutum*; E= EcoR I; H= HindIII. Molecular size (kb) of DNA ladder is indicated on the left.

Northern blot analysis revealed pghD12-1 to be specifically induced during seed development in *G. hirsutum* and as such is likely to play a major role in controlling conversion of oleic to linoleic acid during seed development. In contrast, pghD12-2 showed a low level constitutive expression in leaf tissue and throughout seed development (Fig.2). This is similar to the situation reported for soybean (6).

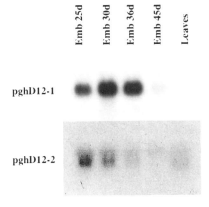

Fig. 2. RNA gel blot analysis of the developmental and tissue-specific expression of cotton (*G. hirsutum*) microsomal ω-6 desaturase genes. The autoradiograph with pghD12-1 (upper) was exposed for 16 hours, while pghD12-2 (lower) was exposed for 7 days. Emb25d, Emb30d, Emb36d, Emb45d indicate RNA isolated from developing embryos of 25, 30, 36, and 45 DAF respectively.

Transient GUS expression studies indicated that the soybean lectin promoter (4) retains its seed specificity and strength (when compared to 35S) in developing cotton embryos (Fig.3). This suggests that an antisense pghD12-1 gene behind a lectin promoter when introduced into *G. hirsutum* may lead to higher oleic acid levels in cottonseed oil. Therefore, an antisense construct of pghD12-1 with a lectin promoter (Fig.4) has been made and is currently being transformed into elite cotton cultivars.

Fig. 3. Histochemical localisation of transient GUS expression in developing embryos of cotton (*G. hirsutum*) bombarded with lectin-GUS constructs.

Fig. 4 Diagram showing the Lectin-ω6-desaturase antisense construct. Arrow shows the direction of normal transcription of pghD12-1.

4. Refererences

1. Okuley, J., Lightner, J., Feldmann, K., Yadav, N., Lark, E. and Browse, J.: *Arabidopsis* FAD2 gene encodes the enzyme that is essential for polyunsaturated lipid synthesis. *The Plant Cell* **6**(1994), 147-158.
2. Singh, S.P., van der Heide, T., McKinney, S. and Green, A.G.: Nucleotide sequence of a cDNA (accession No. X91139) from *Brassica juncea* encoding a microsomal omega-6 desaturase, *Plant Physiol.* (1995) PGR95-107.
3. Paterson, A.H., Brubaker, C.L. and Wendel, J.F.: A rapid method for extraction of cotton (*Gossypium* spp.) genomic DNA suitable for RFLP or PCR analysis, *Plant Molecular Biology Reporter* **11**(1993), 122-127.
4. Cho, M.-J., Widholm, J.M. and Vodkin, L.O.: Cassettes for seed-specific expression tested in transformed embryogenic cultures of soybean, *Plant Molecular Biology Reporter* **13**(1995), 255-269.
5. Jefferson, R.A.: Assaying chimeric genes in plants: the GUS gene fusion system, *Plant Molecular Biology Reporter* **5**(1987), 387-405.
6. Heppard, E.P., Kinney, A.J., Stecca, K.L. and Miao, G.H.: Developmental and growth temperature relation of two different microsomal ω-6 desaturase genes in soybeans, *Plant Physiol.* **110**(1996), 311-319.
7. Hitz, W.D., Carlson, T.J., Booth, J.R.Jr., Kinney, A.J., Stecca, K.L. and Yadav, N.S.: Cloning of a higher-plant plastid omega-6 fatty acid desaturase cDNA and its expression in a *cyanobacterium*, *Plant Physiol.* **105**(1994), 635-641.
8. Arondel, V., Lemieux, B., Hwang, I., Gibson, S., Goodman, H.M., Somerville, C.R.: Map-based cloning of a gene controlling Omega-3 fatty acid desaturation in *Arabidopsis*, *Science* **258**(1992): 1353-1355.

CAN E. *COLI* β-HYDROXYDECANOYL-ACP DEHYDRASE AND β-KETO-ACYL-ACP SYNTHASE I INTERACT WITH *BRASSICA NAPUS* FATTY ACID SYNTHASE TO ALTER OIL SEED COMPOSITION?

J.-A. CHUCK[+], I. VERWOERT[*], E. VERBREE[*], M. SIGGAARD-ANDERSEN[+], P. VON WETTSTEIN-KNOWLES[+,‡], A. STUITJE[*]

[+] *Dept. of Physiology, Carlsberg Laboratory, Gamle Carlsberg Vej 10, DK-2500 Copenhagen, Denmark.* [*] *Dept. of Genetics, Inst. of Molecular and Biological Science, Vrije Universiteit, 1081 HV The Netherlands.* [‡] *Dept. of Genetics, University of Copenhagen, DK-1353 Copenhagen, Denmark.*

1. Introduction

A key difference between bacterial and plant fatty acid biosynthesis is the origin of monounsaturated fatty acids. In plants elongation of the fatty acid occurs before the introduction of double bonds by desaturases. In bacteria a double bond can be introduced during elongation at an acyl chain length of C10. That is, the FAS intermediate β-hydroxydecanoyl-ACP is used as a substrate for a specialized dehydrase, β-hydroxydecanoyl-ACP dehydrase (HDD, encoded by *fabA*). In addition to dehydrase activity, HDD isomerizes the *trans* β-γ double bond to the adjacent position giving a *cis* α-β double bond. The latter is maintained while the fatty acid is elongated to the bacterial products, palmitoleic (C16:1 Δ9) and *cis*-vaccenic (C18:1 Δ11) acid. Can this pathway for bacterial unsaturated fatty acid synthesis be engineered into the seeds of *Brassica napus*? This would result in the appearance or elevation of palmitoleic and *cis*-vaccenic acid in the transgenic plants. It has been reported that the introduction of HDD into plants has no effect on lipid composition (Saito et al., 1995; Verwoert et al., 1995), however, it was not known whether the plant KASes were able to elongate the bacterial unsaturated fatty acid intermediates. In bacteria this elongation is carried out by KAS I (encoded by *fabB*) which is able to use both unsaturated and saturated fatty acids as substrates. We have therefore introduced both the *fabA* and *fabB* genes into *B. napus* to see if the bacterial fatty acids can then be synthesized.

2. Methods

The genes encoding KAS I (Kauppinen et al 1988) and HDD (PCR using sequence information from GenBank J03186) were introduced into an expression cassette in frame behind the sequence encoding the targeting peptide of *B. napus* enoyl-ACP reductase. The cassette also included napin promoter and chalcone synthase transcription terminator sequences. The introduced genes were fully sequenced, and *B. napus* plants transformed using standard *Agrobacterium* techniques. Transformants were identified using antibodies specific for the bacterial enzymes. The primary transformants exhibiting the highest expression (A, 12 µg HDD/mg of soluble protein; B1 and B2, 1.0 and 0.9 µg of KAS I/mg soluble protein) were selfed and crosses made. Western analyses revealed that five out of five seeds on two progeny plants (B1A and AB2) derived from the crosses B1 x A and A x B2, respectively, were double transformants.

3. Results

Fig. 1 demonstrates that both *E. coli* genes were expressed in *B. napus* seeds. Western analyses revealed their parallel expression during seed development with maxium heterologous protein five weeks after anthesis. All data infer that both were targeted into the plastid. Activities of the heterologous proteins present in crude soluble extracts of mature seeds were determined (Kass 1969; Siggaard-Andersen

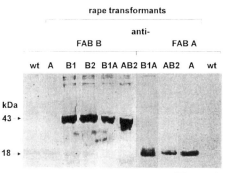

Figure 1

et al., 1995). Using β-hydroxydecanoyl-NAC as a substrate, plant A exhibited 29.8 nmoles/min/mg of protein of dehydrase activity versus 1.8 for non transformed plants. Fig. 2 illustrates the results of complementation assays in which seed extracts were added to crude *E. coli* protein preparations in which endogenous KAS I activity had been inhibited using cerulenin. Quantitation of the long chain acyl-ACPs revealed that seeds on plants B, B1A and AB2 showed up to eight times higher elongation activity than did those on wild type and plant A (lanes 4 + 5 vs 6-10). These results demonstrate that the expressed *E. coli* proteins had dehydrase and elongation activity. Fatty acid methyl esters were analysed by GC-MS. Careful analyses detected *cis*-vaccenic acid in all seeds. This is a known *B. napus* seed lipid albeit in minor amounts. Analyzing single seeds revealed that neither the primary nor double transformants showed any elevation in the levels of *cis*-vaccenic or palmitoleic acid. The amount of lipid per seed and the seed morphology were also unchanged.

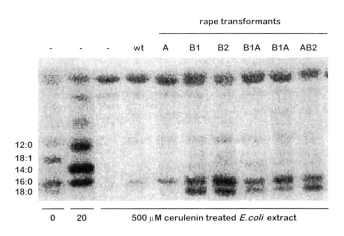

Figure 2

4. Plants do it better

Many possibilities exist to explain why the bacterial lipids did not accumulate, among which is the inability of one or more of the other plant FAS components to use an unusual intermediate as substrates. We investigated this aspect by carrying out the KAS I assays in the presence of the full complement of bacterial FAS components. The *E. coli* extracts were clearly capable of unsaturated fatty acid synthesis (Fig. 2 lane 1). Addition of a seed extract, regardless of whether it was derived from control or transgenic seeds, redirected synthesis to produce C16 and C18 saturated fatty acids (not shown). The extra KAS I or HDD in the transgenic seeds had no effect on increasing the amount of unsaturated fatty acid synthesis. In these assays at least three dehydrases compete for β-hydroxydecanoyl-ACP: (1) HDD from the *E. coli* extracts, (2) the broad range *E. coli* dehydrase involved in other steps of fatty acid biosynthesis and (3) the dehydrase(s) from the plant. The two bacterial enzymes have similar K_m values for the β-hydroxydecanoyl-ACP (Birge et al., 1967, Volpe & Vagelos, 1973). If the plant enzyme has a much higher affinity for the substrate when compared with HDD then the observed pattern of saturated fatty acid synthesis would be consistent. The spinach dehydrase K_m is circa 30 fold lower than that of HDD using crotonyl-ACP as substrate (Shimakata and Stumpf, 1982). The potential exists that the HDD can not compete for substrate in the plant to make unsaturated fatty acids. Our results demonstrate the poor ability of some bacterial enzymes to compete with their analogues in the plant FAS. In fact, Shimakata and Stumpf reported that the K_m value of each spinach FAS enzyme tested was lower than than that of its *E. coli* counterpart. To delineate the basis for the inability of plastids with active heterologous HDD and KAS I to synthesize the characteristic bacterial unsaturated fatty acids, pertinent plant FAS components can be eliminated by antisense technology and/or additional bacterial components added. This knowledge will be fundamental if major redirection of the rape seed FAS is to be achieved.

5. References

Birge, C.H., Silbert, D.F. and Vagelos, P.R. (1967) A β-hydroxydecanoyl-ACP dehydrase specific for saturated fatty acid biosynthesis in *E. coli*. Biochem. Biophys. Res. Commun. 29: 808-814.

Kauppinen, S.K., Siggaard-Andersen, M. and Wettstein-Knowles, P. von (1988) β-Ketoacyl ACP synthase I from Escherichia coli: nucleotide sequence of the *fabB* gene and identification of the cerulenin binding residue. Carlsberg Res. Commun. 53:357-370.

Kass, L.R. (1969) β-hydroxydecanoyl thioester dehydrase from *Escherichia coli*. Meth. Enzymol. 14:73-80.

Saito, K., Hamajima, A., Ohkuma, M., Murakoshi, I., Ohmori, S., Kawaguchi, A., Teeri, T.H. and Cronan, J.E. (1995) Expression of the *Escherichia coli fabA* gene encoding β-hydroxydecanoyl thioester dehydrase and transport to chloroplasts in transgenic tobacco. Transgenic Res. 4:60-69.

Shimakata, T. and Stumpf, P.K. (1982) Fatty acid synthetase of *Spinicia oleracea* leaves. Plant Physiol. 69:1257-1262.

Siggaard-Andersen, M., Wissenbach. M., Chuck, J.-A., Svendsen, I., Olsen, J.G., and Wettstein-Knowles, P. von (1994) The *fabJ*-encoded β-ketoacyl-[acyl carrier protein] synthase IV from *Escherichia coli* is sensitive to cerulenin and specific for short-chain substrates. Proc. Natl. Acad. Sci. USA 91:11027-11031.

Verwoert, I.I.G.S., Verbree, E., Linden, K.H. van der, Nijkamp, J.J. and Stuitje, A.R. (1995) Transgenic expression of bacterial FAS components in rapeseed. In *Plant Lipid Metabolism*. Kader, J.-C. and Mazliak, P. eds., Kluwer Acad. Publ., The Netherlands, 476-478.

ISOLATION OF CYTOCHROME P-450 GENES FROM *VERNONIA GALAMENSIS*

CRAIG SEITHER, S. AVDIUSHKO, AND D. HILDEBRAND
University of Kentucky, Lexington, KY 40546

Introduction

Vernolic (*cis*-12,13-epoxy-*cis*-9-octadecenoic) acid is an epoxy fatty acid which accumulates in the triglyceride of a few plant species such as *Vernonia galamensis* and *Euphorbia lagascae*. Vernolic acid has industrial applications such as in paints, plastic formulations, and protective coatings (1).

Bafor et al. (2) have provided evidence in *E. lagascae* for the enzyme being a cytochrome P-450 based on reductant specificity (NADPH) and inhibition of vernolic acid synthesis by CO. The preferred substrate was linoleic acid esterified to phosphatidylcholine at the *sn*-2 position.

With some modifications in the procedure outlined by Bafor et al., we have detected vernolic acid synthesis from microsomes of developing *V. galamensis* seeds. Vernolic acid synthesis was dependent upon reductant (NADH or NADPH) and inhibited by CO. Liu et al. (3) reported vernolic acid synthesis within one hour of feeding developing *V. galamensis* seeds [^{14}C]acetate. They also reported linoleoyl phosphatidylcholine as the putative substrate for epoxidation (3). These findings and the results presented below support the notion that the epoxygenase in *V. galamensis*, as in *E. lagascae*, is a cytochrome P-450.

Our objective is to clone the gene(s) encoding the enzyme(s) responsible for vernolic acid synthesis in *V. galamensis*. Various molecular techniques were employed to clone the epoxygenase including RT-PCR and heterologous probes. Two cDNA clones have been isolated from a cDNA library constructed at a stage in seed development postulated to have abundant transcripts for the epoxygenase. These clones share highest homology to plant cytochrome P-450s.

Materials and Methods

Microsome Assays. The epoxygenase assays were essentially the same as reported by Bafor et al. [2] with the following modifications. Developing *Vernonia* seeds (10%-30% vernolic acid detected by gas chromatography) were prepared by removing the seed coat and placing approximately 1 g seed tissue into 5 mL buffer A (0.1 M potassium phosphate buffer, pH 7.2, 0.33 M sucrose, 0.1% BSA, 0.1% PVP, 1000 units catalase/mL) on ice. The seed was ground (Tissuemizer) and filtered through Miracloth™. The microsomal pellet was resuspended in 1 ml buffer A. The resuspended microsomal prep was labeled with 2 µCi [^{14}C]18:2-CoA and 4 µmole

CoASH for 15 minutes. Labeled microsomes were separated into equal aliquots for assays. Assay reactions contained 1mM reduced dinucleotide and were allowed to proceed 1-3 h prior to terminating by extracting with 2 vol chloroform:methanol (2:1) and addition of a few drops of 1 M oxalic acid. Chlorofom extracts were blown to dryness with N_2 and lipids were methylated by first adding 0.5 ml sodium methoxide (18 mg/ml) in methanol for 30 min and then adding ethereal diazomethane. In order to ensure complete recovery of lipids, methyl esters were extracted with 2 vol of chloroform; water was added for phase separation. Chloroform extracts were concentrated to a volume of 50 µl for analysis on silica gel TLC in hexane:diethyl ether:acetic acid (85:15:1).

RT-PCR. Total RNA was isolated at a stage in seed development postulated to have abundant transcripts for the epoxygenase. The RNA was reverse transcribed and then PCR was performed by using a highly degenerate 5' primer (YWIHTICCITTYDSINNIGG) designed for the heme binding motif shared among cytochrome P-450s and 3' oligo dT primer ($GGGAGGCCCCT_{16}$) for the poly-A tail as described in [6]. The products were TA cloned and sequenced. A putative candidate was used for screening 1.6 x 10^5 pfu from a cDNA library made from *V. galamensis* at the same stage as above.

Heterologous Probing. The same cDNA library was screened (4 x 10^5 pfu) using a probe for allene oxide synthase (AOS) from flax kindly provided by A. Brash.

Results

Vernolic acid synthesis could be detected from the microsomes of *V. galamensis* developing seeds (Figure 1). The synthesis of vernolic acid was dependent upon reductant and was inhibited by CO.

The RT-PCR products generated were in the range of 200-500 bps. Only one cloned PCR product fit the criterion for a P-450 described in [5]. This fragment was used to screen a cDNA library whereupon a fuller length clone (1.6 Kb) was obtained. This clone has highest homology to several plant cytochrome P-450s at the 3' end of the cDNA and also shares homology to one plant P-450 at the 5' end. The cDNA was excised and used to probe a Northern blot of total RNA from leaf and three stages of seed development. The expression appears to be temporally and spatially regulated with no expression in leaf and most abundant expression in the mid-mature seed.

The heterologous probe made from AOS was used to isolate a cDNA which shows highest homology to several plant cytochrome P-450s at both the 5' and 3' ends. This cDNA was excised and used for probing a Northern blot. The gene appears to be expressed in both leaf and seed with maximal expression early in seed development.

Discussion

The inhibition of vernolic acid synthesis in the presence of CO demonstrates a cytochrome P-450 involvement in the reaction (4). Since our objective is to clone the gene encoding the enzyme responsible for vernolic acid synthesis, our primary interest was whether we could detect activity in the presence of reductants and inhibition of activity in the presence of CO, each of which was demonstrated.

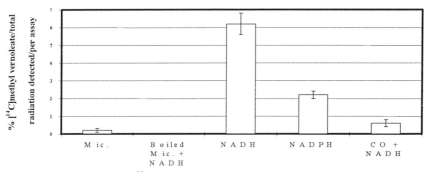

Figure 1. The percentage of [^{14}C]-methyl vernoleate detected from microsome assays conducted with *V. galamensis*. The percentage reports the amount of product ([^{14}C]-methyl vernoleate) relative to the total amount of radioactivity detected per lane on TLC. Each percentage is the mean of three assays ± SE.

The complementary molecular approaches reported constitute a strategy for cloning the epoxygenase gene. The RT-PCR approach was successful in isolating one putative clone which was used for retrieving a fuller length cDNA clone. Partial sequence data at the 5' and 3' ends of the cDNA revealed homology to several P-450s, predominantly at the 3' end. The gene expression pattern best fits our hypothesis that the epoxygenase gene is temporally and spatially regulated with greatest expression prior to and at the onset of vernolic acid synthesis in developing seeds.

The heterologous screening of a cDNA library using the AOS probe was considered less potentially fruitful based on the low degree of sequence homology shared among P-450s. By use of this approach, a cDNA clone was isolated from a cDNA library made at a stage in seed development postulated to have abundant transcripts for the epoxygenase. This clone had highest homology to plant P-450s at both the 5' and 3' ends of the cDNA. The expression pattern showed constitutive expression in leaf and seed. This cDNA does not appear to be a strong candidate.

Both cDNAs are being used for transforming tobacco and *Arabidopsis* under the control of a 35S promoter. The clone isolated via RT-PCR has been cloned behind a phaseolin promoter for subsequent transformation to study its seed-specific expression.

References
1.) Perdue, R.E., Carlson, K.D., and Gilbert, M.G. (1986) *Vernonia galamensis*, potential new crop source of epoxy fatty acid, Econ Bot **40**, 54-68.
2.) Bafor, M., Smith, M.A., Jonsson, L., Stobart, K., and Stymne, S. (1993) Biosynthesis of vernoleate (*cis*-12-epoxyoctadeca-*cis*-9-enoate) in microsomal preparations from developing endosperm of *Euphorbia lagascae*, Arch Biochem Biophys **303**, 145-151.
3.) Liu, L., Hammond, E.G., and Nikolau, B.J. (1995) The biosynthesis of vernolic acid in the seed of *Vernonia galamensis*, Inform **6**, 524.
4.) Bolwell, G.P., Bozak, K., and Zimmerlin, A (1994) Plant cytochrome P450, Phytochem **37**, 1491-1506.
5.) Meijer, A.H., Souer, E., Verpoorte, R., and Hoge, J.H.C. (1993) Isolation of cytochrome P-450 cDNA clones from the higher plant *Catharanthus roseus* by a PCR strategy, Plant Mol Bio **22**, 379-383.

ENGINEERING TRIERUCIN INTO OILSEED RAPE BY THE INTRODUCTION OF A 1-ACYL-*SN*-GLYCEROL-3-PHOSPHATE ACYLTRANSFERASE FROM *LIMNANTHES DOUGLASII*

CLARE L. BROUGH[1], JANE COVENTRY[2], WILLIAM CHRISTIE[3], JOHAN KROON[1], TINA BARSBY[2] & ANTONI SLABAS[1].
[1]Department of Biological Sciences, University of Durham, DH1 3LE, UK, [2]Nickerson BIOCEM Ltd., Cambridge Science Park, Milton Road, Cambridge CB4 4GZ, UK and [3]Scottish Crop Research Institute, Invergowrie, Scotland, UK

Introduction

Currently available cultivars of high erucic acid rape (HEAR) have a theoretical maximum of 66% erucic acid (22:1, $\Delta 13$) in their seed oil due to the specificity of the membrane-bound 1-acyl-*sn*-glycerol-3-phosphate acyltransferase (LPA-AT) enzyme. In HEAR the LPA-AT does not incorporate erucic acid at the *sn*-2 position of triacylglycerols (TAG) but preferentially incorporates oleic acid (18:1, $\Delta 9$), even if this is only a minor component of the total fatty acid pool (1). However, some plant species, e.g *Limnanthes*, can utilise erucoyl-CoA as a substrate and effectively incorporate erucic acid at the *sn*-2 position (2, 3).

A cDNA (pLAT2) has been isolated from *Limnanthes douglasii* developing embryos by complementation of an *E. coli* mutant (JC201) (4) deficient in LPA-AT activity (5, 6). Assays on membranes from complemented JC201 have demonstrated the ability to utilize erucoyl-CoA with a substrate specificity similar to that of *Limnanthes* microsomes and distinctly different to that of wild-type *E. coli*.

In an attempt to engineer trierucin and increased erucic acid levels into HEAR, constructs have been made for expression of either the full length or a truncated open reading frame (ORF) from pLAT2 under the control of a napin promoter. Both constructs were introduced into a HEAR line and seeds from transgenic plants were analysed.

Results and Discussion

The polymerase chain reaction (PCR) was used to introduce suitable restriction sites for cloning the full length open reading frame (ORF) of pLAT2 between the napin promoter and chalcone synthase terminator sequences in a binary vector to produce pSCV144 (7). In pLAT2 the full length ORF is not in frame with the *lacZ* promoter of pBluescript SK and so expression of the complementing polypeptide in the *E. coli* mutant JC201 was proposed to initiate at Met-35, due to the presence upstream of a purine-rich sequence which could act as a Shine-Dalgarno sequence (5). Therefore, to determine if this had an effect on the function of the encoded LPA-AT, a second construct (SCV146) was made which would initiate translation at Met-35 of the coding region of pLAT2. Both constructs were introduced into a HEAR line by *Agrobacterium*-mediated transformation and regenerated plants were self-pollinated and seed collected.

The presence of the transgene and its expression in the embryo was determined by Southern and Northern hybridization (data not shown). To determine the presence of trierucin, triacylglycerols were extracted from the seed of untransformed and transformed plants and analysed by reversed-phase HPLC. A peak which co-chromatographed with trierucin was detected in the transgenic seed TAG extracts and its identity confirmed as trierucin by mass spectrometry. Levels of trierucin ranged from 0.5 to 2.9% in SCV144-transformed plants and from 0.1 to 2.1% in SCV146-transformed plants, indicating that the N-terminal 35 amino acids have no effect on the function of the LPA-AT. No trierucin was detected in the untransformed seed.

Detailed positional analysis of fatty acids in TAG was performed on seed of two selected plants (SCV144-2 and SCV144-9 containing 2.8% and 0.4% trierucin, respectively). Total fatty acid composition was determined by gas chromatography (GC) of methyl esters. To determine the identity of the fatty acids at position 2 the TAG was initially treated with pancreatic lipase, which removes the acyl groups from sn-1 and sn-3, and then the desired 2-monoacylglycerol products were isolated by micro-preparative HPLC, methylated and examined by GC (Table 1). The analyses demonstrated the presence of significant amounts of 22:1 at the sn-2 position (SCV144-9: 9 mol% and SCV144-2: 28.3 mol%) which was not detected in the untransformed control plant. Under the growth conditions used the level of erucic acid in the transgenic seed was not significantly higher than that of the starting population.

TABLE 1. Fatty acid composition of intact TAG and at sn-2 position of control and transgenic seed oil extracted from pooled seed samples.

Plant	Fraction	Acyl group (mol % of total)											
		16:0	16:1	18:0	18:1	18:2	18:3	20:0	20:1	20:2	22:0	22:1 Δ13	22:2
untransformed	TAG	5.2	0.4	1.1	26.0	14.6	9.6	0.6	9.3	0.4	0.4	31.7	0.3
B. napus	sn-2	1.4	0.9	0.4	44.4	32.1	19.9	0.0	0.8	0.0	0.0	0.0	0.0
SCV144-2	TAG	4.2	0.2	1.4	26.0	13.3	8.7	0.8	11.0	0.5	0.6	32.1	0.4
	sn-2	1.0	0.5	0.3	30.5	20.7	12.8	0.2	5.3	0.1	0.0	28.3	0.3
SCV144-9	TAG	4.8	0.3	1.4	27.5	13.2	6.0	0.9	11.6	0.4	0.6	32.2	0.4
	sn-2	0.9	0.7	0.2	45.4	28.8	13.6	0.0	1.3	0.0	0.0	9.0	0.0

Further HPLC analysis of TAG from the T2 seed of selected lines has identified SCV144-9 plants with up to 6.0% trierucin. Detailed fatty acid analysis of TAG from these lines is being performed.

The results obtained are in agreement with those reported by Lassner et al. (1995) using the LPA-AT from *Limnanthes alba* and demonstrate the feasibility of altering the acyl composition at specific positions of TAG in one plant species by introducing an acyltransferase with a particular selectivity from another plant species. Further modifications are required to increase the level of trieucin and erucic acid and these are the subject of current investigations.

References

1. Bernerth, R., Frentzen, M. (1990) Utilisation of erucoyl-CoA by acyltransferases from developing seeds of *Brassica napus* (L.) involved in triacylglycerol biosynthesis. *Plant Science* **67**, 21-28.
2. Cao, Y.-Z., Oo, K.-C., Huang, A.H.C. (1990) Lysophosphatidate acyltransferase in microsomes from maturing seeds of meadowfoam (*Limnanthes alba*). *Plant Physiol.* **94**, 1199-1206.
3. Löhden, I., Frentzen, M. (1992) Triacylglycerol biosynthesis in developing seeds of *Tropaeolum majus* L. and *Limnanthes douglasii* R. Br. *Planta* **188**, 215-224.
4. Coleman J. (1990) Characterization of *Escherichia coli* cells deficient in 1-acyl-*sn*-glycerol-3-phosphate acyltransferase activity. *J. Biol. Chem.* **265**, 17215-17221.
5. Brown, A.P., Brough, C.L., Kroon, J.T.M., Slabas, A.R. (1995) Identification of a cDNA that encodes a 1-acyl-*sn*-glycerol-3-phosphate acyltransferase from *Limnanthes douglasii*. *Plant Mol Biol* **29**, 267-278.
6. Hanke, C., Wolter, F.P., Coleman, J. Peterek, G., Frentzen, M. (1995) A plant acyltransferase involved in triacylglycerol biosynthesis complements an *Escherichia coli sn*-1-acylglycerol-3-phosphate acyltransferase mutant. *Eur J Biochem* **232**, 806-810.
7. Brough, C.L., Coventry, J.M.,Christie, W.W., Kroon, J.T.M., Brown, A.P., Barsby, T.L., Slabas, A.R. (1996) Towards the genetic engineering of triacylglycerols of defined fatty acid composition: major changes in erucic acid content at the *sn*-2 position affected by the introduction of a 1-acyl-*sn*-glycerol-3-phosphate acyltransferase from *Limnanthes douglasii*. *Molecular Breeding* (in press).
8. Lassner, M.W., Levering, C.K., Davies, H.M, Knutzon, D.S. (1995) Lysophosphatidic acid acyltransferase from meadowfoam mediates insertion of erucic acid at the *sn*-2 position of triacylglycerol in transgenic rapeseed oil. *Plant Physiol.* **109**, 1389-1394.

BRASSICA NAPUS CTP:PHOSPHOCHOLINE CYTIDYLYLTRANSFERASE: MOLECULAR CLONING AND SUBCELLULAR LOCALIZATION IN YEAST, AND ENZYME CHARACTERIZATION

I. NISHIDA[1], Y. KITAYAMA[1], R. SWINHOE[2], A.R. SLABAS[2], N. MURATA[3], AND A. WATANABE[1]
[1]*Department of Biological Sciences, Graduate School of Science, University of Tokyo, 7-3-1 Hongo, Bunkyoku, Tokyo, 113 Japan;* [2]*Department of Biological Sciences, University of Durham, Durham DH1 3LE, UK;* [3]*National Institute for Basic Biology, Okazaki, 444 Japan*

1. Introduction

Cellular levels of phosphatidylcholine (PC) have been known to be influenced by plant hormones [1] as well as by environmental stimuli such as low temperature [2]. However, little is known about the mechanism of regulation, and biological meanings are open to conjecture. For better understanding of the phenomena, it is essential to clone the gene of interest.

PC is synthesized in plant cells via the CDP-choline pathway, which includes three enzymatic steps catalyzed by choline kinase (EC 2.7.1.32; CKI), CTP:phosphocholine cytidylyltransferase (EC 2.7.7.15; CCT) and CDP-choline:diacylglycerol choline-phosphotransferase (EC 2.7.8.2; CPT). Among them, CCT has been reported to be a rate-limiting enzyme [3].

Recently, we have cloned four *Brassica napus* CCT cDNAs by complementation in a yeast *cct* mutant [4]. In the present studies, we have further extended our studies of *B. napus* CCTs and have updated our knowledge of the catalytic domain of CCT by comparing deduced amino-acid sequences of CCTs from various sources including rat [5], yeast [6] and *Plasmodium falciparum* [7]. We have also characterized enzymological properties of *B. napus* CCT1 overexpressed in *Escherichia coli* .

2. Results and Discussion

2.1 MOLECULAR CLONING AND FEATURES OF AMINO-ACID SEQUENCES

B. napus root cDNA library was constructed on the plasmid pFL61 [8], and a yeast cct mutant, INY103 [4] (*MATa, CHO1Δ::HIS3, CCTΔ::LEU2*, YCpGPSS [9]), was transformed with the cDNA library. Transformants were screened for phenotypic choline auxotrophs [4]. pFL61 plasmids were isolated and nucleotide sequences of cDNA inserts were determined. Deduced amino-acid sequences were grouped in four independent polypeptides containing 329, 331, 324 and 326 amino-acid residues, which are then designated CCT1, CCT2, CCT3 and CCT4, respectively. CCT1 and CCT2 showed an identity of 95.5% and CCT3 and CCT4 showed that of 99.4%. By contrast, in other combinations identity values was less than 82.4%, indicating that *B. napus* CCTs are subdivided in two subsets.

We have reported that *B. napus* CCTs resembled rat and yeast CCTs in the central domain of polypeptides (see Fig. 1), which are thus considered as the catalytic domain [4]. Recently, Yeo et al. [7] have cloned a gene for CCT from *P. falciparum* and have identified the catalytic domain conserved among CCTs from mammals, yeast and *P. falciparum*. Together with their data, Fig. 1 summarized updated data of sequence identity within the catalytic domain among various CCTs. Interestingly, one of CCT1 cDNA clones (designated p621) had a truncated nucleotide sequence starting from just behind the first ATG codon. We are currently confirming if this clone could synthesize an enzymatically active CCT polypeptide translated from the second Met, i.e., ^{85}Met.

2.2 ENZYME CHARACTERIZATION

CCT1 cDNA was subcloned in the *Bam* HI site of the plasmid pET3b, and the resultant cDNA was overexpressed in *E. coli* BL21(DE3). After incubation with 0.4 mM IPTG for 3 h, a CCT1 polypeptide accumulated as a 41 kDa band to a level of 6% relative to the total cellular proteins. With repeated experiments, overexpressed protein was always recovered in soluble fraction but not in insoluble faction.

E. coli cells were disrupted by passing through a French Pressure cell at 10,000 psi and the CCT1 polypeptide was fractionated by the FPLC protein purification system equipped with a HiTrap Q column (5 ml). The activity was measured in 20 μl of standard reaction mixtures containing 50 mM Tris-HCl (pH 8.0), 5 mM CTP, 25 mM MgCl$_2$, 5 mM DTT and 4 mM [*methyl*-^{14}C]phosphorylcholine (1,000 dpm/nmol). After incubation for 60 min at 25 °C, the reaction was stopped by adding 10 μl of 10% TCA containing 10 μg CDP-choline. Ten-microliter aliquots of the resultant solution were developed on TLC (ethanol : 2% ammonia = 1:1, by volume) and radioactive bands were quantified with a BAS2000 Image Analizer (Fujifilm).

With 80 mM Tris-HCl buffers, the activity was measured at comparable rates between pH 7.5 and pH 9.0, although some inhibitory effects were observed below pH 7.0, probably due to the lower buffering activity of the Tris buffer. Indeed, no inhibitory effect was observed with 80 mM BisTris-HCl buffer between pH 6.5 and pH 7.0, and the maximum activity was measure at pH 7.0 with this buffer system. However, at pH 6.0 with both 80 mM BisTris-HCl and 60 mM Histidine-HCl buffers the activity was

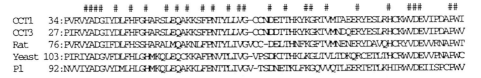

Figure 1. Deduced amino-acid sequence of various CCTs within the catalytic domain. Identical amino-acid residues among *B. napus*, rat, yeast and *Plasmodium* (Pl) CCTs are marked with #.

decreased to 40 % relative to the maximum activity, and the activity was totally inhibited at pH 5.4 with 60 mM Histidine-HCl buffer. The results are consitent with the optimum values of pH for CCTs purified from castor bean endosperm [12] and pea seedling [13], which are reported to be 7.5 at 30 °C and 7.0 at 25 °C, respectively.

The K_m values of *B. napus* CCT for CTP and phosphorylcholine were 0.19-0.24 and 0.43-0.52 mM, respectively. These values are similar to K_m values of purified castor bean CCT for CTP (0.20 mM) and phosphorylcholine (0.37 mM) [12] and K_m values of various rat CCT constructs for CTP (0.41-0.51 mM) and phosphorylcholine (0.39-0.47 mM) [14], and are slightly different from K_m values of purified pea CCT for CTP (0.55 mM) and phosphorylcholine (2.1 mM) [13].

INY103 transformed with a purified pFL61/CCT1 cDNA (p241) restored the CCT activity in both membrane and soluble fractions. This results indicated that *B. napus* CCT is compatible with a yeast cellular machinery which is responsible for targeting yeast CCT to membranes.

Acknowledgements:This work was supported, in part, by the Grant-in-Aid for Scientific Research (C) (No. 08640821) from The Ministry of Education, Science, Sports and Culture, Japan, to IN.

3. References

1. Moore, Jr., T.S., Price-Jones, M.J. and Harwood, J.L. (1983) The effect of indoleacetic acid on phospholipid metabolism in pea stems, *Phytochemistry* 22, 2421-2425.
2. Sikorska, E. and Kacperska-Palacz, A. (1980) Phospholipid involvement in frost tolerance, Physiol. Plant. 48, 201-206.
3. Price-Jones, M.J. and Harwood, J. L. (1983) Hormonal regulation of phosphatidylcholine synthesis in plants. The inhibition of cytidylyltransferase activity by indol-3-ylacetic acid, Biochem. J. 216, 627-631.
4. Nishida, I., Swinhoe, R., Slabas , A.R. and Murata, N. (1996) Cloning of *Brassica napus* CTP:phosphocholine cytidylyltransferase cDNAs by complementation in a yeast *cct* mutant, *Plant Mol. Biol.*, in press.
5. Kalmer, G.B., Kay, R. J., Lachance, A., Aebersold R. and Cornell R. B. (1990) Cloning and expression of rat liver CTP:phosphocholine cytidylyltransferase: An amphipathic protein that controls phosphatidycholilne synthesis, Proc. Natl. Acad. Sci. USA 87, 6029-6033.
6. Tsukagoshi, Y, Nikawa, J. and Yamashita, S. (1987) Molecular cloning and characterization of the gene encoding cholinephosphate cytidylyltransferase in *Saccharomyces cerevisiae*, Eur. J. Biochem. 169, 477-486.
7. Yeo, H. J. , Sri Widada, J., Mercereau-Puijalon, O. and Vial, H.J. (1995) Molecular cloning of CTP:phosphocholine cytidylyltransferase from Plasmodium falciparum, Eur. J. Biochem. 233, 62-72.
8. Minet, M., Dufour, M.-E. and Lacroute, F. (1992) Complementation of *Saccharomyces cerevisiae* auxotrophic mutants by *Arabidopsis thaliana* cDNAs, Plant J. 2, 417-422.
9. Hamamatsu, S., Shibuya, I., Takagi, M. and Ohta, A. (1994) Loss of phosphatidylserine synthesis results in aberrant solute sequestration and vacuolar morphology in *Saccharomyces cerevisiae*, FEBS Lett. 348, 33-36.
12. Wang, X. and Morre, Jr., T.S. (1989) Phosphatidylcholine biosynthesis in castor bean endosperm, *Plant Physiol.* 93, 250-255.
13. Price-Jones, M. and Harwood, J. L. (1985) Purification of CTP:cholinephosphate cytidylyltransferase from pea stem, *Phytochemistry* 24, 2523-2527.
14. Seitzer, T.D. and Kent C. (1994) Expression of wild-type and mutant rat liver CTP:phosphocholine cytidylyltransferase in a cytidylyltransferase-deficient Chinese hamster ovary cell line, *Arch. Biochem. Biophys.* 311, 107-116.

CLONING OF A PUTATIVE PEA CHOLINEPHOSHATE CYTIDYLYLTRANSFERASE AND EVIDENCE FOR MULTIPLE GENES

PHILIPPA L. JONES[1], DAVID L. WILLEY[1], PETER GACESA[2] AND JOHN L. HARWOOD[1].
1 School of Molecular and Medical Biosciences, University of Wales Cardiff, PO Box 911, Cardiff, CF1 3US, UK.
2 Dept. of Appl. Biol., University of Central Lancashire, Preston, PR1 2HE, UK.

1. Introduction

In plants phosphatidylcholine is the major phospholipid in extra-plastid membranes and is synthesised mainly by the CDP-choline pathway. There are three steps in the pathway catalysed by the enzymes choline kinase, cholinephosphate cytidylyltransferase (CPCT) and cholinephosphotransferase (Harwood, 1989). Evidence from studies in animals (Pelech and Vance, 1984) and in plants (Price-Jones and Harwood, 1983) suggests that the intermediate step catalysed by CPCT is the main regulatory step. The cloning of a full length and partial cDNA clones for pea cholinephosphate cytidylyltransferase is described.

2. Results and Discussion

Degenerate oligonucleotides were designed using a comparison of the CPCT genes from rat (Kalmar et al., 1990) and yeast (Tsukagoshi et al., 1987). The central part of the pea gene designated CPCT 1 was amplified by PCR from pea root cDNA using these primers, cloned and sequenced. The partial clone is 81 amino acids in length (Fig. 1) and is 80% and 62% identical at the amino acid level to the rat and yeast sequences, respectively.

An *Arabidopsis* expressed sequence tag (EST) clone was obtained with a significant similarity of 77% at the amino acid level to the rat CPCT gene. This clone was used to screen 400 000 clones in a pea seedling λgt10 cDNA library and 25 positive clones were identified. Three clones were purified and found to have the same nucleotide sequence. All the clones were truncated at the 3' end due to an *Eco*R I site in the DNA sequence. This part of the gene was obtained by PCR using the technique of rapid amplification of cDNA ends (RACE). Pea RNA was reverse transcribed using an oligo-dT adaptor primer (Kille *et al.*, 1991). A primer specific to the pea gene and the adaptor primer were used to amplify the 3' end of the gene designated CPCT 2. The full length cDNA (Fig. 1) codes for a protein 286 amino acids in length. CPCT 2 is 48%, 43% and 76% identical at the amino acid level to the rat, yeast and *Arabidopsis* sequences, respectively.

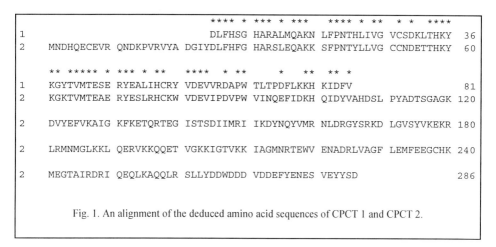

Fig. 1. An alignment of the deduced amino acid sequences of CPCT 1 and CPCT 2.

A Southern blot of pea genomic DNA (30μg per lane) digested with *Hin*d III or *Eco*R I (Fig. 2) was probed sequentially with CPCT 1 and 2. There are one or two copies of the CPCT 1 gene in pea. The CPCT 2 probe hybridised to multiple bands implying there are more than two copies of this gene or closely related sequences in the pea genome.

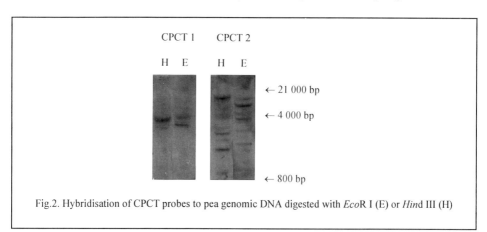

Fig.2. Hybridisation of CPCT probes to pea genomic DNA digested with *Eco*R I (E) or *Hin*d III (H)

Short pieces of pea stems were treated in a phosphate buffer with different concentrations of indoleacetic acid for 2.5 hours as described in Moore *et al.* (1983). Total RNA was isolated from the treated stems and 5μg of each sample was electrophoresed on a denaturing agarose gel and blotted. The CPCT 2 probe was hybridised to the northern blot and a small increase in the RNA expression of CPCT 2 was observed on treatment with indoleacetic acid (Fig. 3). An increase in enzyme activity was observed in similarly treated pea stems over a period of 5 hours (Moore *et al.*, 1983). This effect of indoleacetic acid on enzyme activity may be partly explained by the increase in mRNA levels.

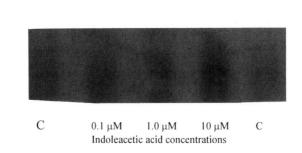

C   0.1 μM   1.0 μM   10 μM   C
Indoleacetic acid concentrations

Fig. 3. Hybridisation of CPCT2 to total RNA isolated from pea stems treated with different concentrations of indoleacetic acid. Controls (C) were not treated with hormone.

3. Acknowledgements

We are grateful to J. S. Gantt for providing the λgt10 pea cDNA library and to P. Kille for providing the oligo-dT adaptor and the adaptor primers. The *Arabidopsis* EST clone was obtained from the *Arabidopsis* Biological Resource Center, Columbus, OH 43210-1002, USA. This work was funded by the BBSRC.

4. References

Harwood, J.L. (1989) Lipid metabolism in plants. *Critical Reviews in Plant Sciences* **8**, 1-43.
Kalmar, G.B., Kay, R.J., Lachance, A., Aebersold, R. and Cornell, R.B. (1990) Cloning and expression of rat liver CTP: phosphocholine cytidylyltransferase: an amphipathic protein that controls phosphatidylcholine synthesis. *Proc. Natl. Acad. Sci., USA* **87**, 6029-6033.
Kille, P., Stephens, P.E. and Kay, J. (1991) Elucidation of cDNA sequences for metallothioneins from rainbow trout, stone loach and pike liver using the polymerase chain reaction. *Biochim. Biophys. Acta* **1089**, 407-410.
Moore, T.S., Price-Jones, M.J. and Harwood, J.L. (1983) The effect of indoleacetic acid on phopholipid metabolism in pea stems. *Phytochem.* **22**, 2421-2425.
Pelech, S.L. and Vance, D.E. (1984) Regulation of phosphatidylcholine biosynthesis. *Biochim. Biophys. Acta* **779**, 217-251.
Price-Jones, M. and Harwood, J.L. (1983) Hormonal regulation of phosphatidylcholine synthesis in plants: the inhibition of cytidylyltransferase activity by indolyl-3-acetic acid. *Biochem. J.* **216**, 627-631.
Tsukagoshi, Y., Nikawa, J. and Yamashita, S. (1987) Molecular cloning and characterisation of the gene encoding cholinephosphate cytidylyltransferase in Saccharomyces cerevisiae. *Eur. J. Biochem.* **169**, 477-486.

IMMUNOLOGICAL IDENTIFICATION OF SUNFLOWER (*HELIANTHUS ANNUUS* L.) SPHEROSOMAL LIPASE

[1]Sellema BAHRI, [2]Raja MARRAKCHI, [1]Ahmed LANDOULSI,
[1]Jeannette BEN HAMIDA and [3] Jean Claude KADER.

1 Laboratoire de Biochimie, Faculté des Sciences de Tunis, Tunisie
2 Laboratoire d'Immunologie, Faculté des Sciences de Tunis, Tunisie.
3 Laboratoire de Physiologie Cellulaire et Moléculaire, Université Pierre et Marie Curie, France.

Synopsis

The spherosomal lipase from sunflower (*Helianthus annuus* L.) seedlings partially purified presented 2 protein bands with an approximate molecular weight of 61 and 66 kDa, in a characteristic zone of plant lipases. In order to determine which one corresponds to the lipase, we prepared specific antibodies against each protein bands. The polyclonal antibodies obtained have been tested by western blotting on spherosomal membrane proteins and also on a pure lipase from hog pancreas. The immunserum directed against the lower band (61 kDa) reacted specifically with the spherosomal lipase protein and is also reactive on hog pancreatic lipase. An enzymatic study showed that only the lower immunserum specifically inhibited sunflower lipolytic activity.

Introduction

It is well known that lipase catalysed the first step of lipids' catabolism, in which they were converted to sugars and other metabolits. Information on these enzymes' action was important especially during agricultural product storage. True lipase plant notion has been neglected until recently (Huang et al 1987) so that two plant lipases have been first purified to homogeneity : the lipase from the scutella of corn (*Zea mays*) associated to "spherosomes" has a molecular weight of 65 kDa (Lin and Huang, 1984) and the lipase from castor bean (*Ricinus communis*) associated to another type of organelles, the "glyoxysomes". This enzyme has a Mr of 62 kDa (Maeshima and Beevers, 1985).

In this paper we report the immunological identification of sunflower spherosomal lipase.

Materials and Methods

SUBCELLULAR FRACTION.

The cotyledons from sunflower seedlings (*Helianthus annuus* L var Albena) of 3 days old were used. The spherosomes were extracted using Qu and al. method (1986) adapted to the sunflower (Bahri and al 1992). All operations were performed at 0 - 4°C.

LIPASE ACTIVITY.

Lipase activity was measured by a colorimetric method (adapted from Nixon and Chan. 1979). The substrate (sunflower oil) was purified. The reaction was carried out at 30°C in a shaker water bath.

IMMUNOLOGICAL PROCEDURE

In collaboration with the immunological laboratory of Tunis University, we prepared polyclonal antibodies against each protein band. After a preparative electrophoresis, the 66 and the 61 kDa bands were extracted from the gel and injected to different rabbits.

Spherosomal membrane proteins were transferred on to nitrocellulose membrane by electroblotting for 2 H at 1 A.The nitrocellulose membrane was blocked by incubation overnight (4 °C) in 3 % w/v dried milk powder in a saline buffer phosphate (PBS, pH 7.2) then washed with PBS tween 20, incubated with a 1 / 100 dilution of the different anti lipase antibodies during 2 h (37 ° C), washed again with PBS tween 20. The membrane was incubated for 2 h with 1/1000 dilution of peroxidase - conjugated goat anti - rabbit antibody (Sigma), washed with PBS tween 20, then incubated with the substrate : Diamine - benzidine + H_2O_2. Finally after 20 mn , the reaction was stopped with water.

Anti - lipase stained bands were localized with reference to molecular weight markers run on the same electrophoretic gel and blotted on to nitrocellulose membrane.

Results and Discussion

True lipase activity was found in unbound final chromatography fractions (Bahri and al 1992). By SDS PAGE we showed the presence of two proteic bands (61 and 66 kDa) in a characteristic zone of lipase molecular band. This result raised the question of the functional and structural relation ship between the two bands. In order to answer this question, we tried to prepare specific antibodies against each protein band. The polyclonal antibodies obtained have been tested by western blotting on delipidated spherosomal membrane proteins.

The immunserum directed against the upper band did not show any reaction , on the contrary the one prepared against the 61 kDa band specifically reacted with a single 61 kDa present in spherosomal membranes (figure 1).

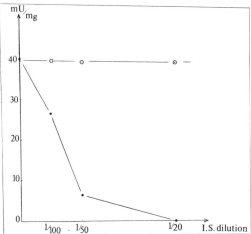

Figure 1. Immunological determination of spherosomal lipase by Western blot .
1, 2 Control antisera ; 3 Upper (66 kDa) antibody ; 4 Lower (61 kDa) antibody; 5 Hog pancreatic lipase (50 kDa).

We also tested lipolytic activity in presence of different dilutions of the 61 kDa immunserum (Figure 2).

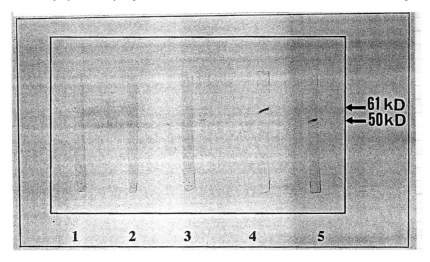

Figure 2 . Lipolytic activity of the spherosomal lipase in presence of different lower immunserum dilutions (•) and immunserum control (o).

We observed that lipolytic activity was 100% inhibited in presence of the lower 1/20 diluted immunserum indicating that the lower band is involved in the lipolytic activity.
Non specific inhibition was controlled for by repeating the experience with the immuncontrol serum The upper immunserum didn't show any incidence on this activity. This result could be in favor of the fact that this molecule is not implicated in lipase activity, but does not exclude the possibility that epitopes recognized by the upper immunserum could not be directly involved in the active site.

Finally we tested the two immunsera by western blot on a pure hog pancreatic lipase. Only the lower immunserum was reactive with this lipase. The cross reactivity with an animal lipase involves primary sequence epitopes rather than conformational epitopes. These two molecules have a sequence homology. Probably the consensus sequence Gly - X - Ser - X - Gly commonly present in animal and microbial lipases (Antonian, 1988; Blow, 1990).

References

Antonian, E. (1988) *Lipids* **23** ,1101 - 1106
Bahri, S., Oursel, A., Ben Hamida, J. and Kader J. C. (1992) *10 th International Symposium on The Metabolisme Structure and Utilization of Plant Lipids* . eds Cherif and al. Tunisia.
Blow, D (1990) *Nature* **343** , 6944 - 695.
Chua, N. H. (1980) *Method Enzymol.* **69**, 434 - 446.
Huang, A. H. C. , Qu, R. , Wang, S., Vance, V. B. Cao, Y., Lin , Y. (1987) In : *The Metabolisme Structure and Function of Plant Lipids* . eds Stumpf P. K, Mudd, J. B. New - York. 239 - 246
Lin , Y. H. , Huang AHC (1984) *Plant Physiol.* **76**, 719 - 722.
Maeshima , M., Beevers, H. (1985) *Plant Physiol.* **79**, 489 - 493.
Qu, R., Wang, S. H., Lin, Y. H. , Vance, V. B. , Huang, A. H. C.*Biochem* . J. **235**, 57 - 65

PURIFICATION AND IMMUNOLOGICAL ANALYSIS OF PHOSPHOLIPASE D FROM *BRASSICA NAPUS* (RAPE SEED)

Z. NOVOTNÁ, O. VALENTOVÁ, J. DAUSSANT[1], J. KÁŠ

Institute of Chemical Technology, Department of biochemistry and microbiology, Technická 3, 166 28 Prague 6, Czech Republic,
[1]*CNRS, 4 ter route des Gardes, 92 190 Meudon, France*

Synopsis

Phospholipase D has been purified to apparent homogeneity from the seeds of *Brassica napus*. Enzyme characteristics are discussed including the molecular weight, isoelectric point, pH optimum, effects of calcium ionts and SDS. The immunochemical cross-reactivity with proteins of soya beans and sunflower seeds was tested.

Introduction

Rape seed (*Brassica napus*) represents one of the most important oil seeds used in food and chemical industry of Central Europe.

Phospholipase D (PLD, EC 3.1.4.4) hydrolyses phospholipids, liberating phosphatidic acid (PA) and the alcohol moiety which participates in the ester linkage (choline, ethanolamine, glycerol or inositol). The enzyme was found to be widespread in the plant kingdom (1). A number of recent studies which investigated the putative roles of these phospholipases suggest common structural features between PLDs of different plant origins.

The studies started with rape seed PLD (2) provided preliminary results on some enzymatic and biochemical features of that enzyme. The study of one oilseed PLD was aimed at enhancing knowledge of an enzyme which may play a negative role during rape seed oil manufacturing.

The detailed study of the enzyme characteristics reported here provides the bakground for the further investigations on the possible regulation of PLD activity in rape seed metabolism.

Materials and methods

Winter rape seed variety Lirajet (double zero) was obtained from the Setuza Inc. The mature dry seeds were stored at $-70\,^0$ C until used.

After extraction by the procedure described earlier (2), the obtained supernatant was extracted with hexane to remove the lipids and centrifuged at 48000g for 20 min.

Proteins were precipitated with ammonium sulphate (30%-50%), desalted and loaded on Octyl-Sepharose CL-4B column (11 × 1.3 cm) equilibrated with the 0.03 M MES buffer pH 6.0 with 120mM $CaCl_2$. Unbound proteins were eluted with 0.01 M MES pH 6.0 with 50 mM $CaCl_2$. The PLD activity was found in the fractions eluted with 0.2 mM EDTA. Active fractions were pooled, concentrated on Centricon 30 and applied on the native PAGE (8%) according to Wang (3). Finally the active PLD protein was electroeluted from polyacrylamide gel.

The PLD activity assay was based on the determination of choline released from emulsified phosphatidylcholine substrate. The choline content was determined by a choline biosensor (4).

Results and Discussion

The enzyme was purified to electrophoretic homogeneity using a three step procedure involving ammonium sulphate fractionation, chromatography on the column of Octyl-Sepharose and non-denaturing PAGE followed by electroelution. The procedure resulted in over 700 fold purification with 4,6% yield of the activity (TABLE I).

TABLE I Summary of phospholipase D purification from winter rape seed (var.Lirajet, 1995)

Purification step	Total activity (U)	Total proteins (mg)	Specific act.ivity (U/mg)	Purification (-fold)	Yield (%)
Extract	58.4	307.4	0.19	1	100
Crude enzyme[a]	33.6	1.44	1.1	5.6	58
Octyl-Sepharose	18.9	0.19	99.5	523	32
PAGE + elution	2.7	0.020	133.3	700	4.6

Analysis by SDS-PAGE revealed a single polypeptide of MW 90 000 Da (*Figure 1*). The native enzyme had a MW of 90-100 kDa when analysed by size-exclusion chromatography (TSK column) suggesting that PLD from rape seed is a monomer.

Biochemical characteristics of purified PLD are alike to most of the plant soluble phospholipase D (TABLE II) concerning the pH optimum, pI and mol. wt. (denaturing conditions) and differes in native mol.wt. and higher concentration of activators is demanded. In the contrast with results published for PLD from Castor bean endosperm (5), rape seed PLD seems to be a monomeric protein.

TABLE II Characteristics of purified PLD from *Brassica napus*

Mol.Wt. native (kDa)	Mol.Wt. denatured (kDa)	K_m PtdCho (mM)	pH optimum	pI	Ca^{2+} (mM)	SDS (mM)	specific activity (U/mg)
90-100	105	2.5	6.0	4,85	120	8.3	133

Immunoprecipitation provided with the extracts of other oil seeds showed crossreactivity of the anti rape seed PLD immunoserum with sunflower seed and soybean PLD.

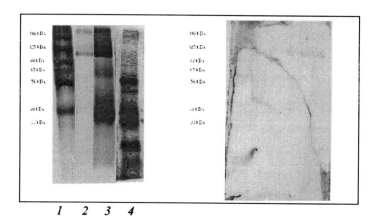

Figure 1. Protein profiles resolved on SDS-PAGE (10%), **A**: lane 1 - molecular weight markers, lane 2 - preparation obtained by native PAGE and electroelution, lane 3 - fraction after chromatography on the Octyl-Sepharose, lane 4 - crude enzyme. **B**: immunoblot analysis of finally purified rape seed PLD.

References

1. Heller, M. (1978) Phospholipase D: A rewiev. *Adv. Lipid. Res.* **16**, 267 - 325.
2. Valentová, O., Novotná, Z., Káš, J., Daussant, J. and Thévenot, C. (1994) Rape seed Phospholipase D, *Potrav. Vedy (Food Sciences)* **12**, 297 - 308.
3. Wang, X., Dyer, J. H. and Zheng, L. (1993) Purification and immunological analysis of phospholipase D from Castor bean endosperm, *Arch. Biochem. Biophys.* **306**, 486 - 494.
4. Vrbová, E., Kroupová, I., Valentová, O., Novotná, Z., Káš, J. and Thévenot, C. (1993) Determination of phospholipase D activity with choline biosensor, *Anal.Chim. Acta* **280**, 43-48.
5. Dyer, J.H., Ryu, S.B. and Wang, X. (1994) Multiple forms of phospholipase D following germination and during leaf developmnet of Castor bean, *Plant. Physiol.* **105**, 715 - 724.

MODIFICATION OF SEED OIL CONTENT AND ACYL COMPOSITION IN THE BRASSICACEAE UTILIZING A YEAST sn-2 ACYLTRANSFERASE (*SLC1-1*) GENE

J.-T. ZOU, V. KATAVIC, E.M. GIBLIN, D.L. BARTON, S.L. MACKENZIE, W.A. KELLER and D.C. TAYLOR
NRC Canada, Plant Biotechnology Institute, 110 Gymnasium Place, Saskatoon, SK, Canada S7N 0W9.

Introduction

The *SLC1* gene was originally cloned from a yeast mutant lacking the ability to make sphingolipids (1). *SLC1* is homologous to the *E. coli plsC* gene, which has been claimed to encode *lyso*-phosphatidic acid acyltransferase (LPAT; EC 2.3.1.51, (2)). The *SLC1* gene was able to complement the growth defect in JC201 (an *E. coli* strain mutated in *PLSC*). It was reported that *SLC1* encodes a yeast *sn*-2 acyltransferase. However, the authors were unable to detect LPAT activity in the complemented *E. coli* JC201 mutant (1). The sequence of a suppressor allele designated *SLC1-1* was also reported and shown to encode a protein which suppresses the genetic defect in sphingolipid long chain base biosynthesis (3). In *SLC1-1*, nucleotide 131 is a T instead of an A, resulting in amino acid 44 being leucine rather than glutamine. The working hypothesis is that the *SLC1-1* gene encodes a variant acyltransferase with an altered substrate specificity, which enables it to use a very long-chain fatty acid (26:0) to acylate the *sn*-2 position of inositol-containing glycerolipids. The authors have not, to date, provided conclusive evidence of activity encoded by *SLC1-1* or *SLC1*.

Based on our initial assessment of the relatively broad acyl specificity of yeast LPAT, and our interest in modifying the very long-chain fatty acid (VLCFA) content of *Brassicaceae* seed oils (4), we expressed the *SLC1-1* gene in the model oilseed *Arabidopsis thaliana*, and in a high erucic acid *B. napus* cultivar. It will be shown that the yeast *SLC1-1* gene can be used to change the content and composition of plant seed oils, via increased seed LPAT activity.

Experimental Methods

Construction of vectors for SLC1-1 transformation: Two primers with 5' BamHI restriction site extensions, (AGAGAGAGGGATCCATGAGTGTGATAGGTAGG) & (GAGGAAGAAGGATCCGGGTCT ATATACTACTCT) designed according to the 5' and 3' end sequences of the *SLC1* gene, respectively, were used in a Polymerase Chain Reaction (PCR) with plasmid p411ΔB/C (obtained from Dr. R. Dickson, University of Kentucky, KY), harboring the *SLC1-1* gene as template, to generate the *SLC1-1* PCR fragment with a BamHI site at both ends. The fragment was digested with BamHI and ligated into the BamHI cloning site located between the tandem 35S promoter and NOS terminator in vector pBI524. The orientation of *SLC1-1* was verified by digestion with BglII. The translation initiation codon of *SLC1-1* is maintained, and hence the construct is a transcriptional fusion. The HindIII and EcoRI fragment containing a tandem 35S promoter, AMV enhancer, *SLC1-1* encoding sequence and NOS terminator was freed from *SLC1-1*-pBI524, and cloned into the

EcoRI-HindIII site of vector pRD400. The final vector p*SLC1-1*/pRD400 (ATCC 97545) was introduced into *Agrobacterium tumefaciens* strain GV3101 (bearing helper plasmid pMP90) by electroporation.

Plant growth conditions: All *A. thaliana* plants were grown in controlled growth chambers, under continuous fluorescent illumination (150-200 $\mu E \cdot m^{-2} \cdot sec^{-1}$) at 22°C. All *B. napus* plants were grown under natural light supplemented with high pressure sodium lamps with a 16 hour photoperiod (16 h light/8 h dark), at 22°C, and a relative humidity of 25-30%.

Plant transformation: Wild type (WT) *A. thaliana* plants of ecotype Columbia were transformed *in planta* by wound inoculation (5) with a suspension of *A. tumefaciens* strain GV3101 bearing helper nopaline plasmid pMP90 and binary vector p*SLC1-1* /pRD400. *B. napus* cv. Hero (high erucic acid variety) was transformed by co-cultivation of hypocotyl explants using modifications of a published protocol (6).

Lipid Analyses and Acyltransferase (LPAT) Assays: Fatty acid analyses, stereospecific analyses and LPAT assays were conducted as described previously (7).

Results and Discussion

A. thaliana SLC1-1 Transformant Seed Lipid Analyses: T_2 seed from a large number of *A. thaliana SLC1-1* transgenic lines (21 of 48) showed significantly increased oil contents (11- 49% increase in μg total fatty acids / 50 seeds) compared to non-transformed controls (n-WT), and pBI121 controls (without *SLC1-1* insert, but with KAN^r selectable marker). Several lines exhibited dramatic increases in total VLCFA content, especially 20:1 and 22:1, and elevated proportions of VLCFAs.

The processes of transformation and kanamycin selection alone had no significant effect on seed oil content or fatty acid composition, since these values were essentially identical in seed from n-WT and pBI121 control plants. For example, the fatty acid content for equivalent numbers of T_2 seed from pBI121 controls was 97% of the value for n-WT seeds. The proportions of polyunsaturated C_{18} fatty acids in n-WT controls and in pBI121 controls were 46.4 ± 2.3 % and 46.1 ± 2.3 %, respectively, while the corresponding proportions of VLCFAs were 27.5 ± 1.3 % and 26.3 ± 1.3 %.

Those T_2 lines showing the greatest increases in seed oil content and proportions of VLCFAs were propagated to give T_3 progeny lines. Several *SLC1-1* T_3 seed lines contained significantly increased lipid content (μg total fatty acids/100 seeds) compared to pBI121 controls (Fig. 1).

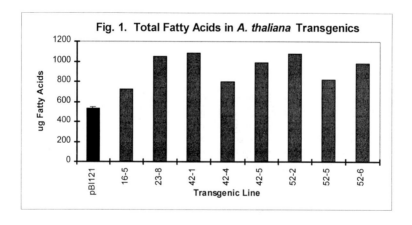

The amounts of VLCFAs (Table 1) and of VLCFA-containing C_{56}-C_{60} TAGs were greatly enhanced.

TABLE 1. Total VLCFA content of T_3 seed of pBI121 controls and selected *SLC1-1* transgenic lines of *A. thaliana*. Averages based on 100-seed samples. For pBI121 controls, values in parentheses = ± SEM. For transgenics, values are average of 2 determinations.

Line	pBI121 n=5	16-5	23-8	42-1	42-4	42-5	52-2	52-4	52-6
VLCFAs (ug)	130 (15)	200	310	315	255	310	315	245	270

T_3 seed TAGs from selected independent *SLC1-1* transgenics contained increased proportions of VLCFAs (e.g. 20:1) at the *sn*-2 position (Table 2). Furthermore, these increases were correlated with decreases in the proportions of polyunsaturated C_{18} fatty acids at this position.

TABLE 2. Proportions of VLCFAs and polyunsaturated fatty acids at the *sn*-2 position of TAGs in T_3 seed of pBI121 controls (± SEM), and *SLC1-1* transgenic lines of *A. thaliana*. (Wt%)

Line	pBI121, n=4	16-5	23-8	42-1	42-4	42-5	52-2	52-5	52-6
sn-2 VLCFAs	1.8 ± 0.8	17.8	13.3	8.9	14.5	9.4	10.6	11.8	13.4
sn-2 [18:2+18:3]	81 ± 1.3	72	62	73	66	72	70	66	64

B. napus cv. Hero SLC1-1 Transformant Seed Lipid Analyses: *B. napus* cv. Hero *SLC1-1* T_2 seed lines exhibited 10-25% increases in oil content and 4-10% increases in erucic acid content.

LPAT Analyses of Transformant Lines: LPAT analyses of developing seed from n-WT and *SLC1-1* transgenics of *A. thaliana* indicated that 20:1-CoA:LPAT activity was increased 15-200% in the transgenics. Similarly, in *B. napus* cv. Hero, both 18:1-CoA:LPAT and 22:1-CoA:LPAT specific activities were increased 2 to 10-fold in the transgenics. Thus, we provide, for the first time, direct evidence that the yeast *SLC1-1* gene product encodes an enzyme which possesses *sn*-2 acyltransferase activity, and which can exhibit LPAT activity *in vitro*.

References

1. Lester, R.L., Wells, G.B., Oxford, G. and Dickson, R.C. (1993) Mutant strains of *Saccharomyces cerevisiae* lacking sphingolipids synthesize novel inositol glycerophospholipids that mimic sphinglipid structure. *J. Biol. Chem.* **268**: 845-856.
2. Coleman, J. (1992) Characterization of the *Escherichia coli* gene for 1-acyl-*sn*-glycerol-3-phosphate acyltransferase (*plsC*) *Mol. Gen. Genet.* **232**:295-303.
3. Nagiec, M.M., Wells, G.B., Lester, R.L. and Dickson, R.C. (1993) A suppressor gene that enables *Saccharomyces cerevisiae* to grow without making sphingolipids encodes a protein that resembles an *Escherichia coli* fatty acyltransferase. *J. Biol. Chem.* **268**:22156-22163.
4. Taylor, D.C., Magus, J.R., Bhella, R., Zou, J., MacKenzie, S.L., Giblin, E.M., Pass, E.W., and Crosby, W.L. (1993). Biosynthesis of triacylglycerols in *Brassica napus* L. cv. Reston; Target: trierucin, in S.L. MacKenzie and D.C. Taylor, (eds), *Seed Oils for the Future*, Am. Oil Chem. Soc., Champaign, Illinois, Chapter 10, pp 77-102.
5. Katavic, V., Haughn, G.W., Reed, D.W., Martin, M. and Kunst, L. (1994) *In planta* transformation of *Arabidopsis thaliana*. *Mol. Gen. Genet.* **245**:363-370
6. Moloney, M.M., Walker, J. and Sharma, K. (1989) High efficiency transformation of *Brassica napus* using *Agrobacterium*. *Plant Cell Reports* **8**:238-242.
7. Taylor, D.C., Barton, D.L., Giblin, E.M., MacKenzie, S.L., van den Berg, C.G.J. and McVetty, P.B.E. (1995) Microsomal *lyso*-phosphatidic acid acyltransferase from a *Brassica oleracea* cultivar incorporates erucic acid into the *sn*-2 position of seed triglycerides. *Plant Physiology*, **109**:409-420.

AUTHOR INDEX

ABIGOR, R.D. 278
ABUSHITA, A. 281
ADLERSTEIN, D. 90, 93, 218
ALBAN, C. 17, 35
ÁLVAREZ-ORTEGA, R. 322
AVDIUSHKO, S. 377, 389
AZUMA, M. 137

BAFOR, M. 57
BAHRI, S. 401
BALDOCK, C. 38
BANAS, A. 57, 230, 244
BARNES, B.J. 236
BARSBY, T. 392
BARTON, D.L. 319, 407
BEILINSON, V. 29
BEN HAMIDA, J. 401
BEREMAND, P.D. 351
BEUTELMANN, P. 253
BIACS, P.A. 215, 281
BIGOGNO, C. 90, 93, 218
BITHEL, S. 38
BROOKER, N.L. 236
BROUGH, C.L. 392
BROUN, P. 342
BROWN, A.P. 368
BROWSE, J. 166, 200, 203
BRUBAKER, C.L. 383
BRYANT, D.A. 380
BURTIN, D. 20
BYERS, S.D. 357

CAHOON, E.B. 6, 374
CAMPBELL, D. 172
CANTISÁN, S. 322
CHAPMAN, K.D. 107
CHERVIN, D. 227
CHO, S.H. 371
CHOI, J.-K. 363
CHOI, S.-B. 371
CHOJNACKI, T. 192
CHRISTIE, W. 392
CHUCK, J.-A. 386
COHEN, Z. 90, 93, 218
COLE, D.J. 17
COUTURIER, D. 195

COVELLO, P.S. 63
COVENTRY, J. 392
CRANMER, A. 304
CROFT, K.P.C. 268
CROTEAU, R. 51

DA SILVA, P.M.R. 307
DALPÉ, Y. 195
DAOOD, H.G. 215, 281
DAUSSANT, J. 404
DAVIES, M. 304
DAVORAN, J.M. 357
DE BOER, G.-J. 360
DEKA, R.K. 48
DELLEDONNE, M. 336
DEWEY, R.E. 110
DIETRICH, C. 336
DIEZ, T. 363
DINSMORE, P.K. 110
DISCH, A. 177, 180
DOUCE, R. 35
DOUMA, A.C. 284
DZIEWANOWSKA, K. 209

EASTMOND, P.J. 66, 307
EICHENBERGER, W. 116
EK, B. 244
ELBOROUGH, K.M. 11, 14, 23
EMES, M.J. 307
ENARI, D. 84
EVANS, I.M. 23
EVENSON, K.J. 45

FAWCETT, T. 81, 360
FERRI, S.R. 137
FEUSSNER, I. 250
FOGLIA, T.A. 96, 259, 265
FRASER, T. 122
FRENTZEN, M. 119
FROESE, C.D. 154
FUJII, T. 212
FURUKAWA-STOFFER, T.L. 257

GACESA, P. 398
GARCÉS, R. 322
GASSER, A. 169

GAWER, M. 227
GIBLIN, E.M. 319, 407
GOMBOS, Z. 87
GOODE, J.H. 110
GOODRICH-TANRIKULU, M. 60
GRANDMOUGIN-FERJANI, A. 195
GRAY, G.R. 206
GREEN, A.G. 383
GREENWAY, D. 239
GRENIER, G. 163
GRIFFITHS, G. 122
GROSBOIS, M. 128
GUAN, X. 363
GUERBETTE, F. 128
GUERN N. 227
GUERRA, D. 209

HAMBERG, M. 99
HAMMERLINDL, J.K. 63
HANSEN, J.D. 336
HAPSORO, D. 268
HARAMATY, E. 157
HARRISON, P.J. 256
HARTMANN, M.-A. 195
HARWOOD, J.L. 17, 104, 262, 398
HATANAKA, A. 247
HAWKER, N. 342
HEBSHI, E. 281
HELSPER, J.P.F.G. 301
HERBERT, D. 17
HERMAN, E. 289
HICKS, K.B. 189
HILDEBRAND, D.F. 268, 377, 389
HILLS, M.J. 311, 368
HLOUSEK-RADOJCIC, A. 45, 339
HODGES, D.M. 357
HOFMANN, M. 116
HOLTMAN, W.L. 284
HONG, Y. 154
HRISTOVA, M.K. 51
HSU, A.-F. 265
HUANG, W. 6
HUANG, A.H.C. 292, 295
HUDAK, K.A. 154
HUNER, N.P.A. 206
HUTCHINGS, D. 307

IMAI, H. 224
ISHIZAKI-NISHIZAWA, O. 137, 212
ITO, S. 224
IWAMATSU, A. 354

JACKSON, F. 122

JAYAKUMAR, P.S. 339
JAWORSKI, J.G. 45
JOB, D. 17
JOHNSON, P.E. 368
JOLLIOT, A. 128
JONES, A. 304
JONES, H. 141
JONES, P.L. 398

KADER, J.-C. 128, 275, 401
KAIJWARA, T. 247, 348
KAMISAKA, Y. 125
KANG, F. 307
KÁŠ, J. 275, 404
KATAVIC, V. 407
KE, J. 363
KELLER, W.A. 63, 407
KERESZTES, Á. 215
KHAN, M.U. 206
KHOZIN, I. 90, 93, 218
KIM, J.C. 371
KIM, M.R. 271
KIM, Y. 160
KINNEY, A.J. 298
KITAGAWA, F. 354
KITAYAMA, Y. 395
KNUTZON, D. 304
KOJIMA, M. 224
KONISHI, T. 32
KROL, M. 206
KROON, J. 392
KROUMOVA, A.B. 42, 54
KÜHN, H. 250
KUNST, L. 72
KUSANO, T. 84

LACEY, D.J. 310
LAGOUCHE, O.J. 113
LANDOULSI, A. 401
LAROCHE, A. 357
LARSON, T. 256
LARUELLE, F. 195
LEE, K.-W. 371
LEIKIN-FRENKEL, A.I. 233
LENMAN, M. 57, 230, 244
LESHEM, Y.Y. 157
LI, C. 313
LICHTENTHALER, H.K. 177, 180
LIN, J.T. 113
LINDQVIST, Y. 6
LIU, D. 339
LIU, Q. 383
LIVNE, A. 131

LOS, D.A. 87

MACKENZIE, S.L. 319, 407
MALIK, Z. 157
MANCHA, M. 322
MARIER, J.-P. 163
MARKHAM, J.E. 11, 23
MARRAKCHI, R. 401
MARSHALL, D.R. 383
MARTÍNEZ-FORCE, E. 322
MASTERSON, C. 75, 78
MASUDA, T. 354
MATSUI, K. 247, 348
MAZLIAK, P. 151, 227
MCANDREW, R.S. 107
MCFERSON, J.R. 319
MCKEAN, A.L. 363
MCKEON, T.A. 60, 113
MENZEL, K. 253
MICHAELSON, L. 122
MIERNYCK, J.A. 75, 78
MIERSCH, O. 99
MILLAR, A. 72
MIQUEL, M. 166
MONKS, D.E. 110
MOON, H. 377
MOORE, T.S. 134
MOREAU, R. 189
MORITA, N. 84, 160
MURATA, N. 87, 395
MURPHY, D.J. 289
MUSTARDY, L. 87

NAKAZATO, H. 160
NIELSEN, N.C. 29
NIJKAMP, H.J.J. 360
NIKOLAU, B.J. 336,363
NISHIDA, I. 87, 395
NORBERG, P. 227
NORMAN, H.A. 236
NOVOTNÁ, Z. 275,404
NUNBERG, A.N. 351
NUÑEZ, A. 96, 259

OHLROGGE, J.B. 3, 26
OHNISHI, M. 224
OHTA, H. 354
OHTANI, T. 212
OKIY, D.A. 278
OKUYAMA, H. 84, 160
OVEREND, C. 81
OZERININA, O. 331

PALLETT, K.E. 17
PALMER, C.E. 63
PAPPAN, K. 345
PEARCE, M. 140
PIAZZA, G.J. 96, 259, 265
POMEROY, M.K. 357
PONGDONTRI, P. 310
POST-BEITTENMILLER, D. 45, 48, 339
POWELL, M.J. 189
PRICE, L.J. 17
PRUSKY, D. 233
PUGH, C.E. 104

QUINN, P.J. 148

RADDATZ, S. 169
RADUNZ, A. 169
RAFFERTY, J. 38
RAHIER, A. 183, 186
RAMLI, U.S. 69
RAWSTHORNE, S. 20, 66, 307
REDDY, A.S. 351
REED, D.W. 63
REVERDATTO, S.V. 29
RHEE, Y. 339
RICE, D. 38
RICHES, C. 239
RÖBBELEN, G. 316
ROBERTSON, D.S. 336
ROBINSON, P. 239
ROHMER, M. 177, 180
ROLPH, C. 239
ROSS, J.H.E. 289
ROUGHAN, P.G. 3
ROUTABOUL, J.-M. 200, 203
ROY, A.B. 104
RÜCKER, B. 316

SAKAMOTO, T. 380
SALAS, J. 325, 328
SAMBANTHAMURTHI, R. 26, 69
SÁNCHEZ, J. 325, 328
SANCHOLLE, M. 195
SARMIENTO, C. 289
SASAKI, Y. 32
SAVAGE, T.J. 51
SCHMID, G.H. 169
SCHMIDT, H. 6
SCHMITT, N. 284
SCHNABLE, P.S. 336
SCHNEIDER, G. 6
SCHWENDER, J. 177, 180
SCOWBY, M. 377

SEDEE, N.J.A. 284
SEEMAN, M. 177
SEITHER, C. 389
SELSTAM, E. 172
SHANKLIN, J. 6, 374
SHARP, P.J. 383
SHIBATA, Y. 348
SHIBUTANI, Y. 348
SHIMOJIMA, M. 354
SHIOI, Y. 354
SIGGAARD-ANDERSEN, M. 386
SIMON, J.W. 38
SINGH, S.P. 383
SITBON, F. 230
SJÖDAHL, S. 244
SKORUPINSKA, K. 192
SLABAS, A.R. 11, 14, 23, 38, 140, 360, 392, 395
SMITH, A.M. 307
SMITH, M.D. 154
SOFER, Y. 157
SOK, D.-E. 271
SOMERVILLE, C. 342
STAFFORD, A.E. 60
STÅHL, U. 57, 244
STENLID, G. 230
STIREWALT, V.L. 380
STOBART, K. 122
STUITJE, A.R. 38, 360, 386
STYMNE, S. 57, 230, 244
SUKENIK, A. 131
SUN, J. 363
SWIEZEWSKA, E. 192
SWINHOE, R. 395
SZYMANSKA, M. 192

TAKAMIYA, K.-I. 354
TANG, F. 134
TATON, M. 183, 186
TAYLOR, D.C. 63, 319, 407
TENASCHUK, D. 319
THOMAS, N. 38
THOMAS, T.L. 351
THOMPSON, G.A. 313
THOMPSON, G.A. Jr 160
THOMPSON, J.E. 254
TING, J.T.L. 295
TISSOT, G. 35
TOGURI, T. 137, 212
TSVETKOVA, N. 87
TSYDENDAMBAEV, V.D. 331

VALENTOVÁ, O. 275, 404
VAN DER PLAS, L.H.W. 301
VAN MECHELEN, J.R. 284
VAN DUIJN, G. 284
VEBREE, E. 386
VERESHCHAGIN, A.G. 331
VERWOERT, I. 386
VIJAYAN, P. 200, 203
VOELKER, T. 304
VON WETTSTEIN-KNOWLES, P. 386

WAGNER, G.J. 42
WALLIS, J. 209
WANG, C. 268
WANG, X. 345
WASTERNAK, C. 250
WATANABE, A. 395
WEAVER, L.M. 363
WEIER, D. 119
WEN, T.-J. 336
WEN, T.-N. 363
WESELAKE, R.J. 357
WHITAKER, B.D. 143
WHITE, A.J. 14, 23
WHITE, G.F. 104
WHITTLE, E. 6
WIBERG, E. 57
WILLEMOT, C. 221
WILLEY, D. 398
WILLIAMS, J.P. 206
WILLIAMS, M. 262
WILMER, J.A. 301
WINZ, R. 23
WOOD, C. 75, 78
WOODRUFF, C.L. 113
WURTELE, E.S. 363

XIA, Y. 336
XU, X. 336

YANIV, Z. 227
YU, F. 363
YU, H. 221
YU, H.Z. 90

ZHENG, L. 345
ZHENG, S. 345
ZIEGLER, J. 99
ZIMAFUALA, A.M. 163
ZOU, J.-T. 407

SUBJECT INDEX

α-keto acid elongation 42, 54
α-oxidation 54
abscisic acid 227, 301
acetyl-CoA carboxylase 3, 11, 14, 17, 20, 23, 26, 29, 32, 357, 363
acetyl-CoA synthetase 3
acetylcarnitine 75
acetylenic fatty acids 57, 253
acyl-ACP thioesterase 304
acyl-ACP desaturase 6, 374
acyl-CoA binding protein 128, 368
acyl-CoA synthetase 45
acylglycerol 259
acylglycerol-3-phosphate acyltransferase 119
acyl lipids 239
acyltransferase 310, 407
alcohol synthesis 247, 328
alfalfa 48
algae 93, 131, 180, 239
alkane hydroxylase 6
alkanes 51
allene oxide cyclase 99
Allium porrum 45, 195
Anacystis nidulans 212
anti-freeze protein 209
antifungal activity 236
antifungal diene 233
antioxidants 215, 281
apple 143
Arabidopsis thaliana 35, 54, 72, 87, 148, 166, 200, 203, 342, 345, 360, 371, 407
arbuscular mychorrhizal fungi 195
avocado 140, 233

β-hydroxydecanoyl-ACP dehydrase 386

β-keto acyl-ACP synthase I 386
β-keto acyl-ACP synthase II 69
β-keto acyl reductase 14, 336
β-oxidation 250
barley 75, 284
betaine lipids 116
biotin 11, 29, 363
biotin carboxylase 11, 20, 23
biotin carboxyl carrier protein 11, 23, 29
biotin holocarboxylase synthetase 35
borage 351
Brassica napus 11, 14, 20, 23, 38, 45, 66, 81, 275, 301, 307, 310, 316, 357, 368, 386, 395, 404, 407
Brassica oleracea 319
Brassica rapa 319

callus 262
canola 313
carbenoxolone 236
carbon metabolism 325
carbon partitioning 307
carnitine acetyltransferase 75
carnitine palmitoyltransferase 78
carotenoids 169, 177, 215, 281
castor bean 113, 134, 345, 342
CDP-choline pathway 398
ceramides 151
cerebrosides 143, 224
chilling injury 221
chilling sensitive 224
chilling tolerance 137, 203
Chlorella 96, 177, 180
chloroplasts 3, 29, 75, 78, 104, 119, 148, 154, 157
choline kinase 110
choline phosphotransferase 110

chondrillasterol 180
cis-trans isomerase 84
coconut 42, 54
cold stress 227
cold tolerance 209
Combretaceae 192
corn 99, 189
corn fiber oil 189
cotton 38, 107, 383
crepenynic acid 57
Crepis alpina 57

CTP:ethanolaminephosphate
 cytidylyltransferase 134
CTP:phosphocholine cytidylyltransferase
 371, 395, 398
cucumber 354
Cuphea 42, 54
cuticular waxes 48, 336, 339
cyanobacteria 87, 172, 212, 380
cytochrome 348, 389
cytosolic lipid-protein particles 154

D1-core peptide 169
dehydrocholesterol reductase 186
deoxycholate 259
diacylglycerol acyltransferase 125, 357,
diacylglycerol 140, 259
diacylglyceryl-N,N,N-trimethyl-homoserine
 116
diatom 256

eicosapentaenoic acid 90, 93
Elaeis oleifera 26
Elaeis guineensis 26, 69
elm seed 244
endoplasmic reticulum 289
endosperm 278
enoyl-ACP reductase 38, 81, 360
epoxygenase 389
erucic acid 301, 319, 392
Escherichia coli 38, 119, 386, 354
ethylcholesterol 195

fatty acids 42, 45, 48, 54, 60, 63, 107, 163,
 169, 195, 200, 203, 215, 259, 316,
 322
fatty acid biosynthesis 29, 42, 63, 66, 75
fatty acid desaturation 63, 78, 122, 209,
 374
fatty acid desaturase 7, 87, 183, 230, 289,
 298, 342
fatty acid desaturase (Δ^6) 8, 122
fatty acid desaturase (Δ^9) 8, 212, 377, 380
fatty acid desaturase (ω6) 383
fatty acid elongase 45, 48, 339
fatty acid elongation 42, 45, 72
fatty acid synthase 3, 38, 60, 386
ferulate 189
fungal inhibition 236

G protein 151
galactolipid 90, 218, 221, 354
geranoyl-CoA carboxylase 363
Gloeobacter violaceus 172
Glomales 195
glycerol-3-phosphate acyltransferase 137,
 392
glycerolipid biosynthesis 137, 221
Glycine max 259, 268
glycosylphosphatidylinositol-anchored
 protein 160
glyoxylate cycle 256
Gossypium hirsutum 107, 383
Gramineae 32

heavy metal 239
herbicides 17
hexadecatrienoic acid 166, 200
hexadecenoic acid 84, 331
hexadecenoic acid (Δ^{11}) 195
hexadecenoic acid (*trans*-Δ^3) 206
high saturate canola 313
Hippophae rhamnoides 331
HPLC 143
hydroxylase 6, 342
hydroperoxide 247, 259, 265
hydroperoxide lyase 96, 247, 348

idioblast oil cells 233
immunocytochemistry 87
inositol triphosphate 151
inositol 227
isopentenyl diphosphate 177, 180
isoprenoids 180, 192

jasmonic acid 99, 151

leek 195, 339
Lemna 163, 177
Lesquerella 63, 342
light-harvesting chlorophyll protein 169
light-harvesting complex II 206
Limnanthes douglasii 392
linoleic acid 230, 247, 259, 351
linolenic acid (α) 200, 172, 247
linolenic acid (γ) 122, 247, 351
linseed 230
lipase 278, 401
lipid antisera 169
lipid bodies 125, 268, 292
lipid protein particles 154
lipid transfer protein (LTP) 128
liposomes 253
lipoxygenase 157, 247, 250, 259, 262, 265, 268, 271, 281, 284
low-temperature resistance 212
Lycopersicon esculentum 221, 281

maize 17, 128
membrane ultrastructure 148
membranes 87, 154, 157, 212, 253, 310, 313
methylcrotonyl-CoA carboxylase 363
mevalonate 177, 180
microsomes 57, 113
microspore-derived embryos 301
monogalactosyl diacylglycerol (MGDG) 90, 93, 131
monogalactosyl diacylglycerol (MGDG) synthase 354,
monounsaturates 298
Mortierella ramanniana 125

moss protonema 253
muskmelon 143

N-acylphosphatidylethanolamine 107
Nannochloropsis 131
Neurospora crassa 60
Nicotiana tabacum 119, 169

octadecatetraenoic acid 351
oil bodies 154, 289, 292, 295
oil oxidation 298
oil processing 275
oil palm 26, 69, 278
oil seed 57, 63, 298, 304, 401
Olea europaea 325, 328
oleic acid 230, 298
oleoresin 51
oleosin 289, 292, 295
oleosin genes 295
olive 262, 325, 328
oryzanol 189
oxylipins 99
oxo-carboxylic acid 96

palmitic acid 60
pea 3, 32, 35, 75, 78, 104, 398
pentose phosphate pathway 309
pepper 212, 348
peptide hydrophobicity 292
Persea americana 140
Phaeodactylum tricornutum 256
Phaseolus vulgaris 154
phosphatidate phosphatase 140
phosphatidic acid 125, 140
phosphatidylcholine 93, 110, 371, 395
phosphatidylethanolamine 107, 134
phosphatidylglycerol 119
phosphatidylinositol-4, 5-bisphosphate 151
phosphoglyceride 259
phospholipase A_2 244
phospholipase D 151, 256, 275, 345, 404
phospholipids 113, 227, 345
photoinhibition 203
photosynthesis 163, 200, 203, 325

photosystem II 169, 200, 203
phytol 177
Pinus jeffreyi 51
Pisum sativum 157
plastids 66, 307, 386
plastoquinone-9 177
Poa annua 17
polyprenols 192
polyunsaturated fatty acid (PUFA) 90, 218, 230, 298
Porphyridium cruentum 90, 93, 218
potato 209
prenylated proteins 192
prenyllipids 177
Pseudomonas 6, 84

rapeseed 289, 392, 404, 407
reductase pathway 250
Rhodobacter sphaeroides 116
rice bran oil 189
ricinoleic acid 113, 342

salicyclic acid 230
salicylhydroxamic acid 230
Scenedesmus 177, 180
sea buckthorn 331
Secale cereale 206
secondary oxygenation 271
seed germination 322
seed oil composition 316, 407
seed maturation 275
Selenastrum capricornutum 239
sitosterol 180
soybean 29, 42, 54, 236, 268, 271, 289, 298, 377
sphingoids 143
sphingolipids 151, 224
sphingosine-1-phosphate 151
Spirodela oilgorrhiza 160

stearoyl-ACP desaturase 6
sterols 183, 186, 189, 195, 239
sterol desaturase 183
storage lipids 256
sulpholipid (SQDG) 90, 104, 172
sunflower 322, 401
Synechococcus 380
Synechocystis 87, 351

Tenera 278
Thea sinensis 247
thermotolerance 200
thioesterase 298, 304
thylakoids 203
tobacco 119, 227, 345, 377
transgenic plants 119, 209, 212, 345
traumatic acid 151
triacylglycerol 90, 244, 253, 278, 289, 310, 319, 331, 357, 401
trienoic fatty acid 200, 203
trierucin 392
triterpenoids 236

UDP-galactose:diacylglycerol galactosyltransferase (UDGT) 131
Ulmus glabra 244
unsaturated hydroxy fatty acids 224
UV-B irradiation 32

Vernonia galamensis 389

wheat 230

X-ray crystallography 38

yeast 395

Zea mays 128, 298